Molecular Ecology of Aquatic Communities

Developments in Hydrobiology 138

Series editor
H. J. Dumont

Molecular Ecology of Aquatic Communities

Edited by

J.P. Zehr and M.A. Voytek

Reprinted from Hydrobiologia, volume 401 (1999)

Springer-Science+Business Media, B.V.

Library of Congress Cataloging-in-Publication Data

A C.I.P. Catalogue record for this book is available from the Library of Congress.

ISBN 978-0-7923-5933-3 ISBN 978-94-011-4201-4 (eBook)
DOI 10.1007/978-94-011-4201-4

Printed on acid-free paper

Hydrobiologia **401:** v–vi, 1999.
J.P. Zehr & M.A. Voytek (eds.) Molecular Ecology of Aquatic Communities.

Molecular Ecology of Aquatic Communities

Hydrobiologia **401:** vii, 1999.
J.P. Zehr & M.A. Voytek (eds), Molecular Ecology of Aquatic Communities.

Preface

Over the past decade, molecular biology approaches have had a significant impact on many areas of biological sciences, including ecology. In 1997, a special session on the application of molecular techniques to aquatic communities was held at the American Society for Limnology and Oceanography Aquatic Sciences Meeting in Santa Fe, New Mexico. The focus of that session, and the collection of papers presented here, is that molecular information can be used to study the concepts involved in the interactions of species and individuals that are the basis for the features that we observe as aquatic communities.

In this volume, papers present approaches and perspectives that address interactions and relationships involved in community level characteristics. Molecular approaches have provided information on organisms at all trophic levels from prokaryotic microbes to fish and mammals, and including important ecosystem components such as viruses and plasmids. Researchers have applied these techniques over the globe, in diverse environments from hot springs to Antarctic lakes and Arctic ocean basins, from tropical and temperate seas to lakes and rivers.

It is hoped that this volume will integrate studies across subdisciplines, and provide a useful research and educational reference. More importantly, it is hoped that the philosophy of looking forward from what we have done with molecular tools, to what we can hope to do in the field of aquatic community ecology, will stimulate molecular ecology students and researchers to pursue new approaches and ask new questions, at the community level.

J.P. ZEHR
M.A. VOYTEK
Guest Editors

Hydrobiologia **401:** 1–8, 1999.
J.P. Zehr & M.A. Voytek (eds), Molecular Ecology of Aquatic Communities.

Molecular ecology of aquatic communities: reflections and future directions

J. P. Zehr[1] & M. A. Voytek[2]
[1]*Department of Biology, Rensselaer Polytechnic Institute, 110 8th St., Troy, NY 12180-3590, U.S.A.*
Current address for J.P. Zehr: Ocean Sciences Department, Earth and Marine Sciences Building, University of California, Santa Cruz, CA 95064, U.S.A.
[2]*U.S. Geological Survey, MS430, 12201 Sunrise Valley Drive, Reston, VA 20192, U.S.A.*

Key words: aquatic ecology, molecular techniques, molecular ecology

Abstract

During the 1980s, many new molecular biology techniques were developed, providing new capabilities for studying the genetics and activities of organisms. Biologists and ecologists saw the promise that these techniques held for studying different aspects of organisms, both in culture and in the natural environment. In less than a decade, these techniques were adopted by a large number of researchers studying many types of organisms in diverse environments. Much of the molecular-level information acquired has been used to address questions of evolution, biogeography, population structure and biodiversity. At this juncture, molecular ecologists are poised to contribute to the study of the fundamental characteristics underlying aquatic community structure. The goal of this overview is to assess where we have been, where we are now and what the future holds for revealing the basis of community structure and function with molecular-level information.

Introduction

Studies of freshwater and marine communities have played an integral role in the history and development of the science of ecology (Lindeman, 1942; Hutchinson, 1957; Paine, 1980). Ecology has matured during the past quarter century, with theoretical and quantitative developments in the description and modeling of populations, communities and ecosystems (Jones & Lawton, 1995). In parallel, the development of molecular biological techniques has spawned new ways of looking at organisms in the environment, assessing biological processes and activities (Zehr, 1998; Zehr & Hiorns, 1998), and studying population genetics and species distributions (Medlin et al., 1995; Vanoppen et al., 1995; Palumbi, 1996; Geller, 1998; Graves, 1998; Parker et al., 1998).

The trajectories of ecological theory and molecular biology technique development have converged during this decade, and the application of molecular techniques has begun to provide information relevant to ecological questions. Ecological studies have focused on different levels and scales ranging from individual organisms to species, populations and ecosystems, and these different perspectives are now being integrated (Grimm, 1995). Given the complexity of ecosystems and ecological interactions, it could be questioned whether the extension of these studies to the scale of molecules has anything to offer the study of community and ecosystem ecology. Nonetheless, aquatic biology and ecology have already benefited from molecular approaches (for reviews, see Falkowski & LaRoche, 1991; Joint, 1995; Burton, 1996; Cooksey, 1998; Parker et al., 1998). The objective of this discussion is to develop a framework for integrating molecular biology into community ecology and community structure studies, thus making a link from spatial scales of molecules to ecosystems that may foster new avenues of ecological research.

Molecular biology contributions to aquatic ecology

Some of the fundamental concepts that have driven studies in aquatic ecology at the community and ecosystem levels are:

1. Energy flow and trophic dynamics (Lindeman, 1942),
2. Biogeochemical cycling of elements,
3. The 'niche' as the ecological hyperdimensional 'space' of an organism (Hutchinson, 1957),
4. Competition for resources (Tilman, 1982),
5. Food web structure including the 'microbial loop' (Pomeroy, 1974; Steele, 1974; Paine, 1980; Azam et al., 1983; Carpenter et al., 1985; Carpenter & Kitchell, 1988; Azam, 1998),
6. Interactions between species including herbivory, predation and symbiotic relationships, and
7. Community properties including diversity, stability and succession (MacArthur, 1955; Connell, 1961; May, 1972).

Although traditional ecological approaches have provided means to investigate these characteristics of communities, molecular biology has injected a new vitality into studies of some of these concepts. Molecular techniques provide information on the genetics, activities and capabilities of organisms at the most fundamental level. In the following discussion, we will provide some examples of areas where molecular approaches have contributed, and are likely to make contributions to ecological studies.

Biodiversity

A major contribution of molecular techniques has been to provide real measures of biodiversity of organisms at the species, population and community levels. Particularly with respect to microbial assemblages that were previously difficult to study due to constraints of culturability and nondescript morphology, nucleic acid sequence information obtained directly from natural communities has provided a new perspective on diversity in aquatic microbial communities and has led to the identification of major new groups of microorganisms (Murray et al., 1996; Ferrari & Hollibaugh, 1999; Nold & Zwart, 1998).

Molecular sequence information has provided for a number of new approaches for microbial ecology, by facilitating the design of oligonucleotide probes for determining the composition of natural assemblages with fluorescent *in situ* hybridization, and primers for polymerase chain reaction based approaches (Muyzer et al., 1993; Amann et al., 1995; Vanhannen et al., 1998).

Molecular information makes it possible to catalogue the distribution of 'species' and 'populations' (Medlin et al., 1995). This information is essential for determining biological diversity and providing a framework for conservation strategies (Haig, 1998; Palumbi & Cipriano, 1998). At the microbial level, information on species-level diversity would be virtually nonexistent if not for the surveys of terrestrial and aquatic environments that have dominated molecular microbial ecology for the past decade (Pace et al., 1986; Pace, 1997; DeLong, 1998; Head et al., 1998; Methé et al., 1998).

Molecular techniques have also provided information on gene transfer among microorganisms in the environment (Ashelford et al., 1997; Williams et al., 1997; Jiang, 1998), with implications for their evolution, as well as the effects of introductions of new species and genetically-engineered organisms. Molecular approaches have provided means to investigate the ecological roles of viruses (Proctor, 1997; Scanlan & Wilson, 1999; Short & Suttle, 1999) and plasmids (Sobecky & Mincer, 1998; Sobecky, 1999). Much of the biodiversity efforts have remained at the cataloguing stage, with studies only recently beginning to detail the dynamics of individual species or phenotypes, or to use the information to ask classical ecological questions.

It is now possible to use the molecular sequence information and databases to develop probes for studying the dynamics of individual species or phylotypes (DiChristina & DeLong, 1993; Amann et al., 1995; Gordon and Giovannoni, 1996; Methé and Zehr, 1999), to use sequence information to calculate diversity indices (Watve & Gangal, 1996; Nubel et al., 1999), and to investigate relationships between microbial diversity and ecosystem attributes such as community stability. The sequence information can also be used as markers to aid in cultivation of specific groups, which ultimately is critical for understanding the physiological ecology of these organisms in the environment (Palleroni, 1997).

Population biology, biogeography and gene flow

The application of molecular approaches to studies of eukaryotes or macroorganisms has focused on population structure and evolutionary questions, on organisms ranging from picoeukaryotes to whales (DeLong, 1998). Molecular information has provided markers for identifying individuals, determining population structure and studying parentage (Coffroth & Lasker, 1998; Zuccarello et al., 1999), as well as documenting the dispersion of species and larvae in the ocean (Bucklin, 1995; France & Kocher, 1996; Bucklin et al., 1999). Population structure data can be used to assess the effects of disturbances, such as the introduction of toxins and contaminants, on population diversity (Guttman 1994; Depledge 1996; Hebert & Murdoch 1996; Guttman & Berg 1998). The expression of stress proteins and other proteins provide the potential to identify environmental stressors prior to shifts in populations (Chen et al., 1999a; Karouna & Zehr, 1999). Molecular techniques have facilitated the identification of the larvae of species that are otherwise too small or nondescript to identify by traditional means (Burton, 1996), facilitating studies of gene flow and population dynamics (DeLong, 1998). This type of information can ultimately be used to study linkages in aquatic communities, such as the effects of predation and competition on population genetic structure. Currently, these studies are usually descriptive in that they generally do not relate the genetics of populations to the environmental basis for selection or fitness in the environment. However, this may be a rewarding, yet difficult, objective of future studies.

Productivity

Productivity and energy flow are the common measures of the performance of aquatic communities. Measures of microbial productivity are currently constrained to measuring 'community' rates, thus information is lost on the contribution of individual species to community productivity. Molecular approaches that target RNA or protein can provide specific assessments of productivity, growth or gene expression in specific groups of microorganisms (Kramer & Singleton, 1993; Pichard et al., 1996), sometimes at the single cell level (Chen et al., 1999b; Orellana & Perry, 1995). Measurements of phytoplankton primary productivity are made in bulk, whereas macrophyte primary productivity assays use individual plants. Molecular techniques provide the means to assay individual phytoplankton for proteins involved in carbon fixation (Orellana and Perry, 1995), growth and cell division (Lin et al., 1995) and to interrogate cells for nutritional or physiological status (LaRoche et al., 1993; Palenik & Koke, 1995; LaRoche et al., 1999; Scanlan & Wilson, 1999) and study the photosynthetic apparatus (Geider et al., 1993). This type of information can also be obtained from macroalgae or macrophytes, providing better information on their physiological status, growth and metabolism. It may be possible to obtain growth information for nonphotosynthetic eukaryotic organisms, including invertebrates, by targeting developmental genes or measuring RNA/DNA ratios (Smerdon, 1998; Buckley et al., 1999). These tools now provide the potential for integrated community studies, to determine the effects of community structure on growth and productivity of species and individuals in populations.

Competition

Competition is one of the classic concepts in ecology. In contradiction to the prediction of basic competitive exclusion principles, the plankton of oligotrophic systems is more diverse than would be expected if the best competitor for the limiting nutrient grew the fastest and outcompeted other species. This diversity was described over thirty years ago as the "Paradox of the Plankton" (Hutchinson, 1961), and various explanations have been offered since then (Richerson et al., 1970; Siegel, 1998). A recent modeling study suggested that one possible explanation is that the outcome of competition is not predictable at the population level, but only by considering the effects of competition at the individual level (Siegel, 1998). Testing this conclusion requires analyses at the level of the individual and the use of molecular tools. As discussed above, several approaches have been developed for investigating the growth (Lin & Carpenter, 1995), productivity (Orellana & Perry, 1995) and physiological status (Palenik & Wood, 1998) of individual phytoplankton cells using microscopy or flow cytometry (Urbach & Chisholm, 1998; Collier & Campbell, 1999). Thus, molecular biology provides a tool for attempting such studies, even in microscopic species.

Biogeochemical cycles

Many of the critical steps in biogeochemical cycles are catalyzed by very specific groups of microorganisms, using specific enzymes. Molecular approaches have provided important inroads for the detection and characterization of microbes involved in biogeochemical processes, from natural elemental cycles such as nitrification and denitrification (Voytek & Ward, 1995; Voytek et al., 1999), nitrogen fixation (Zehr & Capone, 1996), sulfate reduction (Kane et al., 1993) or sulfur oxidation (Schramm et al., 1996; Gray & Head, 1999), to environmentally important transformations of anthropogenic xenobiotics such as metal compounds (Neilson et al., 1992; Nazaret et al., 1994; Sayler et al., 1995; Langworthy et al., 1998). Probes for specific metabolic pathways are particularly useful since they can be used to determine redundancy within guilds involved in biogeochemical cycling, which may be an important factor in community or ecosystem stability.

Food web structure

The pathways of energy and nutrient transfer through different trophic levels is a fundamental characteristic of communities and ecosystems. Molecular and immunological techniques provide markers that can be used to determine the fate of individual organisms and to identify groups such as the heterotrophic nannoflagellates (Caron et al., 1999). Molecular tracers can provide information on trophic pathways (who eats whom). Immunological techniques were used to identify major food of fish larvae (Ohman et al., 1991); tracing the ingestion of species may be an important, yet unexploited contribution of molecular tools to community ecology. Molecular markers can also be used to evaluate the effects of predation on microbial communities (Pernthaler et al., 1997; Suzuki, 1997).

Symbiosis

Symbiotic relationships span a wide range of interactions between host and symbiont, from loose associations to relationships that provide substantial mutual benefit. Molecular techniques that provide high resolution at the species level, as well as the ability to identify individual organisms on the basis of immunoassays or nucleic acid probe hybridization have greatly facilitated the investigation of symbioses (Hackstein, 1997). Previously unidentified symbiotic relationships that are uncovered with molecular techniques may have important implications for biodiversity (Hackstein, 1997). Molecular probes have been used to identify organisms in association with cells, or determine the specific localization of microorganisms within cells or tissues (Cary et al., 1993). Symbiotic organisms can often be identified (Distel & Wood, 1992; Polz et al., 1994), and the interactions between host and symbiont studied at the molecular level. Signals between hosts and symbionts and their effects on gene expression can be studied, providing a model of symbiotic interactions at the molecular level (Weis et al., 1998). Mechanisms of symbiont transfer from generation to generation can be explored (Cary & Giovannoni, 1993). Furthermore, the relationships between diversity of hosts and symbionts can be evaluated (Rowan, 1998).

Adaptation

Important determinants of the distribution of species are the physiological, biochemical and behavioral characteristics that allow an individual species to compete in its unique niche. Studies of how organisms are adapted to their environment, including extreme environments, are enhanced by the use of molecular tools that allow the direct examination of the molecular basis for adaptation and provide information on evolution as well. Ecologically important molecules can be identified and characterized, such as the antifreeze protein in fish (Wang et al., 1995). Molecules involved in damage or responses to environmental factors such as UV, can be assayed by molecular techniques (Lyons et al., 1998) and the effects of factors such as UV-stress have implications for competitive interactions (Miller et al., 1998). The identification and understanding of the expression of these molecules is fundamental to understanding adaptation and selection, which determine the distribution of organisms in time and space, and the outcome of competitive interactions.

Summary

The application of molecular tools was initiated with an exploratory, developmental phase that has blossomed and provided new insights into structure, function, diversity and ecology. Perhaps during this phase the traditional ecologist has been disappointed in the products of molecular biology, but the understanding that has been obtained now poises the ecologist to merge molecular approaches with more traditional experimental techniques to exploit the full potential of molecular level understanding. The molecular approach has perhaps made the most revolutionary impact on microbial ecology, which previously had been limited by the technological ability to identify, characterize and study natural populations. Perhaps the most profound insights are yet to come, when ecological information on rate processes and biomass are routinely collected with molecular information, and when molecular approaches are better integrated into experimental ecology to directly address ecological questions. A number of pioneering studies have shown the potential payoff of using molecular techniques and recombinant organisms in ecological experiments (Sobecky et al., 1996; Pernthaler et al., 1997; Gonzalez et al., 1999).

The next step will be to address questions regarding the specific physiological properties that constitute ecological success under certain nutrient conditions or that characterize populations that are nutrient-limited (controlled by bottom-up forces) or external factors (e.g. predation, top-down mechanisms). There may be molecular markers that define r vs. K strategists, or that characterize the populations at different stages in community succession. Molecular markers that provide indications of disturbance can be used to assess stresses that may be useful for predicting long-term impacts of environmental effects on biodiversity.

At this juncture, we have not yet seen the complete maturation of molecular ecology in the aquatic sciences, but the fusion of molecular approaches with the classical concerns of the ecologist are on the horizon.

Acknowledgements

Many people have contributed to the development of this paper, and to the application of molecular techniques to aquatic ecology. We would like to particularly thank A. Bucklin, J. T. Hollibaugh and J. Collier for encouragement, insight and for reviewing the manuscript.

References

Amann, R. I., W. Ludwig & K.-H. Schleifer, 1995. Phylogenetic identification and *in situ* detection of individual microbial cells without cultivation. Microb. Rev. 59: 143–169

Ashelford, K. E., J. C. Fry, M. J. Day, K. E. Hill, M. A. Learner, J. R. Marchesi, C. D. Perkins & A. J. Weightman, 1997. Using microcosms to study gene transfer in aquatic habitats. FEMS Microb. Ecol. 23: 81–94.

Azam, F., 1998. Microbial control of oceanic carbon flux: The plot thickens. Science 280: 694–696.

Azam, F., T. Fenchel, J. G. Field, J. S. Gray & L. A. T. F. Meyer-Reil, 1983. The ecological role of water-column microbes in the sea. Mar. Ecol. Progr. Ser. 10: 257–263

Buckley, L., E. Caldarone & T. L. Ong, 1999. RNA:DNA ratio and other nucleic acid-based indicators for growth and condition of marine fishes. Hydrobiologia 401 (Dev. Hydrobiol. 138): 269–281.

Bucklin, A., 1995. Molecular markers of zooplankton dispersion in the ocean. Reviews in Geophysics 33: 1165–1175.

Bucklin, A., M. Guarnieri, R. S. Hill, A. M. Bentley & S. Kaartvedt, 1999. Taxonomic and systematic assessment of planktonic copepods using mitochondrial COI sequence variation and competitive, species-specific PCR. Hydrobiologia 401 (Dev. Hydrobiol. 138): 241–257.

Burton, R. S., 1996. Molecular tools in marine ecology. J. exp. mar. Biol. Ecol. 200: 85–101.

Caron, D. A., R. J. Gast, E. L. Lim & M. R. Dennett, 1999. Protistan community structure: molecular approaches for answering ecological questions. Hydrobiologia 401 (Dev. Hydrobiol. 138): 217–229.

Carpenter, S. R. & J. F. Kitchell, 1988. Consumer control of lake productivity. Bioscience 38: 764–769.

Carpenter, S. R., J. F. Kitchell & J. R. Hodgson, 1985. Cascading trophic interactions and lake productivity. Bioscience 35: 634–639.

Cary, S. C. & S. J. Giovannoni, 1993. Transovarial inheritance of endosymbiotic bacteria in clams inhabiting deep-sea hydrothermal vents and cold seeps. Proc. natn. Acad. Sci. U. S. A. 90: 5695–5699.

Cary, S. C., W. Warren, E. Anderson & S. J. Giovannoni, 1993. Identification and localization of bacterial endosymbionts in hydrothermal vent taxa with symbiont-specific polymerase chain reaction amplification and *in situ* hybridization techniques. Molec. mar. Biol. Biotechnol. 2: 51–62.

Chen, C. Y., K. B. Sillett, C. L. Folt, S. L. Whittemore & A. Barchowsky, 1999a. Molecular and demographic measures of arsenic stress in *Daphnia pulex*. Hydrobiologia 401 (Dev. Hydrobiol. 138): 229–238.

Chen, F., W. A. Dustman & R. E. Hodson , 1999b. Microscopic detection of the toluene dioxygenase gene and its expression inside bacterial cells in seawater using prokaryotic *in situ* PCR. Hydrobiologia 401 (Dev. Hydrobiol. 138): 231–240.

Coffroth, M. A. & H. R. Lasker, 1998. Population structure of a conal gorgonian coral – the interplay between clonal reproduction and disturbance. Evolution 52: 379–393.

Collier, J. L. & L. Campbell, 1999. Flow cytometry in molecular aquatic ecology. Hydrobiologia 401 (Dev. Hydrobiol. 138): 34–54.

Connell, J. H., 1961. Effects of competition, predation by *Thais lapillus* and other factors on natural populations of barnacles. Ecol. Monogr. 31: 61–104.

Cooksey, K. E., 1998. Molecular Approaches to the Study of the Ocean. Chapman and Hall, London, 549 pp.

DeLong, E. F., 1998. Molecular phylogenetics: new perspective on the ecology, evolution and biodiversity of marine organisms. In Cooksey K. E. (ed.), Molecular Approaches to the Study of the Ocean. Chapman and Hall, London: 1–28.

DiChristina, T. J. & E. F. DeLong, 1993. Design and application of rRNA-targeted oligonucleotide probes for the dissimilatory iron- and manganese-reducing bacterium *Shewanella putrefaciens*. Appl. envir. Microbiol. 59: 4152–4160.

Distel, D. L. & A. P. Wood, 1992. Characterization of the gill symbiont of *Thyasira flexuosa* (Thyasiridae: Bivalvia) by use of polymerase chain reaction and 16S rRNA sequence analysis. J. Bact. 174: 6317–6320.

Falkowski, P. G. & J. LaRoche, 1991. Molecular biology in studies of ocean processes. Int. Rev. Cytology. 128: 261–303

Ferrari, V. C. & J. T. Hollibaugh, 1999. Distribution of microbial assemblages in the central arctic ocean basin studied by PCR/DGGE: analysis of a large data set. Hydrobiologia 401 (Dev. Hydrobiol. 138): 55–68.

France, S. C. & T. D. Kocher, 1996. Geographic and bathymeteric patterns of mitochondrial 16S rRNA sequence divergence among deepsea amphipods, *Eurythenes gryllus*. Mar. Biol. 126: 633–643.

Geider, R. J., J. LaRoche, R. M. Greene & M. Olaizola, 1993. Response of the photosynthetic apparatus of *Phaeodactylum tricornutum* (bacillariophyceae) to nitrate, phosphate or iron starvation. J. Phycol. 29: 755–766.

Geller, J. B., 1998. Molecular studies of marine invertebrate biodiversity: status and prospects. In K. E. Cooksey (ed.), Molecular Approaches to the Study of the Ocean. Chapman and Hall, London: 359–376.

Gonzalez, J. M., R. E. Hodson & M. A. Moran , 1999. Bacterial populations in replicate marine enrichment cultures: assessing variability in abundance using 16S rRNA-based probes. Hydrobiologia 401 (Dev. Hydrobiol. 138): 69–75.

Gordon, D. A. & S. J. Giovannoni, 1996. Detection of stratified microbial populations related to *Chlorobium* and *Fibrobacter* species in the Atlantic and Pacific Oceans. Appl. envir. Microbiol. 62: 1171–1177.

Graves, J. E., 1998. Molecular insights into the population structures of cosmopolitan marine fishes. J. Heredity 89: 427–437.

Gray, N. D. & I. M. Head, 1999. New insights on old bacteria: diversity and function of morphologically conspicuous sulfur bacteria in aquatic systems. Hydrobiologia 401 (Dev. Hydrobiol. 138): 97–112.

Grimm, N. B., 1995. Why link species and ecosystems: A perspective from ecosystem ecology. In Jones C. G. & J. H. Lawton (ed.), Linking Species and Ecosystems. Chapman and Hall, New York: 5–15.

Hackstein, J. H. P., 1997. Eukaryotic molecular biodiversity: systematic approaches for the assessment of symbiotic associations. Antonie Van Leeuwenhoek 72: 63–76.

Haig, S. M., 1998. Molecular contributions to conservation. Ecology 7: 413–425.

Head, I. M., J. R. Saunders & R. W. Pickup, 1998. Microbial evolution, diversity and ecology – A decade of ribosomal RNA analysis of uncultivated microorganisms. Microbiol. Ecol. 35:1–21.

Hutchinson, G. E., 1957. A treatise on limnology. I. Geography, physics and chemistry. John Wiley and Sons, Inc., New York, 1015 pp.

Hutchinson, G. E., 1961. The paradox of the plankton. Am. Nat. 95: 137–145.

Jiang, S. C. P. J. H., 1998. Gene transfer by transduction in the marine environment. Appl. envir. Microbiol. 64: 2780–2787.

Joint, I., 1995. Molecular Ecology of Aquatic Microbes. Springer, Berlin, 415 pp.

Jones, C. G. & J. H. Lawton, 1995. Linking Species and Ecosystems. Chapman and Hall, New York, 387 pp.

Kane, M. D., L. K. Poulsen & D. A. Stahl, 1993. Monitoring the enrichment and isolation of sulfate-reducing bacteria by using oligonucleotide hybridization probes designed from environmentally derived 16S rRNA sequences. Appl. envir. Microbiol. 59: 682–686.

Karouna, N. K. & J. P. Zehr, 1999. Effects of stress on freshwater invertebrate populations of Chironomus tentans: assaying sublethal stress using heat shock protein 70 (HSP-70) expression. Hydrobiologia 401 (Dev. Hydrobiol. 138): 259–268.

Kramer, J. G. & F. L. Singleton, 1993. Measurement of rRNA synthesis variations in natural communities of microorganisms on the southeastern U. S. continental shelf. Appl. envir. Microbiol. 59: 2430–2436.

Langworthy, D. E., R. D. Stapleton, G. S. Sayler & R. H. Findlay, 1998. Genotypic and phenotypic responses of a riverine microbial community to polycyclic aromatic hydrocarbon contamination. Appl. envir. Microbiol. 64: 3422–3428.

LaRoche, J., M. L. McKay & P. Boyd, 1999. Immunological and molecular probes to detect phytoplankton responses to environmental stress in nature. Hydrobiologia 401 (Dev. Hydrobiol. 138): 179–200.

LaRoche, J., R. J. Geider, L. M. Graziano, H. Murray & K. Lewis, 1993. Induction of specific proteins in eukaryotic algae grown under iron-, phosphorus- or nitrogen-deficient conditions. J. Phycol. 29: 767–777.

Lin, S. & E. J. Carpenter, 1995. Growth characteristics of marine phytoplankton determined by cell cycle proteins: The cell cycle of *Ethmodiscus rex* (Bacillariophyceae) in the southwestern North Atlantic Ocean and Caribbean Sea. J. Phycol. 31: 778–785.

Lin, S., J. Chang & E. J. Carpenter, 1995. Growth characteristics of phytoplankton determined by cell cycle proteins: PCNA immunostaining of *Dunaliella tertiolecta* (Chlorophyceae). J. Phycol. 31: 388–395.

Lindeman, R. L., 1942. The trophodynamic aspect of ecology. Ecology 23: 399–418.

Lyons, M. M., P. Aas, J. D. Pakulski, L. Vanwaasbergen, R. V. Miller, D. L. Mitchell & W. H. Jeffrey, 1998. DNA damage induced by ultraviolet radiation in coral-reef microbial communities. Mar. Biol. 130: 537–543.

MacArthur, R. H., 1955. Fluctuations of animal populations and a measure of community stability. Ecology 36: 533–536.

May, R. M., 1972. Will a large complex system be stable? Nature 238: 413–414.

Medlin, L. K., M. Lange, G. L. A. Barker & P. K. Hayes, 1995. Can molecular techniques change our ideas about the species concept? In Joint I. (ed.), Molecular Ecology of Aquatic Microbes. Springer, Berlin: 133–170.

Methé, B. A., W. D. Hiorns & J. P. Zehr, 1998. Contrasts between marine and freshwater bacterial community composition: analyses of communities in Lake George, NY and six other Adirondack lakes. Limnol. Oceanogr. 43: 368–374.

Methé, B. A. & J. P. Zehr, 1999. Diversity of bacterial communities in Adirondack lakes: do species assemblages reflect lake water chemistry? Hydrobiologia 401 (Dev. Hydrobiol. 138): 77–96.

Miller, S. R., C. E. Wingard & R. W. Castenholz, 1998. Effects of visible light and UV radiation on photosynthesis in a population of a hot spring cyanobacterium, a Synechococcus sp., subjected to high-temperature stress. Appl. envir. Microbiol. 64: 3893–3899.

Murray, A. E., J. T. Hollibaugh & C. Orrego, 1996. Phylogenetic compositions of bacterioplankton from two California estuaries compared by denaturing gradient gel electrophoresis of 16S rDNA fragments. Appl. envir. Microbiol. 62: 2676–2680.

Muyzer, G., E. C. De Waal & A. G. Uitterlinden, 1993. Profiling of complex microbial populations by denaturing gradient gel electrophoresis analysis of polymerase chain reaction-amplified genes coding for 16S rRNA. Appl. envir. Microbiol. 59: 695–700.

Nazaret, S., W. H. Jeffrey, E. Saouter, R. Von Haven & T. Barkay, 1994. *merA* gene expression in aquatic environments measured by mRNA production and Hg(II) volatilization. Appl. envir. Microbiol. 60: 4059–4065.

Neilson, J. W., K. L. Josephson, S. D. Pillai & I. L. Pepper, 1992. Polymerase chain reaction and gene probe detection of the 2,4-dichlorophenoxyacetic acid degradation plasmid, pJP4. Appl. envir. Microbiol. 58: 1271–1275.

Nold, S. C. & G. Zwart, 1998. Patterns and governing forces in aquatic microbial communities. Aquat. Ecol. 32: 17–35.

Nübel, U., F. Garcia-Pinchel, M. Kuhl & G. Muyzer , 1999. Spatial scale and the diversity of benthic cyanobacteria and diatoms in a salina. Hydrobiologia 401 (Dev. Hydrobiol. 138): 201–208.

Ohman, M. D., G. H. Theilacker & S. E. Kaupp, 1991. Immunochemical detection of predation on ciliate protists by larvae of the Northern Anchovy (*Engaulis mordax*). Biol. Bull. 181: 500–504.

Orellana, M. V. & M. J. Perry, 1995. Optimization of an immunofluorescent assay of the internal enzyme ribulose-1,5-bisphosphate carboxylase (RUBISCO) in single phytoplankton cells. J. Phycol. 31: 785–794.

Pace, N. R., 1997. A molecular view of microbial diversity and the biosphere. Science 276: 734–740.

Pace, N. R., D. A. Stahl, D. J. Lane & G. J. Olsen, 1986. The analysis of natural microbial populations by ribosomal RNA sequences. Adv. microb. Ecol. 9: 1–55.

Paine, R. T., 1980. Food webs: linkage, interaction strength and community infra-structure. J. anim. Ecol. 49: 667–686.

Palenik, B. & J. A. Koke, 1995. Characterization of a nitrogen-regulated protein identified by cell surface biotinylation of a marine phytoplankton. Appl. envir. Microbiol. 61:3311–3315.

Palenik, B. & A. M. Wood, 1998. Molecular markers of phytoplankton physiological status and their appliation at the level of individual cells. In Cooksey K. E. (ed.), Molecular Approaches to the Study of the Ocean. Chapman and Hall, London: 187–206.

Palleroni, N. J., 1997. Prokaryotic diversity and the importance of culturing. Antonie Van Leeuwenhoek 72: 3–19.

Palumbi, S. R., 1996. What can molecular genetic contribute to marine biogeography? An urchin's tale. J. exp. mar. Biol. Ecol. 203: 75–92.

Palumbi, S. R. & F. Cipriano, 1998. Species identification using genetic tools – The value of nuclear and mitochondrial gene sequences in whale conservation. J. Heredity 89: 459–464.

Parker, P. G., A. A. Snow, M. D. Schug, G. C. Booton & P. A. Fuerst, 1998. What molecules can tell us about populations: choosing and using a molecular marker. Ecology 92: 361–382.

Pernthaler, J., T. Posch, K. Simek, J. Vrba, R. Amann & R. Psenner, 1997. Contrasting bacterial strategies to coexist with a flagellate predator in an experimental microbial assemblage. Appl. envir. Microbiol. 63: 596–601.

Pichard, S. L., L. Campbell, J. B. Kang, F. R. Tabita & J. H. Paul, 1996. Regulation of ribulose bisphosphate carboxylase gene expression in natural phytoplankton communities .1. Diel rhythms. Mar. Ecol. Progr. Ser. 139: 257–265

Polz, M., D. Distel, B. Zarda, R. Amann, H. Felbeck, J. Ott & C. Cavanaugh, 1994. Phylogenetic analysis of a highly specific association between ectosymbiotic, sulfur-oxidizing bacteria and a mine nematode. Appl. envir. Microbiol. 60: 4461–4467.

Pomeroy, L. R., 1974. The ocean's food web, a changing paradigm. BioScience. 24: 499–504.

Proctor, L. M., 1997. Advances in the study of marine viruses. Microsc. Res. Techn. 37: 136–161.

Richerson, P., R. Armstrong & C. R. Goldman, 1970. Contemporaneous disequilibrium, a new hypothesis to explain the "paradox of the plankton". Proc. natn. Acad. Sci. U. S. A. 67: 1710–1714.

Rowan, R., 1998. Diversity and ecology of zooxanthellae on coral reefs. J. Phycol. 34: 407–417.

Sayler, G. S., A. Layton, C. Lajoie, J. Bowman, M. Tschantz & J. T. Fleming, 1995. Molecular site assessment and process monitoring in bioremediation and natural attenuation. Appl. Biochem. Biotech. 54: 277–290.

Scanlan, D. J. & W. H. Wilson, 1999. Application of molecular techniques to addressing the role of p as key effector in marine ecosystems. Hydrobiologia 401 (Dev. Hydrobiol. 138): 151–177.

Schramm, A., L. H. Larsen, N. P. Revsbech, N. B. Ramsing, R. Amann & K. H. Schleifer, 1996. Structure and function of a nitrifying biofilm as determined by *in situ* hybridization and the use of microelectrodes. Appl. envir. Microbiol. 62: 4641–4647.

Short, S. M. & C. A. Suttle, 1999. Use of the polymerase chain reaction and denaturing gradient gel electrophoresis to study diversity in natural virus communities. Hydrobiologia 401 (Dev. Hydrobiol. 138): 19–33.

Siegel, D. A., 1998. Resource competition in a discrete environment: Why are plankton distributions paradoxical? Limnol. Oceanogr. 43:1133–1146.

Smerdon, G. R., 1998. Towards the molecular analysis of copepod production. In K. E. Cooksey (ed.), Molecular Approaches to the Study of the Ocean. Chapman and Hall, London: 319–328.

Sobecky, P. A., 1999. Plasmid ecology of marine sediment microbial communities. Hydrobiologia 401 (Dev. Hydrobiol. 138): 9–18.

Sobecky, P. A., T. J. Mincer, M. C. Chang, A. Toukdarian, & D. R. Helinski, 1998. Isolation of broad-host-range replicons from marine sediment bacteria. Appl. envir. Microbiol. 64: 2822–2830.

Sobecky, P. A., M. A. Schell, M. A. Moran & R. E. Hodson, 1996. Impact of a genetically engineered bacterium with enhanced alkaline phosphatase activity on marine phytoplankton communities. Appl. envir. Microbiol. 62: 6–12.

Steele, J. H., 1974. The structure of marine ecosystems. Harvard University Press, Cambridge, Massachusetts, 128 pp.

Suzuki, M., 1997. The effect of protistan bacterivory on bacterioplankton community structure. PhD. Thesis. Oregon State University, Corvallis, Oregon.

Tilman, D., 1982, Resource competition and community structure. Princeton University Press, Princeton, New Jersey, 296 pp.

Urbach, E. & S. W. Chisholm, 1998. Genetic diversity in *Prochlorococcus* populations flow cytometrically sorted from the Sargasso Sea and Gulf Stream. Limnol. Oceanogr. 43:1615–1630.

Vanhannen, E. J., M. P. Vanagterveld, H. J. Gons & H. J. Laanbroek, 1998. Revealing genetic diversity of eukaryotic microorganisms in aquatic environments by denaturing gradient gel electrophoresis. J. Phycol. 34: 206–213.

Vanoppen, M. J. H., J. L. Olsen & W. T. Stam, 1995. Genetic variation within and among North Atlantic and Baltic populations of the benthic alga *Phycodrys rubens* (Rhodophyta). Eur. J. Phycol. 30: 251–260.

Voytek, M. A., J. C. Priscu & B. B. Ward, 1999. The distribution and relative abundance of ammonia-oxidizing bacteria in lakes of the McMurdo Dry Valley, Antarctica. Hydrobiologia 401 (Dev. Hydrobiol. 138): 113–130.

Voytek, M. A. & B. B. Ward, 1995. Detection of ammonium-oxidizing bacteria in the beta-subclass of the class *Proteobacteria* in aquatic samples with the PCR. Appl. envir. Microbiol. 61: 1444–1450.

Wang, X., A. L. DeVries & C. C. Cheng, 1995. Antifreeze peptide heterogeneity in an Antarctic eel pout includes an unusually large major variant comprised of two 7 kDa type III AFPs linked in tandem. Biochim.Biophys. Acta 1247:163–172.

Watve, M. G. & R. M. Gangal, 1996. Problems in measuring bacterial diversity and a possible solution. Appl. envir. Microbiol. 62: 4299–4301.

Weis, V. M., Kampen, J. V. & R. P. Levine, 1998. Techniques for exploring symbiosis-specific gene expression in cnidarian/algal associations. In Cooksey K. E. (ed.), Molecular Approaches to the Study of the Ocean. Chapman and Hall, London: 435–448.

Williams, H. G., J. Benstead, M. E. Frischer & J. H. Paul, 1997. Alterations in plasmid DNA following natural transformation to populations of marine bacteria. Molecular Marine Biology and Biotechnology 6: 238–247.

Zehr, J. P., 1998. Molecular approaches to the study of the activities of marine organisms. In Cooksey K. E. (ed.), Molecular Approaches to the Study of the Ocean. Chapman and Hall, London: 91–112.

Zehr, J. P. & D. G. Capone, 1996. Problems and promises of assaying the genetic potential for nitrogen fixation in the marine environment. Microb. Ecol. 32: 263–281.

Zehr, J. P. & W. D. Hiorns, 1998. Molecular approaches for studying the activities of marine organisms. In Cooksey K. E. (ed.), Molecular Approaches to the Study of the Ocean. Chapman and Hall, London: 91–112.

Zuccarello, G. C., J. A. West, M. Kamiya & R. J. King, 1999. A rapid method to score plastid haplotypes in red seaweeds and its use in determining parental inheritance of plastids in the red alga *Bostrychia* (Ceramiales). Hydrobiologia 401 (Dev. Hydrobiol. 138): 209–216.

Hydrobiologia **401:** 9–18, 1999.
J.P. Zehr & M.A. Voytek (eds), Molecular Ecology of Aquatic Communities.

Plasmid ecology of marine sediment microbial communities

P. A. Sobecky
School of Biology, Georgia Institute of Technology, 310 Ferst Drive, Atlanta, GA 30332-0230, U.S.A.
Tel: [+1]404-894-5819; Fax: [+1]404-894-0519; E-mail: patricia.sobecky@biology.gatech.edu

Key words: plasmids, marine bacteria, molecular ecology, diversity, replication, DNA probes

Abstract

It is well documented that bacteria can readily exchange genetic information under artificial conditions typically used in most laboratory studies as well as to some extent in nature. The three mechanisms by which such genetic exchange can occur are transformation, transduction and conjugation. Transformation is the uptake of free DNA into a cell from the surrounding environment, while bacterial viruses mediate the exchange of genetic material during transduction and conjugation involves the direct transfer of DNA during cell-to-cell contact. In most cases, plasmids mediate the transfer of DNA during conjugation events, although chromosomal transfer can also occur. This review will focus mainly on plasmids and the role of conjugation in marine sediment microbial communities. Plasmids, although often dispensible, provide a unique plasticity to an individual host cell or to an entire microbial community 'genome'. Specifically, plasmid-encoded traits mobilized throughout microbial communities can provide a means of rapid adaptation to changing environmental conditions. Examples of such adaptation can be seen in the increased frequencies of catabolic plasmids and antibiotic and heavy metal resistance plasmids within microbial populations upon exposure to selective pressures. Presently, the view of plasmid diversity and horizontal transfer dynamics is predominantly based on broad- and narrow-host-range plasmids isolated from bacteria of clinical and animal origins. While the exchange of plasmids is most likely an important mechanism by which bacterial populations in clinical environments can evolve and adapt, there remains a general lack of information regarding the role of plasmid-mediated transfer in marine ecosystems and how indigenous plasmids impact the microbial community structure and function. The combined application of molecular biology and microbial ecology techniques is providing new approaches to address the ecological role of plasmids in marine environments.

Introduction

Plasmids are autonomously replicating extrachromosomal elements ranging in size from a few kilobases (2–3 kb) to greater than 500 kb. Plasmids typically occur as circular DNA molecules but linear plasmids have been isolated from *Borrelia*, *Streptomyces* and *Rhodococcus* species. The occurrence of plasmids has been well documented among the majority of gram-negative and gram-positive isolates from the Eubacteria, and recently in an hyperthermophilic Archaeon (Erauso et al., 1996). In many instances, the presence of such accessory elements confers a novel or advantageous trait to the host cell. Examples of some typical plasmid-encoded traits include protection from UV light damage (Rochelle et al., 1989), resistance to heavy metals (Hansen et al., 1984; Schutt, 1989), proliferation in the presence of antibiotics (Aviles et al., 1993) and catabolism of xenobiotic compounds (Hada & Sizemore, 1981; Sayler et al., 1990).

In addition to the phenotypic traits described, some plasmids contain a region which encodes a complex transfer (*tra*) system that promotes plasmid movement, (i.e. horizontal transfer), during cell-to-cell contact. Such plasmids, classified as conjugative or self-transmissible, encode a set of genes (usually on a minimum of 20 kb of DNA) that specifies a conjugative pilus and functions for entry exclusion and DNA processing. The initiation of DNA transfer is believed to begin at the *oriT* (origin of transfer) site located at one end of the *tra* region after a single-strand break (nick) is generated by a plasmid-encoded endonuclease. DNA synthesis is associated with the single strand transfer to the recipient cell (Willetts & Wilkins,

1984). Plasmids catagorized as non-conjugative or non-self-transmissible though lacking a set of functional genes required for conjugal transfer may contain *mob* and *oriT* regions which facilitate their mobilization by conjugative plasmids. For plasmid mobilization to occur both regions must be present since the *mob* region encodes a specific nuclease which acts on the *bom* (*oriT*) site to produce a nick in the DNA. While a suitable Mob nuclease may be provided by a related conjugative plasmid, the *bom* site must be present on the non-conjugative plasmid for mobilization to proceed. The nicked DNA can then be transferred via the transfer machinery encoded by a co-resident conjugative plasmid.

The ability of plasmids to either self-transfer or be mobilized means plasmid-encoded genes represent a considerable pool of mobile DNA that may contribute to the genetic adaptation of microbial communities. The unique plasmid-encoded ability to readily transfer DNA between cells serves to promote the movement of genes between both cells of related and cells of diverse genetic backgrounds thereby providing a mechanism for bacterial evolution. The inter- and intragenic transfer of DNA is thus postulated to be a key process that determines the structure and function of marine microbial communities. Numerous studies have demonstrated that genetic exchange by conjugation as well as transduction and transformation occurs between bacteria in the environment (O'Morchoe et al., 1988; Ogunseitan et al., 1990; Saye et al., 1990; Paul et al., 1991; Kinkle et al., 1993). All three transfer mechanisms have been shown to occur in marine systems (Maruyama et al., 1993; Goodman et al., 1993; Hermansson & Linberg, 1994; Frisher et al., 1994; Barkay et al., 1995). This review will focus on plasmids and the importance of conjugal transfer of plasmid-encoded traits within marine (sediment) bacterial populations.

Although numerous studies have reported on the incidence of plasmids in bacteria isolated from marine sediments, estuarine, and pelagic ecosystems (Sizemore & Colwell, 1977; Kobori et al., 1984; Hermansson et al., 1987; Wortman & Colwell, 1988; Belliveau et al., 1991; Aviles et al., 1993; Dahlberg et al., 1997), there remains a general lack of knowledge regarding the diversity (e.g. replicon types) and transfer capabilities of plasmids in indigenous marine bacterial assemblages. However, studies to determine naturally occurring plasmid distribution and diversity and to assess the potential to transfer plasmid-encoded genes within microbial communities require molecular-based methods for typing and classifying plasmids. Previous studies (Benson & Shapiro, 1978; Fry, 1994) have focused on classifying plasmids, mainly from freshwater systems, according to their transfer activity. While transfer abilities are important for predicting the potential for horizontal transfer, such attributes are not sufficient for characterizing plasmid diversity or predicting maintenance of a plasmid in a new host. Molecular-based plasmid classification (i.e. replicon typing) by using DNA sequences of replication origins and incompatibility loci of well-characterized plasmids originally isolated from clinical and animal environments has been shown to be useful in typing or classifying plasmids from bacterial isolates of medical importance (Davey et al., 1984; Couturier et al., 1988). Although many plasmids of medical importance have been well studied, general information is lacking on the host range, maintenance requirements, conjugal abilities and incompatibility groupings of most plasmids present in bacteria isolated from marine environments.

The remarkable array of marine microbial diversity being revealed by nucleic acid-based methods such as 16S rRNA phylogenetic analysis continues to indicate the presence of novel and as yet uncharacterized microbial types (DeLong, 1997). The considerable biodiversity currently being detected in marine systems may also extend to accessory elements such as plasmid populations occurring in marine microbial communities. These plasmids could prove to be a rich and valuable source of biotechnologically important genes.

Plasmid incidence and distribution in marine bacteria

Two general methods are used for the determination of plasmid incidence and abundance in natural microbial communities. One method, sometimes referred to as endogenous plasmid isolation, requires the initial cultivation of bacterial isolates for the screening and confirmation of plasmids. Numerous procedures employing various cell lysis and extraction conditions have been developed for the isolation of plasmids from bacteria ranging in size from 5 kb to > 400 kb. An obvious drawback of the endogenous isolation procedure is the necessity to cultivate the bacterial host. Such a reliance on isolation and cultivation of plasmid-containing hosts likely results in a skewed or biased collection of plasmid populations, since the vast majority of bacteria from aquatic and terrestrial environments are resistant to standard laboratory isolation procedures. A more recent approach, exogenous isolation, eliminates the need to cultivate specific bacterial hosts by isolating plasmid populations based on either a selectable phenotypic trait or on the ability of indigenous plasmids to either self-transfer or to mobilize a nonconjugative, broad-host-range plasmid to a selected recipient rather than the isolation and direct screening of bacterial isolates (Hill et al., 1992). Thus, in theory, the exogenous isolation method provides a means by which to obtain plasmids from non-culturable bacterial populations.

The ubiquity of plasmids in bacteria isolated from diverse environments is well documented. The vast majority of these studies have relied on the endogenous plasmid isolation procedure previously described. A high percentage of plasmid-bearing isolates has been cultivated from freshwater and marine water column and sediment samples with studies detecting one or more plasmids in as much as 50% of the isolates screened (Kobori et al., 1984; Pickup, 1989). In earlier studies to determine plasmid incidence, reported frequencies varied from 27%, for more than 400 putative marine *Vibrio* spp. (Hada & Sizemore, 1981), to 43% of bioluminescent bacteria (Simon et al., 1982) and 60% of nearshore and open ocean isolates (Sizemore & Colwell, 1977). In a similar study, Glassman & McNicol (1981) reported 46% of estuarine bacteria from Chesapeake Bay carried plasmids. A later study by Baya et al. (1986) reported that bacterial isolates containing plasmids ranged from 17% for open ocean samples to 48% for bacteria isolated near a sewage outfall diffuser which released pharamaceutical and industrial wastes.

While a high percentage of bacteria from marine environments have been shown to carry plasmids, attributing specific traits and functions to these plasmids has proven difficult. Baya et al. (1986) demonstrated that the frequency of plasmid DNA and resistance to antibiotics and toxic chemicals increased in bacterial isolates in closest proximity to the outfall diffuser, but the authors were unable to demonstrate a direct correlation between plasmid presence and the observed phenotypes. In a similar study, Hada & Sizemore (1981), reported a 1.5-fold increase in plasmid-containing isolates from a Gulf of Mexico oil field relative to a control site located 8 km from the impacted sampling sites. Although attempts were made to assign phenotypic traits such as hydrocarbon utilization and heavy metal resistance to the plasmid-bearing isolates obtained during the study, no correlation between plasmid content and the resistance determinants was observed. Similarly, Leahy et al. (1990) were unable to detect a direct correlation between hydrocarbon degradation capabilities of marine sediment microbial communities and plasmid incidence in 242 heterotrophic sediment bacterial isolates obtained from an offshore site in the Gulf of Mexico chronically contaminated with varying concentrations of petroleum hydrocarbons. The inability to correlate plasmid content and antibiotic and heavy metal resistance traits was also reported for more than 30 plasmid-containing *Bacillus* isolates, representing a total of 102 plasmids obtained from Canadian coastal marine sediments (Belliveau et al., 1991).

The inability to correlate or assign a particular function to plasmids occurring in natural bacterial isolates appears to be a common feature of endogenously isolated plasmids, and the terms cryptic and genotypically barren have often been used to describe such plasmids. The exact nature of the function(s) encoded on many naturally occurring plasmids may be difficult to ascertain using standard laboratory conditions, and a lack of readily assigned traits may simply reflect a lack of suitable methods to assay plasmid-encoded traits in environmental bacteria. Several studies have reported on the prolonged persistence of plasmid types in terrestrial, freshwater and marine systems, suggesting an ecological importance to these natural microbial community even in the absence of apparent selection (Pickup, 1989; Lilley et al., 1996; Sobecky et al., 1998). In such instances, a direct molecular approach such as sequencing the various persistent plasmid types may help to shed light on plasmid-encoded functions.

Molecular properties used in plasmid identification and classification

Incompatibility is a heritable trait of plasmids and is defined as the inability of two coresident plasmids to be stably maintained in the same host in the absence of selection (Novick et al., 1976; Datta, 1979; Novick, 1987). Plasmids belonging to the same incompatibility group will share similar or identical replication functions which prevent them from being stably maintained in the same host cell. This incompatibility phenomenon between related plasmids is largely due to stochastic selection for replication and partitioning events. The sharing of any function between plasmids

Table 1. Broad-host-range and narrow-host-range plasmid incompatibility and replication probes from clinical isolates

Inc group	Probe size (bp)	Plasmid source
Broad host range		
N	1000	R46
P	750	RK2
Q	357	R1162
W	1150	RSa
Narrow host range		
B/O	1600	pMU700
FIA	917	F
FII	543	R1*drd-19*
FIB	1202	P307
HI1	2250	TR6
HI2	1800	TP116
I1	1100	R64*drd-11*
L/M	800	pMU407.1
X	942	R6K
U	950	RA3

that is required for the control of plasmid replication is likely to result in loss of one of the co-resident plasmids. A formal scheme of plasmid classification initially proposed by Datta & Hedges (1972) assigned plasmids to specific groups based on incompatibility. Using incompatibility as a means to classify and type plasmids has resulted in the identification of more than 30 different incompatibility groups for plasmids primarily isolated from gram-negative bacteria of medical importance (i.e. primarily enterics) and 7 incompatiblility groups for staphylococcal plasmids (Bukhari et al., 1977).

The traditional method for determining which incompatibility group a plasmid should be assigned to has been either through conjugation, transformation or transduction of the plasmid of interest into a host containing a plasmid belonging to a known incompatibility group. If the resident plasmid is subsequently lost, then the entering plasmid is assigned to the same incompatibility group. However, this approach has some limitations including the lack of suitable marker genes on some plasmids and surface exclusion. Surface exclusion refers to the property that greatly limits or inhibits the host cell containing a resident plasmid to act as a recipient for related plasmids.

In recent years, a molecular-based approach, referred to as replicon typing, has been used to assign plasmids to incompatibility groups using specific DNA probes containing replication control genes from well-characterized plasmids (Couturier et al., 1988). The primary source of the majority of these well-characterized plasmids have been bacteria from clinical and animal origins. This more direct and less time-consuming method for classifying plasmids is possible due to the nature of the basic replicon of plasmids. The basic or minimal replicon of a plasmid consists of the genes and sites necessary to ensure and control autonomous replication. The genes essential for plasmid replication and maintenance are typically clustered on a contiguous segment of DNA usually no more than 2–3 kb in size (Helinski et al., 1996). It is this compact nature of plasmid replication origins that has facilitated the isolation and characterization of replicons from plasmids obtained from bacteria of clinical and animal origins. The bank of replicon probes developed by Couturier et al. (1988) contain unique DNA sequences derived from 19 different basic replicons cloned in high copy number plasmid vectors. This collection of replicon (*inc/rep*) probes have been shown to be suitable for the molecular typing of plasmids from bacteria of medical importance (Table 1).

Interestingly, recent studies that have attempted to use these clinically-based replicon probes to type plasmids from bacterial isolates obtained from terrestrial soils (Kobayashi & Bailey, 1994) as well as sediments (Sobecky et al., 1997), bulk water, air-water interfaces and biofilms of marine environments (Dahlberg et al., 1997) have been unsuccessful. None of the hundreds of plasmid-containing isolates from these different environments shared homology to the *inc/rep* group-specific DNA probes currently available for plasmid typing. Such findings indicate that plasmids isolated from bacterial populations occurring in terrestrial soils and marine aquatic and sediment systems encode novel replication and incompatibility loci that lack homology to clinically-derived plasmid incompatibility groups. Moreover, the extent of plasmid diversity occurring in natural microbial communities, such as marine sediments, cannot be determined using the present molecular classification system based on plasmids of clinical and animal origins. Therefore, *inc/rep* probes specific for replicons isolated from the marine environment are necessary to characterize naturally occurring plasmid distribution and diversity.

Table 2. Plasmid replication origins obtained from marine bacteria

Isolate designation	Phylogenetic affiliation genus (species)[a]	Approximate size of plasmid	Replication origin designation and size (kb)	Cross-reactivity of replication origins
32	*Vibrio (fischeri)*	6.5	repSD32 (2.0)	
41	*V. (fischeri)*	7.0	repSD41 (2.3)	121
121	*V. (splendidus)*	6.0	repSD121 (2.2)	41
172	*V. (alginolyticus)*	30.0	repSD172 (1.8)	
164	*Roseobacter (litoralis)*	6.5	repSD164 (2.1)	

[a]As determined by sequencing of 16S rRNA gene and fatty acid analysis.

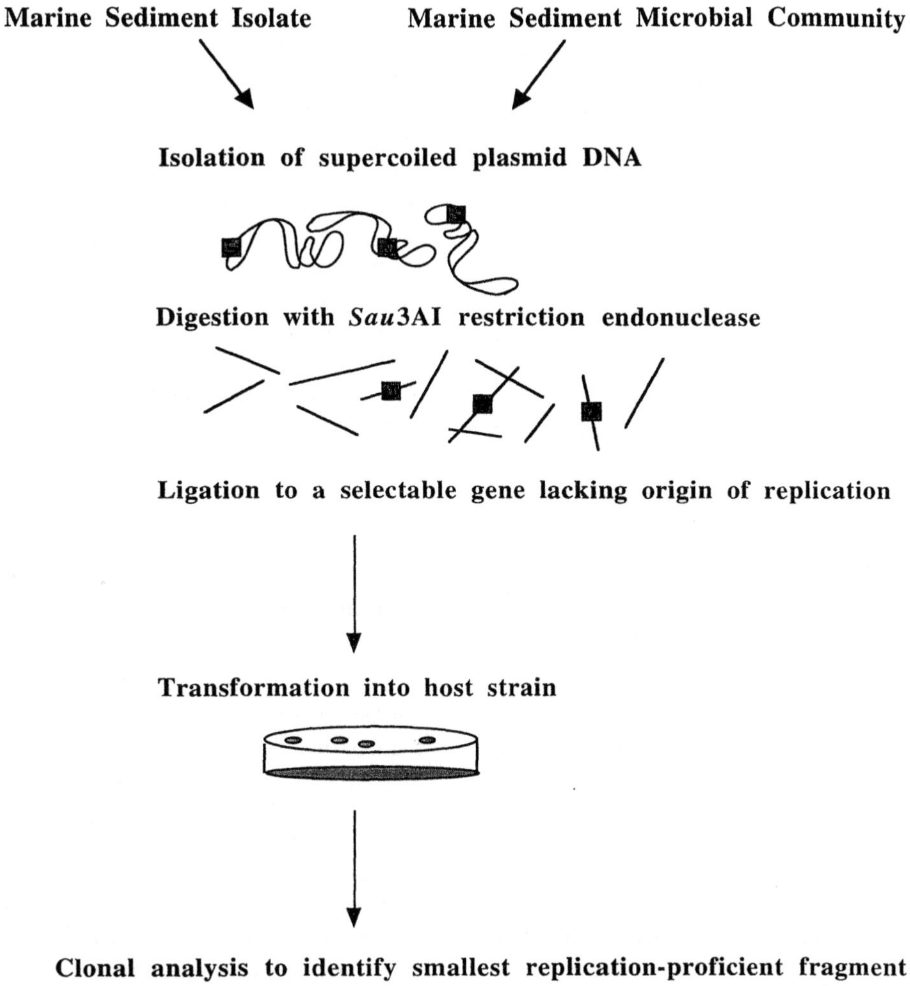

Figure 1. Outline of the 'replicon rescue' protocol for isolation of plasmid replication-proficient fragments. This methodology has been used to isolate replication-proficient fragments from culturable marine bacteria belonging to the α- and γ-Proteobacteria groups. The (■) represents site of replication and incompatibility sequence.

Isolation of plasmid-specific DNA probes (replicon rescue)

To better understand plasmid distribution, diversity and abundance in marine sediment microbial communities, the isolation and characterization of replication sequences from naturally occurring plasmid populations is necessary. Ideally, such information could be used to develop a collection of environmentally-based incompatibility group-specific replicon probes suitable for typing plasmids from non-clinical environments. An increasing body of literature, based largely on the analysis of plasmids from culturable bacteria from diverse environments, supports the existence of new plasmid groups which appear to have evolved along separate lines from plasmid groups occurring in clinical bacterial populations (Kobayashi & Bailey, 1994; Top et al., 1994; Dahlberg et al., 1997; Sobecky et al., 1997; Van Elsas et al., 1998). Therefore, studies designed to isolate and characterize plasmid replication and incompatibility sequences from environmental isolates will aid in the determination of plasmid diversity, as well as to provide more detailed insights into gene movement in microbial communities.

Sobecky et al. (1998) have devised a replicon rescue strategy to isolate replication and incompatibility sequences from gram-negative marine bacteria. This general approach has been used successfully in isolating numerous plasmid replication origins from marine bacteria belonging to the α and γ subclass of the *Proteobacteria* (Table 2). The outlined methodology should be applicable to isolating plasmid replication origins with extended host ranges from a variety of gram-negative marine bacteria (Figure 1). Specifically, plasmid DNA is obtained from a 500 ml cell culture grown in either TSS or half-strength YTSS (Sobecky et al., 1996). To facilitate recovery of large, low copy number plasmids (i.e. >ca. 50 kb; less than 15 copies per chromosome), the volume of the cell culture should be increased several-fold. A modification of the alkaline lysis method of Birnboim & Doly (1979) is used to isolate supercoiled plasmid DNA with subsequent purification of the DNA by cesium chloride gradient centrifugation (Sobecky et al., 1998). Approximately 1 μg of plasmid DNA is partially digested with the restriction endonuclease *Sau3A*I. The partially digested plasmid DNA is ligated to the Tn903 *npt* gene isolated as a *BamH*I fragment from pUC4K (Vieria & Messing, 1982). The ligation mixture is sequentially transformed into the host strains *E. coli* DH5α and the *polA1 E.coli* C2110. Assaying replicons for replication in *E. coli* C2110 confirms the lack of requirement for host DNA polymerase I (Pol I). Clonal analysis is done to identify the transformant(s) containing the smallest replication-proficient fragment. Typically, smaller replication-proficient fragments (2 kb–3 kb) can be generated by increasing the length of *Sau3*AI incubation time which expedites ease and cost of sequencing the plasmid DNA fragment containing the replication origin of interest.

Since culturable bacterial isolates are not representative of the total microbial community in terms of species composition and abundance (Giovannoni et al., 1990; DeLong, 1992; Barns et al., 1994), information on the diversity and abundance of plasmid populations occurring in the non-culturable bacterial community is also needed. Current methods being used to isolate total community DNA are not suitable for the isolation of plasmid-encoded replication and incompatibility sequences, due to the anticipated large size (>50 kb) and low-copy-number of many plasmids which hinders their isolation. In addition, care must be taken to avoid possible contamination of plasmid DNA with chromosomal origins (*oriC*) since *oriC*-containing fragments may be capable of autonomous replication. Attempts to modify existing methods and develop new protocols for the isolation of high quality and quantity supercoiled plasmid DNA from marine sediment microbial communities are in progress (Cook and Sobecky, unpublished).

Plasmid diversity in marine microbial communities

To date, there have been few studies attempting to characterize the molecular diversity and transfer dynamics of plasmid populations encountered in naturally occurring marine bacterial assemblages. While some plasmids confer phenotypes such as antibiotic and heavy metal resistance, colicin production, and virulence traits that can be used to differentiate plasmids into groups, these traits cannot be used to characterize relationships between plasmids. Therefore, a collection of incompatibility sequences (e.g. that are proven to be plasmid-group specific) derived from marine bacteria will greatly facilitate the elucidation of plasmid diversity in naturally occurring marine microbial communities.

Although data is lacking on plasmid diversity in marine bacteria, some information on the extent of diversity is available from studies characterizing plasmids in *Escherichia* and *Bacillus*. Previously, Selander et al. (1987) reported a high degree of plasmid diversity in *E. coli* strains containing numerous plasmids, indicative of the presence of multiple incompatibility groups occurring in the same host. *E. coli* strains containing plasmids conferring antibiotic resistance traits and colicin production also display a high degree of diversity (Novick, 1987; Riley & Gordon, 1992). In contrast to the high levels of *E. coli* plasmid diversity observed, Zawadzki et al. (1996) recently reported a lack of plasmid diversity in *Bacillus* strains. Southern hybridization analysis of thirteen plasmids isolated from the *Bacillus subtilus*, *B. mojavensis* and *B. licheniformis* strains obtained from geographically distant locales, indicated that all but one of the plasmids had extensive regions of homology to each other.

Sobecky et al. (1998) have undertaken preliminary studies to examine the extent of plasmid diversity in marine bacteria obtained from coastal California salt marsh sediments. Replication-proficient fragments were isolated from purified, endonuclease digested plasmid DNA obtained from culturable gram-negative marine bacteria by rescuing in an *E. coli* host background as previously described. Fatty acid determinations and 16S rRNA phylogenetic analysis classified four of the five plasmid-bearing marine isolates to the genus *Vibrio* (Table 2). Analysis of the four replication fragments designated repSD32, repSD41 and repSD172 indicated that these fragments lacked sequence homology, however repSD121 shared considerable regions of homology (77%–94%) to repSD41 but the two origins are compatible in the same host (Sobecky, unpublished). Although the sample size is small, three of the four naturally occurring marine *Vibrio* sp. harbored different replication sequences, indicating a considerable level of plasmid diversity amongst culturable gram-negative marine sediment bacteria, regardless of the phenotypes that they may confer. Continued studies to isolate and characterize replication-proficient fragments from marine bacteria, particularly from phylotypes that are shown to be either numerically dominant or ecologically important, will greatly enhance our understanding of the extent of plasmid diversity in marine microbial communities.

Plasmid transfer in marine microbial communities

Plasmids influence evolutionary events (e.g. adaptations to changing environmental conditions) in microbial populations by their ability to transfer genes to unrelated species. Much of the current understanding of horizontal gene exchange is derived from plasmids occurring in bacteria of medical and agricultural importance. These plasmids, however, represent a rather specific collection of replicons that are not representative of the plasmid populations occurring in marine sediment and water column bacterial isolates (Dahlberg et al., 1997; Sobecky et al., 1997). Perhaps not surprisingly, most studies addressing gene transfer in marine ecosystems to date have used *incP* plasmids (e.g. RK2/RP1/RP4) because of the *incP* vegetative origin ability to replicate in diverse host backgrounds (Thomas & Helinski, 1989) and the versatility of the transfer genes encoded on these plasmids (Guiney, 1993). Nonetheless, this approach has provided valuable insights into the nature and frequency of plasmid-mediated transfer events likely to occur in marine environments.

Previous studies using nutrient-rich conditions have reported the transfer of mercury (Gauthier et al., 1985) and antibiotic resistance determinants (Sandaa et al., 1992) to *E. coli* from gram-negative marine bacteria. Reported transfer frequencies ranged from 10^{-3} to 10^{-8}. The ability of plasmids to transfer between bacteria has been demonstrated under varying abiotic conditions (e.g. nutrient depletion, pH and temperature fluctuations) in both simulated and natural environments (Goodman et al., 1994). For example, Goodman et al. (1993) pre-starved marine *Vibrio* donors containing RP1 and *Vibrio* recipients for prolonged periods (as much as 100 days in the case of the marine recipient) and detected plasmid transfer in the absence of nutrients. Their findings clearly demonstrated that bacteria adapted to oligotrophic nutrient conditions, common for many marine systems, maintain the ability to readily transfer plasmids. Such results provide evidence that plasmid-mediated gene exchange is likely to be an important factor in determining the structure and function of marine microbial communities, even under less than optimal growth conditions.

Biotic factors such as cell densities are also known to affect frequencies of horizontal transfer (Trevors et al., 1987). The higher donor and recipient cell densities and cell-to-cell contact likely to occur in marine sediment environments, and biofilm bacterial communities favor considerably higher frequencies of genetic exchange, relative to water column-based microbial communities. Previously, Angles et al. (1993) have observed 100-fold increases in frequencies of plasmid transfer among marine bacterial isolates introduced into artificial biofilms as compared to the same bacteria present in the water column. Predation, in the form of heterotrophic protozoan grazing, has also been shown to increase the frequency of transfer of the broad-host-range plasmid RK2 between marine *Vibrio* strains by more than 100-fold (Otto et al., 1997).

To date, however, relatively few studies have focused on the role of naturally occurring broad-host-range plasmids in promoting gene transfer in marine ecosystems and how such plasmids determine the structure of bacterial populations. Because of their ability to replicate in diverse bacterial genera, plasmids with broad-host-range capabilities are likely to influence microbial community structure and function. Moreover, broad-host-range plasmids with mobilization and/or self-transfer capabilities will promote the dissemination of advantageous genes throughout the indigenous microbial population. Although conjugation appears to be a primary mode of gene transfer in many environments, transformation and transduction may also be important methods of exchange in soil and sediment systems as well as in aquatic systems (Trevors et al., 1987; Chamier et al., 1993; Frischer et al., 1994). A previous review by Hermansson & Linberg (1994) reported on all three mechanisms of genetic exchange (transformation, transduction and conjugation) in marine environments. Regardless of the mechanism of gene exchange, additional studies are needed to fully determine the potential for and impact of indigenous broad-host-range plasmid transfer on the structure and function of marine sediment bacterial communities.

Conclusion

The application of molecular biology techniques to microbial ecology studies allows innovative approaches to elucidate the structure and function of marine sediment microbial communities. Recently, Stretton et al. (1998) employed laser scanning confocal microscopy (LSCM) to visualize marine bacteria localized in biofilms. The marine bacteria had been tagged with green fluorescent protein (GFP) using a mini-Tn10 transposon delivery system. The *gfp* gene expression was monitored in living cells *in situ* and in real time, thereby providing a unique opportunity to study gene expression in marine bacteria. Dahlberg et al. (1998) also employed GFP to detect plasmid transfer from *Pseudomonas putida* to indigenous marine bacteria in seawater. By tagging a conjugative plasmid with the *gfp* gene, Dahlberg et al. (1998) could monitor plasmid transfer to individual bacterial cells by epifluorescence microscopy.

By promoting the movement of genes throughout bacterial populations, plasmids can exert a direct effect on ecological processses. Presently, additional basic information on the molecular functions (i.e. transfer, maintenance, host range, replication and incompatibility) of indigenous plasmids is needed to assess the role of *in situ* plasmid-mediated gene exchange in marine bacterial populations. Continued efforts to identify and characterize plasmid distribution and diversity in marine ecosystems should provide new insights and understanding of bacterial gene flux mediated by naturally occurring plasmids.

Acknowledgements

The author's work is supported by the Office of Naval Research (N00014-98-1-0076). Thanks are extended to J. Mallonee and N. Reyes for critical review of the manuscript.

References

Angles, M. L., K. C. Marshall & A. E. Goodman, 1993. Plasmid transfer between marine bacteria in the aqueous phase and biofilms in reactor microcosms. Appl. envir. Microbiol. 59: 843–850.

Aviles, M., J. C. Codina, A. Perez-Garcia, F. Cazorla, P. Romero & A. de Vicente, 1993. Occurrence of resistance to antibiotics and metals and of plasmids in bacterial strains isolated from marine environments. Wat. Sci. Tech. 27: 475–478.

Barkay, T., N. Kroer, L. D. Rasmussen & S. J. Sorensen, 1995. Conjugal transfer at natural population densities in a microcosm simulating an estuarine environment. FEMS Microbiol. Ecol. 16: 43–54.

Barns, S. M., R. E. Fundyga, M. W. Jeffries & N. R. Pace, 1994. Remarkable archaeal diversity detected in a Yellowstone National Park hot spring environment. Proc. natl. Acad. Sci. U.S.A. 91: 1609–1613.

Baya, A. M., P. R. Brayton, V. L. Brown, D. J. Grimes, E. Russek-Cohen & R. R. Colwell, 1986. Coincident plasmids and antimicrobial resistance in marine bacteria isolated from polluted and unpolluted Atlantic Ocean samples. Appl. envir. Microbiol. 51: 1285–1292.

Belliveau, B. H., M. E. Starodub & J. T. Trevors, 1991. Occurrence of antibiotic and metal resistance and plasmids in *Bacillus* strains isolated from marine sediment. Can. J. Microbiol. 37: 513–520.

Benson, S. & J. Shapiro, 1978. TOL is a broad-host-range plasmid. J. Bact. 135: 278–280.

Birnboim, H. C. & J. Doly, 1979. A rapid alkaline extraction procedure for screening recombinant plasmid DNA. Nucleic Acids Res. 7: 1513–1523.

Bukhari, A. I., J. A. Shapiro & S. L. Adhya, 1977. DNA insertion elements, plasmids, and episomes. Cold Spring Harbor Laboratory, Cold Spring Harbor, N.Y., U.S.A.

Chamier, B., M. G. Lorenz & W. Wackernagel, 1993. Natural transformation of *Acinetobacter calcoaceticus* by plasmid DNA adsorbed on sand and groundwater aquifer material. Appl. envir. Microbiol. 59: 1662–1667.

Cook, M.A. and P.A. Sobecky, unpublished data.

Couturier, M. F., F. Bex, P.L. Bergquist & W. K. Maas, 1988. Identification and classification of bacterial plasmids. Microbiol. Rev. 52: 375–395.

Dahlberg, C., M. Bergström & M. Hermansson, 1998. *In situ* detection of high levels of horizontal plasmid transfer in marine bacterial communities. Appl. envir. Microbiol. 64: 2670–2675.

Dahlberg, C., C. Linberg, V. L. Torsvik & M. Hermansson, 1997. Conjugative plasmids isolated from bacteria in marine environments show various degrees of homology to each other and are not closely related to well-characterized plasmids. Appl. envir. Microbiol. 63: 4692–4697.

Datta, N., 1979. Plasmid classification: incompatibility grouping. In K. N. Timmis & A. Puhler (ed.), Plasmids of Medical, Environmental and Commercial Importance. Elsevier/North-Holland Biomedical Press, Amsterdam, The Netherlands. 3–11.

Datta, N. & R. W. Hedges, 1972. Host ranges of R factors. J. Gen. Microbiol. 70: 453–460.

Davey, R. B., P. I. Bird, S. M. Nikoletti, J. Prazkier & J. Pittard, 1984. The use of mini-gal plasmids for rapid incompatability grouping of conjugative R plasmids. Plasmid 11: 234–242.

DeLong, E. F., 1992. Archaea in coastal marine environments. Proc. Natl. Acad. Sci. U.S.A. 89: 5685–5689.

Delong, E. F., 1997. Marine microbial diversity: the tip of the iceberg. Trends Biotechnol. 15: 203–207.

Erauso, G., S. Marsin, N. Benbouzid-Rollet, M.-F. Baucher, T. Barbeyron, Y. Zivanovic, D. Prieur & P. Forterre, 1996. Sequence of plasmid pGT5 from the Archaeon *Pyrococcus abyssi*: evidence for rolling-circle replication in a hyperthermophile. J. Bact. 178: 3232–3237.

Frischer, M. E., G. J. Stewart & J. H. Paul, 1994. Plasmid transfer to indigenous marine bacterial populations by natural transformation. FEMS Microbiol. Ecol. 15: 127–136.

Fry, J. C., 1994. Genetic transfer in water. Presented at the Juan March Centre for International Meetings on Biology, Sevilla, Spain, 14–16 February, 1994.

Gauthier, M. J., F. Cauvin & J.-P. Breittmayer, 1985. Influence of salts and temperature on the transfer of mercury resistance from a marine pseudomonad to *Escherichia coli*. Appl. envir. Microbiol. 50: 38–40.

Giovannoni, S. J., T. B. Britschgi, C. L. Moyer & K. G. Field, 1990. Genetic diversity in Sargasso Sea bacterioplankton. Nature 345: 60–63.

Glassman, D. L. & L. A. McNicol, 1981. Plasmid frequency in natural populations of estuarine microorganisms. Plasmid 5: 231.

Goodman, A. E., E. Hild, K. C. Marshall & M. Hermansson, 1993. Conjugative plasmid transfer between bacteria under simulated marine oligotrophic conditions. Appl. envir. Microbiol. 59: 1035–1040.

Goodman, A. E., K. C. Marshall & M. Hermansson, 1994. Gene transfer among bacteria under conditions of nutrient depletion in simulated and natural aquatic environments. FEMS Microbiol. Ecol. 15: 55–60.

Guiney, D. G., 1993. Broad host range conjugative and mobilizable plasmids in gram-negative bacteria. In Don Clewell (ed.), Bacterial Conjugation. Plenum Press, N.Y., U.S.A. 75–103.

Hada, H. S. & R. K. Sizemore, 1981. Incidence of plasmids in marine *Vibrio* spp. isolated from an oil field in the northwestern Gulf of Mexico. Appl. envir. Microbiol. 41: 199–202.

Hansen, C. L., G. Zwolinsk, D. Martin & J. W. Williams, 1984. Bacterial removal of mercury from sewage. Biotechnol. Bioeng. 26: 1330–1303.

Helinski, D. R., A. E. Toukdarian & R. P. Novick, 1996. Replicaton control and other stable maintenance mechanisms of plasmids. In F. C. Neidhardt, R. Curtiss III, J. L. Ingraham, E. C. C. Lin, K. B. Low, B. Magasanik, W. S. Reznikoff, M. Riley, M. Schaechtger & H. E. Umbarger (ed.), *Escherichia coli* and *Salmonella typhimurium*: Cellular and Molecular Biology. American Society for Microbiology, Washington, D.C., U.S.A.: 2295–2324.

Hermansson, M., G. W. Jones & S. Kjelleberg, 1987. Frequency of antibiotic and heavy metal resistance, pigmentation, and plasmids in bacteria of the marine air-water interface. Appl. envir. Microbiol. 53: 2338–2342.

Hermansson, M. & C. Linberg, 1994. Gene transfer in the marine environment (minireview). FEMS Microbiol Ecol. 15: 47–54.

Hill, K. E., A. J. Weightman & J. C. Fry, 1992. Isolation and screening of plasmids from the epilithon which mobilize recombinant plasmid pD10. Appl. envir. Microbiol. 58: 1292–1300.

Kinkle, B. K., Sadowsky, M. J., E. L. Schmidt & W. C. Koskinen, 1993. Plasmids pJP4 and R68.45 can be transferred between populations of bradyrhizobia in soil. Appl. envir. Microbiol. 59: 1762–1766.

Kobayashi, N. & M. J. Bailey, 1994. Plasmids isolated from the sugar beet phyllosphere show little or no homology to molecular probes currently available for plasmid typing. Microbiology 140: 289–296.

Kobori, H., C. W. Sullivan & H. Shizuya, 1984. Bacterial plasmids in Antarctic natural assemblages. Appl. envir. Microbiol. 48: 515–518.

Leahy, J. G., C. C. Somerville, K. A. Cunningham, G. A. Adamantiades, J. J. Byrd & R. R. Colwell, 1990. Hydrocarbon mineralization in sediments and plasmid incidence in sediment bacteria from the Campeche Bank. Appl. envir. Microbiol. 56: 1565–1570.

Lilley, A. K., M. J. Bailey, M. J. Day & J. C. Fry, 1996. Diversity of mercury resistance plasmids obtained by exogenous isolation from the bacteria of sugar beet in three successive years. FEMS Microbiol. Ecol. 2: 211–227.

Maruyama, A., M. Oda & T. Higashihara, 1993. Abundance of virus-sized non-DNase-digestible DNA (coated DNA) in eutrophic seawater. Appl. envir. Microbiol. 3: 712–718.

Novick, R. P., R. C. Clowes, S. N. Cohen, R. Curtiss III, N. Datta & S. Falkow, 1976. Uniform nomenclature for bacterial plasmids: a proposal. Bact. Rev. 40: 168–189.

Novick, R. P., 1987. Plasmid incompatibility. Microbiol. Rev. 51: 381–395.

Ogunseitan, O. A., G. S. Sayler & R. V. Miller, 1990. Dynamic interactions between *Pseudomonas aeruginosa* and bacteriophages in lakewater. Microb. Ecol. 19: 171–185.

O'Morchoe, S., O. Ogunseitan, G. S. Sayler & R. V. Miller, 1988. Conjugal transfer of R68.45 and FP5 between *Pseudomonas aeruginosa* strains in a freshwater environment. Appl. envir. Microbiol. 54: 1923–1929.

Otto, K., D. Weichart & S. Kjelleberg, 1997. Plasmid transfer between marine *Vibrio* strains during predation by the heterotrophic microflagellate *Cafeteria roenbergensis*. Appl. envir. Microbiol. 63: 749–752.

Paul, J. H., M. E. Frisher & J. M. Thurmond, 1991. Gene transfer in marine water column and sediment microcosms by natural plasmid transformation. Appl. envir. Microbiol. 57: 1509–1515.

Pickup, R. W., 1989. Related plasmids found in an English Lake District stream. Microb. Ecol. 18: 211–220.

Riley, M. A. & D. M. Gordon, 1992. A survey of Col plasmids in natural isolates of *Escherichia coli* and an investigation into the stability of Col-plasmid lineages. J. Gen. Microbiol. 138: 1345–1352.

Rochelle, P. A., J. C. Fry & M. J. Day, 1989. Factors affecting conjugal transfer of plasmids encoding mercury resistance from pure cultures and mixed bacterial suspensions of epilithic bacteria. J. gen. Microbiol. 135: 409–424.

Sandaa, R.-A., V. L. Torsvik & J. Goksoyr, 1992. Transferable drug resistance in bacteria from fish-farm sediments. Can. J. Microbiol. 38: 1061–1065.

Saye, D. J., O. Ogunseitan, G. S. Sayler & R. V. Miller, 1990. Transduction of linked chromosomal genes between *Pseudomonas aeruginosa* strains during incubation *in situ* in a freshwater habitat. Appl. envir. Microbiol. 56: 140–145.

Sayler, G. S., S. W. Hooper, A. C. Layton & J. H. King, 1990. Catabolic plasmids of environmental and ecological significance. Microb. Ecol. 19: 1–20.

Schutt, C., 1989. Plasmids in the bacterial assemblage of a dystrophic lake: evidence for plasmid-encoded nickel resistance. Microb. Ecol. 17: 49–62.

Selander, R. K., D. A. Caugant & T. S. Whittam, 1987. Genetic structure and variation in natural populations of *Escherichia coli*. In Neidhardt, F. C., J. L. Ingraham, K. B. Low, B. Magasanik, M. Schaechtger & H. E. Umbarger (eds), *Escherichia coli* and *Salmonella typhimurium*: Cellular and Molecular Biology. American Society for Microbiology, Washington, D.C., U.S.A.: 1625–1648.

Simon, R. D., M. Shilo & J. W. Hastings, 1982. The absence of a correlation between plasmids and luminescence in marine luminous bacteria. Curr. Microbiol. 7: 175–180.

Sizemore, R. K. & R. R. Colwell, 1977. Plasmids carried by antibiotic resistant marine bacteria. Antimicrob. Agents Chemother. 12: 372–382.

Sobecky, P. A., T. J. Mincer, M. C. Chang & D. R. Helinski, 1997. Plasmids isolated from marine sediment microbial communities contain replication and incompatibility regions unrelated to those of known plasmid groups. Appl. envir. Microbiol. 63: 888–895.

Sobecky, P. A., T. J. Mincer, M. C. Chang, A. E. Toukdarian & D. R. Helinski, Isolation of broad-host-range replicons from marine sediment bacteria. Appl. envir. Microbiol. 64: 2822–2830.

Sobecky, P. A., M. A. Schell, M. A. Moran & R. E. Hodson, 1996. Impact of a genetically engineered bacterium with enhanced alkaline phosphatase activity on marine phytoplankton communities. Appl. envir. Microbiol. 62: 6–12.

Stretton, S., S. Techkarnjanaruk, A. M. McLennan & A. E. Goodman, 1998. Use of green fluorescent protein to tag and investigate gene expression in marine bacteria. Appl. envir. Microbiol. 64: 2554–2559.

Thomas, C. M. & D. R. Helinski, 1989. Vegetative replication and stable inheritance of IncP plasmids. In C. M. Thomas (ed.), Promiscuous Plasmids of Gram-Negative Bacteria. Academic Press, London: 1–25.

Top, E., I. De Smet, W. Verstraete, R. Dijkmans & M. Mergeay, 1994. Exogenous isolation of mobilizing plasmids from polluted soils and sludges. Appl. envir. Microbiol. 60: 831–839.

Trevors, J. T., T. Barkay & A. W. Bourquin, 1987. Gene transfer among bacteria in soil and aquatic environments: a review. Can. J. Microbiol. 33: 191–198.

Van Elsas, J. D., B. B. McSpadden-Gardener, A. C. Wolters & E. Smit, 1998. Isolation, characterization, and transfer of cryptic gene-mobilizing plasmids in the wheat rhizosphere. Appl. envir. Microbiol. 64: 880–889.

Vieria, J. & J. Messing, 1982. The pUC plasmids, an M13mp7-derived system for insertion mutagenesis and sequencing with synthetic universal primers. Gene 19: 259–268.

Willetts, N. & B. Wilkins, 1984. Processing of plasmid DNA during bacterial conjugation. Microbiol. Rev. 48: 24–41.

Wortman, A. T. & R. R. Colwell, 1988. Frequency and characteristics of plasmids in bacteria isolated from deep sea Amphipods. Appl. envir. Microbiol. 54: 1284–1288.

Zawadzki, P., M. A. Riley & F. M. Cohen, 1996. Homology among nearly all plasmids infecting three *Bacillus* species. J. Bacteriol. 178: 191–198.

Hydrobiologia **401**: 19–32, 1999.
J.P. Zehr & M.A. Voytek (eds), Molecular Ecology of Aquatic Communities.

Use of the polymerase chain reaction and denaturing gradient gel electrophoresis to study diversity in natural virus communities

S. M. Short & C. A. Suttle*
The Departments of Earth and Ocean Sciences, Microbiology and Immunology, and Botany, University of British Columbia, 6270 University Blvd. , Vancouver, British Columbia, Canada, V6T 1Z4
E-mail: suttle@eos.ubc.ca; Fax: (604) 822-6091

Key words: phytoplankton, viruses, diversity, polymerase, PCR, DGGE

Abstract

Viruses are abundant members of marine and freshwater microbial communities, and are important players in aquatic ecology and geochemical cycles. Recent methodological developments have allowed the use of the polymerase chain reaction (PCR) to examine the diversity of natural communities of viruses without the need for culture. DNA polymerase genes are highly conserved and are, therefore, suitable targets for PCR analysis of microbes that do not encode rRNA. As natural virus communities are largely made up of dsDNA viruses, and as many dsDNA algal viruses encode their own DNA polymerase, PCR primers can be designed to amplify fragments of these genes. This approach has been used to examine the genetic diversity in natural communities of viruses that infect phytoplankton. Algal-virus-specific primers were used to amplify polymerase fragments from natural virus samples, demonstrating the presence of a diverse community of viruses closely related to those that are known to infect phytoplankton. We have modified this approach by using denaturing gradient gel electrophoresis (DGGE) to rapidly analyze PCR products. DGGE will permit rapid and efficient fingerprinting of natural marine viral communities, and allow spatial and temporal differences in viral community structure to be examined. This paper provides a brief overview of how PCR and DGGE can be used to examine diversity in natural viral communities drawing on viruses that infect phytoplankton as an example.

Introduction

Torella & Morita (1979) provided the first evidence that viruses were abundant in seawater, but it was not until a decade later that accurate estimates were obtained of their abundance and ubiquitous distribution in marine and fresh waters (Bergh et al., 1989; Proctor & Fuhrman, 1990; Suttle et al., 1990). Electron and epifluorescence microscopy revealed that abundances of these small (0.02 to 0.2 μm dia.) particles ranged from $<10^4$ to $>10^8$ ml^{-1}, depending on the season and the environment sampled (reviewed in Thingstad et al., 1993; Fuhrman & Suttle, 1993; Børsheim, 1994). In general, viral abundance and distribution parallels that of other biotic factors such as bacteria, Chl *a*, and dissolved and particulate DNA (Paul et al., 1991; Weinbauer et al., 1993).

While viruses have an obvious effect on natural populations as a source of mortality, Suttle (1994) stated that the greatest impact of viruses on aquatic ecosystems might be through non-steady-state processes such as genetic exchange between microbial populations or through alterations of community composition. Viruses are, by definition, obligate pathogens that require host cells in order to replicate. Furthermore, infection of cells by viruses requires direct contact between a virus particle and its host. The probability of successful infection is dependent on the contact rate between viruses and their hosts, and a rapid increase in virus abundance typically occurs only when host cells reach a threshold density (Wiggins & Alexander, 1985). As viruses are known to infect cells in a density dependent manner, it follows that viruses potentially have the greatest impact on the most abundant host cells in a community. Lysis of the most

* Author for correspondence.

abundant host cells may lead to increased diversity through competitive release, or may prevent some species from forming blooms (Fuhrman & Suttle, 1993; Thingstad et al., 1993; Bratbak et al., 1994; Suttle, 1994). The cosmopolitan phytoplankton *Micromonas pusilla* may be an example of the latter (Cottrell & Suttle, 1995a). In this way, viruses may have a direct impact on community composition and species diversity.

Viruses may also infect cells through a lysogenic rather than lytic life cycle and may be directly involved in the exchange of genetic material with their hosts. In the lytic pathway of infection, viruses inject their DNA into a host cell. This DNA is then replicated, transcribed, and translated to produce new virus particles. Host cells eventually lyse and new particles are released. In the lysogenic, or latent, life cycle, virus genetic information is not used to replicate new particles. Instead this DNA is archived in the host cell and can remain inactive through many host-cell generations. Eventually, the virus may be induced by an environmental signal, the virus genome is then replicated, and new infectious particles are formed. When hosts maintain viral DNA, host and viral genomes are often intimately associated; this association often results in increased probability of genetic transformation of hosts. To date there is no experimental evidence for widespread lysogeny in marine eukaryotic phytoplankton; however, there is accumulating evidence that lysogeny occurs in many marine prokaryotes (Jiang & Paul, 1994; Ohki & Fujita, 1996). Furthermore, Chiura (1997) described the process of gene transfer by viruses isolated from marine bacteria. While non-steady-state processes such as competitive release or genetic transformation may be important to phytoplankton communities, there is no direct experimental evidence that virus infection can lead to increased phytoplankton species diversity, or that latent viruses infecting eukaryotic phytoplankton exist.

Brown (1972) wrote the first comprehensive review of algal viruses. At this time there was substantial evidence, based largely on ultrastructural studies, that viruses infected both prokaryotic and eukaryotic algae, yet few of the viruses were isolated. The first marine algal viruses isolated infected the picoflagellate *Micromonas pusilla* (Mayer & Taylor, 1979). Currently, lytic viruses infecting at least six taxa of eukaryotic microalgae are established in culture. The first of these to be characterized in detail are a group of viruses which infect freshwater *Chlorella*-like algae which are found as symbionts in *Hydra viridis* and *Paramecium bursaria* (for review see Van Etten et al., 1991). These viruses have been assigned to a new family, the Phycodnaviridae. The other viruses that have been isolated infect marine phytoplankton. These include viruses that infect the prasinophyte *Micromonas pusilla* (Mayer & Taylor, 1979; Cottrell & Suttle, 1991), the prymnesiophytes *Chrysochromulina* spp. (Suttle & Chan, 1995) and *Phaeocystis pouchetii* (Jacobsen et al., 1996), the pelagophyte *Aureococcus anophagefferens* (Milligan & Cosper, 1994), and the raphidophyte *Heterosigma akashiwo* (Nagasaki & Yamaguchi, 1997). With the exception of the *A. anophagefferens* virus, these viruses share a common morphology; they are large (>100 nm dia.) polyhedral viruses with double-stranded DNA.

An important problem that has been difficult to address is to identify the range of phytoplankton species that are infected by viruses. Suttle et al. (1990) demonstrated that increasing the abundance of naturally occurring viruses in seawater samples significantly reduced photosynthetic rates. Furthermore, photosynthetic rates of specific subsets of the phytoplankton assemblage were affected by the addition of concentrated marine virus communities (Suttle, 1992). While these studies demonstrate that viruses may infect a wide range of marine planktonic algae, determination of the range of phytoplankton species infected has remained elusive. This difficulty stems from the fact that study of the range of organisms infected has relied on culture techniques that tend to be slow and cumbersome. Furthermore, studies have shown that the most abundant phytoplankton strains may be resistant to the most abundant viruses (Waterbury and Valois, 1993). If cultured strains of specific phytoplankton species are not representative of a natural community, or if certain strains are not easily cultured, virus isolation with host cultures allows identification of only a small portion of the viruses present in a given sample. Therefore, the diversity of viruses that infect marine phytoplankton may be greatly underestimated.

PCR amplification of virus-specific sequences allows the rapid determination of genetic diversity in natural samples without the need to purify the organisms in culture. Information on diversity, when combined with data on specific virus isolates can lead to an understanding of the effects of these pathogens. For example, DNA polymerase sequences from cultured algal virus strains form distinct clades dependent on the microalgal host infected (Chen et al., 1996; Figure 1). This information also permits inferences to be made on the number of host taxa in a nat-

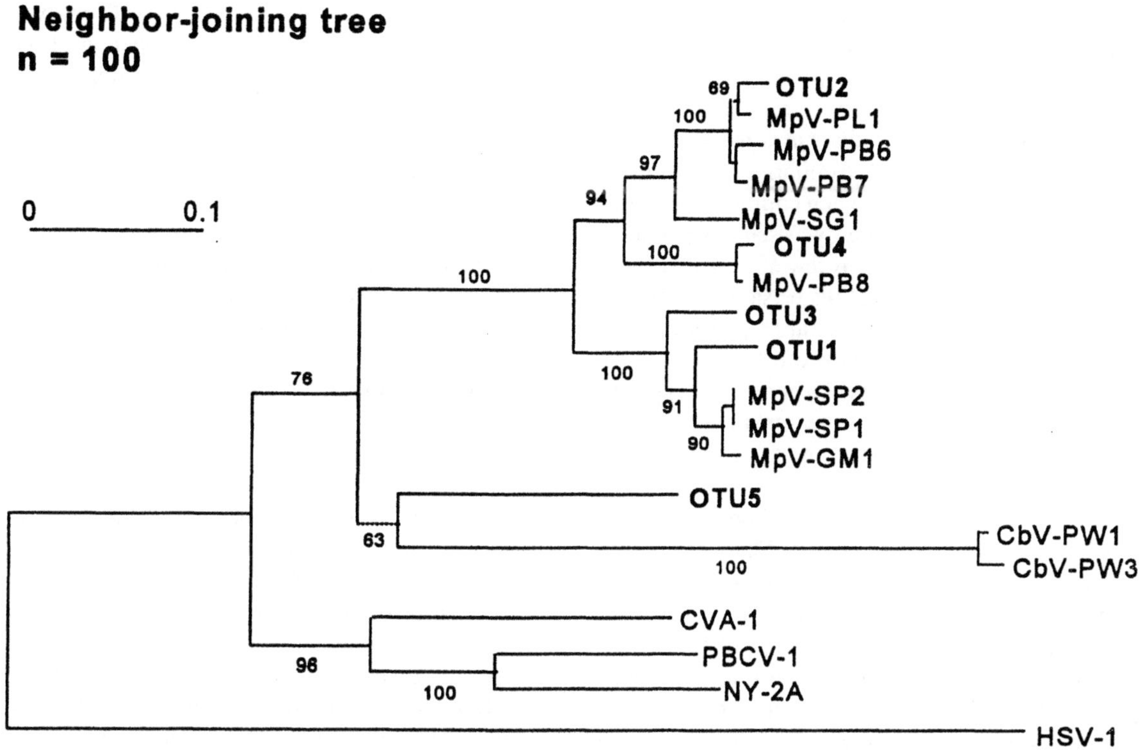

Figure 1. A neighbor-joining tree showing the phylogenetic relationships among five operational taxonomic units amplified from the Gulf of Mexico and 13 other algal viruses based on sequences of DNA *pol* gene fragments. Herpes simplex virus type 1 was used as an out-group. The numbers at the nodes indicate bootstrap values (n =100), and branches with values less than 50 have been collapsed. The scale bar represents 0.1 fixed mutation per nucleotide position. Abbreviations are as follows: MpV = *Micromonas pusilla* virus; CbV = *Chrysochrormulina brevifilum* virus; CVA-1, PBCV-1, and NY-2A = viruses infecting *Chlorella* -like algae; HSV-1 = Herpes simplex virus type 1; OTU = operational taxonomic units.

ural community; i.e. it is reasonable to hypothesize that virus communities composed of many separate clades of virus sequences infect a larger number of taxa than communities with fewer clades. Further, as more viral polymerase gene sequences are collected, it will be possible to identify specific virus assemblages in nature and infer the host organisms that are infected. Such data should allow a more accurate assessment of the role of marine viruses in regulating algal community structure and primary production than is currently available. For example, if knowledge of the community structure of algal viruses can be coupled with data on their abundance and turnover rates, it is possible to determine the effect of viruses on the mortality of specific components of the phytoplankton community. A similar approach was used to estimate that MpV may be responsible for the mortality of 2.0–10.0% of the *M. pusilla* population on a daily basis (Cottrell & Suttle, 1995a) and that cyanophages may remove from 2.0 to 7.0% of *Synechococcus* cells daily (Suttle & Chan, 1995).

Furthermore, PCR amplification of viral polymerase genes provides a tool for screening of phytoplankton genomic DNA for the presence of lysogens. As mentioned, the algal viruses isolated to date appear to be lytic, and there is no direct evidence of lysogenic viruses that infect eukaryotic phytoplankton. Whether this is a reflection of current methods or a biological reality is unknown. Through molecular techniques, we will be better able to assess algal virus diversity and the potential for lysogeny of algal hosts. The following discussion will focus on PCR amplification of DNA polymerase gene sequences as a tool for studying diversity in natural virus communities.

Polymerase Chain Reaction (PCR)

Ever since PCR using a thermostable DNA polymerase (*Taq*) was first described (Mullis and Faloona, 1987; Saiki et al., 1988), the technique has become widely used in studies ranging from forensics to ecology (Innis et al., 1990). The wide spread use of

PCR stems from the fact that it allows the detection and exponential amplification of rare sequences from heterogeneous DNA mixtures. As an example, PCR has permitted the detection of as few as 100 cells against a background of 10^{11} diverse nontarget organisms (Steffan & Atlas, 1991). Similarly, other studies have demonstrated that PCR, when combined with the use of radiolabelled gene probes, was sensitive enough to detect gene sequences from as few as 1 to 5 bacteria cells in 100 ml environmental water samples (Bej et al., 1990). For the detection of viral DNA sequences, PCR has been proven to be equally effective. In a study of enteroviruses in water, Kopecka et al. (1993) found that PCR was $10–10^3$ times more sensitive than cell culture, and was $10^5–10^7$ times more sensitive than direct probe hybridization.

A widely utilized PCR-based method for aquatic studies has been the amplification of DNA sequences which encode small subunit rRNA in marine microorganisms. A long-standing dogma in marine microbiology is that most of the bacteria present in a given water sample are uncultivable. Therefore, estimates of population diversity based on cultivable bacteria greatly underestimate the actual diversity present in a sample (Giovannoni & Cary, 1993). To circumvent this problem many aquatic microbiologists have used molecular techniques to estimate the diversity of microbes in water samples. For example, Giovannoni et al. (1990) used PCR amplified small subunit rRNA sequences to estimate the genetic diversity of bacterioplankton in the Sargasso Sea. Similarly Fuhrman et al. (1992) identified novel archaebacteria using rDNA sequence analysis. Recently, researchers have continued to use these methods for the determination of marine bacterial genetic diversity (Field et al., 1997; Suzuki et al., 1997; Wright et al., 1997). It is worth noting that molecular techniques have also been used to examine the diversity and distribution of many important primary producers such as the toxic dinoflagellate *Alexandrium* spp.and prokaryotic cyanobacteria (Scholin & Anderson, 1994; Nübel et al., 1997). It is apparent that in spite of any shortcomings, PCR has become, and will remain, an important tool for the examination of genetic diversity of organisms from environmental samples.

As noted by Giovannoni & Cary (1993), molecular techniques such as cloning may potentially involve biases that are not fully understood, however, it was concluded that these biases are much less encumbering than those imposed by cell culture methods. Others have also noted potential problems with molecular techniques used to determine genetic diversity in natural samples. For example Reysenbach et al. (1992), noted that PCR primers favored the amplification of rDNA sequences from *Saccharomyces cerevisiae* in mixtures of DNA from this yeast and an extremely thermophilic archaebacterium. However, these workers noted that the addition of a denaturant (5% acetamide) eliminated this primer bias. Further, it has been noted that in some cases the formation of chimeric rDNA may lead to erroneous conclusions about the genetic diversity of certain samples (Lisack et al., 1991). However, these workers concluded that detailed analysis of amplified sequences could prevent spurious results.

As is the case for many experimental methods, critical analysis of results is necessary to prevent inaccurate conclusions. Furthermore, when compared to classical culture methods, determination of sample genetic diversity with PCR-based techniques allows the detection of organisms otherwise missed. Therefore, molecular techniques reveal a more accurate picture of species diversity in microbiological studies and will continue to be favored in coming years. However, as viruses do not contain ribosomal RNA sequences it is necessary to use another conserved gene to make phylogenetic inferences.

Conserved nature of DNA polymerases

DNA polymerases are the enzymes responsible for DNA replication and therefore, are essential to all organisms. The structure and function of DNA polymerase enzymes and the genes which code for them have been extensively studied for several decades (for review see Wang, 1991; Joyce & Steitz, 1994). A result of these studies has been the discovery, in many DNA polymerases, of regions of highly conserved amino acid sequence.

One of the first conserved motifs noted in polymerases was a 14 amino-acid sequence that occurred in polymerases from 15 viruses which infect animals, plants and bacteria (Kamer & Argos 1984). The discovery of this conserved region (YGDTDS i.e. Asp-Asp motif) in RNA-dependent polymerases of viruses lead the authors to suggest that the viruses may be derived from a common ancestor. As more polymerase amino acid sequences were determined, comparison revealed other homologous segments among several other viruses; namely ϕ29, adenoviruses, and Epstein-Barr virus (Argos et al., 1986). However, the similarity of one of the homologous segments to the

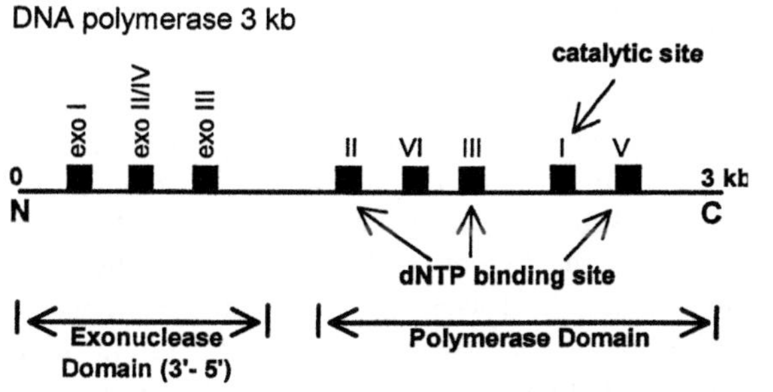

Figure 2. Generic map of B-family DNA polymerases. Conserved motifs are represented by black boxes. Possible functions are shown for the polymerase domain motifs I, II, III and V. The most highly conserved amino acid sequence YGDTDS is located in motif I and is a proposed catalytic site for B-family polymerases. dNTP = deoxynucleoside triphosphate.

Asp-Asp motif described for RNA-dependent polymerases was not noted at this time. It was not until the human DNA-dependent DNA polymerase sequence was determined that the similarities among these motifs were recognized. Furthermore, it was noted that the most highly conserved amino acid sequence was the Asp-Asp motif (Wong et al., 1988), which has been observed in DNA and RNA-dependent RNA and DNA polymerases of humans, yeast and viruses (e.g. Argos, 1988; Pizzagalli et al., 1988; Iwasaki et al., 1991; Braithwaite & Ito, 1993; Illana et al., 1996). Argos (1988) went further and predicted that this amino acid motif would form a strand-loop-strand structure essential to the function of the enzyme.

The deduced amino acid sequence of human DNA polymerase α also shared six regions of homology with yeast DNA pol I, bacteriophages T_4 and $\phi 29$, herpes viruses, vaccinia virus and adenovirus (Wong et al, 1988). It has also been noted by several researchers that these regions of homology are always in the same linear spatial arrangement (Wong et al., 1988; Pizzagalli et al., 1988; Damagnez et al., 1991; Figure 2). The discovery of this segmental homology in many polymerases eventually led to the formation of a family of polymerases with similar conserved sequences. Initially this group of polymerases was informally classified as α-like polymerases based on homology to the human DNA polymerase α. However, it was also noted that the α-like polymerases were homologous to *E. coli* DNA polymerase II (Iwasaki et al., 1991). Subsequently, based on homology with *E. coli* DNA polymerase II, this group of DNA dependent DNA polymerases have been classified as B-family DNA polymerases which includes polymerases from viruses, bacteria, yeast and higher eukaryotes (Ito & Braithwaite, 1991). Furthermore, the B-family DNA polymerases are subdivided into two groups depending on whether they are protein or nucleic acid primed (Braithwaite & Ito, 1993).

DNA polymerase sequences as genetic markers

As a result of the essential and conserved nature of the α-like polymerases, these enzymes can potentially be used as markers for evolutionary relationships among organisms. It was postulated that because of the high degree of amino-acid sequence conservation, α-like polymerases from both prokaryotic and eukaryotic organisms share a common ancestor and differences are the result of divergent evolution (Kamer & Argos, 1984; Jung et al., 1987). Since this initial hypothesis was formulated, many other workers have used different polymerase gene and amino acid sequences to determine the phylogenetic relationships of these enzymes (e.g. Ito & Braithwaite, 1990; Iwasaki et al., 1991; Braithwaite & Ito, 1993). It is apparent that there is considerable utility in the use of polymerase sequences for the determination of phylogenetic relationships of organisms, especially viruses that have genomes of limited complexity when compared to cellular organisms.

Phylogenies based on polymerase genes, or any single copy gene, are not always incontrovertible. For example, Meyer et al. (1997) noted that phylogenies based on polymerase genes differed slightly from those based on other essential genes. Some workers have also found differences in the homology of different polymerases from two organisms; 45% and 32% for Pol δ and Pol α respectively when comparing sequences from *Plasmodium falciparum* and *Saccharomyces cerevisiae* (Ridley et al., 1991). Information such as this has lead to criticism of single-gene-based taxonomy. For example, Calisher et al. (1995) noted several drawbacks to virus phylogenies based on polymerases. Among these were the following:

1. Polymerase based taxonomic schemes are not inclusive and viruses which do not encode a polymerase will be excluded.
2. Genes that occur in a wide range of organisms may have moved horizontally between species leading to erroneous taxonomy.
3. Recombination events in the evolutionary relationships of some viruses may also lead to inaccurate taxonomic relationships.

These authors feel that the International Committee on Taxonomy of Viruses (ICTV) data driven 'bottom up' strategy of classification is the best approach. This strategy is based on the approach of clustering entities into higher taxa when accumulated data demonstrates relationships between lower taxa (Calisher et al., 1995).

While the arguments of Calisher et al. (1995) are noteworthy, aspects of them are, as the authors acknowledge, open to debate. As many different viruses lack distinguishing morphological characteristics, such as those that infect unicellular algae, taxonomic schemes must be based on life histories or genetic information. Therefore, if virus and gene relationships are the result of horizontal gene transfer, or recombination between viruses coinfecting the same organism, this is biologically relevant information and is useful for taxonomic inferences. Indeed, coinfection of a single host suggests that two viruses are very similar, at least in host specificity and life history. Calisher et al. (1995) state that it is not known with certainty if viruses are monophyletic or polyphyletic in origin and that any taxonomic scheme must be contrived. If virus taxonomic schemes are contrived, it is appropriate to base these schemes on biologically relevant information such as host range. While polymerase genes may not be appropriate for the determination of phylogenetic relationships of all viruses, i.e. DNA-dependent DNA polymerases are only useful for taxonomy of viruses with DNA genomes, this does not negate the information that can be gained from the study of these genes. When very little information is known for a specific group of viruses, genetic markers such as polymerase genes are a useful tool and preliminary relationships can be established based on sequence homology of these markers.

The initial evidence for phycodnaviridae-coded α-like DNA polymerases was from a study of viruses that infect a unicellular, exsymbiotic, eukaryotic *Chlorella*-like green alga, strain NC64A (Rohozinski & Van Etten, 1989). This study demonstrated that it was possible to distinguish three different DNA polymerase activities in virus infected *Chlorella* strain NC64A. The authors also demonstrated that the activities of these enzymes differed between infected and uninfected cells suggesting that the enzymes in virus-infected cells could be host polymerases modified by virus-encoded protein(s), or entirely virus encoded. Further study on the DNA polymerase (*pol*) of two viruses (PBCV-1 and NY-2A) that infect this alga demonstrated that the polymerase gene sequences of these viruses were more homologous to each other than to polymerase sequences of other organisms (Grabherr et al., 1992). Interestingly, different regions of the gene had highest homology to different viruses. For example, the amino-acid sequence of the 3'→5' exonuclease domain had highest identity (23% identical amino acids) with phages PRD1 and ϕ29 and herpes simplex virus, while the polymerase domain had highest identity (43% identical amino acids) with two herpes viruses, HSV and EBV. Nonetheless, these studies of viruses infecting a unicellular algae provided the first evidence that some algal viruses encode their own DNA polymerase and this polymerase is closely related to members of the B-family (α-like) DNA polymerases.

PCR amplification of algal virus pol

Following characterization of DNA polymerase genes of viruses infecting a *Chlorella*-like alga, a DNA polymerase gene from a virus (MpV-SP1) which infects the unicellular alga *Micromonas pusilla* was sequenced (Zhang & Suttle, 1994). Analysis of sequences from PBCV-1, NY-2A and MpV-SP1 revealed that the DNA polymerase genes of these viruses contain regions both universally conserved in B-family DNA polymerase genes and conserved only among these viruses. This information was used to design, from inferred amino acid sequences, three highly degenerate primers that can be used for amplification of algal virus *pol* sequences by nested PCR. Two of the primers are complementary to conserved regions of algal virus polymerase genes, and one is complementary to a universally conserved sequence (Chen & Suttle, 1995a,b).

The nested PCR involves two separate PCR reactions using three primers that are complementary to the three regions of the polymerase gene (Chen & Suttle, 1995b; Figure 3). The first round of PCR is carried out with the two primers specific to the algal virus polymerase sequence. A second round of PCR is carried out using one of the primers from the first reaction and a third nested *pol* gene specific primer. This round of PCR is used to confirm the identity of any amplified products from the first reaction (Chen & Suttle, 1995a,b, 1996). Following amplification, products are cloned into plasmid vectors and the resulting library can be screened by restriction fragment length polymorphism (RFLP). Unique clones identified in this way can then be sequenced, and phylogenetic relationships can be determined. It was demonstrated that this method allowed the successful amplification of algal virus polymerase sequences from several cul-

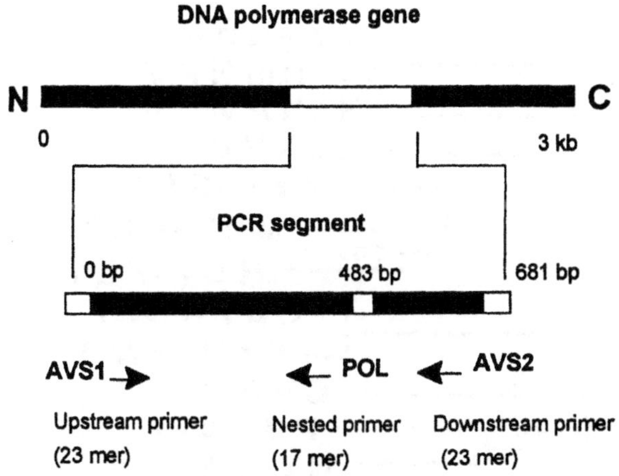

Figure 3. Relative positions of PCR primers on the DNA polymerase gene. The upstream and downstream primers (AVS1 and AVS2) are specific for algal viruses, and the nested primer (POL) is universal for B-family DNA polymerases. The amplified segment shown in white is contained within the polymerase domain of the DNA polymerase gene.

tured algal viruses and viruses from natural samples (Chen & Suttle, 1995b; Chen et al., 1996; Figure 2).

Analysis of amino acid sequences deduced from the polymerase genes of algal viruses and 16 other DNA viruses showed that algal viruses clustered as a distinct group (Chen & Suttle, 1996; Figure 4). As well, when compared to data from total viral genome hybridization (Cottrell & Suttle, 1995b), polymerase sequence analysis produced similar phylogenetic trees thereby demonstrating that DNA pol sequences are reflective of the whole genome of MpV (Chen & Suttle, 1996). Analysis of the virus-*pol*-based phylogeny indicated that DNA polymerase sequences of the Phycodnaviridae are more closely related to each other than to other dsDNA virus polymerases, suggesting that they share a common ancestor (Chen & Suttle, 1996). Furthermore, the viruses fall into three clades, which correlate with their hosts. For example, the sequence from the *Chlorella* virus PBCV-1 is 77% similar to those from two other *Chlorella* viruses, NY-2A and CVA-1, while CVA-1 and NY-2A are 64% similar. Among viruses that infect the members of the Prasinophyceae, including MpV-SP1, the proportion of identical nucleotides for all pairwise combinations ranged from 78 to 99%. The two viruses that infect members of the Prymnesiophyceae, CbV-PW1 and CbV-PW3 shared a high sequence similarity (98%). Overall, sequence similarities among the DNA polymerasegene fragments of three *Chlorella* viruses and all Prasinovirus and Prymnesiovirus clones ranged from 49 to 99%. Furthermore, the Phycodnaviridae clones shared higher sequence similarity to each other than to the most closely related viruses, the Herpesviridae (Chen & Suttle, 1996).

DNA polymerase genes in the study of marine virus communities

Many virus genomes contain DNA polymerase genes. For example, marine phages in the family Styloviridae and Myoviridae both encode their own polymerases. Therefore, the approach taken to examine algal virus communities is applicable to other natural virus populations including bacteriophages and cyanophages. The following section will briefly outline the approach to design algal-virus-specific PCR primers and will summarize the results of specific field studies.

The first algal-virus *pol* genes sequenced were from the viruses NY-2A and PBCV-1. Details of the methods used to obtain these sequences have been previously described (Grabherr et al., 1992). This information was combined with information from the DNA polymerase sequence of MpV-SP1 to design algal-virus-specific PCR primers. Obtaining sequence information from MpV-SP1 first involved isolating and amplifying cultivable viruses; details of this, as well as purification of MpV genomic DNA have been previously described (Cottrell & Suttle, 1991; Cottrell & Suttle, 1995b). Once purified genomic MpV DNA was obtained, restriction digests of the DNA were used in Southern blot hybridizations with a radiolabeled probe according to the protocol of Sambrook et al. (1989). The 5' end-labeled probe used to locate the polymerase gene was a degenerate oligonucleotide based on the amino acid sequence YGDTDS (universally conserved among B-family DNA polymerases). After probing several digests of MpV, a fragment of appropriate size that hybridized to the probe was selected for cloning. The fragment selected was approximately 5kb in size; the fragment selected was larger than the 2800 bp genes of NY-2A and PBCV-1 to increase the chance of cloning the entire gene. Subsequent digestions and subcloning permitted sequencing the entire DNA polymerase gene. Once the sequence was obtained, inferred amino acid sequences were aligned with other DNA *pol* sequences available in GenBank. Sequences unique to the algal viruses PBCV-1, NY-2A, and MPV-SP1 were used to design two degenerate PCR primers to amplify DNA from these closely related viruses. Furthermore, the oligonucleotide based

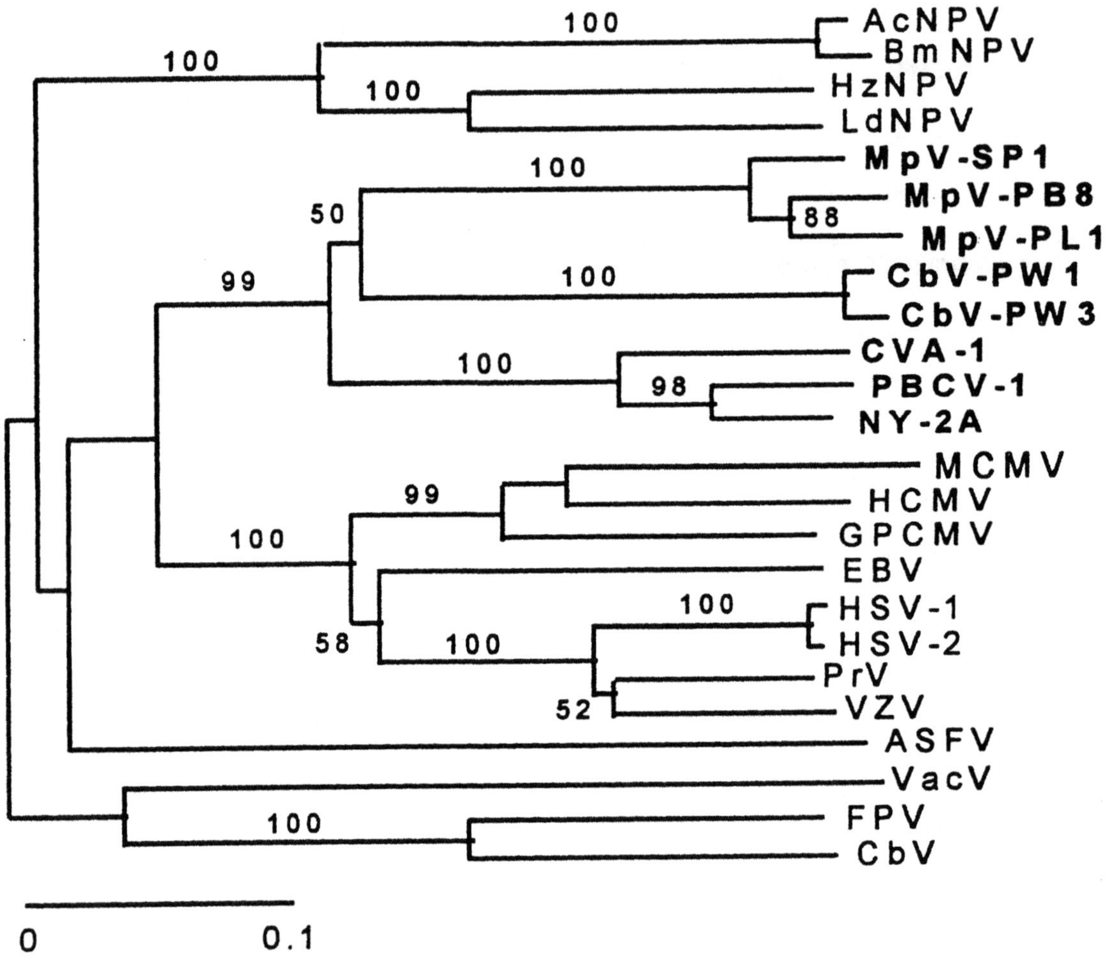

Figure 4. Neighbor-joining tree of 100 bootstrap analyses. The phylogenetic tree was constructed by using 217 of 335 amino acid sites. Virus clones infecting microalgae are indicated by boldface type. The numbers at the nodes indicate bootstrap values. The scale bar represents 0.1 fixed mutations per amino acid position. Abbreviations are as follows: AcNPV = *Autographa californica* nuclear polyhedrosis virus; BmNPV = *Bombyx mori* nuclear polyhedrosis virus; HzNPV = *Helicoverpa zea* nuclear polyhedrosis virus; LdNPV = *Lymantria dispar* nuclear polyhedrosis virus; MpV = *Micromonas pusilla* virus; CbV = *Chrysochormulina brevefilum* virus; CVA-1, PBCV-1, and NY-2A = viruses infecting *Clorella* -like algae; MCMV = Murine cytomegalovirus; HCMV = Human cytomegalovirus; GPCMV = Guinea pig cytomegalovirus; EBV = Epstein-Barr virus; HSV-1 = Herpes simplex virus type 1; HSV-2 = Herpes simplex virus type 2; PrV = Pseudorabies virus; VZV = Varicella-Zoster virus; ASFV = African swine fever virus; VacV = Vaccinia virus; FPV = Fowlpox virus; CbV = *Choristoneura biennis* poxvirus.

on the amino acid sequence YGDTDS was used as a third primer for nested PCR (Zhang & Suttle, 1994, Figure 5). This third primer is of particular importance as it allows confirmation that the desired target sequence, in this case DNA *pol*, was amplified in the first round of PCR. The motivation for this work was to design PCR primers suitable for the amplification of DNA from eukaryotic phytoplankton viruses (Chen & Suttle, 1995b). However, it should be possible to use the same methods to develop PCR primers specific to other groups of related viruses.

Use of this PCR method resulted in the amplification of *pol* sequences from 13 clonal isolates of different viruses. The viruses with amplifiable sequences included 2 viruses that infect an endosymbiotic *Chlorella*-like algae, 9 viruses that infect *M. pusilla*, and 2 viruses that infect *C. brevifilum*. Furthermore, *pol* sequences were amplified from virus communities concentrated from the Gulf of Mexico, indicating that the PCR method was suitable for the study of natural virus communities (Chen & Suttle, 1995b). Subsequently, PCR with the algal-

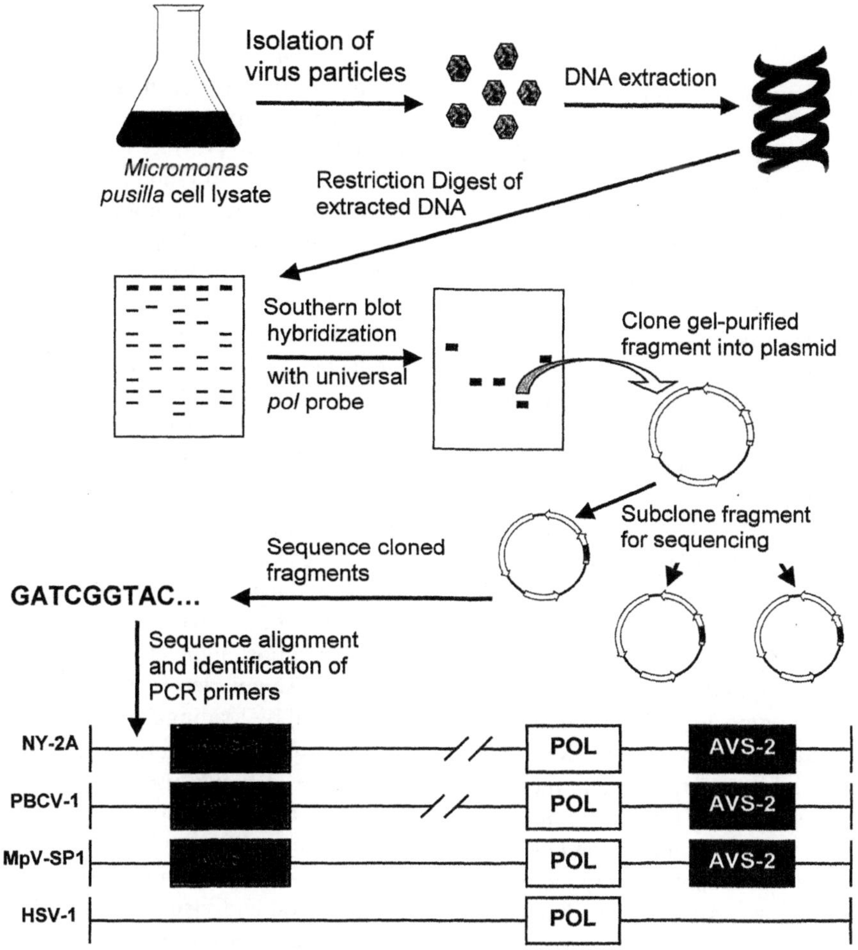

Figure 5. Flowchart of PCR primer design. Chart shows the order of steps taken to clone and sequence the polymerase gene from the *Micromonas pusilla* virus MpV-SP1. Following sequencing, the pol gene of MpV-SP1 was aligned with other known sequences including two *Chlorella*-like alga viruses (NY-2A and PBCV-1) and Herpes simplex virus type 1 (HSV-1). The primers AVS-1 and AVS-2 are conserved only among algal viruses while the primer Pol is conserved among all B-family DNA polymerases.

virus-specific primers was used to examine the genetic diversity of a concentrated natural virus community collected in the Gulf of Mexico. It was hypothesized that PCR with algal-virus-specific primers would result in the production, from a single natural virus sample, of several different DNA products of approximately the same size (Chen et al., 1996). Therefore, amplified *pol* fragments from a single PCR reaction were cloned to facilitate separation of different sequences. RFLP analysis of cloned fragments revealed that five different sequences were amplified from a single natural virus concentrate. This demonstrated that algal virus polymerase genes, closely related to those of cultured virus strains, could be amplified and sequenced from a natural marine virus sample. Moreover, a single unknown algal virus *pol* fragment was recovered which did not cluster with any sequences of known marine algal viruses (Chen et al., 1996). These results indicate that the PCR-based method is useful for examining the genetic diversity of naturally occurring marine viruses.

While the primers used in these studies were specific to algal-viruses, it should be possible to use similar methods to study other natural virus communities. As previously mentioned, many marine cyanophages and bacteriophages may encode DNA polymerases and it should be possible to design primers specific for these groups. These molecular techniques allow examination of virus community diversity without the need to isolate or culture specific members of these communities.

DGGE as a tool for comparing virus communities

In addition to direct-sequence or RFLP analysis of pol gene fragments, DGGE (denaturing gradient gel electrophoresis) can be used to examine natural virus communities. This method allows the resolution of DNA fragments of the same size that differ in as little as a single base pair. DGGE has been used to analyze the diversity of complex microbial communities ranging from biofilms in wastewater treatment reactors to hot-spring microbial mats (Muyzer et al., 1993; Ferris et al., 1996; Santegoeds et al., 1996). We have begun to use this method to examine the diversity of algal viruses in natural samples.

DGGE makes use of the fact that denaturation (separation) of double-stranded (ds) DNA depends on temperature, concentration of denaturant (e.g. urea and/or formamide), and interactions of adjacent base pairs stacked in a helix (Myers et al., 1987). Therefore, lengths of dsDNA separate under different conditions according to their base composition. Furthermore, as a result of the formation of DNA strands with three ends, partial denaturation inhibits electrophoretic mobility. Resolution of dsDNA samples of the same length, but differing in sequence, occurs by running the DNA in a polyacrylamide gel that has a continuously increasing concentration of denaturant. As each strand that differs in sequence partially denatures at a different concentration of denaturant, migration halts at a different position in the gel. Following gel separation of heterogeneous PCR products, bands can be excised from the gel and sequenced, and phylogenetic relations can be determined.

A major advantage of DGGE is that it eliminates cloning PCR products, and offers a rapid means of detecting predominant populations. As the products are separated by electrophoresis, it also eliminates the laborious screening of redundant clones in a library (Ferris et al., 1996). Furthermore, as DGGE is a highly sensitive method for detecting sequence differences, the possibility of mistaking chimeric sequences produced during PCR amplification for sample genetic diversity can be eliminated (Muyzer et al., 1993).

As a preliminary investigation of the utility of DGGE for the analysis of virus polymerase fragments, several PCR products were analyzed using this technique. Samples of clonal viruses or natural virus communities were amplified using only the outer primer pair (the algal-virus-specific PCR primers AVS-1 and AVS-2) according to previously established protocols (Chen et al., 1996). Briefly, colony PCR was performed with stabs from E. coli colonies containing cloned *pol* fragments from *Micromonas pusilla* viruses MpV-SP1, MpV-PB5, *Chlorella*-like algal viruses PBCV-1 and CVA-1, and *Chrysochromulina* spp. virus CbV-PW3. Natural communities of viruses that had been concentrated from the Gulf of Mexico (GOM-A1) and British Columbia, Canada, coastal waters (BC-268, BC-271, BC-277, BC-288, BC-308) using ultrafiltration were pelleted by ultracentrifugation and subjected to hot/cold treatments. Samples BC-268, BC-271, and BC-308 were collected from surface waters at the Jericho Beach pier in the Burrard Inlet while samples BC-277 and BC-288 were collected in the Strait of Georgia. While the amplified products from natural samples are not confirmed algal virus DNA polymerases, negative control experiments have been conducted using DNA templates from a wide range of sources. These have included phage DNA, bacterial DNA, and host algae DNA itself (data not shown).

Agarose gel electrophoresis of PCR products was performed by loading 5μl of clonal samples and 10μl of natural samples into a gel composed of 1.5% SeaKem® LE agarose (FMC BioProducts) in 0.5x TBE (45 mM Tris-borate, 1 mM EDTA, pH 8.0). The gel was run at 90 v for 1.5 hours. After electrophoresis, the gel was stained in a 50 μg/ml ethidium bromide solution and visualized on a UV transilluminator. Examination of the gel revealed the expected single bands from all clonal viruses, a single 700 bp band from the Gulf of Mexico sample, and two predominant bands of 700 and 550 bp in all samples from British Columbia coastal waters (Figure 6a). It is worth noting that the amplicon from the virus PBCV-1 is larger than the other amplicons as the DNA polymerase of this virus contains a 100 bp intron (Grabherr et al., 1992).

Further analysis of these PCR products was conducted using a 0–60% (100% denaturant is defined as 7 M urea and 40% deionized formamide) linear denaturing gradient 6% polyacrylamide gel, cast with 1x TAE (40 mM Tris-acetate, 2 mM EDTA, pH 8.5). For the clonal virus samples 10μl was loaded into wells while 20μl was loaded for the natural samples. Electrophoresis was carried out for 6 hours at 135 volts using the D-code™ electrophoresis system (Bio-Rad laboratories). The gel was stained in a SyBr® Green I (Molecular Probes, Inc.) solution overnight and was visualized on a UV transilluminator.

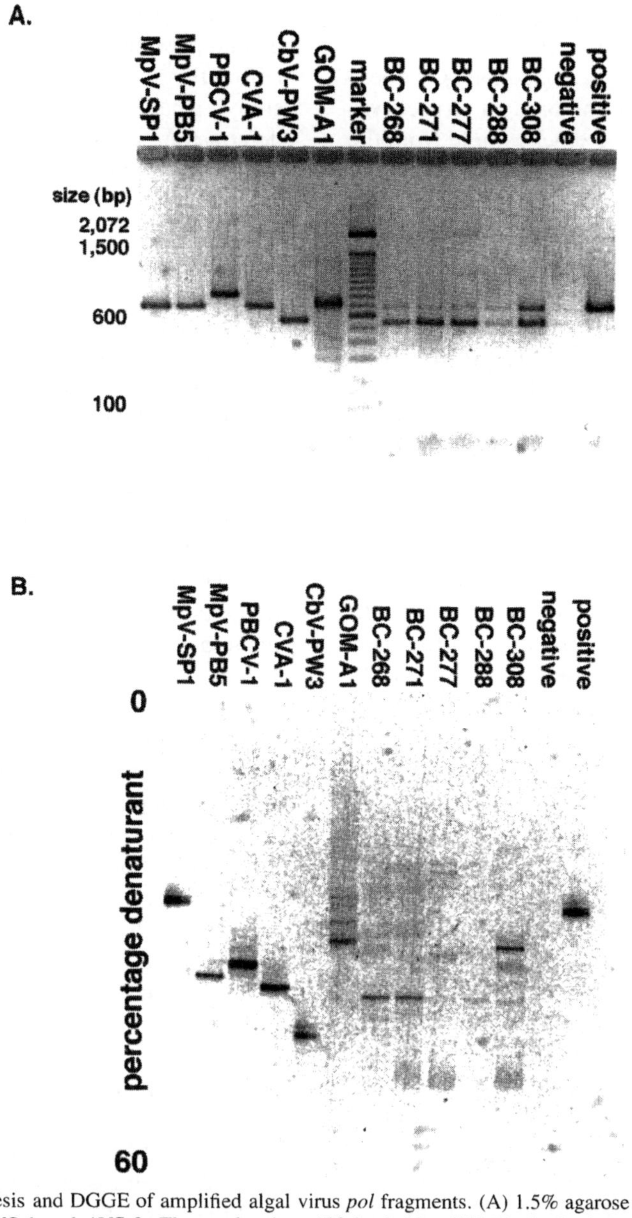

Figure 6. Agarose gel electrophoresis and DGGE of amplified algal virus *pol* fragments. (A) 1.5% agarose gel of PCR amplified DNA using the algal-virus-specific primers AVS-1 and AVS-2. The marker is a 100 bp ladder (Life Technologies). (B) 6% polyacrylamide gel with 0–60% linear denaturing gradient loaded with PCR amplified DNA using the algal-virus-specific primers AVS-1 and AVS-2. Abbreviations are as follows: MpV-SP1 and MpV-PB5 = *Micromonas pusilla* virus strains SP1 and PB5; PBCV-1 and CVA-1 = *Chlorella*-like algae viruses; GOM-A1 = Gulf of Mexico sample A1; BC-268, BC-271, BC-277, BC-288, BC-308 = British Columbia, Canada, coastal water samples. See text for further detail.

Algal virus PCR products of the same size could be resolved using DGGE. For example, PCR products of cloned virus samples MpV-SP1, MpV-PB5, and CVA-1 cannot be resolved in the agarose gel, yet are clearly separated in the gradient gel (Figure 6). Similarly, PCR products from natural samples cannot be resolved by agarose gel electrophoresis, yet several different banding patterns were revealed using DGGE (Figure 6). The bands observed in the sample from the Gulf of Mexico are not present in any of the other samples implying that virus communities in the Gulf of Mexico are genetically different than viral communities of coastal British Columbia. Interestingly, samples BC-268 and BC-271 appear very similar, with the same dominant band. These two samples were collected from the same location but 10 weeks apart, implying that during the time of sample collection (October–January) the genetic composition of algal viruses was relatively stable. However, the other sample from the same location (BC-308) collected in July has a different dominant band demonstrating that the genetic composition of algal virus communities can differ seasonally. Furthermore, samples BC-277 and BC-288 from the Strait of Georgia have very different banding patterns implying that virus communities differ over relatively small spatial scales. The results from this preliminary study using DGGE imply that the genetic structure of algal virus communities differs with time and geographic location and that DGGE is useful for investigating the genetic composition of Phycodnaviridae communities.

Recent modifications of DGGE have increased its sensitivity and range of application, and will allow for further optimization of the technique for examining marine virus communities. For example, attachment of a 40-base-pair GC-rich sequence to one end of the DNA fragments prevents complete denaturation of short amplicons, and allows for better resolution of similar sequences. Attachment of a 'GC-clamp' to DNA fragments is easily achieved by adding the sequence to one of the PCR primers. All amplified products from the reaction will have the sequence incorporated at one end (Sheffield et al., 1989; Sheffield et al., 1992). This modification of DGGE has become standard in many applications of this method. Incorporation of a 'GC clamp' has been used to examine rDNA sequence diversity in many diverse microbial communities (e.g. Muyzer et al., 1993; Ferris et al., 1996, Rölleke et al., 1996; Heuer et al., 1997; Nübel et al., 1997). The amplified rDNA fragments in these studies commonly ranged in size from 100 to 400 bp. In our study the amplicons were longer (>550 bp), making a GC clamp unnecessary. Even after 6 h of electrophoresis the amplified fragments remained within the polyacrylamide gel. While most studies employing DGGE have examined rDNA sequence diversity, the method is applicable to any PCR-based study of genetic diversity.

DGGE is a powerful method for studying diversity in natural virus communities, and will allow rapid examination of viral diversity in different environments. This will permit researchers to obtain 'fingerprints' of natural virus communities. Comparisons of fingerprints obtained from different environments will lead to the formulation and testing of hypotheses on the relationship between virus diversity and other environmental parameters. Although much work is necessary before this method will become an optimized tool for studying natural virus communities, we hope to further demonstrate its usefulness for examining diversity in algal virus communities in the near future.

Conclusion

Viruses are currently recognized as numerically and ecologically significant members of aquatic ecosystems (see reviews by: Fuhrman & Suttle, 1993; Thingstad et al., 1993; Suttle, 1994; Bratbak et al., 1994). They are also regulators of carbon flow in the marine environment and are probably important controlling agents of bacterial and algal community composition. Currently, very little information on the diversity of natural virus communities, or the range of phytoplankton infected by viruses is available. Molecular techniques such as PCR and DGGE should provide a means for rapid, efficient determination of viral genetic diversity in environmental samples. Data on diversity, when combined with sequence information from cultured viruses, will permit further estimation of the range of phytoplankton species infected by viruses. Furthermore, these molecular techniques, through examination of host genomes, permit alternate means to examine the possibility of latent replication in these viruses. Such information is essential to understanding the impact of marine viruses on global primary production and carbon flow.

Acknowledgement

We would like to thank Egidio Spinelli and Derek Yuen for their assistance preparing viral concentrates, and the crew of the CCGS Vector for their assistance with sample collection. This research was supported by NSERC Research & Ship Operations Grants to C.A.S. and a NSERC postgraduate fellowship to S.M.S.

References

Argos, P., A. D. Tucker, L. Philipson (1986). Primary structural relationships may reflect similar DNA replication strategies. Virology 149: 208–216.

Argos, P. (1988). A sequence motif in many polymerases. Nucl. Acids Res. 16(21): 9909–9916.

Bej, A. K., R. J. Steffan, J. DiCesare, L. Haff, R. M. Atlas (1990). Detection of coliform bacteria in water by polymerase chain reaction and gene probes. Appl. envir. Microbiol. 56: 307–314.

Bergh, O., K. Y. Børsheim, G. Bratbak, M. Heldal (1989). High abundance of viruses found in aquatic environments. Nature 340: 467–468.

Børsheim, K. Y., G. Bratbak, M. Heldal (1990). Enumeration and biomass estimation of planktonic bacteria and viruses by transmission electron microscopy. Appl. envir. Microbiol. 56(2): 352–356.

Braithwaite, D. K., J. Ito (1993). Compilation, alignment, and phylogenetic relationships of DNA polymerases. Nucl. Acids Res. 21(4): 787–802.

Bratbak, G., F. Thingstad, M. Heldal (1994). Viruses and the microbial loop. Microb. Ecol. 28: 209–221.

Brown R. M., Jr. (1972). Algal viruses. Advances in Virus Res. 17: 243–277.

Calisher, C. H., M. C. Horzinek, M. A. Mayo, H. -W. Ackermann, J. Maniloff. (1995). Sequence analyses and a unifying system of viral taxonomy: consensus via consent. Arch. Virol. 140: 2093–2099.

Chen, F., C. A. Suttle (1995a). Nested PCR with three highly degenerate primers for amplification and identification of DNA from related organisms. BioTechniques 18: 609–612.

Chen, F., C. A. Suttle (1995b). Amplification of DNA polymerase gene fragments from viruses infecting microalgae. Appl. envir. Microbiol 61: 1274–1278.

Chen, F., C. A. Suttle (1996). Evolutionary relationships among large double stranded DNA viruses that infect microalgae and other organisms as inferred from DNA polymerase genes. Virology 219: 170–178.

Chen, F., C. A. Suttle, S. M. Short (1996). Genetic diversity in marine algal virus communities as revealed by sequence analysis of DNA polymerase genes. Appl. envir. Microbiol. 62(8): 2869–2874.

Chiura, H. X. (1997). Generalized gene transfer by virus-like particles from marine bacteria. Aquat. microb. Ecol. 13: 75–83.

Cottrell, M. T., C. A. Suttle (1991). Wide-spread occurrence and clonal variation in viruses which cause lysis of a cosmopolitan, eukaryotic marine phytoplankter, *Micromonas pusilla*. Mar. Ecol. Prog. Ser. 78: 1–9.

Cottrell, M. T., C. A. Suttle (1995a). Dynamics of a lytic virus infecting the marine picoflagellate *Micromonas pusilla*. Limnol. Oceanogr. 40(4): 730–739.

Cottrell, M. T., C. A. Suttle (1995b). Genetic diversity of algal viruses which lyse the photosynthetic picoflagellate *Micromonas pusilla* (Prasinophyceae). Appl. envir. Mirobiol. 61(8): 3088–3091.

Damagnez, V., J. Tillit, A-M. de Recondo, G. Baldacci (1991). The *POL I* gene from fission yeast, *Schizosaccharomyces pombe*, shows conserved amino acid blocks specific for eukaryotic DNA polymerases alpha. Mol. Gen. Genet. 226: 182–189.

Field, K. G., D. Gordon, T. Wright, m Rappé, E. Urbach, K. Vergin, S. J. Giovannoni (1997). Diversity and Depth-specific distribution of SAR11 cluster rRNA genes from marine planktonic bacteria. Appl. envir. Microbiol. 63(1): 63–70.

Ferris, M. J., G. Muyzer, D. M. Ward (1996). Denaturing gradient gel electrophoresis profiles of 15S rRNA-defined populations inhabiting a hot spring microbial mat community. Appl. envir. Microbiol. 62: 340–346.

Fuhrman, J. A., K. McCallum, A. A. Davis (1992). Novel major archaebacterial group from marine plankton. Nature 356: 148–149.

Fuhrman, J. A., C. A. Suttle (1993). Viruses in marine planktonic systems. Oceanography 6(2): 51–63.

Giovannoni, S. J., T. B. Britschgi, C. L. Moyer, K. G. Field (1990). Genetic diversity in Sargasso Sea bacterioplankton. Nature 345: 60–63.

Giovannoni, S., S. C. Cary (1993). Probing marine systems with ribosomal RNAs. Oceanography 6(3): 95–104.

Grabherr, R., P. Strasser, J. L. Van Etten (1992). The DNA polymerase gene from Chlorella viruses PBCV-1 and NY-2A contains and intron with nuclear splicing sequences. Virology 188: 721–731.

Heuer, H., M. Krsek, P. Baker, K. Smalla, E. M. H. Wellington (1997). Analysis of Actinomycete communities by specific amplification of genes encoding 16s rRNA and gel-electrophoretic separation in denaturing gradients. Appl. envir. Microbiol. 63(8): 3233–3241.

Illana, B., L. Blanco, M. Salas (1996). Functional characterization of the genes coding for the terminal protein and DNA polymerase from bacteriophage GA-1. Evidence for a sliding-back mechanism during protein-primed GA-1 DNA replication. J. mol. Biol. 264: 453–464.

Innis, M. A., D. H. Gefland, J. J. Sninsky, T. J. White (1990). PCR protocols: A Guide to Methods and Applications. Academic Press, Inc., San Diego. 482 pp.

Ito, J., D. K. Braithwaite (1990). Yeast mitochondrial DNA polymerase is related to the family A DNA polymerases. Nucl. Acids Res. 18(22): 6716.

Ito, J., D. K. Braithwaite (1991). Compilation and alignment of DNA polymerase sequences. Nucl. Acids. Res. 19(15): 4045–4057.

Iwasaki, H., Y. Ishino, H. Toh, A. Nakata, H. Shinagawa (1991). *Escherichia coli* DNA polymerase II is homologous to a-like DNA polymerases. Mol. Gen. Genet. 226: 24–33.

Jacobsen, A., G. Bratbak, M. Heldal (1996). Isolation and characterization of a virus infecting Phaeocystis pouchetii (Prymnesiophyceae). J. Phycol. 23: 923–927.

Jiang, S. C., J. H. Paul (1994). Seasonal and diel abundance of viruses and the occurrence of lysogeny/bacteriocinogy in the marine environment. Mar. Ecol. Prog. Ser. 104: 163–172.

Joyce C. M., Steitz, T. A. (1994). Function and Structure Relationships in DNA Polymerases. Annu. Rev. Biochem. 63: 777–822.

Jung, G., M. C. Leavitt, J.-C., Hsieh, J. Ito (1987). Bacteriophage PRD1 DNA polymerase: Evolution of DNA polymerases. Proc. natl. Acad. Sci. 84: 8287–8291.

Kamer, G., Argos P. (1984). Primary structural comparison of RNA-dependent polymerases from plant, animal and bacterial viruses. Nucl. Acids Res. 12(18): 7269–7282.

Kopecka, H. S., S. Dubrou, J. Prevut, J. Marechal, J. M. Lopez-Pila (1993). Detection of naturally occurring enteroviruses in waters by reverse transcription, polymerase chain reaction, and hybridization. Appl. envir. Microbiol. 59: 1213–1219.

Liesack, W., H. Weyland, E. Stackebrandt (1991). Potential risks of gene amplification by PCR as determined by 16s rDNA analysis of a mixed-culture of strict barophilic bacteria. Microb. Ecol. 21: 191–198.

Mayer, J. A., F. J. R. Taylor (1979). A virus which lyses the marine nanoflagellate *Micromonas pusilla*. Nature 281: 299–301.

Meyer, G., C. Vlcek, V. Paces, M. K. O'Hara, P.-P. Pastoret, E. Thiry, M. Schwyzer (1997). Sequence analysis of the bovine herpesvirus type 1 genes homologous to the DNA polymerase (UL30), the major DNA-binding protein (UL29) and ICP18.5 assembly protein (UL28) genes of herpes simplex virus. Arch Virol 142: 89–102.

Milligan, K. L., E. M. Cosper (1994). Isolation of virus capable of lysing the brown tide microalga, *Aureococcus anaphagefferens*. Science 266: 805–807.

Mullis, K. B., Faloona, F. A. (1987). Specific synthesis of DNA in vitro via a polymerase-catalyzed chain reaction. Methods Enzymol 155: 335–351.

Muyzer, G., E. C. De Waal, A. G. Uitterlinden (1993). Profiling of complex microbial populations by denaturing gradient gel electrophoresis analysis of polymerase chain reaction-amplified genes coding for 16S rRNA. Appl. envir. Microbiol. 59: 695–700.

Myers, R. M., T. Maniatis, L. S. Lerman (1987). Detection and localization of single base changes by denaturing gradient gel electrophoresis. Meth. Enzymol. 155: 501–527.

Nagasaki, K., M. Yamaguchi (1997). Isolation of a virus infectious to the harmful bloom forming alga *Heterosigma akashiwo* (Raphidophyceae). Aquat. microb. Ecol. 13: 135–140.

Nübel, U., F. Garcia-Pichel, G. Muyzer (1997). PCR primers to amplify 16s rRNA genes from cyanobacteria. Appl. envir. Microbiol. 63(8): 3327–3332.

Ohki, K., Y. Fujita (1996). Occurrence of a temperate cyanophage lysogenizing the marine cyanophyte *Phormidium persicinium*. J. Phycol. 32: 365–370.

Paul, J. H., S. C. Jiang, J. B. Rose (1991). Concentration of viruses and dissolved DNA from aquatic environments by vortex flow filtration. Appl. envir. Microb. 57: 2197–2204.

Pizzagalli, A., P. Valsasnini, P. Plevani, G. Lucchini (1988). DNA polymerase I gene of *Saccharomyces cerevisiae*: Nucleotide sequence, mapping of a temperature-sensitive mutation, and protein homology with other DNA polymerases. Proc. natl. Acad. Sci. 85: 3772–3776.

Proctor, L. M., J. A. Fuhrman (1990). Viral mortality of marine bacteria and cyanobacteria. Nature 343: 60–62.

Reysenbach, A.-L., L. J. Giver, G. S. Wickham, N. R. Pace (1992). Differential amplification of rRNA genes by polymerase chain reaction. Appl. envir. Microbiol. 58(10): 3417–3418.

Ridley, R. G., J. H. White, S. M. McAleese, M. Goman, P. Alano, E. de Vries, B. J. Kilbey (1991). DNA polymerase δ: gene sequences from *Plasmodium falciparum* indicate that this enzyme is more highly conserved that DNA polymerase α. Nucl. Acids Res. 19(24): 6731–6736.

Rohozinski, J., J. L. Van Etten (1989). Characterization of DNA polymerases in an uninfected and virus PBCV-1 infected green algae – Chlorella strain NC64A. Intervirology 30: 156–162.

Rölleke, S., G. Muyzer, C. Wawer, G. Wanner, W. Lubitz (1996). Identification of bacteria in a biodegraded wall painting by denaturing gradient gel electrophoresis of PCR-amplified gene fragments coding for 16s rRNA. Appl. envir. Microbiol. 62(6): 2059–2065.

Sambrook, J., E. F. Fritsch, T. Maniatis (1989). Molecular Cloning: A Laboratory Manual, 2nd ed. Cold Spring Harbour Laboratory Press, Cold Spring Harbour, N. Y.

Saiki, R. K., D. H. Gelfand, S. Stoffel, S. J. Scharf, R. G. Higuchi, T. T. Horn, K. B. Mullis, H. A. Erlich (1988). Primer-directed enzymatic amplification of DNA with a thermostable DNA polymerase. Science 239: 487–491.

Santegoeds, C. M., S. C. Nold, D. M. Ward (1996). Denaturing gradient gel electrophoresis used to monitor the enrichment culture of aerobic chemoorganotrophic bacteria from a hot spring cyanobacterial mat. Appl. envir. Microbiol. 62: 3922–3928.

Sheffield, V. C., D. R. Cox, L. S. Lerman, R. M. Meyers (1989). Attachment of a 40-base-pair G +C rich sequence (GC-clamp) to genomic DNA fragments by the polymerase chain reaction results in improved detection of single-base changes. Proc. natl. Acad. Sci 86: 232–236.

Sheffield, V. C., J. S. Beck, E. M. Stone. (1992). A simple efficient method for attachment of a 40-base pair, GC-rich sequence to PCR-amplified DNA. BioTechniques 12(3): 386–387.

Scholin, C. A., D. M. Anderson (1994). Identification of group- and strain-specific genetic markers for globally distributed *Alexandrium* (Dinophyceae). I. RFLP analysis of SSU rRNA genes. J. Phycol. 30: 744–754.

Steffan, R. J., R. M. Atlas (1991). Polymerase chain reaction: applications in environmental microbiology. Annu. Rev. Microbiol. 45: 137–161.

Suttle, C. A., A. M. Chan, M. T. Cottrell (1990). Infection of phytoplankton by viruses and reduction of primary productivity. Nature 347: 467–469.

Suttle, C. A. (1992). Inhibition of photosynthesis in phytoplankton by the submicron size fraction concentrated from seawater. Mar. Ecol. Prog. Ser. 87: 105–112.

Suttle, C. A. (1994). The significance of viruses to mortality in aquatic microbial communities. Microb. Ecol. 28: 237–243.

Suttle, C. A., A. M. Chan (1995). Viruses infecting the marine Prymnesiophyte *Chyrsochromulina* spp.: isolation, preliminary characterization and natural abundance. Mar. Ecol. Prog. Ser. 118: 275–282.

Suzuki, M. T., M. S. Rappé, Z. W. Haimberger, H. Winfield, N. Adair, J. Ströbel, S. J. Giovannoni (1997). Bacterial diversity among small-subunit rRNA gene clones and cellular isolates from the same seawater sample. Appl. envir. Microbiol. 63(3): 983–989.

Thingstad, T.F., M. Heldal, G. Bratback, I. Dundas (1993). Are viruses important partners in pelagic food webs? Trends ecol. Evol. 8(6): 209–213.

Torella, F., R. Y. Morita (1979). Evidence by electron micrographs for a high incidence of bacteriophage particles in the waters of Yaquina Bay, Oregon: ecological and taxonomic implications. Appl. envir. Microbiol. 37: 774–778.

Van Etten, J. L., L. C. Lane, R. H. Meints (1991). Viruses and virus like particles of eukaryotic algae. Microbiol. Rev. 55: 586–620.

Wang, S.-F. (1991). Eukaryotic DNA Polymerases. Annu. Rev. Biochem. 60: 513–552.

Waterbury, J. B., F. W. Valois (1993). Resistance to co-occurring phages enables marine *Synechococcus* communities to coexist with cyanophages abundant in seawater. Appl. envir. Microbiol. 59(10): 3393–3399.

Weinbauer, M. G., D. Fuks, P. Peduzzi (1993). Distribution of viruses and dissolved DNA along a coastal trophic gradient in the northern Adriatic Sea. Appl. envir. Microbiol. 59: 4074–4082.

Wiggins, B. A., M. Alexander (1985). Minimum bacterial density for bacteriophage replication: implications for significance of bacteriophages in natural ecosystems. Appl. envir. Microbiol. 49(1): 19–23.

Wong, S. W., A. F. Wahl, P-M. Yuan, N. Arai, B. E. Pearson, K. Arai, D. Korn, M. W. Hunkapiller, T .S. F. Wang (1988). Human DNA polymerase a gene expression is cell proliferation dependent and its primary structure is similar to both prokaryotic and eukaryotic replicative DNA polymerases. EMBO 7(1): 37–47.

Wright, T. D., K. L. Vergin, P. W. Boyd, S. J. Giovannoni (1997). A novel δ-subdivision proteobacterial lineage from the lower ocean surface layer. Appl. envir. Microbiol. 63(4): 1441–1448.

Zhang, Y. P., C. A. Suttle (1994). Design and use of PCR primers for B-family DNA polymerase genes to detect and identify viruses and microbes (abstract), p. a-85. ASLO/PSA Joint Meeting, Florida, June 12–16, 1994.

Hydrobiologia **401:** 33–53, 1999.
J.P. Zehr & M.A. Voytek (eds), Molecular Ecology of Aquatic Communities.

Flow cytometry in molecular aquatic ecology

J. L. Collier[1] & L. Campbell[2]
[1]*Biology Department, MRC 345, Rensselaer Polytechnic Institute, 110 8th Street, Troy, NY 12180, U.S.A. Tel: [+1]518-276-2178; E-mail: collij3@rpi.edu*
[2]*Department of Oceanography, Texas A&M University, College Station, TX 77843-3146, U.S.A. E-mail: lcampbell@ocean.tamu.edu*

Key words: fluorescent whole-cell hybridization, immunofluorescence, microbial ecology, flow cytometry, probes

Abstract

In working towards understanding ecosystems that are often dominated by microorganisms, aquatic ecologists have historically relied on measuring bulk, community-level properties and synecological processes. However, developing a mechanistic and predictive explanation for the factors structuring aquatic ecosystems will require understanding the roles that individual microorganisms play in these higher-order phenomena. The application of molecular biological techniques to examine nucleic acids extracted in bulk from microbial communities can provide information about the taxonomic structure of microbial communities and the physiological ecology of particular types of organisms at various levels of specificity. Yet, even if accomplished at the 'species' level, these data still represent bulk parameters because they can reveal only an average value for the organisms and community of interest. A more detailed view may be gained by investigations performed at a single-cell level. Flow cytometry allows the measurement of one cell at a time, at a rate of thousands of cells per second. When combined with fluorescent stains, including nucleic acid and antibody-based molecular probes, flow cytometry permits rapid analysis of cell-specific information for particular types of microbes within complex microbial assemblages. Thus, the autecology of microbial populations and structure of microbial communities can be examined from the vantage point of the individual cells comprising them. By bringing the level of analysis closer to the relevant scale of the organisms being investigated, the combination of molecular tools and flow cytometry will bring powerful new insights into the autecology of aquatic microorganisms.

Introduction

A long-term goal of ecological research is to develop a mechanistic understanding of biological processes that will not only allow the explanation of how an ecosystem is currently operating, but will also provide the capacity to predict its response to perturbations. One major focus of this endeavor lies in gaining a more detailed understanding of the factors regulating the fluxes of materials and energy through ecosystems. Given the importance of microorganisms in driving these fluxes, the relevant controls may well be found within the realm of microbial ecology. This is especially true in aquatic ecosystems, where microorganisms play more of the critical roles than they do in terrestrial ecosystems. The classic works of Pomeroy (1974) and Azam et al. (1983) have prompted the recent focus on the importance of microbial communities in aquatic ecosystems. However, largely because of technical constraints, the inclusion of microorganisms in our understanding of ecosystems often remains synecological and process-oriented; that is, we consider the contribution of the microbial community as a whole to a process of interest (e.g. primary production, nitrogen cycling, decomposition) and measure rates on a bulk scale, without extensive consideration of the specific organisms involved. Yet it appears likely that the explicit consideration of the autecology of the microbes involved will be required to gain a greater predictive, mechanistic understanding of ecosystem structure and function (Tate, 1986) than is attainable from bulk community measurements alone. Where it has been possible to do so, aquatic ecologists have already taken a more organism-centered approach (generally complemented by synecological measurements), tracking the abundances and interactions of particular functional groups or of identifiable species. Successful examples of this autecological approach exist for aquatic microbes such as phytoplankton that can both be grown in the laboratory for detailed physiological examination and also be relatively easily identified and enumerated in natural samples. These characteristics allow comparative and experimental investigations to be performed, and permit the description of factors regulating phytoplankton community structure and of the roles particular phytoplankton species play within aquatic ecosystems (Sandgren, 1988). Extending this type of mechanistic and autecological approach to defining the taxonomic structure and the physiological ecology of other groups of microbes must be one of the most important goals of modern aquatic ecology.

As the papers in this volume show, molecular techniques will aid in the attainment of these goals because they not only reveal unprecedented detail about the taxonomic composition of microbial communities, but also permit a targeted physiological approach to be taken more widely in microbial ecology. Techniques such as fluorescent *in situ* hybridization (FISH) and *in situ* PCR are being developed to bring molecular analyses to the single-cell level, with detection often accomplished by fluorescence microscopy. For some systems, particularly the planktonic component of aquatic ecosystems, the adaptation of these same approaches to detection by flow cytometry allows the simultaneous collection of five or more parameters while improving the speed and objectivity of data collection. Our intent in this review is to consider how flow cytometry in combination with molecular tools will improve the capability of ecologists to understand the roles of microbes in aquatic ecosystems. First, we provide brief general summaries of flow cytometric and molecular techniques already established in aquatic microbial ecology. Next, we examine how flow cytometry and molecular tools have been combined in the pioneering efforts to utilize this new technology. Finally, we discuss the limitations that must be overcome before aquatic ecologists can take full advantage of this approach to understand the diversity of microbes and their role in aquatic ecosystems.

The fundamentals of flow cytometry

There have been a number of informative reviews published recently that discuss the development and design of modern optical flow cytometers (Davey & Kell, 1996), detail technical aspects of their application to specific problems (Phinney & Cucci, 1989; Ault, 1991; Steen, 1991; Button & Robertson, 1993; Olson et al., 1993), and summarize recent applications of flow cytometry in microbial ecology (Olson et al., 1989, 1991; Fouchet et al., 1993; Lloyd, 1993; Troussier et al., 1993; Davey & Kell, 1996; Edwards et al., 1996; J. Porter et al., 1996; Hauer & Eipel, 1997). A number of excellent overall introductions to the theory and protocols of flow cytometry are also available (H. Shapiro, 1995). It is not our intent to provide a complete overview; rather, we will focus on aspects most relevant to the detection of molecular probes by flow cytometric analysis.

The internal workings of all modern optical flow cytometers are fundamentally similar. The instruments consist of a pressurized fluid system that delivers and hydrodynamically focuses the sample, an illumination source, and a detection system to acquire and record data. Commercially available flow cytometers use lasers or arc lamps as light sources. For flow cytometers that use an arc lamp as a light source, excitation wavelength is selected by excitation filters or a monochromator. To provide flexibility in excitation wavelength, some laser-based flow cytometers may employ multiple lasers or a single tunable laser, but in most bench-top models only a single, fixed wavelength of excitation light is available. The blue (488 nm) line of an Ar laser is probably the most commonly used light source.

Flow cytometers are able to record data at a rate of thousands of events per second. From two up to eight different parameters can be collected for each particle that passes through the interrogation point, i.e. where the light source is focused on the sample stream and the light collection and detection optics are arrayed. Together, these parameters provide an optical characterization of each particle, which typically includes two measures of light scattering and up to four measures of fluorescence. As a particle crosses the beam, it scatters light. The amount and direction of the light scattered by a particle are largely related to properties such as its size, surface and internal structure, and refractive index. Most of the light scattered is deflected to small forward angles and measured as forward scatter (FSC or FALS) by a photodiode. Light that is either scattered at higher angles or emitted as fluorescence from particle-associated fluorochromes is collected by a lens placed orthogonally to both the light beam and the sample stream. A series of dichroic mirrors and optical filters divides this light by wavelength and directs it to photomultiplier tubes (PMTs) for detection and quantitation. Scattered excitation light collected in this manner is known as side scatter, right angle scatter or wide angle scatter (SSC, RALS). Typically, fluorescence emitted in the green, orange, and red regions of the spectrum is also measured.

Many flow cytometers are equipped to sort particles with specific optical characteristics out of the flow stream and into a collection tube; this capability offers the opportunity to collect particles of interest and then to reanalyze them by other methods. Most sorting instruments have employed a piezoelectric transducer–vibrational mechanism to generate droplets, which are then electrostatically charged and deflected for sorting. More recently, a design in which a mechanical arm is injected into the fluid stream to isolate cells of interest has been introduced. Advantages of this design include ease of use and lack of aerosol production; however, sorting speed is limited (to a maximum of 300 cells s^{-1}) and the sample is diluted many-fold as it is sorted.

General limitations of flow cytometry for microbial ecology

Although it is possible to build a flow cytometer to broader specifications (Dubelaar et al., 1989; Peeters et al., 1989), most bench-top, commercially available flow cytometers are designed for the analysis of cells in the size range of 1–100 microns. Particles must be smaller than the physical size limit of orifices within the cytometer flow path, yet large enough to produce signals detectable above the background noise of the instrument. Particles of interest in the sample must also be able to remain suspended in solution and be reasonably abundant. Since smaller organisms tend to occur at higher densities, bacterial abundances in natural samples may be sufficient (see below). For larger or less abundant organisms, the utility of flow cytometry can be limited. Epifluorescence microscopy or scanning confocal laser microscopy (SCLM) may be better suited for investigating organisms larger than 100 μm in diameter, symbioses (Lloyd et al., 1996), rhizospheres (Assmus et al., 1995) or biofilms (Rothemund et al., 1996; Moller et al., 1998). Flow cytometry using commercially available instruments is particularly attractive for examining abundant planktonic organisms ($>10^2$ cells ml^{-1}, although less abundant organisms can be concentrated to levels appropriate for analysis) that are smaller than 50 μm in their largest dimension. In investigations of natural microbial communities, flow cytometry has been used extensively to monitor the dynamics of photosynthetic microorganisms by their autofluorescence. Nonautofluorescent microbes, particularly heterotrophic bacteria, have been enumerated via flow cytometry after staining with fluorescent dyes specific for a particular macromolecule or enzymatic process. More specific staining has been achieved with fluorescent antibody or nucleic acid probes. Flow cytometry has been applied in a similar manner to monitor the growth and physiological status of laboratory cultures of microorganisms subjected to various experimental conditions.

Nonmolecular applications of flow cytometry in aquatic ecology

Autofluorescence

The utility of flow cytometry in aquatic microbial ecology was predicted 20 years ago and has since been clearly established (Legendre & Yentsch, 1989; Yentsch & Horan, 1989; Olson et al., 1991). Since light-scattering characteristics alone are usually not sufficient to reveal much detail about either the taxonomic affinities or the physiological status of microorganisms (Davey & Kell, 1996), the flow cytometric measurement of fluorescence is a key aspect of its utility in microbial ecology. Not surprisingly then, one of the most successful applications of flow cytometry in aquatic microbial ecology has been in analyzing the structure of phytoplankton assemblages (Olson et al., 1993). Phytoplankton make ideal subjects for flow cytometric analysis because they are naturally autofluorescent by virtue of their complement of photosynthetic pigments. These molecules make fine fluorochromes for flow cytometry because most of them can absorb the blue light of the 488 nm line of an Ar laser, yet they can be distinguished because of their unique fluorescence emission spectra. Standard filter arrangements can distinguish and quantify chlorophyll fluorescence (red, >650 nm) and phycoerythrin (PE) fluorescence (orange, 575 nm $\pm$ 20 nm). Thus, flow cytometric data collected from natural

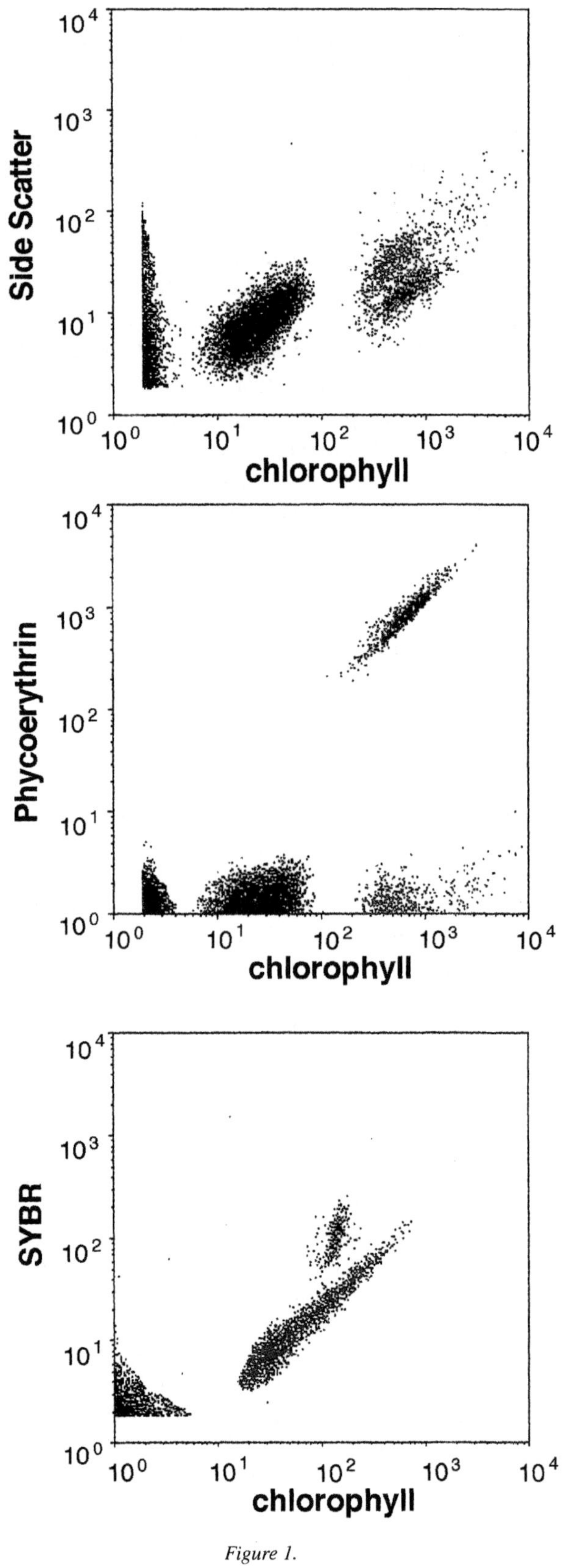

Figure 1.

phytoplankton assemblages can be used to identify and classify phytoplankton based on scattering characteristics (size) and fluorescence (pigmentation) (for an example, see Figure 1). Based on flow cytometric analyses, biological oceanographers typically divide the phytoplankton into *Synechococcus*, *Prochlorococcus*, small (pico-) eukaryotes, and large eukaryotes, and are able to define the distributions and dynamics of each group (e.g. Olson et al., 1990; Campbell et al., 1994a; Li, 1995; Lindell & Post, 1995; Partensky et al., 1996; Campbell et al., 1997). The PE-containing *Synechococcus* can be distinguished from the similarly-sized *Prochlorococcus* by the presence of 'orange' fluorescence from PE. Eukaryotic phytoplankton are distinguished based on their larger scatter and chlorophyll fluorescence signals. In some cases, the larger cells may be further distinguished based on their scattering characteristics (coccolithophorids) or the presence of both PE and chlorophyll (cryptophytes) (Olson et al., 1989). Autofluorescence-based flow cytometric analysis of phytoplankton is also used to study community interactions such as selective grazing both in natural phytoplankton assemblages (Landry et al., 1995; Reckermann & Veldhuis, 1997) and in artificial

←
Examples of flow cytometric analyses from an experiment conducted at the Bermuda Atlantic Time Series site (31°50′N, 64°10′W) in March, 1997. Samples were collected from the chlorophyll maximum depth (~ 100 m), fixed with paraformaldehyde and analyzed using a Becton-Dickinson FACSCalibur equipped with a 15 mW 488nm laser. Data were stored in a list mode file, analyzed with CYTOWIN software developed by D. Vaulot (Roscoff, France), and plotted with Win MDI software (© Joseph Trotter, La Jolla, CA). Standard beads, SPHERO™ 2 μm Purple-Yellow fluorescent particles (Spherotech, Inc., Libertyville, IL), were added to the sample and are plotted as yellow events in top and middle panels.
Top panel: Scatter plot of chlorophyll fluorescence vs. 90° light scatter (side scatter). Populations of *Prochlorococcus* (in red), *Synechococcus* (in orange), picoeukaryotes (in green) can be distinguished by autofluorescence.
Middle panel: Plotting the same data as a scatter plot of chlorophyll fluorescence vs. phycoerythrin fluorescence permits the cyanobacteria *Synechococcus* to be distinguished easily from the other picophytoplankton (populations color coded as in top panel).
Bottom panel: To the same sample, the DNA-stain SYBR (Molecular Probes, Oregon) was added to distinguish and enumerate heterotrophic bacteria (in green) and *Prochlorococcus* (in red). Higher PMT high voltage settings were used to detect chlorophyll fluorescence in this analysis, thus the *Prochlorococcus* population appears brighter here than in the top and middle panels. Evidence of spillover from the SYBR fluorescence into the chlorophyll channels for the heterotrophic bacteria can be an artifact when using this stain (see text). Abundances of *Synechococcus* and picoeukaryotes are much lower than either heterotrophic bacteria or *Prochlorococcus*, thus they do not appear in this plot.

assemblages created from cultured organisms (Cucci et al., 1989; Bougrier et al., 1997; Weisse & Kirchhoff, 1997).

Fluorescent Stains

For non-autofluorescent organisms ranging from heterotrophic bacteria to heterotrophic nanoflagellates, the utility of flow cytometry for analyzing natural assemblages (or even laboratory cultures) is limited unless the organisms of interest can be fluorescently labeled. This limitation is not unique to flow cytometry, and epifluorescence microscopy-based techniques employing fluorochromes to identify, enumerate and determine the physiological status of heterotrophic microorganisms are also amenable to flow cytometric analysis.

For example, enumeration of bacteria is often accomplished microscopically by capturing bacteria on a filter and staining them with nucleic acid (DNA)-specific fluorochromes such as the UV-excited dyes 4′,6-diamidino-2-phenylindone (DAPI) or Hoechst 33342, or the blue-excited dyes acridine orange (AO) or propidium iodide (PI), and then counting stained cells visualized by epifluorescence microscopy (Hobbie et al., 1977; K. G. Porter & Feig, 1980). These dyes can also be detected by appropriately equipped flow cytometers, and so these techniques have been transferred directly to flow cytometric analysis both for monitoring natural bacterial populations (del Giorgio et al., 1996; DeLeo & Baveye, 1996; Lange et al., 1997) and for measuring effects of antibiotics and toxins on cells grown in culture (Walberg et al., 1996; Walberg et al., 1997), often with significant improvement in speed and accuracy and without significant increases in cost. In fact, in marine planktonic samples it can be difficult to distinguish dimly autofluorescent *Prochlorococcus* from heterotrophic bacteria by microscopic techniques. Accurate counts of each group can be achieved by staining samples with a DNA-specific dye such as Hoechst 33342 and using dual-beam flow cytometry (488 nm and UV) to determine whether cells have only Hoechst fluorescence and should be counted as heterotrophic bacteria, or have both Hoechst and chlorophyll fluorescence and should be counted as *Prochlorococcus* (Campbell et al., 1994a). In combination with flow cytometry, DNA-specific dyes also allow the quantification of groups of heterotrophic bacteria with differing DNA content (Li et al., 1995; Button et al., 1996). DNA-specific dyes can also be used to provide estimates of the growth rates of microorganisms by cell cycle analysis (Carpenter & Chang, 1988). Both DAPI and Hoechst 33342 have been used successfully to measure cell cycle parameters and to estimate growth rates of natural populations of the marine picophytoplankter *Prochlorococcus* (Vaulot & Partensky, 1992; Vaulot et al., 1995; Liu et al., 1997). In efforts to make these types of analyses possible on more popular bench-top flow cytometers that provide only 488 nm excitation, several of the new DNA-specific dyes such as YOYO-1, YOPRO-1, and PicoGreen (Marie et al., 1996) along with SYBR Green I (Marie et al., 1997), have also been tested on a variety of marine picoplankton both from laboratory cultures and from natural samples. Although most these dyes are limited by the high ionic strength of sea water and transfer to TE buffer is recommended for marine samples (Marie et al., 1996), Veldhuis and co-workers reported that both PicoGreen and SYTOXGreen provided coefficients of variation for DNA histograms acceptable for cell cycle analysis of marine phytoplankton (Veldhuis et al., 1997). SYBR has been used successfully directly in sea water samples (Marie et al., 1997). An example of field samples enumerated by autofluorescence and after SYBR staining is shown in Figure 1. The *Prochlorococcus* and heterotrophic bacteria can be distinguished in a plot of chlorophyll vs. SYBR fluorescence. As demonstrated by the appearance of 'chlorophyll' fluorescence in SYBR-stained heterotrophic bacteria, fluorescence from a dye added to a sample is not always neatly confined to one variable, but may also affect others. This problem is not limited to DNA-binding dyes, and must be considered in any application of flow cytometry involving multiple fluorochromes. Interpretation of stained samples therefore requires some caution. In some cases, electronic compensation has been used to reduce the spillover between channels (Marie et al., 1997), but appropriate controls are not always possible for field samples. In cases like that shown in Figure 1, accurate chlorophyll fluorescence measurements should be obtained from unstained samples.

Fluorochromes can also be used in combination with flow cytometry to identify other biological molecules and activities with varying degrees of specificity. In addition to the nucleic acid-specific dyes discussed above, some fluorochromes naturally bind to biological macromolecules such as protein (e.g. fluorescein isothiocyanate (FITC) and rhodamine 101 isothiocyanate (Texas Red)) or lipid (e.g. Nile Red), while nonfluorescent binding agents such as lectins can be conjugated to a fluorochrome. These stains have been used to monitor cellular protein and poly-β-hydroxybutyrate content and to trace the fates of labeled cells (Davey & Kell, 1996; Porro et al., 1997; Barcina et al., 1997). Several DNA-binding fluorochromes (e.g., ethidium bromide, propidium iodide (PI), SYTOX Green) are unable to penetrate living cells and have been used to detect and enumerate cells that have lost membrane integrity (Roth et al., 1997; Magarinos et al., 1997; Faber et al., 1997) during exposure to toxins, antibiotics or other unfavorable conditions. In contrast, fluorochromes such as Rhodamine 123 and Oxonol (DiBAC) dyes accumulate in cells that have an actively energized membrane system (Kaprelyants et al., 1996; Barcina et al., 1997). Simultaneous application of both a 'dead' and 'live' stain can provide information on changes in the physiological status and/or viability of bacteria exposed to starvation (Magarinos et al., 1997; Joux et al., 1997a,b). Enzyme activities may be detected when they convert nonfluorescent substrates into fluorescent products that are then trapped in a cell (e.g., fluorescein diacetate (FDA) or carboxyFDA converted to fluorescein, or 5-cyano-2,3-ditolyl tetrazolium chloride (CTC) converted to its fluorescent formazan). These techniques have been used both in monitoring physiological changes in cultured cells exposed to starvation or toxins (Joux et al., 1997a, b; Faber et al., 1997) and in determining the distribution of metabolically active cells in natural populations (Porter et al., 1995b; del Giorgio et al., 1997).

Sorting

Flow cytometers equipped for fluorescence-activated cell sorting (FACS) permit any identifiable group of organisms to be separated from the rest of the community for further analysis. Li (1994) was successful in using the sorting capacity of a FACSort to estimate group-specific rates of ^{14}C fixation for four subgroups of picophytoplankton. Using an EPICS V, Lipschultz (1995) sorted different size-classes of larger phytoplankton (>3 μm) to measure their incorporation of ^{15}N-ammonium or ^{15}N-nitrate without the interference of heterotrophic bacteria or detrital particles. While in both cases enough material was collected to permit analysis, the practical limits of current types of flow cytometric sorting were reached. At the lowest productivity site examined, Li (1994) found that it took 3 h per sample to sort the tens of thousands of cells required to make each measurement. Lipschultz (1995) found that by preconcentrating his samples 200-fold, enough material (2.5×10^5 cells) could be sorted in 30 minutes for an exceptionally sensitive ^{15}N analysis, but hours were required to collect sufficient material for bulk CHN analysis.

Although flow cytometric analysis based on the autofluorescence of photosynthetic pigments may give the impression of greater distinction among phytoplankton, it is widely acknowledged that within each of the operationally defined groups outlined above (*Synechococcus*, *Prochlorococcus*, 'small' and 'large' eukaryotes) lies a great deal of taxonomic diversity (Campbell & Vaulot, 1993; Goericke & Repeta, 1993; Partensky et al., 1993; Palenik, 1994; Scanlan et al., 1996; Potter et al., 1997; Toledo & Palenik, 1997, Urbach et al., 1998). Some of this diversity can be observed in autofluorescence-based flow cytometric profiles (e.g. Wood et al., 1985; Olson et al., 1990; Moore et al., 1998), but flow cytometry simply cannot distinguish morphologically similar cells with similar pigmentation by endogenous characters, no matter how taxonomically distinct the organisms might be. Although Simon et al. (1994) were able to use scattering characteristics in combination with binding of two DNA-specific dyes (Hoechst 33342 and Chromomycin A3, which bind AT- versus GC-rich regions, respectively) to discriminate among small marine eukaryotes grown in culture and to conclude that these strains are not representative of the types actually dominating in the ocean, further progress in defining the details of phytoplankton community structure will require the application of tools capable of more specific identification of phytoplankton cells. Similar constraints clearly apply to most problems in microbial ecology, as most heterotrophic microorganisms offer even fewer distinguishing characteristics that are naturally detectable by flow cytometry than do phytoplankton. The information required to design tools to overcome these constraints, and to use flow cytometry to gather information from natural microbial communities at any required level of taxonomic or physiological specificity, will undoubtedly come from the realm of molecular ecology.

Approaches to molecular microbial ecology

Two fundamentally different approaches have been used in molecular microbial ecology (Falkowski & LaRoche, 1991; Atlas et al., 1992; Amann et al.,

1995; Amann et al., 1996). The first approach requires the extraction of nucleic acids from environmental samples for subsequent detailed analysis. The second employs probes to detect nucleic acids in their native state ('*in situ*') with minimal manipulation. Successful application of the second approach requires the availability of considerable sequence data to guide the design and application of each new probe (Amann et al., 1995; Medlin & Simon, 1998).

Analysis of Community Structure from Extracted Nucleic Acids

Perhaps the most important example of the first approach is the explosion of taxonomic information now being derived via the amplification, cloning and sequencing of genes directly from bulk community DNA purified from environmental samples (for reviews, see Amann et al., 1995; Pace, 1997). 16S rRNA has been the most widely used target for this type of analysis because it is universal among living organisms, generally believed to provide a reliable record of the real evolutionary relationships among living things, and flexible as to the level of taxonomic information revealed. 16S rRNA genes are often amplified using universal primers that target sites common to every known organism (Giovannoni et al., 1990; Britschgi & Giovannoni, 1991; Schmidt et al., 1991; DeLong, 1992; Fuhrman et al., 1993). Specific primers can also be designed for a finer taxonomic scale so that amplification is limited to one or a few groups of organisms (Voytek & Ward, 1995; M. Lange et al., 1996; Nubel et al., 1997; Ward et al., 1997). The same approach can also be applied to other genes, either for the purpose of gaining more detailed taxonomic information about one particular group of microorganisms (Palenik, 1994; Chen & Suttle, 1995; Palenik & Swift, 1996) or with the goal of detecting all organisms capable of a certain enzymatic activity, regardless of their taxonomic affinities (Paul et al., 1990; Kirshtein et al., 1991; Sinigalliano et al., 1995; Zehr et al., 1995; Zehr & Capone, 1996; Lechner & Conrad, 1997). The use of process-specific probes may allow the identification of all microorganisms that can perform a particular function (Wagner et al., 1996), but low similarity of homologous sequences in different taxa and the existence of multiple enzymatic pathways performing the same function will make it difficult to assure that all potential contributors to a process are detected (Barkay et al., 1989).

Another potential advantage of using genes encoding enzymes with essential functions for molecular ecological work is that they permit questions about physiological ecology to be addressed in addition to questions about taxonomic community structure. For example, analysis of total mRNA extracted from natural communities allows the determination of whether the gene of interest is being expressed and how its expression is regulated in the field (Pichard & Paul, 1991, 1993; Pichard et al., 1997; Lindell et al., 1998b).

One outcome of this work has been the recognition that in many cases, 'the great plate count anomaly' arises because the dominant bacteria in natural communities are inefficiently cultivated by standard techniques (Rehnstam et al., 1993; Giovannoni et al., 1995). Instead, the organisms brought into pure culture from natural environments are not necessarily representative of those actually dominating natural microbial communities (Wagner et al., 1993; Ferris et al., 1996; Kampfer et al., 1996; Pizarro et al., 1996; Suzuki et al., 1997). As a result, while some sequences isolated directly from the environment appear closely related to known groups, many others are not closely related to any microorganism that has been examined in culture. The ability to bypass a cultivation step offers the advantage of faster and more complete descriptions of microbial diversity, but does not make the cultivation of microorganisms any less essential to microbial ecologists. There is simply not yet enough information available to confidently deduce the physiological characteristics of the cell from which a novel 16S rRNA sequence was derived. Neither can we yet predict the pattern of expression of an enzyme represented by a gene recovered from a novel organism. The establishment of cultures of the organisms from which novel sequences are derived would provide basic biochemical and physiological information that should aid greatly in connecting them with their role in their natural surroundings.

In addition, care must be taken to confirm that sequences recovered from bulk community DNA actually represent organisms that are present and active within the original environment. The importance of this concern is demonstrated by the danger of amplifying and cloning a relatively high frequency of chimeric molecules that do not represent real organisms (Wang & Wang, 1996), and by the potential for one organism to contain multiple different rRNA sequences (Nubel et al., 1996). Confirmation that recovered sequences represent real organisms may be accomplished by a thorough comparison of sequences and strains independently isolated from the same sample (Chandler et al., 1997), by bulk extractive approaches in which a specifically designed probe is hybridized against total community DNA or rRNA (Moran et al., 1993; Wagner et al., 1993; Ferris et al., 1996), by quantitative PCR (S.-Y. Lee et al., 1996), or by cell-specific approaches such as fluorescent *in situ* hybridization (FISH) or whole-cell hybridization (Amann et al., 1995, 1997b).

Analysis of microbial populations at the single-cell level

The isolation of sequences from bulk community DNA reveals the presence of particular organisms, and in combination with hybridization to bulk nucleic acids, can give information on the relative abundance of particular types of organisms. There is, however, potentially more information that can be gained from single-cell level analyses, particularly where molecular probes providing physiological insight are employed. At least two routes for achieving this goal are available: the application of specific antibodies and the application of nucleic acid probes. In both cases, detection of probe binding is often achieved either by fluorescently labeling the probes directly or by using an indirect fluorescent detection system. Epifluorescence or scanning laser confocal microscopy are commonly used tools for gathering such data.

Immunofluorescence is a long-established autecological technique in microbial ecology (Bohlool & Schmidt, 1980; Vrieling & Anderson, 1996). Relative to DNA-based studies, it has three major drawbacks: organisms must already be in culture in order for antibodies to be raised against them; the specificity of probes is less predictable and must be defined empirically; and the flow of information is only one-way. In efforts to develop antibodies that can be used to specifically identify phytoplankton species by immunofluorescence, polyclonal antisera have been raised against cell-surface antigens for several marine cyanobacterial isolates (Campbell et al., 1983) and representatives from other major algal groups (L. P. Shapiro et al., 1989). These antisera proved reasonably effective as probes of the serogroups present in cyanobacterial assemblages from several coastal and open-ocean sites (Campbell et al., 1983; Campbell & Carpenter, 1987; Campbell & Iturriaga, 1988). Species- and genus-specific antibodies directed against representative eukaryotic phytoplankton were able to detect a small percentage of cells in a natural population (Campbell et al., 1994b). The observation that these antisera did not react with a larger percentage might be explained either by the tremendous diversity among the picophytoplankton or by the insufficiency of our culture collections (hence, antigen source) to represent the actual community.

Immunofluorescence approaches have also been used in efforts to quantify processes. Molecular detection of ammonium oxidizers via immunofluorescence was combined with the incorporation of ^{14}C (Ward, 1984) to estimate the metabolic activity of nitrifiers in open ocean waters. In cases where the quantitation of a particular antigen directly reveals either actual or potential physiological activity, such an approach should be valuable in investigating microbial physiological ecology. For example, Scanlan et al. (1997) raised an antibody against a protein expressed only in phosphorus-limited *Synechococcus* and showed that it could be combined with FITC-labeled secondary antibody to detect the phosphorus status of cells in natural populations by epifluorescence microscopy. La Roche et al. (1995, 1996) raised an antibody that recognized flavodoxin in diatoms and used total protein extracts and Western blots along with single-cell assays to determine the iron status of both cultures and natural populations of diatoms. Other workers are in the process of developing new molecular probes that should give information about the physiological status of the organisms of interest (e.g. Palenik & Koke, 1995; Dyrhman & Palenik, 1997; Lindell et al., 1998a).

Since the first demonstrations of the feasibility of using *in situ* hybridization of single cells to 16S rRNA-directed oligonucleotides to identify and enumerate bacteria (Giovannoni et al., 1988; DeLong et al., 1989; Amann et al., 1990a), fluorescent *in situ* hybridization (FISH) of cells fixed to slides followed by epifluorescent microscopic analysis has become a well-established tool in microbial ecology. rRNA provides a target for the design of oligonucleotide probes with varying degrees of taxonomic resolution, and with appropriate hybridization conditions single-basepair mismatch discrimination can be achieved (Manz et al., 1992). Examples include the identification and enumeration of particular types of bacteria in food (Schleifer et al., 1995), animal gut flora (Fischer et al., 1995; Santo Domingo et al., 1998), activated sludge (Amann et al., 1996; Kampfer et al., 1996; Manz et al., 1996; Erhart et al., 1997; Snaidr et al., 1997), water-treatment biofilms (Wagner et al., 1995,1996; Neef et al., 1996; Schramm et al., 1996), lake 'snow' (Weiss et al., 1996) and winter ice cover (Alfreider et al., 1996), soils (Hahn et al., 1992), hydrothermal vents (Harmsen et al., 1997), and in bacterioplankton (Hicks et al., 1992; Ramsing et al., 1996). FISH has also been used to describe the impacts of grazing on bacterial community structure in experimental microcosms (Pernthaler et al., 1997; Simek et al., 1997).

A similar approach can be taken towards identifying eukaryotic microorganisms based on 18S rRNA sequences (Lischewski et al., 1996). In fact, Lim et al. (1996) found that rRNA-targeted biotin-labeled oligonucleotides visualized with avidin-FITC gave higher counts of heterotrophic nanoflagellates in natural samples than did either DAPI or AO staining, probably because the oligonucleotide probes gave less nonspecific background staining, making the small eukaryotes more easily identified among debris.

Flow cytometry and the taxonomic structure of microbial communities

The goal of molecular flow cytometric analyses is basically the same as that of non-flow cytometric molecular analyses performed at the single-cell level; that is, to allow microbial ecologists to determine how many of which specific types of microorganisms are present in a particular sample and what they are doing. Even though it is in an earlier stage of development, the flow cytometric approach offers the advantages of higher throughput and sensitivity than epifluorescence microscopy-based analyses, and should be capable of using basically the same protocols while producing data that is more objective and amenable to statistical analysis. Here we summarize the progress to date in using both antibody and nucleic acid-based probes in combination with flow cytometry.

Immunological Detection

A major practical application of flow cytometric detection of molecular probes is in monitoring water sources and foods for the presence of pathogenic microorganisms (Donnelly & Baigent, 1986; Meier et al., 1997). Tyndall et al. (1985) used FITC-labeled antibodies and PI staining to identify *Legionella* in cooling tower water by flow cytometry. Flow cytometry improved the detection limits for *Giardia* and *Cryptosporidium* identification by immunofluorescence from 2 to 20 fold over epifluorescence microscopy (Vesey et al., 1994; Dixon et al., 1997), although the low abundances at which these organisms are of concern for water quality makes effective preconcentration strategies essential. Vrieling et al. (1997) developed FITC-labeled monoclonal antibodies that reliably identify the toxic dinoflagellates *Gyrodinium aureolum* and *Gymnodinium nagasakiense* in laboratory cultures grown under several different conditions.

Immunofluorescence and flow cytometry have also been used by Volsch et al. (1990) to examine the abundance of two serotypes of *Nitrosomonas* in sewage treatment plants. In this study, the proportion of the total bacterial community (as determined by PI staining) reacting with a *Nitrosomonas*-specific antibody (visualized with FITC-labeled secondary antibodies) varied from below the detection limit of 0.1% up to 2%.

Orellana & Perry (1995) developed a procedure for binding an anti-ribulose-1,5-bisphosphate carboxylase/oxygenase (Rubisco) antibody to both cultured and field-collected phytoplankton samples and quantitating the extent of labeling via a FITC-labeled secondary antibody and flow cytometry. These authors extensively tested various fixation/permeabilization protocols in search of a method that kept cells intact yet allowed sufficient access of the antibodies to their intracellular targets. They found ethanol fixation to be the best method tested on cultured cells, although both ethanol fixation and permeabilization by sodium hypochlorite oxidation were required for natural samples. This study showed that, in addition to identification based on extracellular antigens, immunofluorescence approaches can be taken to quantitate intracellular antigens of biogeochemical significance at a single-cell level.

Nucleic Acid Probes

In 1990, Amann et al. (1990b) demonstrated the capacity of fluorescent whole-cell hybridization combined with flow cytometry to quantify defined mixtures of target and nontarget bacteria. They designed and tested four tetramethylrhodamine isothiocyante (TRITC)-labeled oligonucleotide probes directed against 16S rRNA with various degrees of taxonomic specificity (eubacteria, sulfate-reducing bacteria, *Desulfobacter* and eukaryotes) on laboratory-grown cultures of *Escherichia coli*, *Desulfovibrio gigas* and *Desulfobacter hydrogenophilus*. Specific labeling signals varied from 4- to 7-fold higher than the 'noise' generated by nonspecific probe binding. The signal to noise ratio could not be increased by labeling each oligonucleotide with multiple fluorochrome molecules, but could be improved by applying more than one singly-labeled oligonucleotide probe at a time.

Several groups have been working to develop flow cytometric protocols for identifying eukaryotic microorganisms by fluorescent whole-cell hybridization to oligonucleotides complementary to 18S rRNA. Based on sequence information from 50 strains representing 12 classes of eukaryotic algae, Simon et al. (1995) designed 18S rRNA-targeted FITC-labeled oligonucleotides to distinguish members of the Chlorophyceae from nonchlorophytes. The two primers constructed were targeted to the same stretch of 18S rRNA, and differed only by one out of 18 nucleotides. Testing on 11 eukaryotic marine pico- and nano-phytoplankton from four different classes revealed that although the specificity of binding by the FITC-labeled probes themselves was low, it could be improved greatly by the inclusion of the unlabeled form of the alternate probe as a competitor. The results were consistent with the existing classification of the test organisms, and signal intensity was related to cell volume for actively growing cultures, suggesting that fixation and permeabilization were efficient in all cases. However, the specificity of labeling decreased as cultures entered stationary phase, due both to a decline in signal from the specific probe and to an increase in nonspecific, background autofluorescence. Picophytoplankton cells were much harder to identify, particularly in stationary phase, than were nanophytoplankton cells. This difficulty was increased by the fact that the original flow cytometric signatures of the organisms were altered by the fixation and hybridization procedure, so that scattering and chlorophyll autofluorescence could no longer be used to aid in detection and identification. For organisms like phytoplankton, the application of molecular tools may come at the cost of the endogenous characteristics that make them amenable to flow cytometric analysis without manipulation.

In an effort to develop probes that allow the fast and certain identification of members of the class Prymnesiophyceae in general and of the genus *Phaeocystis* in particular, M. Lange et al. (1996) designed two new probes targeted to different regions of the 18S rRNA to detect each of these taxa specifically. In whole-cell hybridization, their FITC-labeled probes were of limited utility for epifluorescent microscopic observations because of interference by autofluorescence resulting from chlorophyll degradation. The probes were more easily detected when analyzed by flow cytometry. The specificity of binding of the Prymnesiophyte- and *Phaeocystis*-specific probes was demonstrated by the ability of competition from the same probe in unlabeled form to reduce the signal, although the extent of this reduction varied among strains. Conversely, the lack of specificity in nontarget strains was confirmed by the absence of any significant effect of unlabeled competitor. Additional species-specific probes have been developed for studies of the bloom dynamics of *Chrysochromulina* and *Prymnesium* species, and these have been tested by whole-cell hybridization (Simon et al., 1997). Specific probes for major picoeukaryote groups (the classes Pelagophyceae and Prymnesiophyceae and the division Chlorophyta) have also been designed and tested by whole cell hybridization along with dot blots (Simon et al., 1996). Although signals were two to sevenfold higher than controls, the signal must be further amplified to successfully distinguish field populations.

Others have been working towards using whole-cell hybridization with 18S rRNA-targeted oligonucleotides combined with flow cytometry for detecting and identifying heterotrophic marine nanoplanktonic protists. Lim et al. (1993) showed that a FITC-labeled oligonucleotide probe directed against a conserved region of the 18S rRNA of eukaryotes hybridized to laboratory cultures of the ciliate *Uronema* sp. The signal was easily detectable by flow cytometry for actively growing cells, but declined to near background noise levels as the cells entered stationary phase. Using quantitative epifluorescence microscopy along with four 18S rRNA eukaryote-specific probes, a negative-control probe directed against the opposite strand, and a eubacterial-specific probe, they then tested several methods for increasing the signal to noise ratio. Attaching multiple fluorochrome molecules to each oligonucleotide was ineffective due to increased non-specific binding of these probes. In contrast, the simultaneous application of up to three probes recognizing different sites in 18S rRNA successfully increased the intensity of specific labeling, although not linearly. The signal was increased up to five-fold by utilizing biotin-labeled oligonucleotides for detection by an indirect method with FITC-labeled avidin, although nontarget binding of the avidin itself increased background fluorescence. Similar results were obtained for the flagellate *Cafeteria* sp. Probes were also tested on natural samples ($\sim 10^3$ cells ml^{-1}) following concentration or enrichment culturing, and examined by epifluorescent microscopy. In both cases the signal was low, as in stationary-phase cultures, but cells could easily be detected by the biotin-avidin or multiple-probes labeling strategies.

In their efforts to develop species-specific oligonucleotide probes for marine heterotrophic nanoflagellates, Rice et al. (1997a, b) found flow cytometry to be superior to epifluorescence microscopy in providing the quantification needed to optimize their techniques. Rice et al. (1997a, b) used five FITC or TRITC labeled probes targeted to the variable regions of *Paraphysomonas vestita* 18S rRNA along with their complements as controls, a general eukaryotic probe, and one probe targeting *P. imperforata*. While most investigators have used paraformaldehyde as a fixative before ethanol dehydration, Rice et al. (1997a,b) found that fixation with trichloroacetic acid (TCA) was superior in terms of cell preservation, low background autofluorescence, and high specific hybridization signals obtained. Specific and nonspecific labeling, as well as the signal-to-noise ratio, were higher in growing than in stationary phase *P. vestita*. Signal levels were also correlated with cell size. Both probe concentration and hybridization duration had significant effects on the signal-to-noise ratio. Simultaneous application of multiple probes targeted to different sequences increased the signal obtained in a manner similar to increasing the concentration of a single probe. However, these investigators were not able to use the biotin/avidin system for signal amplification due to high backgrounds generated by the avidin stain. Probes that bound specifically could be outcompeted by their own unlabeled forms, but not by other probes; conversely, competition had no effect on nonspecifically-binding probes. The *P. vestita*-specific probes were shown to label *P. vestita* more strongly than the *Paraphysomonas* congeners tested, although the relative intensity yielded by each of the five probes varied with the region of the 18S rRNA molecule targeted. The probe designed for *P. imperforata* failed to produce a signal above background levels.

Techniques allowing the application of whole-cell hybridization and flow cytometry to soil and sediment samples have also been developed and should be of great value to ecologists investigating the role of benthic processes in aquatic ecosystems. Wallner et al. (1995) hybridized seven FITC-labeled oligonucleotide probes of varying taxonomic specificity to cells from laboratory cultures as well as from samples of activated sludge, and counterstained with the DNA-specific dye Hoechst 33342. Simultaneous detection of this combination of fluorochromes requires the use of a dual-laser flow cytometer (488 nm and UV excitation). They found that a DNA counterstain was essential to help separate bacterial cells from debris, but that the choice of DNA dye was limited by spectral overlap and quenching interactions with the FITC used for the oligonucleotide probes. The probes used to detect beta- versus gamma-proteobacteria differed only by one nucleotide out of 17; sufficiently stringent hybridization conditions to distinguish between the two were achieved by including with each an equimolar concentration of the other probe in unlabeled form as competitor. Using six group-specific probes, Wallner et al. (1995) were able to quantitate groups comprising from 79 to 0.2% of the total Hoechst-stained counts with coefficients of variation ranging from 4% to 24%, respectively. For most groups, labeled percentages were similar to but more reproducible than those derived from epifluorescent microscopy counts performed on the same samples. Exceptions occurred with chain-forming or floc-bound cells, which were not completely dispersed by the fixation procedure and were therefore undercounted by flow cytometry. Within each group identified by the rRNA probes, multiparameter analysis based on forward scatter and DNA content (determined from the intensity of Hoechst staining) permitted the identification of distinct subgroups. This study of natural microbial populations via whole-cell hybridization and flow cytometry probably benefited technically from the relatively high activity of microbes found in the rich habitat of activated sludge.

Whole-cell hybridization in combination with flow cytometry has also been applied to identifying bacteria in bioaerosols (J. L. Lange et al., 1997) and soil samples (Thomas et al., 1997). The application of flow cytometry to samples such as these is complicated by the difficulties of distinguishing the organisms of interest from noise and debris, releasing bacteria from flocs or aggregates (Wallner et al., 1995), and by the low signal often obtained from natural bacterial samples (J. L. Lange et al., 1997). Thomas et al. (1997) circumvented these problems by using a multi-parameter approach, in which light-scattering properties, ethidium bromide staining, and whole-cell hybridization with 16S rRNA-based fluorescently labeled probes together serve to identify the bacteria of interest amid soil particles and other organisms.

Combining Probes

A good example of the multiparameter capacity of flow cytometry was provided by Wallner et al. (1996), who combined FLUOS (fluorescein)-labeled oligonucleotide probes and biotin-labeled antibodies (detected with PE-conjugated streptavidin) together with DAPI counterstaining to examine laboratory cultures of *Escherichia coli*, *Comomonas testosteroni* and *Pseudomonas aeruginosa*. These investigators noted that both SDS and TritonX-100 caused nonspecific binding of negative control antibodies, while Tween 20 and formamide did not. They found no significant difference in the results whether whole-cell hybridization or immunolabeling was performed first, although for some antibodies an additional deblocking step after immunolabeling was needed to recover the normal efficiency of subsequent whole-cell hybridization. While staining of the *E. coli* cells was not completely homogeneous, all three cell types could clearly be distinguished in mixtures based on various combinations of antibody and oligonucleotide probe labeling. In this case, spectral overlap between the fluorescent probes was corrected by compensation performed after data collection.

Sorting

Fluorescence activated cell sorting may provide a powerful general tool for targeting components of a microbial community for further analysis. For example, Moore et al. (1998) sorted 'bright' from 'dim' *Prochlorococcus* cells co-occurring in the same freshly collected sample at sea. They succeeded in establishing cultures of each type from the sorted cells, and showed that each type retained its unique flow cytometric and genetic signatures during growth in the lab. Wallner et al. (1997) presented several examples of the use of the sorting capacity of flow cytometry in combination with molecular analyses. A magnetic field enrichment derived from the sediment of a freshwater lake was dominated by a morphologically conspicuous bacterium, yet the 16S rDNA for this organism could not be obtained directly from the enrichment. Wallner et al. (1997) used flow cytometric sorting, based on scattering characteristics, to enrich the cells of interest. They were then able to use this sorted sample to amplify and clone the desired gene. Similarly, morphologically distinctive cells observed in samples from a freshwater lake were sorted, based on scatter and DAPI staining, achieving enrichment from about 5% of the cells present to nearly 100% purity. Wallner et al. (1997) also performed flow cytometric analysis on activated sludge samples hybridized with oligonucleotide probes directed against either *Acinetobacter* or *Leptothrix* 16S rRNA. They were able to discriminate morphologically distinct subpopulations among the cells that hybridized to each probe and to significantly (5–280-fold) purify each of these subpopulations by sorting, even though each subpopulation accounted for as little as 0.3% of the total DAPI-stained cells detected. Most remarkably, they were able to use as templates for PCR amplification both cells grown in culture and cells in activated sludge samples that had been fixed with paraformaldehyde or ethanol, hybridized to a fluorescent oligonucleotide, and sorted by flow cytometry. Similar treatments should be useful in field work with natural populations from many different sources.

Reporter strains

In addition to allowing the detection of endogenous markers, flow cytometry can be used to monitor the expression of exogenous genes introduced into genetically engineered reporter organisms by the immunological detection of the gene product (Mouawad et al., 1997), by the measurement of its activity against a fluorescent substrate, or by the fluorescence of the reporter itself. Tombolini et al. (1997) have shown that the gene encoding the green fluorescent protein (GFP, which absorbs blue light and emits green) can be introduced into *Pseudomonas fluorescens*, and that a single copy of *gfp* allows the engineered cells to be detected by flow cytometry, epifluorescence microscopy, or SCLM, regardless of their metabolic state. J. Porter et al. (1995a) used PCR to simultaneously amplify and FITC-label a plasmid-borne gene (*xylE*) inside fixed, intact *E. coli* and *P. putida* cells. Although only about half of the cells survived the thermal cycling procedure intact, this *in situ* PCR approach shows promise for increasing the signal to noise ratio for virtually any gene, whether introduced or endogenous.

Current limitations and needs for improvement

Currently, both the techniques for applying molecular probes at a single-cell level and the flow cytometric technology available to detect them face some limitations. Here we will point to some of the most important problems that must be solved in order to generalize the applicability of flow cytometry to molecular aquatic ecology and describe additional potential solutions to these problems that were not mentioned above.

Fixation and permeabilization

Optimizing fixation and permeabilization protocols remains a major issue in the application of molecular tools at a cellular level, just as it does in preserving any sample for flow cytometric analysis (Lepasteur et al., 1993). Paraformaldehyde fixation followed by ethanol dehydration/permeabilization is probably the most commonly used protocol. Following this treatment, a wide variety of organisms ranging from bacteria to protists appear to be freely permeable to oligonucleotides on the order of 15–20 residues in length (e.g. Hicks et al., 1992; Lim et al., 1996). However, several investigators have observed that gram-positive bacteria fixed with paraformaldehyde are much less accessible to oligonucleotide probes than are paraformaldehyde-fixed gram-negative or archaebacterial cells, and therefore require additional enzymatic permeabilization steps prior to hybridization (Wallner et al., 1993; Manz et al., 1994; Roller et al., 1994; Lloyd et al., 1996; Snaidr et al., 1997; Meier et al., 1997), although others have encountered less difficulty (Frischer et al., 1996; Santo Domingo et al., 1998). Gaining entry for the much larger molecules required in many enzyme-based indirect detection and signal amplification strategies while preserving cell integrity for flow cytometric analysis is even more difficult (see 'Sensitivity', below). Clearly, different preservation and permeabilization protocols may be required to allow equal access of probes to organisms with differing cell walls and other aspects of cell structure, and identifying appropriate protocols may require extensive experimentation (Orellana & Perry, 1995; Erhart et al., 1997). The optimal protocol can even vary for the same organism growing under different conditions. For example, Hahn et al. (1992) found that lysozyme treatment was required for the detection of the gram-positive bacterium *Streptomyces scabies* grown in culture, but not when grown in soil.

Another problem arising in the application of whole-cell hybridization techniques is the alteration or loss of endogenous properties used in the detection, identification and enumeration of untreated cells. Optical properties related to scatter measurements may be dramatically altered by fixation and permeabilization, while pigments may be extracted or destroyed. Microbial ecologists will have to carefully develop and evaluate these techniques to retain or replace essential native characteristics while adding informative new markers to the organisms of interest.

Sensitivity

Sensitivity, i.e. the signal to noise ratio, is the second major limitation of current single-cell molecular ecological and flow cytometric technology. 16S rRNA is one of the most abundant molecular targets available in bacterial cells; a rapidly growing cell can contain 10^4–10^5 ribosomes, and ribosome content is often directly related to growth rate (DeLong et al., 1989). The importance of this natural amplification is shown by the common observation that fast-growing cells are easily detected by FISH but slowly growing cells are not (DeLong et al., 1989). Generating sufficient signal from small or slowly growing organisms is one of the most significant technical problems facing the application of this technique in natural environments (Hahn et al., 1992). Fluorescent *in situ* protocols for flow cytometry will probably benefit from many of the same solutions now being developed for fluorescent microscopy protocols (Amann et al., 1995, 1997a). These include the use of more stable and more hydrophilic fluorochromes and of better filter blocks for choosing excitation and emission wavelengths (Alfreider et al., 1996; Glockner et al., 1996). While the incorporation of multiple fluorochromes into oligonucleotides has generally been found only to increase background noise due to nonspecific dye binding (e.g. Wallner et al., 1993), the use of multiple labels in polynucleotide probes approximately 100 residues in length may improve signal levels for some applications (Timm & Stewart, 1992; Trebesius et al., 1994). For oligonucleotides, signal may be increased by using multiple different singly-labeled probes with the same taxonomic specificity (Hicks et al., 1992; S. Lee et al., 1993; S. Lee & Kemp, 1994), by carefully optimizing hybridization conditions (Wallner et al., 1993), and by considering the secondary structure of rRNA while designing primers (Assmus et al., 1995; Frischer et al., 1996). Manipulations such as chloramphenicol treatment to increase the number of rRNA targets per cell (Ouverney & Fuhrman, 1997) may also be useful, but for DNA or mRNA targets other signal amplification strategies will be needed.

One approach to signal amplification involves labeling oligonucleotide probes with nonfluorescent moieties that will allow secondary detection systems to increase the signal output. In the most direct form of this approach, oligonucleotides labeled with biotin or digoxigenin (DIG) are detected with streptavidin or anti-DIG Fab antibody fragments that are themselves conjugated to one or more fluorochrome molecules.

Lim et al. (1996) reported good results from the biotin approach, but Zarda et al. (1991) found no improvement of DIG-labeled oligonucleotides over oligonucleotides directly labeled with a single fluorochrome. Timm & Stewart (1992), using DIG-labeled polynucleotides, found that high nonspecific background arose from the anti-DIG detection step.

Taking the system to the next level of amplification, Zarda et al. (1991) reported promising results using anti-DIG conjugated to either horseradish peroxidase (HRP) or alkaline phosphatase (APH), combined with non-fluorescent enzymatic detection reactions. However, Fisher et al. (1995) found that non-fluorescent detection reactions, driven either directly by HRP-labeled oligonucleotides or indirectly by DIG-labeled oligonucleotides detected with APH conjugated to anti-DIG, were inferior to directly FITC- or Cy3-labeled oligonucleotides due to technical difficulties with fixation, permeabilization and visualization. Schonhuber et al. (1997) used HRP-labeled oligonucleotides to catalyze the deposition of fluorescein-tyramide and were successful in increasing the sensitivity of detection of some bacteria by FISH on slides with this indirect fluorescence technique. However, they found that many bacterial strains tested were not permeable to the system by FISH. Additionally, even strains that labeled well by FISH yielded heterogeneous populations in flow cytometric analyses. Lebaron et al. (1997) found that biotin-labeled oligonucleotides detected with HRP-streptavidin and visualized with tyramide-TRITC gave higher signals than did a mixture of 5 monolabeled oligonucleotides for a variety of cultured bacteria and natural samples analyzed by FISH on slides. Again, however, the technique did not work in combination with flow cytometry. Yamaguchi et al. (1996) developed an indirect detection technique by which DIG-labeled oligonucleotides are bound with APH-anti-DIG. APH then dephosphorylates 2-hydroxy-3-napthoic acid-2'-phenylanilide phosphate (HNPP) to HNP, which combines with Fast Red TR to make a bright and stable fluorochrome, HNP-TR. This system gave up to an 8-fold signal amplification over singly fluorochrome-labeled oligonucleotides.

For detecting targets present at lower copy numbers, amplification techniques such as *in situ* PCR, perhaps used in combination with other signal-enhancement strategies, promise effective results (Hodson et al., 1995; Tolker-Nielsen et al., 1997, 1998; Tani et al., 1998).

Efforts to improve the sensitivity of flow cytometric detection of molecular probes should not ignore the flow cytometer itself (Steen, 1992). All of the flow path, optical path, and electronic components of a flow cytometer can impact its sensitivity, reliability and suitability for a particular application. While some operators will benefit from making their own modifications to commercial flow cytometers to suit their specific needs (Dusenberry & Frankel, 1994), it is likely that most operators will benefit from simply using a machine with design features that best suit their particular application, maintaining their equipment in good order, and replacing or upgrading components as needed.

Rare events

In microbial communities, important processes may be driven by organisms that are very rare relative to the total community. Therefore, the detection of a sufficient number of cells to perform interesting analyses may not be a simple matter. For microscopy-based detection, Amann et al. (1995) argue that target cells need to be present at abundances greater than 10^3 per cm^2 area examined (translating to initial densities of 10^5 cells ml^{-1} if a typical liquid sample of 10 μl is spotted onto a 1 cm^2 area) to be detectable with reasonable effort. The FISH-epifluorescence microscopy technique of Wagner et al. (1996) had a detection limit of about 10^4 cells ml^{-1} in the initial sample. Microscopic investigations using fluorescently-labeled rRNA probes of fairly broad taxonomic specificity can detect positive cells accounting for about 1% of the total population (that is, an initial density above 10^4 labeled cells ml^{-1} in a total community of 10^6 cells ml^{-1}), but these numbers are also very close to the counts obtained with nonspecific, negative-control probes (Glockner et al., 1996). Ramsing et al. (1996) reported a higher detection limit, 10^5 cells ml^{-1} or 10% of total cells, except when specific morphological parameters could simultaneously be used to lower the effective detection limit to 10^4 cells ml^{-1} or 1% of the total cells. Similarly, the bulk nucleic acid extraction and hybridization method of Pichard & Paul (1991) also had a detection limit of around 10^4 cells ml^{-1}.

Flow cytometry can be used to monitor laboratory cultures growing at densities much more relevant to those found in natural environments (10^2 cells ml^{-1}, Sorensen & Jakobsen, 1997) than the $>10^6$ cells ml^{-1} more typical of classical microbiological laboratory investigations. Volsch et al. (1990) reported a detection limit of 0.1% of total cells for their immunofluorescent technique analyzed by flow cytometry. Limits of detection as low as 10 cells ml^{-1} have been cited for flow cytometers (Troussellier et al., 1993). However, in natural samples this limit may be difficult to reach. Due to both volume throughput and electronic constraints, the flow cytometric analysis of a 1 ml sample can require from 10 to 100 min. Whether the data obtained from an organism present at 10 cells ml^{-1} will be recognizable as a distinct group depends on the counting efficiency and reproducibility of the signal from each event, both of which may be maximized by lower flow rates (longer run times). Although an easily identifiable group may in fact be detectable at natural abundances near 10 cells ml^{-1}, in many cases a more practical detection limit may be between 100 and 1000 cells ml^{-1}, which is still 10–100-fold better than current microscopic protocols and more easily achieved. Preconcentration of samples may relieve volume limitations and improve the detection of less abundant particles, but will still be subject to other constraints.

Depending on the proportion at which a specific type of organism is present (one cell in one thousand or one cell in one million), flow cytometric analysis may range from difficult to impossible. For example, the calculations of Rosenblatt et al. (1997) show that detecting 1000 signals from cells representing 0.1% of the total events detected will require examining 10^7 events. At a rate of 5000 events per second, that analysis would require less than 4 min. However, for organisms present at only one per million, acquiring data from the same 1000 cells would require examining 10^9 events. At the same event rate, this analysis would require over 58 h. Although it is possible to ignore non-target particles to some extent by excluding them from the data that are recorded, physical and electronic limitations on the rate at which flow cytometers can examine particles are not as easily circumvented. Concentrating a sample to increase the abundance of rare particles only makes the analysis easier if most other particles can also be selectively removed.

A comparison of these limits with some abundance estimates currently in hand is informative. Ward (1984) found the abundance of ammonium-oxidizing bacteria to be approximately 10^4 cells l^{-1} (10 cells ml^{-1}) in marine samples. K.-H. Lee & Ruby (1995) estimated the abundance of *Vibrio fischeri* at 10^2–10^3 cells ml^{-1} (mostly viable but nonculturable) in seawater from Hawaii. In the Mariager Fjord, Ramsing et al. (1996) estimated that delta-proteobacteria accounted for 10^5 cells ml^{-1} in a total bacterial population of 10^6 cells ml^{-1}. Within that fraction, cells hybridizing to *Desulfovibrio* and *Desulfobacter* probes were present at less than 10^3 cells ml^{-1}, while a *Desulfobacterium* probe hybridized to 10^3–10^4 cells ml^{-1}. Using different techniques, Teske et al. (1996) detected only 10^2 sulfate-reducing cells ml^{-1} at the same site. Clearly most of these examples are below the limit of detection for microscopy-based protocols without further sample concentration. Although many of these examples are near or above the expected detection limits of flow cytometers, the precise limit for any given application will be influenced by many factors, including volume and electronic throughput limitations and the signal-to-noise ratio achieved for the target organism.

Conclusions

Flow cytometry offers several advantages that appear to be essential for advancing aquatic microbial ecology. It is a rapid technique for collecting multiparameter data on large numbers of individual cells, and its intrinsic scale of analysis is much closer to the relevant physical scale of microorganisms than is the scale of most traditional bulk techniques. Additionally, flow cytometry offers the ability to sort organisms of interest out of a sample for further independent analysis. However, flow cytometric analysis based on endogenous characteristics of microbes is limited in its applications. New ways of putting exogenous labels on organisms of interest are required to expand its utility. To accomplish this task by utilizing flow cytometry in combination with molecular probes, several current methodological limitations must be overcome. First, methods to manipulate samples in a quantitative and reproducible manner are needed, particularly for rare organisms. Additionally, preservation and permeabilization protocols that are suitable for a wide phylogenetic range of organisms must be identified. An essential component of that technical effort must focus on the need for signal-amplification protocols that generate little or no background noise. More sequence data from a variety of organisms are also needed to aid in the design of oligonucleotide probes. Beyond these technical issues, flow cytometry offers microbial ecologists the opportunity to collect much more data at the single-cell level than has previously been possible. Hence, the temporal and/or spatial coverage of a system can be performed at scales much more complete and relevant to microbial processes than has been possible historically. However, as a result microbial ecologists will soon have to deal explicitly with statistical concerns about the representativeness, reproducibility and appropriateness of their sampling schemes in order to scale organismal-level information back up to ecosystem-level questions.

References

Alfreider, A., J. Pernthaler, R. Amann, B. Sattler, F.-O. Glockner, A. Wille & R. Psenner, 1996. Community analysis of the bacterial assemblages in the winter cover and pelagic layers of a high mountain lake by *in situ* hybridization. Appl. envir. Microbiol. 62: 2138–2144.

Amann, R., F.-O. Glockner & A. Neef, 1997a. Modern methods in subsurface microbiology: *in situ* identification of microorganisms with nucleic acid probes. FEMS Microbiol. Rev. 20: 191–200.

Amann, R., J. Snaidr, M. Wagner, W. Ludwig & K.-H. Schleifer, 1996. *In situ* visualization of high genetic diversity in a natural microbial community. J. Bact. 178: 3496–3500.

Amann, R., N. Springer, W. Schonhuber, W. Ludwig, E. N. Schmid, K.-D. Muller & R. Michel, 1997b. Obligate intracellular bacterial parasites of Acanthamoebae related to *Chlamydia* spp. Appl. envir. Microbiol. 63: 115–121.

Amann, R. I., L. Krumholz & D. A. Stahl, 1990a. Fluorescent-oligonucleotide probing of whole cells for determinative, phylogenetic, and environmental studies in microbiology. J. Bact. 172: 762–770.

Amann, R. I., W. Ludwig & K.-H. Schleifer, 1995. Phylogenetic identification and *in situ* detection of individual microbial cells without cultivation. Microbiol. Rev. 59: 143–169.

Amann, R. I., B. J. Binder, R. J. Olson, S. W. Chisholm, R. Devereux & D. A. Stahl, 1990b. Combination of 16S rRNA-targeted oligonucleotide probes with flow cytometry for analyzing mixed microbial populations. Appl. envir. Microbiol. 56: 1919–1925.

Assmus, B., P. Hutzler, G. Kirchhof, R. Amann, J. R. Lawrence & A. Hartmann, 1995. *In situ* localization of *Azospirillum brasilense* in the rhizosphere of wheat with fluorescently labeled, rRNA-targeted oligonucleotide probes and scanning confocal laser microscopy. Appl. envir. Microbiol. 61: 1013–1019.

Atlas, R. M., G. Sayler, R. S. Burlage & A. K. Bej, 1992. Molecular approaches for environmental monitoring of microorganisms. BioTechniques 12: 706–717.

Ault, K. A., 1991. Calibration and quantitative analysis. In S. Demers (ed.), Particle Analysis in Oceanography, NATO ASI Series, G27. Springer-Verlag, Berlin: 47–58.

Azam, F., T. Fenchel, J. G. Field, J. S. Gray, L. A. Meyer-Reil & F. Thingstad, 1983. The ecological role of water-column microbes in the sea. Mar. Ecol. Prog. Ser. 10: 257–263.

Barcina, I., P. Lebaron & J. Vives-Rego, 1997. Survival of allochthonous bacteria in aquatic systems: a biological approach. FEMS Microbiol. Ecol. 23: 1–9.

Barkay, T., C. Liebert & M. Gillman, 1989. Hybridization of DNA probes with whole-community genome for detection of genes that encode microbial responses to pollutants: *mer* genes and Hg^{2+} resistance. Appl. envir. Microbiol. 55: 1574–1577.

Bohlool, B. B. & E. L. Schmidt, 1980. The immunofluorescence approach in microbial ecology. Adv. microb. Ecol. 4: 203–241.

Bougrier, S., A. J. S. Hawkins & M. Heral, 1997. Preingestive selection of different microalgal mixtures in *Crassostrea gigas* and *Mytilus edulis*, analyzed by flow cytometry. Aquaculture 150: 123–134.

Britschgi, T. B. & S. J. Giovannoni, 1991. Phylogenetic analysis of a natural marine bacterioplankton population by rRNA gene cloning and sequencing. Appl. envir. Microbiol. 57: 1707–1713.

Button, D. K. & B. R. Robertson, 1993. Use of high-resolution flow cytometry to determine the activity and distribution of aquatic bacteria. In Kemp P. F. (ed.), Handbook of Methods in Aquatic Microbial Ecology. CRC Press, Inc., Boca Raton, Florida: 163–173.

Button, D. K., B. R. Robertson & F. Juttner, 1996. Microflora of a subalpine lake: bacterial populations, size and DNA distributions, and their dependence on phosphate. FEMS Microbiol. Ecol. 21: 87–101.

Campbell, L. & E. J. Carpenter, 1987. Characterization of phycoerythrin-containing *Synechococcus* spp. populations by immunofluorescence. J. Plankton Res. 9: 1167–1181.

Campbell, L. & R. Iturriaga, 1988. Identification of *Synechococcus* spp. in the Sargasso Sea by immunofluorescence and fluorescence excitation spectroscopy performed on individual cells. Limnol. Oceanogr. 33: 1196–1201.

Campbell, L. & D. Vaulot, 1993. Photosynthetic picoplankton community structure in the subtropical North Pacific Ocean near Hawaii (Station ALOHA). Deep Sea Res. 40: 2043–2060.

Campbell, L., E. J. Carpenter & V. J. Iacono, 1983. Identification and enumeration of marine chroococcoid cyanobacteria by immunofluorescence. Appl. envir. Microbiol. 46: 553–559.

Campbell, L., H. A. Nolla & D. Vaulot, 1994a. The importance of *Prochlorococcus* to community structure in the central North Pacific Ocean. Limnol. Oceanogr. 39: 954–961.

Campbell, L., L. P. Shapiro & E. Haugen, 1994b. Immunochemical characterization of eukaryotic ultraplankton from the Atlantic and Pacific Oceans. J. Plankton Res. 16: 35–51.

Campbell, L., H. Liu, H. A. Nolla & D. Vaulot, 1997. Annual variability of phytoplankton and bacteria in the subtropical North Pacific Ocean and Station ALOHA during the 1991–1994 ENSO event. Deep Sea Res. 44: 167–192.

Carpenter, E. J. & J. Chang, 1988. Species-specific phytoplankton growth rates via diel DNA synthesis cycles. I. Concept of the method. Mar. Ecol. Prog. Ser. 43: 105–111.

Chandler, D. P., S.-M. Li, C. M. Spadoni, G. R. Drake, D. L. Balkwill, J. K. Fredrickson & F. J. Brockman, 1997. A molecular comparison of culturable aerobic heterotrophic bacteria and 16S rDNA clones derived from a deep subsurface sediment. FEMS Microbiol. Ecol. 23: 131–144.

Chen, F. & C. A. Suttle, 1995. Amplification of DNA polymerase gene fragments from viruses infecting microalgae. Appl. envir. Microbiol. 61: 1274–1278.

Cucci, T. L., S. E. Shumway, W. S. Brown & C. R. Newell, 1989. Using phytoplankton and flow cytometry to analyze grazing by marine organisms. Cytometry 10: 659–669.

Davey, H. M. & D. B. Kell, 1996. Flow cytometry and cell sorting of heterogeneous microbial populations: the importance of single-cell analyses. Microbiol. Rev. 60: 641–696.

del Giorgio, P. A., Y. T. Prairie & D. F. Bird, 1997. Coupling between rates of bacterial production and the abundance of metabolically active bacteria in lakes, enumerated using CTC reduction and flow cytometry. Microb. Ecol. 34: 144–154.

del Giorgio, P. A., D. F. Bird, Y. T. Prairie & D. Planas, 1996. Flow cytometric determination of bacterial abundance in lake plankton with the green nucleic acid stain SYTO 13. Limnol. Oceanogr. 41: 783–789.

DeLeo, P. C. & P. Baveye, 1996. Enumeration and biomass estimation of bacteria in aquifer microcosm studies by flow cytometry. Appl. envir. Microbiol. 62: 4580–4586.

DeLong, E. F., 1992. Archaea in coastal marine environments. Proc. natn. Acad. Sci. U.S.A. 89: 5685–5689.

DeLong, E. F., G. S. Wickham & N. R. Pace, 1989. Phylogenetic stains: ribosomal RNA-based probes for the identification of single cells. Science 243: 1360–1363.

Dixon, B. R., M. Parenteau, C. Martineau & J. Fournier, 1997. A comparison of conventional microscopy, immunofluorescence microscopy and flow cytometry in the detection of *Giardia lamblia* cysts in beaver fecal samples. J. imm. Meth. 202: 27–33.

Donnelly, C. W. & G. J. Baigent, 1986. Method for flow cytometric detection of *Listeria monocytogenes* in milk. Appl. envir. Microbiol. 52: 689–695.

Dubelaar, G. B. J., A. C. Greonewegen, W. Stokdijk, G. J. van den Engh & J. W. M. Visser, 1989. Optical plankton analyser: a flow cytometer for plankton analysis, II: specifications. Cytometry 10: 529–539.

Dusenberry, J. A. & S. L. Frankel, 1994. Increasing the sensitivity of a FACScan flow cytometer to study oceanic picoplankton. Limnol. Oceanogr. 39: 206–209.

Dyhrman, S. T. & B. P. Palenik, 1997. The identification and purification of a cell-surface alkaline phosphatase from the dinoflagellate *Prorocentrum minimum* (Dinophyceae). J. Phycol. 33: 602–612.

Edwards, C., J. Diaper & J. Porter, 1996. Flow cytometry for the targeted analysis of the structure and function of microbial populations. In Pickup R. W. & J. R. Saunders (eds), Molecular Approaches to Environmental Microbiology. Ellis Horwood Limited, Cambridge: 137–162.

Erhart, R., D. Bradford, R. J. Seviour, R. Amann & L. L. Blackall, 1997. Development and use of fluorescent *in situ* hybridization probes for the detection and identification of "*Microthrix parvicella*" in activated sludge. System. apl. Microbiol. 20: 310–318.

Faber, M. J., L. M. J. Smith, H. J. Boermans, G. R. Stevenson, D. G. Thompson & K. R. Solomon, 1997. Cryopreservation of fluorescent marker-labeled algae (*Selenastrum capricornutum*) for toxicity testing using flow cytometry. Envir. Tox. Chem. 16: 1059–1067.

Falkowski, P. G. & J. LaRoche, 1991. Molecular biology in studies of ocean processes. Int. Rev. Cytol. 128: 261–303.

Ferris, M. J., A. L. Ruff-Roberts, E. D. Kopczynski, M. M. Bateson & D. M. Ward, 1996. Enrichment culture and microscopy conceal diverse thermophilic *Synechococcus* populations in a single hot spring microbial mat habitat. Appl. envir. Microbiol. 62: 1045–1050.

Fischer, K., D. Hahn, R. I. Amann, O. Daniel & J. Zeyer, 1995. *In situ* analysis of the bacterial community in the gut of the earthworm *Lumbricus terrestris* L. by whole-cell hybridization. Can. J. Microbiol. 41: 666–673.

Fouchet, P., C. Jayat, Y. Hechard, M.-H. Ratinaud & G. Frelat, 1993. Recent advances of flow cytometry in fundamental and applied microbiology. Biol. Cell. 78: 95–109.

Frischer, M. E., P. J. Floriani & S. A. Nierzwicki-Bauer, 1996. Differential sensitivity of 16S rRNA targeted oligonucleotide probes used for fluorescence *in situ* hybridization is a result of ribosomal higher order structure. Can. J. Microbiol. 42: 1061–1071.

Fuhrman, J. A., K. McCallum & A. A. Davis, 1993. Phylogenetic diversity of subsurface marine microbial communities from the Atlantic and Pacific Oceans. Appl. envir. Microbiol. 59: 1294–1302.

Giovannoni, S. J., T. D. Mullins & K. G. Field, 1995. Microbial diversity in oceanic systems: rRNA approaches to the study of unculturable microbes. In Joint I. (ed.), Molecular Ecology of Aquatic Microbes, NATO ASI Series, G38. Springer-Verlag, Berlin: 217–248.

Giovannoni, S. J., T. B. Britschgi, C. L. Moyer & K. G. Field, 1990. Genetic diversity in Sargasso Sea bacterioplankton. Nature 344: 60–63.

Giovannoni, S. J., E. F. DeLong, G. J. Olsen & N. R. Pace, 1988. Phylogenetic group-specific oligonucleotide probes for identification of single microbial cells. J. Bact. 170: 720–726.

Glockner, F. O., R. Amann, A. Alfreider, J. Pernthaller, R. Psenner, K. Trebesius & K.-H. Schleifer, 1996. An *in situ* hybridization protocol for detection and identification of planktonic bacteria. System. appl. Microbiol. 19: 403–406.

Goericke, R. & D. J. Repeta, 1993. Chlorophylls *a* and *b* and divinyl chlorophylls *a* and *b* in the open subtropical North Atlantic Ocean. Mar. Ecol. Prog. Ser. 101: 307–313.

Hahn, D., R. I. Amann, W. Ludwig, A. D. L. Akkermans & K.-H. Schleifer, 1992. Detection of microorganisms in soil after *in situ* hybridization with rRNA-targeted, fluorescently labelled oligonucleotides. J. gen. Microbiol. 138: 879–887.

Harmsen, H. J. M., D. Prieur & C. Jeanthon, 1997. Group-specific 16S rRNA-targeted oligonucleotide probes to identify thermophilic bacteria in marine hydrothermal vents. Appl. envir. Microbiol. 63: 4061–4068.

Hauer, B. & H. Eipel, 1997. Flow Cytometry. Useful tool for analyzing bacterial populations cell by cell. In Shapiro J. A. & M. Dworkin (eds), Bacteria as Multicellular Organisms. Oxford University Press, New York: 273–291.

Hicks, R. E., R. I. Amann & D. A. Stahl, 1992. Dual staining of natural bacterioplankton with 4′,6-diamidino-2-phenylindole and fluorescent oligonucleotide probes targeting kingdom-level 16S rRNA sequences. Appl. envir. Microbiol. 58: 2158–2163.

Hobbie, J. E., R. J. Daley & S. Jasper, 1977. Use of Nucleopore filters for counting bacteria by fluorescence microscopy. Appl. envir. Microbiol. 33: 1225–1228.

Hodson, R. E., W. A. Dustman, R. P. Garg & M. A. Moran, 1995. *In situ* PCR for visualization of microscale distribution of specific genes and gene products in prokaryotic communities. Appl. envir. Microbiol. 61: 4074–4082.

Joux, F., P. Lebaron & M. Troussellier, 1997a. Changes in cellular states of the marine bacterium *Deleya aquamarina* under starvation conditions. Appl. envir. Microbiol. 63: 2686–2694.

Joux, F., P. Lebaron & M. Troussellier, 1997b. Succession of cellular states in a *Salmonella typhimurium* population during starvation in artificial seawater microcosms. FEMS Microbiol. Ecol. 22: 65–76.

Kampfer, P., R. Erhart, C. Beimfohr, J. Bohringer, M. Wagner & R. Amann, 1996. Characterization of bacterial communities from activated sludge: culture-dependent numerical identification versus *in situ* identification using group- and genus-specific rRNA-targeted oligonucleotide probes. Microb. Ecol. 32: 101–121.

Kaprelyants, A. S., G. V. Mulkamolova, H. M. Davey & D. B. Kell, 1996. Quantitative analysis of the physiological heterogeneity within starved cultures of *Micrococcus luteus* by flow cytometry and cell sorting. Appl. envir. Microbiol. 62: 1311–1316.

Kirshtein, J. D., H. W. Paerl & J. Zehr, 1991. Amplification, cloning, and sequencing of a *nifH* segment from aquatic microorganisms and natural communities. Appl. envir. Microbiol. 57: 2645–2650.

La Roche, J., P. W. Boyd, R. M. L. McKay & R. J. Geider, 1996. Flavodoxin as an *in situ* marker for iron stress in phytoplankton. Nature 382: 802–805.

La Roche, J., H. Murray, M. Orellana & J. Newton, 1995. Flavodoxin expression as as indicator of iron limitation in marine diatoms. J. Phycol. 31: 520–530.

Landry, M. R., J. Constantinou & J. Kirshtein, 1995. Microzooplankton grazing in the central equatorial Pacific during February and August, 1992. Deep Sea Res. 42: 657–671.

Lange, J. L., P. S. Thorne & N. Lynch, 1997. Application of flow cytometry and fluorescent *in situ* hybridization for assessment of exposures to airborne bacteria. Appl. envir. Microbiol. 63: 1557–1563.

Lange, M., L. Guillou, D. Vaulot, N. Simon, R. I. Amann, W. Ludwig & L. K. Medlin, 1996. Identification of the class Prymnesiophyceae and the genus *Phaeocystis* with ribosomal RNA-targeted nucleic acid probes detected by flow cytometry. J. Phycol. 32: 858–868.

Lebaron, P., P. Catala, C. Fajon, F. Joux, J. Baudart & L. Bernard, 1997. A new sensitive, whole-cell hybridization technique for detection of bacteria involving a biotinylated oligonucleotide probe targeting rRNA and tyramide signal amplification. Appl. envir. Microbiol. 63: 3274–3278.

Lechner, S. & R. Conrad, 1997. Detection in soil of aerobic hydrogen-oxidizing bacteria related to *Alcaligenes eutrophus* by PCR and hybridization assays targeting the gene of the membrane-bound (NiFe) hydrogenase. FEMS Microbiol. Ecol. 22: 193–206.

Lee, K.-H. & E. G. Ruby, 1995. Symbiotic role of the viable but nonculturable state of *Vibrio fischeri* in Hawaiian coastal seawater. Appl. envir. Microbiol. 61: 278–283.

Lee, S. & P. F. Kemp, 1994. Single-cell RNA content of natural marine planktonic bacteria measured by hybridization with multiple 16S rRNA-targeted fluorescent probes. Limnol. Oceanogr. 39: 869–879.

Lee, S., C. Malone & P. F. Kemp, 1993. Use of multiple 16S rRNA-targeted fluorescent probes to increase signal strength and measure cellular RNA from natural planktonic bacteria. Mar. Ecol. Prog. Ser. 101: 193–201.

Lee, S.-Y., J. Bollinger, D. Bezdicek & A. Ogram, 1996. Estimation of the abundance of an uncultured soil bacterial strain by a competitive quantitative PCR method. Appl. envir. Microbiol. 62: 3787–3793.

Legendre, L. & C. M. Yentsch, 1989. Overview of flow cytometry and image analysis in biological oceanography and limnology. Cytometry 10: 501–510.

Lepasteur, M., J. M. Martin & A. Fleury, 1993. A comparative study of different preservation methods for phytoplankton cell analysis by flow cytometry. Mar. Ecol. Prog. Ser. 93: 55–63.

Li, W. K. W., 1994. Primary production of prochlorophytes, cyanobacteria, and eucaryotic ultraphytoplankton: measurements from flow cytometric sorting. Limnol. Oceanogr. 39: 169–175.

Li, W. K. W., 1995. Composition of ultraphytoplankton in the central North Atlantic. mar. Ecol. Prog. Ser. 122: 1–8.

Li, W. K. W., J. F. Jellett & P. M. Dickie, 1995. DNA distributions in planktonic bacteria stained with TOTO or TO-PRO. Limnol. Oceanogr. 40: 1485–1495.

Lim, E. L., D. A. Caron & E. F. DeLong, 1996. Development and field application of a quantitative method for examining natural assemblages of protists with oligonucleotide probes. Appl. envir. Microbiol. 62: 1416–1423.

Lim, E. L., L. A. Amaral, D. A. Caron & E. F. DeLong, 1993. Application of rRNA-based probes for observing marine nanoplankton protists. Appl. envir. Microbiol. 59: 1647–1655.

Lindell, D. & A. F. Post, 1995. Ultraphytoplankton succession is triggered by deep winter mixing in the Gulf of Aqaba (Eilat), Red Sea. Limnol. Oceanogr. 40: 1130–1141.

Lindell, D., E. Padan & A. F. Post, 1998a. Regulation of *ntcA* expression and nitrite uptake in the marine *Synechococcus* sp. Strain WH 7803. J. Bact. 180: 1878–1886.

Lindell, D., E. Padan, B. Lazar & A. Post, 1998b. *ntcA*, an indicator gene for the N status of marine *Synechococcus*. EOS Supplement 79(1).

Lipschultz, F., 1995. Nitrogen-specific uptake rates of marine phytoplankton isolated from natural populations of particles by flow cytometry. Mar. Ecol. Prog. Ser. 123: 245–258.

Lischewski, A., R. I. Amann, D. Harmsen, H. Merkert, J. Hacker & J. Morschhauser, 1996. Specific detection of *Candida albicans* and *Candida tropicalis* by fluorescent *in situ* hybridization with an 18S rRNA-targeted oligonucleotide probe. Microbiology 142: 2731–2740.

Liu, H., H. A. Nolla & L. Campbell, 1997. *Prochlorococcus* growth rate and contribution to primary production in the equatorial and subtropical North Pacific Ocean. Aquat. microb. Ecol. 12: 39–47.

Lloyd, D., 1993. Flow Cytometry in Microbiology. Springer Verlag, London, 188 pp.

Lloyd, D., A. G. Williams, R. Amann, A. J. Hayes, L. Durrant & J. R. Ralphs, 1996. Intracellular prokaryotes in rumen ciliate protozoa: detection by confocal laser scanning microscopy after *in situ* hybridization with fluorescent 16S rRNA probes. European J. Protistol. 32: 523–531.

Magarinos, B., J. L. Romalde, A. Cid & A. E. Toranzo, 1997. Viability of starved *Pasteurella piscicida* in seawater monitored by flow cytometry and the effect of antibiotics on its resuscitation. Lett. appl. Microbiol. 24: 122–126.

Manz, W., M. Wagner, R. Amann & K.-H. Schleifer, 1994. *In situ* characterization of the microbial consortia active in two wastewater treatment plants. Wat. Res. 28: 1715–1723.

Manz, W., R. Amann, W. Ludwig, M. Wagner & K.-H. Schleifer, 1992. Phylogenetic oligonucleotide probes for the major subclasses of proteobacteria: problems and solutions. System. appl. Microbiol. 15: 593–600.

Manz, W., R. Amann, W. Ludwig, M. Vancanneyt & K.-H. Schleifer, 1996. Application of a suite of 16S rRNA-specific oligonucleotide probes designed to investigate bacteria of the phylum cytophaga-flavobacter-bacteroides in the natural environment. Microbiology 142: 1097–1106.

Marie, D., D. Vaulot & F. Partensky, 1996. Application of the novel nucleic acid dyes YOYO-1, YO-PRO-1, and PicoGreen for flow cytometric analysis of marine prokaryotes. Appl. envir. Microbiol. 62: 1649–1655.

Marie, D., F. Partensky, S. Jacquet & D. Vaulot, 1997. Enumeration and cell cycle analysis of natural populations of marine picoplankton by flow cytometry using the nucleic acid stain SYBR Green I. Appl. envir. Microbiol. 63: 186–193.

Medlin, L. & N. Simon, 1998. Phylogenetic analysis of marine phytoplankton. In K. E. Cooksey (ed.), Molecular Approaches to the Study of the Ocean. Chapman and Hall, London: 161–186.

Meier, H., C. Koob, W. Ludwig, R. Amann, E. Frahm, S. Hoffmann, U. Obst & K.-H. Schleifer, 1997. Detection of enterococci with rRNA targeted DNA probes and their use for hygenic drinking water control. Wat. Sci. Technol. 35: 437–444.

Moller, S., C. Sternberg, J. B. Andersen, B. B. Christensen, J. L. Ramos, M. Givskov & S. Molin, 1998. *In situ* gene expression in mixed-culture biofilms: evidence of metabolic interactions between community members. Appl. envir. Microbiol. 64: 721–732.

Moore, L. R., G. Rocap & S. W. Chisholm, 1998. Physiology and molecular phylogeny of coexisting *Prochlorococcus* ecotypes. Nature 393: 464–467.

Moran, M. A., V. L. Torsvik, T. Torsvik & R. E. Hodson, 1993. Direct extraction and purification of rRNA for ecological studies. Appl. envir. Microbiol. 59: 915–918.

Mouawad, R., D. Khayat, A. Zerrouqi, A. M. Ghoumari & C. Soubrane, 1997. A flow cytometric method for the rapid detection of β-galactosidase transfected cells: an in vitro and in vivo study. J. imm. Meth. 204: 51–56.

Neef, A., A. Zaglauer, H. Meier, R. Amann, H. Lemmer & K.-H. Schleifer, 1996. Population analysis in a denitrifying sand filter: conventional and *in situ* identification of *Paracoccus* spp. in methanol-fed biofilms. Appl. envir. Microbiol. 62: 4329–4339.

Nubel, U., F. Garcia-Pichel & G. Muyzer, 1997. PCR primers to amplify 16S rRNA gene from cyanobacteria. Appl. envir. Microbiol. 63: 3327–3332.

Nubel, U., B. Engelen, A. Felske, J. Snaidr, A. Wieshuber, R. I. Amann, W. Ludwig & H. Backhaus, 1996. Sequence heterogeneities of genes encoding 16S rRNAs in *Paenibacillus polymyxa* detected by temperature gradient gel electrophoresis. J. Bact. 178: 5636–5643.

Olson, R. J., E. R. Zettler & O. K. Anderson, 1989. Discrimination of eukaryotic phytoplankton cell types from light scatter and autofluorescence properties measured by flow cytometry. Cytometry 10: 636–643.

Olson, R. J., E. R. Zettler & M. D. DuRand, 1993. Phytoplankton analysis using flow cytometry. In Kemp P. F. (ed.), Handbook of Methods in Aquatic Microbial Ecology. CRC Press, Inc., Boca Raton, Florida: 175–186.

Olson, R. J., S. W. Chisholm, E. R. Zettler & E. V. Armbrust, 1990. Pigments, size, and distribution of *Synechococcus* in the North Atlantic and Pacific Oceans. Limnol. Oceanogr. 35: 45–58.

Olson, R. J., E. R. Zettler, S. W. Chisholm & J. A. Dusenberry, 1991. Advances in oceanography through flow cytometry. In Demers S. (ed.), Particle Analysis in Oceanography, NATO ASI Series, G27. Springer-Verlag, Berlin: 351–399.

Orellana, M. V. & M. J. Perry, 1995. Optimization of an immunofluorescent assay of the internal enzyme ribulose-1,5-bisphosphate carboxylase (RUBISCO) in single phytoplankton cells. J. Phycol. 31: 785–794.

Ouverney, C. C. & J. A. Fuhrman, 1997. Increase in fluorescence intensity of 16S rRNA *in situ* hybridization in natural samples treated with chloramphenicol. Appl. envir. Microbiol. 63: 2735–2740.

Pace, N. R., 1997. Molecular view of microbial diversity and the biosphere. Science 276: 734–740.

Palenik, B., 1994. Cyanobacterial community structure as seen from RNA polymerase gene sequence analysis. J. Bact. 60: 3212–3219.

Palenik, B. & J. A. Koke, 1995. Characterization of a nitrogen-regulated protein identified by cell surface biotinylation of a marine phytoplankton. Appl. envir. Microbiol. 61: 3311–3315.

Palenik, B. & H. Swift, 1996. Cyanobacterial evolution and Prochlorophyte diversity as seen in DNA-dependent RNA polymerase gene sequences. J. Phycol. 32: 638–646.

Partensky, F., J. Blanchot, F. Lantoine, J. Neveux & D. Marie, 1996. Vertical structure of picophytoplankton at different trophic sites of the tropical northeastern Atlantic Ocean. Deep Sea Res. 43: 1191–1213.

Partensky, F., N. Hoepffner, W. K. W. Li, O. Ulloa & D. Vaulot, 1993. Photoacclimation of *Prochlorococcus* sp. (Prochlorophyta) strains isolated from the North Atlantic and the Mediterranean Sea. Pl. Physiol. 101: 285–296.

Paul, J. H., L. Cazares & J. Thurmond, 1990. Amplification of the *rbcL* gene from dissolved particulate DNA from aquatic environments. Appl. envir. Microbiol. 56: 1963–1966.

Peeters, J. C. H., G. B. J. Dubelaar, J. Ringelberg & J. W. M. Visser, 1989. Optical plankton analyser: A flow cytometer for plankton analysis, I: design considerations. Cytometry 10: 522–528.

Pernthaler, J., T. Posch, K. Simek, J. Vrba, R. Amann & R. Psenner, 1997. Contrasting bacterial strategies to coexist with a flagellate predator in an experimental microbial assemblage. Appl. envir. Microbiol. 63: 596–601.

Phinney, D. A. & T. L. Cucci, 1989. Flow cytometry and phytoplankton. Cytometry 10: 511–521.

Pichard, S. L. & J. H. Paul, 1991. Detection of gene expression in genetically engineered microorganisms and natural phytoplankton populations in the marine environment by mRNA analysis. Appl. envir. Microbiol. 57: 1721–1727.

Pichard, S. L. & J. H. Paul, 1993. Gene expression per gene dose, a specific measure of gene expression in aquatic microorganisms. Appl. envir. Microbiol. 59: 451–457.

Pichard, S. L., L. Campbell, K. Carder, J. B. Kang, J. Patch, F. R. Tabita & J. H. Paul, 1997. Analysis of ribulose bisphosphate carboxylase gene expression in natural phytoplankton communities by group-specific gene probing. Mar. Ecol. Prog. Ser. 149: 239–253.

Pizarro, J., E. Jedlicki, O. Orellana, J. Romero & R. T. Espejo, 1996. Bacterial populations in samples of bioleached copper ore as revealed by analysis of DNA obtained before and after cultivation. Appl. envir. Microbiol. 62: 1323–1328.

Pomeroy, L. R., 1974. The ocean's food web, a changing paradigm. BioScience 24: 499–504.

Porro, D., E. Martegani, B. M. Ranzi & L. Alberghina, 1997. Identification of different daughter and parent subpopulations in an asynchronously growing *Saccharomyces cerevisiae* population. Res. Microbiol. 148: 205–215.

Porter, J., R. Pickup & C. Edwards, 1995a. Flow cytometric detection of specific genes in genetically modified bacteria using *in situ* polymerase chain reaction. FEMS Microbiol. Lett. 134: 51–56.

Porter, J., D. Deere, R. Pickup & C. Edwards, 1996. Fluorescent probes and flow cytometry: new insights into environmental bacteriology. Cytometry 23: 91–96.

Porter, J., J. Diaper, C. Edwards & R. Pickup, 1995b. Direct measurements of natural planktonic bacterial community viability by flow cytometry. Appl. envir. Microbiol. 61: 2783–2786.

Porter, K. G. & Y. S. Feig, 1980. The use of DAPI for identifying and counting aquatic microflora. Limnol. Oceanogr. 25: 943–948.

Potter, D., T. C. Lajeunesse, G. W. Saunders & R. A. Anderson, 1997. Convergent evolution masks extensive biodiversity among marine coccoid picoplankton. Biodiv. Conserv. 6: 99–107.

Ramsing, N. B., H. Fossing, T. G. Ferdelman, F. Andersen & B. Thamdrup, 1996. Distribution of bacterial populations in a stratified fjord (Mariager Fjord, Denmark) quantified by *in situ* hybridization and related to chemical gradients in the water column. Appl. envir. Microbiol. 62: 1391–1404.

Reckermann, M. & M. J. W. Veldhuis, 1997. Trophic interactions between picophytoplankton and micro- and nanozooplankton in the western Arabian Sea during the NE monsoon 1993. Aquat. microb. Ecol. 12: 263–273.

Rehnstam, A.-S., S. Backman, D. C. Smith, F. Azam & A. Hagstrom, 1993. Blooms of sequence-specific culturable bacteria in the sea. FEMS Microbiol. Ecol. 102: 161–166.

Rice, J., C. D. O'Connor, M. A. Sleigh, P. H. Burkhill, I. G. Giles & M. V. Zubkov, 1997a. Fluorescent oligonucleotide rDNA probes that specifically bind to a common nanoflagellate, *Paraphysomonas vestita*. Microbiology 143: 1717–1727.

Rice, J., M. A. Sleigh, P. H. Burkhill, G. A. Tarran, C. D. O'Connor & M. V. Zubkov, 1997b. Flow cytometric analysis of characteristics of hybridization of species-specific fluorescent oligonucleotide probes to rRNA of marine nanoflagellates. Appl. envir. Microbiol. 63: 938–944.

Roller, C., M. Wagner, R. Amann, W. Ludwig & K.-H. Schleifer, 1994. *In situ* probing of gram-positive bacteria with high DNA G+C content using 23S rRNA-targeted oligonucleotides. Microbiology 140: 2849–2858.

Rosenblatt, J. I., J. A. Hokanson, S. R. McLaughlin & J. F. Leary, 1997. Theoretical basis for sampling statistics useful for detecting and isolating rare cells using flow cytometry and cell sorting. Cytometry 27: 233–238.

Roth, B. L., M. Poot, S. T. Yue & P. J. Millard, 1997. Bacterial viability and antibiotic susceptibility testing with SYTOX Green nucleic acid stain. Appl. envir. Microbiol. 63: 2421–2431.

Rothemund, C., R. Amann, S. Klugbauer, W. Manz, C. Bieber, K.-H. Schleifer & P. Wilderer, 1996. Microflora of 2,4-dichlorophenoxyacetc acid degrading biofilms on gas permeable membranes. System. appl. Microbiol. 19: 608–615.

Sandgren, C. D., 1988. Growth and Reproductive Strategies of Freshwater Phytoplankton. Cambridge University Press, Cambridge, 442 pp.

Santo Domingo, J. W., M. G. Kaufman, M. J. Klug & J. M. Tiedje, 1998. Characterization of the cricket hindgut microbiota with fluorescently labeled rRNA-targeted oligonucleotide probes. Appl. envir. Microbiol. 64: 752–755.

Scanlan, D. J., W. R. Hess, F. Partensky, J. Newman & D. Vaulot, 1996. High degree of genetic variation in *Prochlorococcus* (Prochlorophyta) revealed by RFLP analysis. Eur. J. Phycol. 21: 1–9.

Scanlan, D. J., N. J. Silman, K. M. Donald, W. H. Wilson, N. G. Carr, I. Joint & N. H. Mann, 1997. An immunological approach to detect phosphate stress in populations and single cells of photosynthetic picoplankton. Appl. envir. Microbiol. 63: 2411–2420.

Schleifer, K.-H., M. Ehrmann, C. Beimfohr, E. Brockmann, W. Ludwig & R. Amann, 1995. Application of molecular methods for the classification and identification of lactic acid bacteria. Int. Dairy J. 5: 1081–1094.

Schmidt, T. M., E. F. DeLong & N. R. Pace, 1991. Analysis of a marine picoplankton community by 16S rRNA gene cloning and sequencing. J. Bact. 173: 4371–4378.

Schonhuber, W., B. Fuchs, S. Juretschko & R. Amann, 1997. Improved sensitivity of whole-cell hybridization by the combination of horseradish peroxidase-labeled oligonucleotides and tyramide signal amplification. Appl. envir. Microbiol. 63: 3268–3273.

Schramm, A., L. H. Larsen, N. P. Revsbech, N. B. Ramsing, R. Amann & K.-H. Schleifer, 1996. Structure and function of a nitrifying biofilm as determined by *in situ* hybridization and the use of microelectrodes. Appl. envir. Microbiol. 62: 4641–4647.

Shapiro, H., 1995. Practical Flow Cytometry. 3rd edn. Wiley-Liss, New York, 542 pp.

Shapiro, L. P., L. Campbell & E. M. Haugen, 1989. Immunochemical recognition of phytoplankton species. Mar. Ecol. Prog. Ser. 57: 219–224.

Simek, K., J. Vrba, J. Pernthaller, T. Posch, P. Hartman, J. Nedoma & R. Psenner, 1997. Morphological and compositional shifts in an experimental bacterial community influenced by protists with contrasting feeding modes. Appl. envir. Microbiol. 63: 587–595.

Simon, N., H. Nolla & L. Campbell, 1996. Identification of phytoplankton by *in situ* hybridization combined with flow cytometery. EOS 76(3): OS112.

Simon, N., J. Brenner, B. Edvardsen & L. K. Medlin, 1997. The identification of *Chrysochromulina* and *Prymnesium* species (Haptophyta, Prymnesiophyceae) using fluorescent or chemiluminescent oligonucleotide probes: a means for improving studies on toxic algae. Eur. J. Phycol. 32: 393–401.

Simon, N., R. G. Barlow, D. Marie, F. Partensky & D. Vaulot, 1994. Characterization of oceanic photosynthetic picoeukaryotes by flow cytometry. J. Phycol. 30: 922–935.

Simon, N., N. LeBot, D. Marie, F. Partensky & D. Vaulot, 1995. Fluorescent *in situ* hybridization with rRNA-targeted oligonucleotide probes to identify small phytoplankton by flow cytometry. Appl. envir. Microbiol. 61: 2506–2513.

Sinigalliano, C. D., D. N. Kuhn & R. D. Jones, 1995. Amplification of the *amoA* gene from diverse species of ammonium-oxidizing bacteria and from an indigenous bacterial population from seawater. Appl. envir. Microbiol. 61: 2702–2706.

Snaidr, J., R. Amann, I. Huber, W. Ludwig & K.-H. Schleifer, 1997. Phylogenetic analysis and *in situ* identification of bacteria in activated sludge. Appl. envir. Microbiol. 63: 2884–2896.

Sorensen, B. B. & M. Jakobsen, 1997. The combined effects of temperature, pH and NaCl on growth of *Debaryomyces hansenii* analyzed by flow cytometry and predictive microbiology. Int. J. Food Microbiol. 34: 209–220.

Steen, H. B., 1991. Flow cytometry instrumentation. In Demers S. (ed.), Particle Analysis in Oceanography, NATO ASI Series, G27. Springer-Verlag, Berlin: 3–29.

Steen, H. B., 1992. Noise, sensitivity, and resolution of flow cytometers. Cytometry 18: 822–830.

Suzuki, M. T., M. S. Rappe, Z. W. Haimberger, H. Winfield, N. Adair, J. Strobel & S. J. Giovannoni, 1997. Bacterial diversity among small-subunit rRNA gene clones and cellular isolates from the same seawater sample. Appl. envir. Microbiol. 63: 983–989.

Tani, K., K. Kurokawa & M. Nasu, 1998. Development of a direct *in situ* PCR method for detection of specific bacteria in natural environments. Appl. envir. Microbiol. 64: 1536–1540.

Tate, R. L., III, 1986. Importance of autecology in microbial ecology. In R. L. Tate, III (ed.), Microbial Autecology. John Wiley & Sons, New York: 1–26.

Teske, A., C. Wawer, G. Muyzer & N. B. Ramsing, 1996. Distribution of sulfate-reducing bacteria in a stratified fjord (Mariager Fjord, Denmark) as evaluated by most-probable-number counts and denaturing gradient gel electrophoresis of PCR-amplified ribosomal DNA fragments. Appl. envir. Microbiol. 62: 1405–1415.

Thomas, J.-C., M. Desrosiers, Y. St-Pierre, P. Lirette, J.-G. Bisaillon, R. Beaudet & R. Villemur, 1997. Quantitative flow cytometric detection of specific microorganisms in soil samples using rRNA targeted fluorescent probes and ethidium bromide. Cytometry 27: 224–232.

Timm, E. A., Jr. & C. C. Stewart, 1992. Fluorescent *in situ* hybridization en suspension (FISHES) using digoxigenin-labeled probes and flow cytometry. BioTechniques 12: 362–367.

Toledo, G. & B. Palenik, 1997. *Synechococcus* diversity in the California Current as seen by RNA polymerase (*rpoC1*) gene sequences of isolated strains. Appl. envir. Microbiol. 63: 4298–4303.

Tolker-Nielsen, T., K. Holmstrom & S. Molin, 1997. Visualization of specific gene expression in individual *Salmonella typhimurium* cells by *in situ* PCR. Appl. envir. Microbiol. 63: 4196–4203.

Tolker-Nielsen, T., K. Holmstrom, L. Boe & S. Molin, 1998. Nongenetic population heterogeneity studied by *in situ* polymerase chain reaction. Mol. Microbiol. 27: 1099–1105.

Tombolini, R., A. Unge, M. E. Davey, F. J. de Bruijn & J. K. Jansson, 1997. Flow cytometric and microscopic analysis of GFP-tagged *Pseudomonas fluorescens* bacteria. FEMS Microbiol. Ecol. 22: 17–28.

Trebesius, K., R. Amann, W. Ludwig, K. Muhlegger & K.-H. Schleifer, 1994. Identification of whole fixed bacterial cells with nonradioactive 23S rRNA-targeted polynucleotide probes. Appl. envir. Microbiol. 60: 3228–3235.

Troussellier, M., C. Courties & A. Vaquer, 1993. Recent applications of flow cytometry in aquatic microbial ecology. Biol. Cell. 78: 111–121.

Tyndall, R. L., R. E. Hand, Jr., R. C. Mann, C. Evans & R. Jernigan, 1985. Application of flow cytometry to detection and characterization of *Legionella* spp. Appl. envir. Microbiol. 49: 852–857.

Urbach, E., D. J. Scanlan, D. L. Distel, J. B. Waterbury & S. W. Chisholm, 1998. Rapid diversification of marine picophytoplankton with dissimilar light-harvesting structures inferred from sequences of *Prochlorococcus* and *Synechococcus* (Cyanobacteria). J. mol. Evol. 46: 188–201.

Vaulot, D. & F. Partensky, 1992. Cell cycle distributions of prochlorophytes in the north western Mediterranean Sea. Deep Sea Res. 39: 727–742.

Vaulot, D., D. Marie, R. J. Olson & S. W. Chisholm, 1995. Growth of *Prochlorococcus*, a photosynthetic prokaryote, in the equatorial Pacific Ocean. Science 268: 1480–1482.

Veldhuis M. J. W., T. L. Cucci & M. E. Sieracki, 1997. Cellular DNA content of marine phytoplankton using two new fluorochromes: taxonomic and ecological implications. J. Phycol. 33: 527–541.

Vesey, G., P. Hutton, A. Champion, N. Ashbolt, K. L. Williams, A. Warton & D. Veal, 1994. Application of flow cytometric methods for the routine detection of *Cryptosporidium* and *Giardia* in water. Cytometry 16: 1–6.

Volsch, A., W. F. Nader, H. K. Geiss, G. Nebe & C. Birr, 1990. Detection and analysis of two serotypes of ammonia-oxidizing bacteria in sewage plants by flow cytometry. Appl. envir. Microbiol. 56: 2430–2435.

Voytek, M. A. & B. B. Ward, 1995. Detection of ammonium-oxidizing bacteria of the beta-subclass of the class *Proteobacteria* in aquatic samples with the PCR. Appl. envir. Microbiol. 61: 1444–1450.

Vrieling, E. G. & D. M. Anderson, 1996. Immunofluorescence in phytoplankton research: applications and potential. J. Phycol. 32: 1–16.

Vrieling, E. G., W. H. van den Poll, G. Vriezekolk & W. W. C. Gieskes, 1997. Immuno-flow cytometric detection of the ichthyotoxic dinoflagellates *Gyrodinium aureolum* and *Gymnodinium nagasakiense*: independence of physiological state. Neth. J. Sea Res. 37: 91–100.

Wagner, M., R. Amann, H. Lemmer & K.-H. Schleifer, 1993. Probing activated sludge with oligonucleotides specific for proteobacteria: inadequacy of culture-dependent methods for describing microbial community structure. Appl. envir. Microbiol. 59: 1520–1525.

Wagner, M., G. Rath, R. Amann, H.-P. Koops & K.-H. Schleifer, 1995. *In situ* identification of ammonia-oxidizing bacteria. System. appl. Microbiol. 18: 251–264.

Wagner, M., G. Rath, H.-P. Koops, J. Flood & R. Amann, 1996. *In situ* analysis of nitrifying bacteria in sewage treatment plants. Wat. Sci. Tech. 34: 237–244.

Walberg, M., P. Gaustad & H. B. Steen, 1996. Rapid flow cytometric assessment of mecillinam and ampicillin bacterial susceptibility. J. antimicrob. Chemo. 37: 1063–1075.

Walberg, M., P. Gaustad & H. B. Steen, 1997. Rapid assessment of ceftazidime, ciprofloxacin, and gentamicin susceptibility in exponentially-growing *E. coli* cells by means of flow cytometry. Cytometry 27: 169–178.

Wallner, G., R. Amann & W. Beisker, 1993. Optimizing fluorescent *in situ* hybridization with rRNA-targeted oligonucleotide probes for flow cytometric identification of microorganisms. Cytometry 14: 136–143.

Wallner, G., R. Erhart & R. Amann, 1995. Flow cytometric analysis of activated sludge with rRNA-targeted probes. Appl. envir. Microbiol. 61: 1859–1866.

Wallner, G., I. Steinmetz, D. Bitter-Suermann & R. Amann, 1996. Combination of rRNA-targeted hybridization probes and immuno-probes for the identification of bacteria by flow cytometry. System. appl. Microbiol. 19: 569–576.

Wallner, G., B. Fuchs, S. Spring, W. Beisker & R. Amann, 1997. Flow sorting of microorganisms for molecular analysis. Appl. envir. Microbiol. 63: 4223–4231.

Wang, G. C.-Y. & Y. Wang, 1996. The frequency of chimeric molecules as a consequence of PCR co-amplification of 16S rRNA genes from different bacterial species. Microbiology 142: 1107–1114.

Ward, B. B., 1984. Combined autoradiography and immunofluorescence for estimation of single cell activity by ammonium-oxidizing bacteria. Limnol. Oceanogr. 29: 402–410.

Ward, B. B., M. A. Voytek & K.-P. Witzel, 1997. Phylogenetic diversity of natural populations of ammonia oxidizers investigated by specific PCR amplification. Microb. Ecol. 33: 87–96.

Weiss, P., B. Schweitzer, R. Amann & M. Simon, 1996. Identification *in situ* and dynamics of bacteria on limnetic organic aggregates (lake snow). Appl. envir. Microbiol. 62: 1998–2005.

Weisse, T. & B. Kirchhoff, 1997. Feeding of the heterotrophic freshwater dinoflagellate *Peridiniopsis berolinense* on cryptophytes: analysis by flow cytometry and electronic particle counting. Aquat. microb. Ecol. 12: 153–164.

Wood, A. M., P. K. Horan, K. Muirhead, D. A. Phinney, C. M. Yentsch & J. B. Waterbury, 1985. Discrimination between types of pigments in marine *Synechococcus* spp. by scanning spectroscopy, epifluorescence microscopy and flow cytometry. Limnol. Oceanogr. 30: 1303–1315.

Yamaguchi, N., S. Inaoka, K. Tani, T. Kenzaka & M. Nasu, 1996. Detection of specific bacterial cells with 2-hydroxy-3-napthoic acid-2′-phenylanilide phosphate and Fast Red TR *in situ* hybridization. Appl. envir. Microbiol. 62: 275–278.

Yentsch, C. M. & P. K. Horan, 1989. Cytometry in the aquatic sciences. Cytometry 10: 497–499.

Zarda, B., R. Amann, G. Wallner & K.-H. Schleifer, 1991. Identification of single bacterial cells using digoxigenin-labelled, rRNA-targeted oligonucleotides. J. gen. Microbiol. 137: 2823–2830.

Zehr, J. P. & D. G. Capone, 1996. Problems and promises of assaying the genetic potential for nitrogen fixation in the marine environment. Microb. Ecol. 32: 263–281.

Zehr, J. P., M. Mellon, S. Braun, W. Litaker, T. Steppe & H. W. Paerl, 1995. Diversity of heterotrophic nitrogen fixation genes in a marine cyanobacterial mat. Appl. envir. Microbiol. 61: 2527–2532.

Hydrobiologia **401:** 55–68, 1999.
J.P. Zehr & M.A. Voytek (eds), Molecular Ecology of Aquatic Communities.

Distribution of microbial assemblages in the Central Arctic Ocean Basin studied by PCR/DGGE: analysis of a large data set

V. C. Ferrari & J. T. Hollibaugh*
Department of Marine Sciences, University of Georgia, Athens, GA 30602-3636, U.S.A.

Key words: Arctic Ocean, biogeography, bacteria, DGGE, phylogeny, SCICEX, Cavalla

Abstract

Analysis of the biogeographic distributions of bacteria has been limited by potential biases inherent in the isolations required for classical taxonomy and by the time required for phylogenetic analyses. We have attempted to circumvent both of these limitations by using denaturing gradient gel electrophoresis (DGGE) to resolve the products of polymerase chain reaction (PCR) amplifications of mixed template DNA isolated from microbial communities. DGGE separates DNA fragments based on their denaturation characteristics, which vary with the nucleotide sequence of the fragment. The banding patterns in the electropherograms were then subjected to similarity analysis using pattern matching and band comparison software. Replication experiments tested the robustness of band patterns within and between gels. Samples were collected from the Central Arctic Ocean basin during April of 1995 on the SCICEX 95 cruise of the USS Cavalla. One hundred samples collected from a depth of 59 m are the focus of this biogeographical analysis. The band identification algorithm of the software identified between 12 and 30 bands (operational taxonomic units, OTUs) per sample (mean: 21.5) with minimal editing. This number approximately doubled with more extensive editing. Four OTUs seemed to be common to most samples. The samples grouped into five major clusters with similarities greater than approximately 80%. Twenty nine samples in one of these clusters were in two branches with internal similarities greater than approximately 90%. These samples had relatively nondescript banding patterns (numerous bands with roughly equal intensity). Another cluster contained 15 samples with distinctive banding patterns dominated by one or two intense bands. These samples were collected in the same general area of the Arctic Ocean (Canada Basin) and may reflect a community response to local environmental conditions.

Introduction

The Arctic Ocean water column receives organic matter (OM) from a number of sources including the surrounding land masses, phytoplankton production and ice-algal production. The terrestrial organic matter transported into the Arctic Ocean Basin by several large rivers has distinct chemical characteristics and is a significant component of the DOM found in the central Arctic Ocean Basin (Wheeler et al., 1996, 1997; S. Osphal & R. Benner, pers. comm.). Other markers of riverine inflow (e.g. dissolved Ba, Guay & Falkner, 1997) also indicate substantial riverine contributions, for example of MacKenzie River water to Canada Basin surface waters. Elevated primary production in the surrounding marginal seas (Bering, Chukchi, Barents, Laptev, Kara, Norwegian, Greenland and Beaufort Seas) can also contribute organic matter to the central Arctic Ocean Basin via mixing and advection. Decomposing OM transported into the Arctic Ocean by Pacific Ocean water flowing in through the Bering Straits leaves a clear geochemical signal (Wheeler et al., 1997).

In situ production by phytoplankton represents another source of OM, in this case in the form of fresh algal material. Arctic Ocean primary production was previously thought to be low, yet DOC data collected by Wheeler et al. (1997) suggest that primary production may be contributing more than was previously suspected. Ice algal production may also contribute OM to the sub-ice water column. The magnitude of ice-algal production is difficult to estimate, but it

* Author for correspondence.

may be large. Organic matter derived from ice-algae may be biochemically distinct from OM derived from phytoplankton because of physiological adaptations required for survival in sea-ice brine channels. The relative contribution of these different sources to the organic C budget of the Arctic Ocean is not known.

From our knowledge of bacterial physiology, it is likely that different enzyme systems are required to use these different types of organic matter. If this is true, then as suggested in Hollibaugh (1994), it is also likely that there are phylogenetic differences between bacterial communities using different types of OM because metabolic capabilities define classical bacterial taxonomy. Thus, the phylogenetic composition of microbial communities in the Arctic Ocean might be expected to vary depending on changes in the relative dominance of these sources of OM. Detecting specific suites of organisms could provide information on the spatial or temporal variation of OM quality and its processing.

In general, the taxonomic composition of bacterial communities has been difficult to analyze and to date little is known about the composition of Arctic Ocean bacterial communities. Historically, bacterial taxonomy has been determined functionally, focusing primarily on physiological characteristics determined with pure cultures of bacteria. This necessitates isolating, purifying and culturing bacteria which may lead to biases as a low percentage of the cells in a sample give rise to colonies on non-restrictive, solid media (Ferguson et al., 1984). Recent developments in molecular biology have revolutionized the study of microbial taxonomy and community composition. The ability to compare bacterial DNA using molecular techniques, such as cloning and DNA sequencing, has given microbial ecologists their first glimpse at bacterioplankton community structure (Giovannoni et al., 1990; Fuhrman et al., 1993; DeLong et al., 1993; Pedros-Alio, 1993). Although these methods have been successful, one drawback is that they are very labor intensive: 16S rRNA gene sequences must be determined for a large number of clones to have confidence that a statistically robust species rarefaction curve has been obtained. This limits their utility for analyzing the number of samples needed to characterize biogeographical distributions.

An alternative approach, denaturing gradient gel electrophoresis (DGGE), has been used previously to compare bacterial assemblages in environmental samples. In DGGE, short DNA fragments generated by polymerase chain reaction (PCR, Saiki et al., 1988) amplification of a mixture of template DNA are separated based on denaturation characteristics which vary depending on the nucleotide sequence of the fragment. This approach has been used successfully to study phylogenetically simple communities found in microbial mats, hydrothermal vents, hot springs and for studies of specific groups of organisms in sediments (Ward et al., 1990; Muyzer et al., 1993, 1995; Wawer & Muyzer, 1995; Wawer et al., 1995, 1997; Ferris, et al., 1996). It has also been used to compare relatively complex estuarine bacterioplankton communities (Murray, 1994; Murray et al., 1996). Though this approach does not provide phylogenetic information directly, it allows rapid processing of large numbers of samples, generating profiles of assemblage composition that can be used to compare samples. Phylogenetic information can be obtained by excising bands and sequencing the DNA they contain (e.g. Santegoeds et al., 1996).

Here we explore the potential of PCR/DGGE for analyzing bacterioplankton biogeography in the Arctic Ocean. A more detailed study of the relationship between the composition of Arctic Ocean bacterioplankton assemblages and environmental variables will be presented elsewhere.

Materials and methods

Sample Collection

Samples were collected from a depth of 59 m at 100 stations in the central Arctic Ocean Basin (Figure 1) between March 26th and May 8th, 1995, during the SCICEX 95 cruise of the nuclear submarine *USS Cavalla* (DeLaca et al., 1996). Water was collected from the intake line of the submarine's cooling water system then pressure filtered (at least 10 l at 60 psi) through a Millipore Sterivex® filter cartridge (0.22 μm pore size, mixed cellulose ester). Seawater remaining in the filter cartridge was expelled with pressurized air (syringe), then extraction buffer was added to the cartridge (1.8 ml: 0.75 M sucrose, 40 mM EDTA, 50 mM TRIS; pH 8.3). Filter cartridges were capped and frozen at −20 °C until processed.

Nucleic acid extraction

The filter cartridges were thawed at room temperature, lysozyme (40 μl of 50 mg/ml) added and then incubated at 37 °C for 30–45 min while rolling at 1–2 RPM. Proteinase K (100 μl of 10 mg/ml) and SDS (100 μl

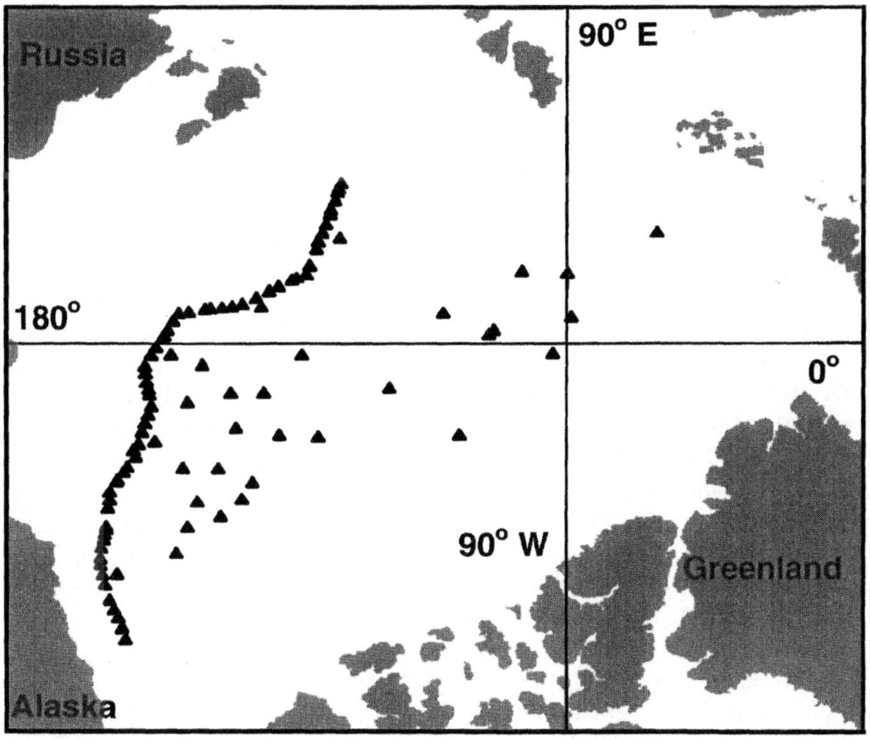

Figure 1. Locations where samples were collected during the SCICEX 95 cruise. The line of stations along the western side of the basin follows the shelf break.

of a 20% solution) were added and the cartridges incubated at 55 °C for 1–2 h. The lysate was extracted a second time with 1 ml of the complete mixture to maximize DNA extraction and recovery. The lysates were pooled and two 800 μl subsamples frozen as archive material.

The lysate was extracted with phenol:$CHCl_3$:isoamyl alcohol (25:24:1), centrifuged and then re-extracted with $CHCl_3$:isoamyl alcohol (24:1). The aqueous layer was extracted with n-butanol. The aqueous layer was combined with TE (10 mM Tris, 1 mM EDTA, pH 8.0) and concentrated to approximately 200 μl using an Amicon Centricon® 100 filter. Successful extraction of high molecular weight DNA was assessed by electrophoresis of the sample on a 0.75% agarose gel.

DNA amplification by PCR

A fragment encompassing the variable 3 region of the 16S rRNA gene (v3 rDNA) was amplified by PCR using primer sequences (17 bp each) complementary to positions 341–358 (341f) and 517–534 (534r) of the *E. coli* 16S rRNA gene. This primer set was chosen to amplify most species of Bacteria (Muyzer et al., 1993; Murray, 1994), although we have found that it also amplifies some chloroplast sequences. A 40 bp GC clamp (Myers et al., 1985; Murray, 1994) was added to the forward primer (341f) to help maintain stable melting behavior during DGGE (Myers et al., 1985; Muyzer et al., 1993). The primer sequences and predicted melting points (Tm) were:

341f :
5′ CCTACGGGAGGCAGCAG 3′,
(‘Bacterial’), Tm = 60.1 °C
534r : 5′ ATTACCGCGGTCGCTGG 3′,
(‘Universal’), Tm = 57.7 °C
GC – clamped 341f :
5′ CGCCCGCCGCGCCCCGCGCCCGTCCCGC
CGCCCCCGCCCCCCTACGGGAGGCAGCAG 3′,
Tm = 116.4 °C

Primers were synthesized by Operon Technologies. Lyophilized primer were reconstituted to a final concentration of 10 μM and stored at −80 °C.

To avoid DNA contamination during PCR, all reaction mixtures were prepared in a molecular biology clean room using a laminar flow cabinet equipped with a bactericidal UV lamp. Other precautions included regular decontamination of the clean room and laminar flow cabinet with 5% bleach, use of filter tips for all PCR reaction pipeting and use of sterilized, virgin plastic reaction tubes. All water used for dilutions and reactions was 0.22 μm filtered molecular biology grade. PCR blanks included extracts of filter cartridges through which no water had been filtered as well as TE buffer. Positive PCR controls used bacterial genomic DNA purchased from Sigma Chemical.

Reactions for subsequent DGGE analysis were performed in 100 μl reaction volumes. Each reaction was prepared using a 'master mix' of Promega 10X Thermophilic Buffer (500 mM KCl, 100 mM Tris, 1% Triton X-100, pH 9.0), $MgCl_2$ (25 mM), Promega DNA Taq polymerase (5 units Taq/μl) and sterile water. The PCR reactions were overlain with mineral oil and (typically) 4 μl of a 1:40 dilution DNA extract template added. The second reaction mix, added after the master mix was heated to 80 °C, ('hot start', Erlich et al., 1991), contained dNTP's, primer and sterile water. Final concentrations in a 100 μl reaction were: 1×Taq DNA polymerase buffer, 2.5 μM MgCl2, 2.5 units of Taq DNA polymerase, 0.75 μM dNTP, 0.5 μM of each primer pair and 1–10 ng of DNA.

To increase primer specificity, a 'touchdown' PCR cycling protocol was used (Don et al., 1991). Thermal cycling parameters were: initial denaturation at 94 °C for 3.5 min., subsequent denaturation at 94 °C for 45 sec.; annealing for 45 sec. at 65–55 °C (stepping down 1 degree every 2 cycles from 65 to 62 °C and 1 degree per cycle from 62 to 55 °C) and extension at 72 °C for 30 sec. This was followed by 30 cycles of: denaturation at 94 °C for 45 sec., annealing at 55 °C for 45 sec., and extension at 72 °C for 30 sec. with a 5 min extension on the final cycle.

DGGE

DGGE analysis of each sample was performed using the combined product of two separate 100 μl PCR reactions, concentrated by ethanol precipitation and resuspended in distilled water (20 μl) and mixed with 10 μl of loading dye. Twelve to 14 μl samples (~1 μg DNA) were analyzed by DGGE using a CBS Scientific DGGE system (Del Mar, CA). A 40%–70% gradient of denaturant was used for all gels. Final concentrations in the 40% denaturant gel were 16.0% deionized formamide, 2.8 M urea, 6.5% acrylamide and 1×TAE gel buffer (100 mM Tris base, 12.5 mM sodium acetate, 1 mM EDTA, pH adjusted to 7.4 with acetic acid). Final concentrations in the 70% solution were 28% deionized formamide, 4.9 M urea, 6.5% acrylamide and 1×TAE gel buffer. All DGGE gels were 0.75 mm thick and run at 60 °C in 1×TAE. Gels were run at 200 volts for 5 h.

Standards

A set of standards consisting of genomic DNA from *Clostridium perfingens* (Standard 1, SIGMA Genomic Ultra Pure, D5139) and *Bacillus thuringiensis* (Standard 2, SIGMA High Molecular Weight, D3171) was run with every gel. These standards were chosen because they were readily and universally available, and because their v3 rDNA PCR products had high melting points and thus traveled most of the length of the gel before denaturing. This allowed us to normalize gel images (see below) without having to extrapolate too far beyond standard bands. Standards were run in the outside and middle lanes of each gel.

Staining and documentation of gels

DGGE gels were stained for 30 min in an ethidium bromide bath and destained for 30 min in deionized water. Gel images were displayed and recorded using a UVP GDS 7500 Gel Documentation System. Gel images were adjusted for brightness and contrast, and viewed at different exposure lengths to adjust for differences in staining intensity and properties of the banding pattern. Gel images were recorded digitally as inverted TIFF files. Only lanes that were uniformly stained and had clear bands were selected for subsequent image analysis. PCR and/or DGGE was repeated until an acceptable image was obtained for each sample.

Each lane in a DGGE gel represents the v3 rDNA signature or fingerprint of one sample and we use the terms 'lane' and 'sample' interchangeably below. In keeping with Muyzer et al. (1995) and Murray et al. (1996), we also use the terms 'operational taxonomic unit' ('OTU'), 'phylotype' or 'band' interchangeably to represent individual bands in the DGGE gel when discussing these data. This convention assumes that each band resolved in the DGGE gel represents a unique v3 rDNA sequence and thus a phylogenetically distinct organism. We know that this assumption is not always valid, for example the presence of comigrating bands or of multiple bands for the same organism may violate this assumption. However, it is a useful convention for discussing changes in banding patterns, and the validity of the assumption for any particular band or pattern can be verified by excising and sequencing the band.

DGGE image processing

DGGE gel images were analyzed using software (Molecular Analyst Fingerprint Plus® marketed by Bio-Rad Laboratories, which appears to us to be identical to GelCompare® marketed by Applied Maths, Kortrijk, Belgium). This software allows the user to extract densitometric curves from banding patterns, to normalize the banding patterns in a gel to a universal standard for comparisons between gels, then to compare banding patterns from different samples statistically (Bio-Rad, 1992).

The first step in processing gel images involves digitizing the banding patterns. Lanes are delineated, band intensity is averaged over 8 pixels across the lane, rescaled against the highest and lowest intensities in the lane, then the resulting densitometric curve is stored as a set of paired position and intensity values. We rescaled band intensities to correct for lane to lane variation in overall intensity due to differences in staining or in the amount of PCR product loaded in a lane.

Next, the positions of major bands in standard lanes are compared to a universal standard (an image of a standard lane recorded earlier, the lane marked 'D 1+2' and indicated by an arrow in Figure 2). The position of the bottom of the well at the top of the gel is used as one reference point and the positions of major bands in standards are used as additional reference points. The program then uses a linear interpolation based on the discrepancy between the positions of these reference points in the gel being analyzed and the corresponding reference points in the universal standard to normalize all of the digitized densitometric curves for that gel to the universal standard. In effect, the program stretches or shrinks the gel image so that the bottoms of the wells and the standard bands all line up with the corresponding markers in the universal standard.

When necessary, internal standard markers (distinctive bands shared by a number of samples) were used to correct for gel distortion. We minimized application of this approach since it requires the assumption that the bands being aligned represent the same OTU. We verified this assumption on a limited number of samples by sequencing DNA in the bands, but we also found instances where the sequence of bands with very similar positions in a gel differed by 17 base pairs. Once normalized, a non-linear background subtraction (rolling ball algorithm) was performed.

Data analysis

Pairs of samples were compared either using point by point regressions of the rescaled, normalized densitometric curves or using the position of individual bands. We could not use the position of individual bands to compare the whole data set because the software could only process a maximum of 50 samples at a time. Regression analysis generally gave results that were more consistent with visual assessments of similarity and thus was preferred for routine analyses.

Regression-based similarity analysis used the set of pairs of intensity values (y_{ia}, y_{ib}) for each position (x_i) on densitometric curves a and b to calculate a Pearson product-moment correlation coefficient for that pair of curves. We limited the regression to a zone in the middle of the curve (between positions 50–100 and 350–400, depending on the sample set being compared) to avoid spurious correlations caused by smears often present at the beginning and ends of the electropherograms (Figure 2a). The matrix of correlation coefficients thus obtained was used with cluster analysis (UPGMA) to generate a similarity dendrogram. This analysis makes no assumptions regarding the significance of band intensity or position, focusing rather on the general patterns of the densitometric curves being compared.

For band-based comparisons, individual bands were identified automatically by the software. We set the program to identify only bands with areas $\geq$0.5% of the total area under the densitometric curve and heights above the local background of $\geq$1%. The band calls for each lane were then manually edited. Comparisons were made with the tolerance for position differences between two bands, expressed as the percentage of the total length of the curve, set at either 0.01 or 0.7%. The 0.01% setting was used for band (OTU) counts since it determined the absolute position in the gel and did not combine bands in close proximity. The 0.7% setting was used to compare subsets of the data because it allowed for residual error in position alignment ($\pm$3 positions) following normalization. The similarity of two samples was assessed using information in banding patterns by a Jaccard index, S_j for each pair of samples. (Magurran, 1988). The Jaccard index is calculated as:

$$S_j = n_{AB}/(n_A + n_B) - n_{AB};$$

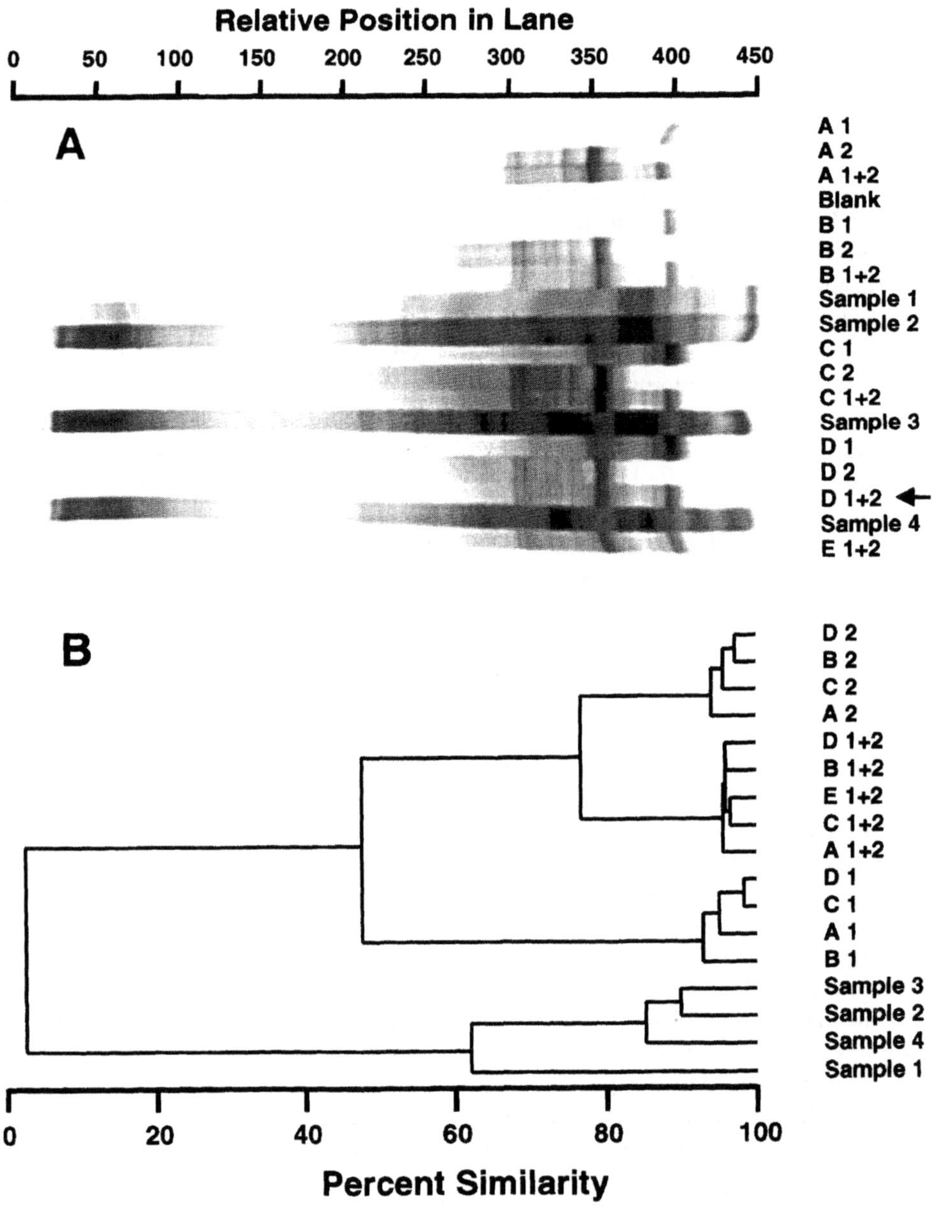

Figure 2. (A) Image of a DGGE gel containing PCR products amplified from genomic DNA of *Clostridium perfringens* (1) or *Bacillus thuringiensis* (2) or both, with Arctic Ocean samples (Sample 1–Sample 4) for comparison; used to test within-gel replication of PCR/DGGE analysis. Each set of standards (A–E) was amplified independently. The arrow indicates the lane (D 1+2) used as a universal standard for normalizing the rest of the gels. (B) Cluster analysis dendrogram of the similarities (regression model) of the lanes in this gel.

where n_{AB} is the number of bands common to both samples (lanes) A and B, n_A is the total number of bands in sample A, and n_B is the total number of bands in sample B. A value of zero indicates that the samples are completely different and a value of one indicates the samples are identical. The matrix of Jaccard similarities was then used to generate a similarity dendrogram as with regression analysis.

Results

Basic DGGE data set

The samples were analyzed in 20 different DGGE gels. Four of these gels were repeats run, because of problems with the first gel (warping, tearing, uneven staining). One of these pairs, in which 2 lanes were lost in the first gel because of a tear, but which was otherwise an acceptable replicate, was used to assess between-gel replication. One additional gel consisting entirely of standards from different PCR reactions was run to assess the reproducibility of the DGGE banding patterns within a gel.

Replication

Replication of PCR amplification and the resulting DGGE pattern on a given gel (Figure 2a) was very good, as indicated by the close clustering (>93% similarity) of standards and combinations of standards shown in Figure 2b. Banding patterns were not affected by the amount of template used or by the amount of PCR product loaded onto the gel, except that bands were fainter (or disappeared) when small quantities of PCR product were loaded or smeared when large quantities were loaded (data not shown). We found 1 μg of PCR product to be the optimum amount to load on our gels.

Our standards consistently yielded multiple bands upon analysis (Figure 2a). The banding pattern for each of the two standards was different, but the patterns and their relative migration down the gel (R_f values) were extremely repeatable as indicated by examination of standard lanes included in all gels (data not shown). However, the absolute position of a given band in the gel varied from gel to gel, presumably due to small differences in gel composition or running conditions. Figure 3 compares the raw images of two replicate gels (Figure 3a,b) with the processed (normalized, rescaled, background subtracted) data from sample lanes of replicate gels which have been combined into one image (Figure 3c). Visually, the agreement between the processed gels (Figure 3c) was good, but small differences in the position of some of the bands (for example, compare lane pairs 2 and 10 in Figure 3c) can be seen. The visual similarity of the gels is reflected in the clustering of pairs of samples in groups at ≥90% similarity in Figure 4a. Gel-to-gel variation was not always completely eliminated by normalization to the universal standard and background subtraction, as replicates were paired together as nearest neighbors in only 2 cases (lanes 2 and 7). In some cases, for example, the other gels we repeated because of staining or warping problems, gel-to-gel variation could be large: these samples clustered strongly by gel rather than as sample pairs (data not shown).

Similarity of samples in the replicated gels was also compared using the Jaccard Index (Figure 4b). This dendrogram was much less tightly clustered than the dendrogram generated by regression analysis and tended to cluster samples by gel. Only one pair of replicates (lane 5) clustered as nearest neighbors, although a second pair (lane 10) clustered closely. When compared by the Jaccard Index, samples typically diverged at 50–70% similarity (none more closely related than 82%), while the dendrogram generated by regression analysis clustered paired samples at >90%. The apparent similarity was also strongly affected by program settings for position tolerance and peak overlap. In general, regression-based analysis provided a more satisfactory, robust, and objective means of grouping similar samples.

The number of bands (OTUs) per lane (sample) in replicates, as identified by the program with minimal editing, ranged from 10 to 29. The number of bands per lane was correlated between gels (r^2=0.44), with a slope of 0.67, which was not significantly different from 1 (p<0.05, Figure 5). Manual identification of bands missed by the program's algorithm (usually either fine bands or bands associated with the shoulder of larger peaks in the densitometric curve) approximately doubled the number of bands (range: 22–50) and improved the correlation (r^2=0.68). The slope of the regression curve (1.23) was not significantly different from 1 (p<0.05). Despite the improved correlation, we relied on minimally edited machine calls to determine the number of OTUs per sample to minimize subjectivity in data analysis. Correlations between the number of OTUs per sample for other pairs of repeat gels were not statistically significant (p>0.05, data not shown).

Assemblage composition

The number of OTUs identified in these samples ranged from 12 to 30, with a mean of 21.5 OTUs were recorded at 290 different positions with position stringency set to 0.01%. A graph of the frequency distribution of bands against position is given in Figure 6. Upon visual inspection of gels, 4 bands appear to

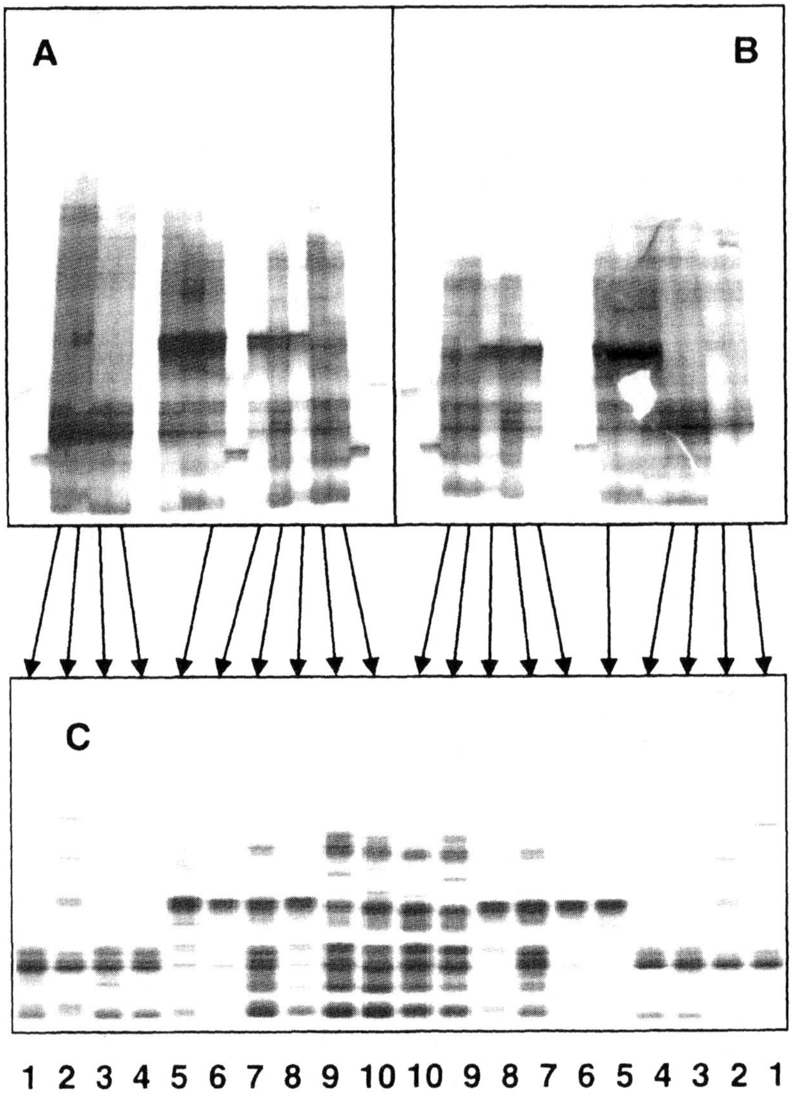

Figure 3. (A and B) Images of gels used to test between-gel replication. (C). Image of both gels after processing.

be almost universal throughout the sample set. These bands correspond to position values of 322, 328, 334 and 339. With band position stringency set at 0.7, subsets of 50 samples (the maximum number of samples the program can use at a time for band-based comparisons) also showed 4 common bands at positions 321.2–324.4, 326.7–330.9, 332.3–336.3 and 338.5–341.2. The position values are slightly different in each subset of 50 samples because the program assigns an average position to bands that it groups together and this varies from sample subset to sample subset when the tolerance is set to be wider than an exact match.

Assemblage similarity

Regression analysis grouped the samples into 5 major clusters with similarities greater than approximately 80% (A–E in Figure 7). Twenty nine samples in one cluster were in 2 branches with similarities >90% (C2 and C3 in Figure 7). These samples were run on 4 (cluster C2) and 11 (cluster C3) different gels, indicating that this association is not an artifact of gel-to-gel variation as discussed above. Visual examination of the gel images (lanes 3 and 4 in Figure 3 and lane 11 in Figure 8) revealed that they were characterized by bands of relatively uniform intensity through most

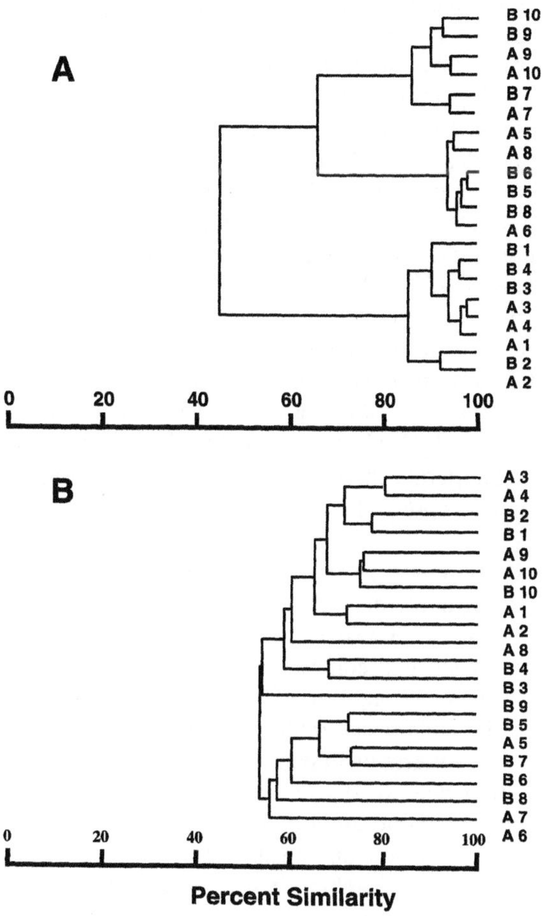

Figure 4. (A) Cluster analysis dendrogram of the similarities (regression model) of the samples (lanes) replicated in the gels in Figure 3. Letter and number identify the gel and lane, respectively, from Figure 3. (B) Cluster analysis dendrogram of the similarities (band-based Jaccard model) of the samples (lanes) replicated in the gels in Figure 3.

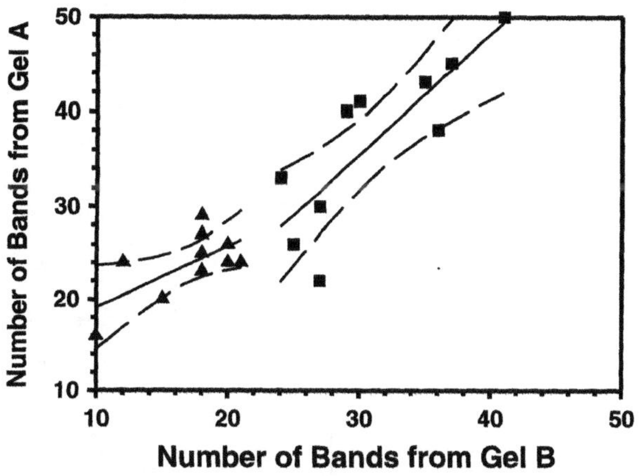

Figure 5. Regressions of the number of bands (OTUs) in lanes representing a given sample in two replicate gels. The two sets of points and regressions represent minimal manual editing (triangles, lower values) and extensive manual editing (squares, higher values).

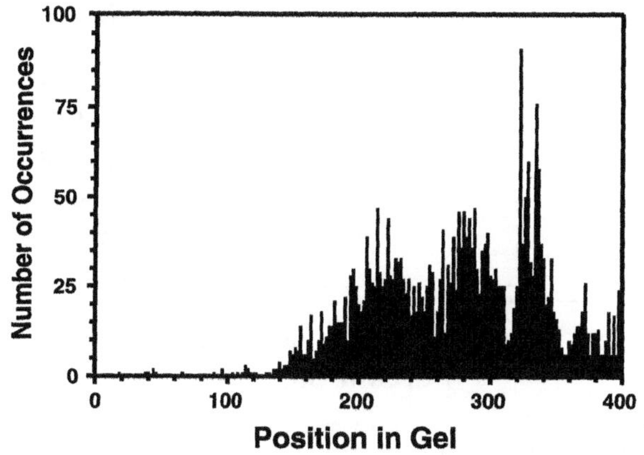

Figure 6. Frequency distribution of bands identified by the program as a function of position.

of the lane with a set of 2–3 more intense bands in the high melting point region of the gel (bottom of the image in these figures).

Another cluster (A in Figure 7) was composed of samples containing one or two dominant bands in the middle of the lane, (lanes 6 and 7 in Figure 3a and lanes 3, 5, 6 and 10 in Figure 8). Most of the samples in this cluster were collected in the Canada Basin, April 23–25, 1995. Assuming that PCR biases haven't affected the relative intensities of the bands, this suggests dominance of these assemblages by one or two OTUs.

Discussion

Number of phylotypes

Analysis of these samples suggested that bacterioplankton assemblages in the Arctic Ocean can be complex, even in late winter. The number of OTUs per sample ranged from 12 to 30, with a mean of 21.5 OTUs per sample. Careful editing of the electropherograms increased this number substantially (Figure 5), indicating that, with the settings we used, the program significantly underestimated the number of OTUs in these samples. These samples were not filtered to remove eucaryotes prior to collection, so some of the OTUs might represent chloroplast sequences. However, chlorophyll concentrations were extremely low in these samples (generally <15 ng/l, Whitledge pers.

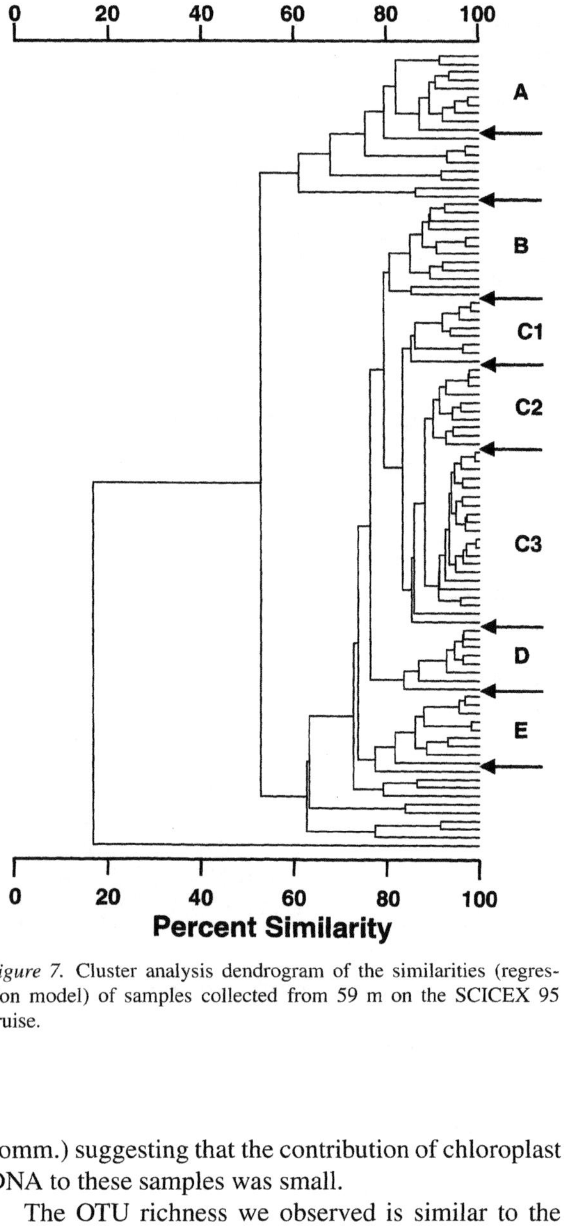

Figure 7. Cluster analysis dendrogram of the similarities (regression model) of samples collected from 59 m on the SCICEX 95 cruise.

comm.) suggesting that the contribution of chloroplast DNA to these samples was small.

The OTU richness we observed is similar to the OTU richness reported by Murray et al. (1996) for California coastal waters. Murray et al. (1996) reported a range of 15–24 OTU/sample (mean of 20.3) in six Tomales Bay samples and a range of 22–30 OTU/sample (mean 25.8) in eight San Francisco Bay samples. Studies of microbial mats and biofilm communities by Muyzer et al. (1993, 1995) revealed fewer phylotypes. Muyzer et al. (1993) detected five phylotypes in a microbial mat community, eight in an anaerobic biofilm community and ten in an aerobic biofilm community. Santegoeds et al. (1996) detected 14 unique 16S rRNA sequence types in their study of an enrichment culture of aerobic chemoorganotrophic bacteria from a hot spring cyanobacterial mat.

The above studies all used the same primer set, producing a fragment of approximately 200 base pairs (Murray, 1994 detected length variants of 169–234 bp in samples from California coastal waters amplified using this primer pair). Not surprisingly, analysis of longer rDNA fragments by other methods yields richer communities, for example Liu et al. (1997) report detecting 45–72 OTUs in samples from sewage sludge, from a contaminated aquifer and from termite guts, by a variant of restriction fragment length analysis of a ca. 900 bp fragment. Giovanonni et al. (1990), DeLong et al. (1993) and Fuhrman et al. (1993) detected large numbers of OTUs in bacterioplankton assemblages from temperate and tropical waters using a clone-and-sequence approach with fragments of ca. 1000 bp length.

Direct comparison of our data to these latter studies is difficult due to differences in protocols and data analyses; however, it is likely that the short sequences we use combine different OTUs, resulting in an underestimate of the richness of Arctic Ocean bacterioplankton assemblages. This could occur either due to co-migration of bands or to significant sequence variation that we didn't detect because it is outside of the region we amplified. Since the major goal of the study was to examine spatial distributions of bacterioplankton assemblages, and since there appears to be enough information in these sequences to accomplish that goal, working with a longer sequence would not increase the power of the analysis. In addition, DGGE can become problematic with long sequences due to inclusions of multiple melting domains in a fragment, although it has been used successfully with longer fragments (Wawer et al., 1997 and others).

Four phylotypes were widely distributed in this sample set. Murray et al. (1996) reported that 5 phylotypes were common to all samples in their comparison of San Francisco and Tomales Bay samples. Visual inspection of the gels suggests that there are more phylotypes in common between samples than identified by the software, as regression analysis clustered a number of samples closely and some bands seem to appear in most gels. Because slight differences between gels in the exact position that a particular band appears are not completely corrected during the normalization procedure, position assignments vary enough so that the software considers them to be different OTUs. Verification that a band at the same position in different lanes or gels represents the same phylotype ultimately

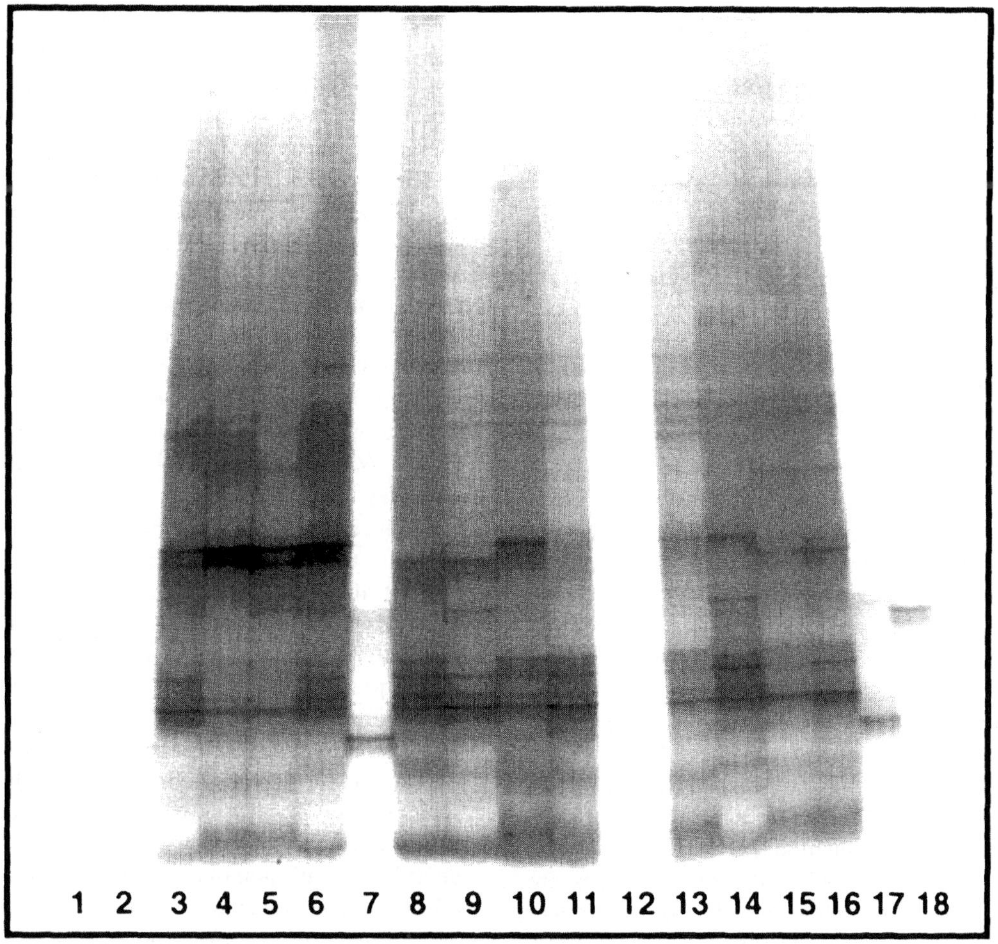

Figure 8. Image of a DGGE gel of Arctic Ocean samples. Lanes 1, 2, 7, 17 and 18 contain standards; lane 12 is empty.

depends on determining the base-pair sequence of the DNA in that band. This can be accomplished readily by excising bands of interest, eluting the DNA and using it as template for cycle sequencing. This analysis is currently underway for this set of samples.

Sensitivity of the PCR/DGGE method

Murray et al. (1996) conclude that this method gives a minimum estimate of the richness of bacterioplankton assemblages. Sensitivity of this method has been estimated by Muyzer et al. (1993) who were able to detect a given rDNA fragment when it is 1% of total DNA present. Murray et al. (1996) report they could easily detect template DNA present as 1.6% of the total DNA in a sample. Although this method is sensitive, it is possible that some members of the bacterioplankton community are excluded due to preferential extraction or amplification (Reysenbach et al., 1992; Suzuki & Giovannoni, 1996; Murray et al., 1996) or other sequence-specific losses during processing. On the other hand, if some organisms give more than one band (as seen with our standards), richness may be overestimated. We have observed that many, but not all, marine isolates produce multiple bands when subjected to PCR/DGGE analysis using this primer set. However, preliminary experiments suggest that this is not a problem with complex mixed template amplifications. An analysis by Santegoeds et al. (1996) showed that each band in DGGE profiles of hot spring communities represented different sequences.

These richness estimates reflect only sequences amplified by the primer set we used. Murray (1994) analyzed primer specificity and determined that most Eubacteria and some plastid sequences were amplified. Archaebacteria are excluded. DeLong (1992) reported widespread occurrence of Archaea in coastal waters off California. More significantly, DeLong et al. (1994) report that Archaea constitute up to 34% of the prokaryotes in coastal Antarctic waters and that they are abundant in other pelagic cold water systems. Including Archaea in our analysis could have increased species richness significantly.

Spatial variation in bacterioplankton assemblages

Spatial variation of assemblage composition was assessed using regression analysis of the similarity of densitometric curves. The dendrogram generated by this analysis contained 5 major clusters, separated at a minimum similarity level of approximately 80% (Figure 7). Cursory examination of ancillary environmental data for these samples revealed no striking correlations between environmental factors and these clusters. Most of the samples in Cluster A were collected over 3 days in a specific region of the Arctic Ocean (Canada Basin). Other clusters did not map strongly with any particular geographic feature, but rather seemed to be associated more with specific phases of the cruise. These correlations may be confounded with temporal variation in the Arctic Ocean over the 44 day period of the cruise. The SCICEX 95 cruise sampled the major seasonal transition from the late Arctic winter to early spring, when dramatic changes are likely to be occurring. However, the spring/summer phytoplankton bloom had not yet begun as chlorophyll concentrations and chlorophyll/phaeopigment ratios were low in these samples (Whitledge, pers comm.). The degree to which clustering reflects a temporal shift *versus* spatial variation is difficult to assess without analyzing additional data from other times of the year.

Since the order in which we analyzed these was essentially random, the clustering we observed is not simply an artifact of sample processing. However, all other things being equal, samples from the same gel tend to cluster more closely than samples from different gels, reflecting the inability of the program to completely remove the 'gel signature' from the processed data. This appears to be due in part to gel-to-gel differences in background staining which are not completely removed during image processing. Variation in the amount of DNA loaded per lane or in exposure time when documenting gels also contribute to this problem, as does physical distortion while handling the gel. Thus, it is imperative that a standard gel processing protocol be implemented for comparisons of large data sets. In our experience, visual discrimination of banding patterns is superior to the program's algorithms. We chose to use the program to identify bands to minimize potential bias of the analysis by the observer and because the data set was too large to analyze otherwise.

Murray et al. (1996) suggest that, unless verified by other analyses, band intensity should not be interpreted as a quantitative measure of relative abundance or used in calculating diversity indices that include a measure of evenness, such as the Shannon Index. Band intensity is related to the amount of DNA in that band and it is tempting to equate band intensity with the relative abundance of an OTU in the original sample. However, Reysenbach et al. (1992) and Suzuki & Giovannoni (1996) suggest that PCR biases may make interpreting band intensity as relative abundance unreliable and the possibility of multiple copies of the 16S rRNA gene in some organisms (Farrelly et al., 1995) further complicates interpretation. Murray et al. (1996) identified co-migration of bands from different phylotypes as another potential problem with equating band intensity and relative abundance. However, Murray (1994) noted a semi-quantitative relationship between relative concentration of a particular template in an artificial assemblage and the relative intensity of the band it represented in the final DGGE gel. Muyzer et al. (1993) also suggested an at least semi-quantitative relationship between relative template concentration and relative band intensity. Riemann et al. (1998) noted a strong correlation between the relative abundance of *Synechococcus* sp. in samples from the Arabian Sea, as determined by epifluorescence microscopy and the intensity of a band in PCR/DGGE analyses of the same sample that could be attributed to this organism by sequencing the band.

We conclude that PCR/DGGE and similar short-cut approaches (for example, see Liu et al., 1997) are capable of providing useful qualitative information on the spatial and temporal variation in the composition of microbial communities and, if combined with sequence information, on the distribution of organisms. Because PCR/DGGE uses a relatively small portion of the 16S rRNA gene as a phylogenetic discriminator, it is likely that several 'species' are combined in some of the OTUs. Nevertheless, patterns emerge from the data as long as care is taken to ensure reproducibility of the analysis so that gel-to-gel variation does not obscure them.

Based on our experience with this approach, we recommend the following. Standards should be included with every gel. Care should be taken to ensure that approximately the same amount of PCR product is loaded into each lane, and gel composition and run conditions must be standardized. Since the banding patterns obtained from PCR/DGGE analysis are much finer than, for example, a normal RFLP analysis, a high resolution gel scanner will improve the subsequent image analysis. The 480 pixel along-lane resolution of the scanner we used is the minimum acceptable because fine bands only 1 pixel wide are considered to be noise by the band-calling algorithms of most gel analysis software. Using a laser-based gel scanner and fluorescently-labeled primers eliminates the problems of staining variation and physical distortion. Careful selection of image analysis software is also important. The software should be able to average intensity across lanes, be able to digitize lanes that are not straight, and produce exportable densitometric curve files or provide a module for analyzing the curves internally. The band identification module in the software we used, while useful for enumerating OTUs, was less suitable than simple regression analysis for testing the similarity of samples.

Acknowledgements

We would like to thank the U.S. Navy and especially the Captain and crew of the USS Cavalla and personnel at the Arctic Submarine Laboratory for their role in making this work possible. We are particularly grateful to Dr Ted DeLaca, Dean Stockwell, Jeff Gossett and the rest of the science program 'riders' for carrying, and then filtering, many liters of seawater for us. This work was funded by the National Science Foundation, NSF OPP 9500237.

References

Bio-Rad Inc., 1992. Molecular Analyst fingerprinting and fingerprinting plus software instruction manual, version 1.0. Bio-Rad Inc. Hercules, CA.

DeLaca, T., B. Coakley, T. Boyd & D. Stockwell, 1996. Cruise Report for the SCICEX95 Mission on-board the USS Cavalla. Arctic Research Consortium of the U.S. Inc., Univ. of Alaska, Fairbanks, AK.

DeLong, E. F, 1992. Archaea in coastal marine environments. Proc. natn. Acad. Sci. U.S.A. 89: 5685–5689.

DeLong, E., D. Franks & A. Alldredge, 1993. Phylogenetic diversity of aggregate-attached vs. free-living marine bacterial assemblages. Limnol. Oceanogr. 38: 924–934.

DeLong, E. F., K. Y. Wu, B. B. Prezelin & R. V. M. Jovine, 1994. High abundance of Archaea in Antarctic marine picoplankton. Nature 371: 695–697.

Don, R. H., P. T. Cox, B. J. Wainwright, K. Baker & J. S. Mattick, 1991. 'Touchdown' PCR to circumvent spurious priming during gene amplification. Nucleic acids Res. 19: 4008.

Erlich, H., D. Gelfand & J. Sninsky, 1991. Recent advances in the polymerase chain reaction. Science 252: 1643–1651.

Farrelly, V., F. A. Rainey & E. Stackebrandt, 1995. Effect of genome size and rrn gene copy number on PCR amplification of 16S rRNA genes from a mixture of bacterial species. Appl. envir. Microbiol. 61: 2798–2801.

Ferguson, R. L., E. N. Buckley & A. V. Palumbo, 1984. Response of marine bacteria to differential filtration and confinement. Appl. envir. Microbiol. 47: 49–55.

Ferris, M., G. Muyzer & D. M. Ward, 1996. Denaturing gradient gel electrophoresis profiles of 16S rRNA-defined populations inhabiting a hot spring microbial mat community. Appl. envir. Microbiol. 62: 340–346.

Fuhrman, J., K. McCallum & A. A. Davis, 1993. Phylogenetic diversity of subsurface marine microbial communities from the Atlantic and Pacific oceans. Appl. envir. Microbiol. 59: 1294–1302.

Giovannoni, S., T. Britschgi, C. L. Moyer & K. G. Field, 1990. Genetic diversity in Sargasso Sea bacterioplankton. Nature 345: 60–63.

Guay, C. K. & K. K. Falkner, 1997. Barium as a tracer of Arctic halocline and river waters, Deep Sea Res. II 44: 1543–1569.

Hollibaugh, J. T., 1994. Relationship between thymidine metabolism, bacterioplankton community metabolic capabilities, and sources of organic matter. Microb. Ecol. 28: 117–131.

Liu, W-T, T. L. Marsh, H. Cheng & L. J. Forney, 1997. Characterization of microbial diversity by determining terminal restriction fragment length polymorphisms of genes encoding 16S rRNA. Appl. envir. Microbiol. 63: 4516–4522.

Magurran, A. E., 1988. Ecological Diversity and its Measurement. Princeton University Press, Princeton, 179 pp.

Murray, A. E., 1994. Community fingerprint analysis: a molecular method for studying marine bacterioplankton diversity. M.A. Thesis, Department of Biology, San Francisco State University. San Francisco, 128 pp.

Murray, A. E., J. T. Hollibaugh & C. Orrego, 1996. Phylogenetic compositions of bacterioplankton from two California estuaries compared by denaturing gradient gel electrophoresis of 16S rDNA fragments. Appl. envir. Microbiol. 62: 2676–2680.

Muyzer, G., E. C. D. Waal & A. G. Uitterlinden, 1993. Profiling of complex microbial populations by denaturing gradient gel electrophoresis analysis of polymerase chain reaction-amplified genes coding for 16S rRNA. Appl. envir. Microbiol. 59: 695–700.

Muyzer, G., A. Teske, C. O. Wirsen & H. W. Jannasch, 1995. Phylogenetic relationships of *Thiomicrospira* species and their identification in deep-sea hydrothermal vent samples by dentauring gradient gel electrophoresis of 16S rDNA fragments. Arch. Mikrobiol. 164: 165–172.

Myers, R. M., S. G. Fischer, L. S. Lerman & T. Maniatis, 1985. Nearly all single base substitutions in DNA fragments joined to a GC-clamp can be detected by denaturing gradient gel electrophoresis. Nucleic Acids Res. 13: 3131–3145.

Pedros-Alio, C., 1993. Diversity of bacterioplankton. Trends Ecol. Evol. 8: 86–90.

Reysenbach, A. L., J. Giver, G. S. Wickham & N. R. Pace, 1992. Differential amplification of rRNA genes by polymerase chain reaction. Appl. envir. Microbiol. 58: 3417–3418.

Riemann, L., L. B. Fandino, G. F. Steward, D. C. Smith, H. W. Ducklow & F. Azam, 1998. Variability of bacterial community composition in the Arabian Sea during the 1995 JGOFS program. EOS 79: OS 68.

Saiki, R. K., D. H. Gelfand, S. Stoffel, S. J. Scharf, R. Higuchi, G. T. Horn, K. B. Mullis & H. A. Erlich, 1988. Primer-directed enzymatic amplification of DNA with a thermostable DNA polymerase. Science 239: 487–491.

Santegoeds, C. M., S. C. Nold & D. M. Ward, 1996. Denaturing gradient gel electrophoresis used to monitor the enrichment culture of aerobic chemoorganotrophic bacteria from a hot spring cyanobacterial mat. Appl. envir. Microbiol. 62: 3922–3928.

Suzuki, M. T. & S. J. Giovannoni, 1996. Bias caused by template annealing in the amplification of mixtures of 16S rRNA genes by PCR. Appl. envir. Microbiol. 62: 625–630.

Ward, D. M., R. Weller & M. M. Bateson, 1990. 16S rRNA sequences reveal numerous uncultured microorganisms in a natural community. Nature 345: 63–65.

Wawer, C. & G. Muyzer, 1995. Genetic diversity of *Desulfovibrio* spp. in environmental samples analyzed by denaturing gradient gel electrophoresis of (NiFe) hydrogenase gene fragments. Appl. envir. Microbiol. 61: 2203–2210.

Wawer, C., H. Ruggerberg, G. Meyer & G. Muyzer, 1995. A simple and rapid electrophoresis method to detect sequence variation in PCR-amplified DNA fragments. Nucleic Acids Res. 23: 4928–4929.

Wawer, C., M. S. M. Jetten & G. Muyzer, 1997. Genetic diversity and expression of the NiFe hydrogenase large subunit gene of *Desulfovibrio* spp in environmental samples. Appl. envir. Microbiol. 63: 4360–4369.

Wheeler, P. A., M. Gosselin, E. Sherr, D. Thibault, D. L. Kirchman, R. Benner & T. E. Whitledge, 1996. Active cycling of organic carbon in the central arctic ocean. Nature 380: 697–699.

Wheeler, P. A., J. M. Watkins & R. L. Hansing, 1997. Nutrients, organic carbon and organic nitrogen in the upper water column of the Arctic Ocean: implications for the sources of dissolved organic carbon. Deep Sea Res. II. 44: 1571–1592.

Hydrobiologia **401:** 69–75, 1999.
J.P. Zehr & M.A. Voytek (eds), Molecular Ecology of Aquatic Communities.

Bacterial populations in replicate marine enrichment cultures: assessing variability in abundance using 16S rRNA-based probes

J. M. González, R. E. Hodson & M. A. Moran*
Department of Marine Sciences, University of Georgia, Athens, GA 30602-3636, U.S.A.
Tel: [+1]706-542-6481; Fax: [+1]706-542-5888; E-mail: mmoran@arches.uga.edu

Key words: microcosms, bacterial populations, 16S rRNA oligonucleotide probes,lignin, coastal seawater

Abstract

Most studies of marine bacterial communities focus on functional attributes of the community, rather than on population or community structure, at least in part, because of the difficulty in enumerating individual species within complex communities. Here, we describe a study in which populations of three bacterial species were followed over time in replicate marine enrichment cultures using 16S rRNA-based oligonucleotide probes. Three identical enrichment microcosms were established with lignin-rich pulp mill waste as a sole carbon source, inoculated with coastal seawater, and transferred at two-week intervals. Population levels were assessed throughout a six-week period using species-specific 16S rRNA-based oligonucleotide probes directed toward three bacterial species that were numerically important (and culturable) members of the enrichments. Substantial differences in the population levels of each bacterial species were found among the triplicate incubations, despite the fact that the enrichments were inoculated and treated identically. Stochastic differences in the composition of the inoculum and/or ecological interactions within the enrichment replicates may have been important in determining final population levels. Functional ability, assessed as rates of degradation of a synthetic lignin preparation, were fairly similar among the three replicate enrichments (within 70%), despite the substantial differences in population levels of the representative lignin-degrading species.

Introduction

Microcosms and other laboratory systems with complex microbial communities have been used as tools for the study of a variety of microbially-mediated biogeochemical processes, including pollutant biodegradation (e.g. Spain et al., 1980), bioremediation (e.g. Bauer & Capone, 1988) and elemental cycling (e.g. Jørgensen & Fenchel, 1974). Laboratory-based microbial systems have also been used in ecotoxicology studies (e.g. Pratt et al., 1988) and for assessing the potential ecological effects of genetically engineered organisms (e.g. Dwyer et al., 1988). Because microorganisms are so abundant in natural waters, soils and sediments, a relatively small laboratory system can harbor a large and diverse microbial community. Thus, these systems can be replicated at low cost and can have multiple parameters manipulated individually in controlled studies.

Although use of laboratory microbial systems is widespread, there are currently few studies that compare microbial populations among replicate systems. Whether microbial species colonize and develop similarly in systems treated identically, or whether they diverge among systems with time, is not well known. How diverging species populations might relate to functional attributes of the microbial communities is similarly unclear. While community composition of replicate microcosms have in some cases been compared at the level of culturable bacteria (Wagner-Döbler et al., 1992), the diversity of bacterial isolates is likely to be low relative to the total community and the composition of isolate collections highly biased (Brock, 1987). Thus molecular techniques that circumvent the necessity of culturing bacteria may be

* Author for correspondence.

valuable in assessing population or community structure of model systems.

In this study, three replicate marine enrichment cultures were established with lignin-rich pulp mill waste as the sole carbon source in a study originally designed to isolate marine lignin-degrading bacteria (González et al., 1996). The replicate cultures were inoculated with coastal seawater and treated identically during a six-week incubation. At three different time points during the incubation, the population levels of three representative bacterial species were assessed using species-specific 16S rRNA oligonucleotide probes. Because significant differences in population levels of the three species were found among the replicates, we also determined whether or not differences in functional attributes were evident, in this case by measuring rates of degradation of synthetic lignin.

Materials and methods

Enrichment microcosms

Marine enrichments consisted of 6-l flasks containing 3 l of filter-sterilized seawater (0.2 μm pore-size) collected from a coastal Georgia marsh, 24 ml of the high molecular weight fraction of pulp mill effluent (obtained from Federal Paper Board Company Inc., Augusta, Georgia) which is rich in lignin and lignin byproducts (20 mg C l^{-1} final concentration) and inorganic nutrients (5 μM N as NH_4NO_3 and 1 μM P as KH_2PO_4). Pulp mill effluents contributed 90% of the carbon in the final enrichment medium and the natural seawater contributed 10%.

The three replicate microcosms (designated Flasks A, B and C) were inoculated with 20-ml subsamples of freshly collected seawater which had been filtered through a 1-μm pore-size Nuclepore filter to obtain a bacterial size fraction; each enrichment was inoculated with approximately 1.1×10^7 bacterial cells, based on epifluorescence microscopy of Acridine Orange-stained cells (Hobbie et al., 1977). Enrichments were incubated in the dark at room temperature with shaking at 100 rpm, and a 20 ml subsample was removed from each replicate and transferred to a new 6 l flask containing 3 l of medium every two weeks during a six-week period (Figure 1). At each transfer and at the end of the six-week period, bacterial cells were harvested from microcosms by filtration onto 0.2-μm pore-size Nuclepore filters and DNA was extracted directly from the filters as previously described (González et al., 1996).

Oligonucleotide probe design

Bacterial strains were isolated from the final enrichment flasks on half-strength YTSS agar plates [4 g yeast extract, 2.5 g tryptone, 20 g sea salts (Sigma, St. Louis, MO) and 18 g agar per liter] or pulp mill waste plates (enrichment medium plus 18 g agar per liter). To identify the isolates with the highest population levels in the final enrichments, the 133 isolated strains were screened using whole-genome colony hybridizations with ^{32}P-labeled community DNA from the final time point of two enrichment cultures (Flasks A and B) as probes. Three of the isolates, designated IRE-31, E-37 and KW-40, were chosen for further study based on indications from the whole-genome screens of significant contributions to enrichment DNA from either Flask A or B (González et al., 1996). All three organisms were isolated from each of the three marine enrichments, as determined by subsequent screening of the isolate collections with species-specific 16S rRNA oligonucleotide probes (see below).

The 16S rRNA genes of isolates IRE-31, E-37 and KW-40 were sequenced as described previously (González et al., 1996). Isolate E-37 is a member of the α subclass of the Proteobacteria and was subsequently named *Sagittula stellata* (González et al., 1997b). Isolates IRE-31 and KW-40 are members of the γ subclass of Proteobacteria and were subsequently named *Marinobacterium georgiense* and *Microbulbifer hydrolyticus* (González et al., 1997a). Species-specific oligonucleotide probes were designed following alignments of the sequences with related sequences in Genbank. Probes were chosen to target regions where variability was high with respect to the closest phylogenetic groups and for which there was no match with any other sequence in the databanks. The sequence 5′-CTTGCAGGATTGACGTTACC was specific for strain IRE-31; 5′-TTCTGTTGAGGATAGCCC was specific for E-37; and 5′-AATACCCGCTTGCTGTG-ACG was specific for KW-40. These probes might also target other, highly related bacteria, if they were present in the marine enrichments, and we use the term 'species-specific' with this caveat in mind. In all cases in which a probe hybridized to an isolate other than the one for which it was designed, that isolate was determined based on sequence and biochemical data to be a duplicate isolation of the same

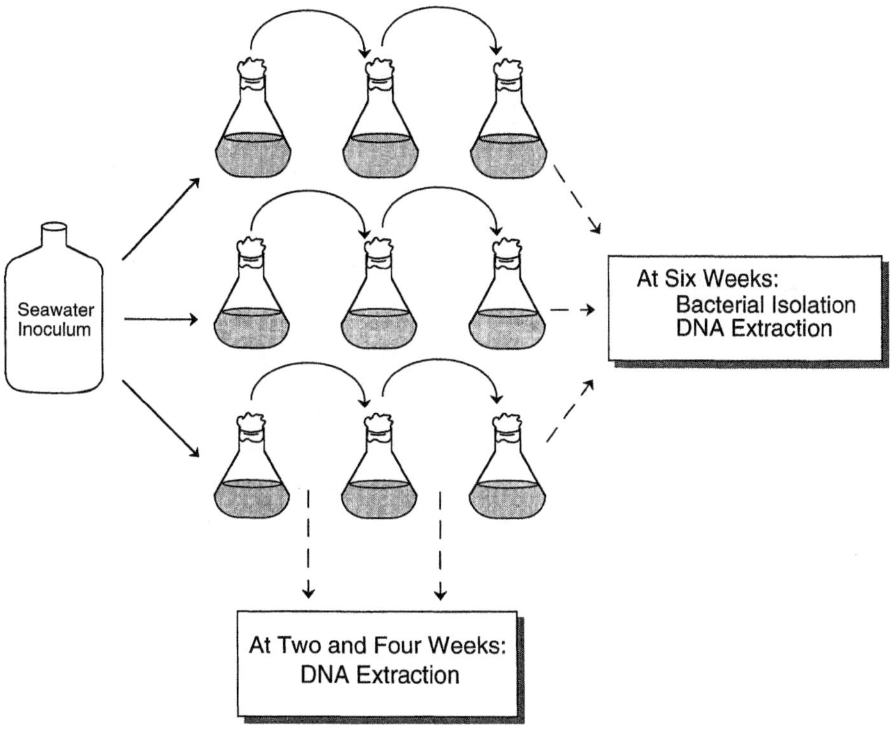

Figure 1. Enrichment and sampling scheme.

strain (González et al., 1996). The Universal probe 5′-ACGGGCGGTGTGTRC served as a positive control in the dot blot hybridizations (Lane et al., 1985).

Enrichment community structure

Relative population levels of each species, expressed as a percent of the total rDNA genes in the enrichment communities, were estimated at three time points: at the two week transfer, at the four week transfer and at the end of the enrichment (at six weeks). Oligonucleotide probes were labeled with ^{32}P and hybridized to community DNA obtained from the enrichment flasks at each time point using methodology described in González et al. (1996). DNA was quantified with Hoescht dye 33258 immediately before each hybridization. DNA from the species to be quantified was spotted on a nylon membrane at four different concentrations from 50 ng to 500 ng to serve as standards. Two different amounts of DNA from the communities (500 ng and 1 μg) were spotted on the same membrane. A standard curve based on a linear regression of the quantity of species' DNA *versus* hybridization signal was used to quantify the amount of DNA from that organism as a percent of community rDNA. The percentage of rDNA attributable to each species was calculated following normalization to hybridization signals with the Universal probe, as described in Giovannoni et al. (1990). Average standard deviation was 2.4% for replicate dots, and only dots for which the hybridization signal was significantly higher than the background (no DNA or non-target DNA controls) were considered positive.

Synthetic lignin mineralization

The ability of the three replicate enrichment communities to mineralize synthetic lignin was tested at the end of the six-week period. An inoculum (100 μl) was removed from each enrichment and added to triplicate milk dilution bottles containing 30 ml BM medium (Baumann & Baumann, 1981) that contained the same concentration of high molecular weight pulp mill effluent as the original enrichments. A radiolabeled synthetic lignin, [^{14}C-β] dehydropolymerisate (DHP; Freudenberg & Neish, 1968; Haider & Trojanowski, 1975) of coniferyl alcohol (400 000 dpm mg^{-1}) was added to the medium at a concentration of 1.4 μg ml^{-1}. Based on direct microscopic counts of Acridine Orange-stained cells (Hobbie et al., 1977), bottles were inoculated with 4.6 $\times$ 10^5 cells. Mineralization of the labeled DHP was measured by trapping $^{14}CO_2$

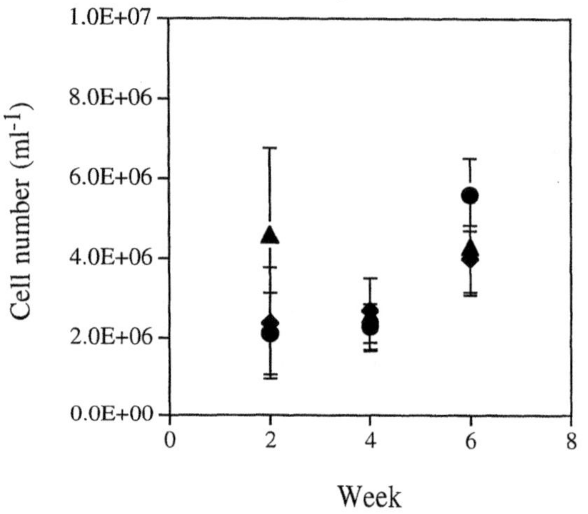

Figure 2. Bacterial cell density at the end of each two-week incubation period in replicate enrichment communities A (♦), B (▲) and C (●). Each flask was originally inoculated with 1.1×10^7 bacterial cells from seawater. A 20-ml volume was transferred to 3 l of new medium after two weeks and four weeks of incubation. Error bars represent 1 S.D. for three AODC slides, with 10 fields counted per slide.

in a 2-ml solution of 0.2 M NaOH placed in a small vial inside each bottle. At three-day intervals, the bottles were opened, the vials removed, and a 1-ml portion of the trapping solution was added directly to 10 ml scintillation cocktail (ScintiSafe Econo 2; Fisher Scientific, Pittsburgh, PA) for counting. The bottles were left open for 5 min before adding new NaOH solution and resealing.

Individual rates of lignin mineralization by the three isolates were also measured. Cells were first grown in 200 ml of BM medium with 0.1% (w/v) ferulic acid as the sole carbon source. When each culture reached exponential growth, the cells were washed twice in BM medium (with no carbon source). The cell pellets were resuspended in BM medium and transferred to the same DHP medium as used for the enrichments. Finally, a mixed culture consisting of all three species (IRE-31, E-37 and KW-40) was also tested for mineralization of synthetic lignin.

Results and discussion

Each of the enrichment cultures was inoculated with 20 ml of seawater to a final density of 3.7×10^3 cells ml^{-1} (S.D. = 1.4×10^3 cells ml^{-1}). Cell densities were not different among the replicate enrichments at the time of the first transfer ($P > 0.15$; ANOVA) or the second transfer ($P > 0.29$). At the end of the enrichment period, cell density in Flask C was about 40% higher than in the other two flasks ($P < 0.01$; Figure 2).

Hybridization with species-specific 16S rRNA oligonucleotide probes indicated that relative population levels of the three species varied substantially among the replicate microcosms, despite the fact that each enrichment was inoculated and treated identically (Figure 3). Species E-37 was numerically important in Flask A throughout the study, contributing 32% of the total rDNA by the end of the six-week experiment. This bacterium was not detected in the other two flasks until the final time point, when it contributed 14% and 4% of the rDNA in Flasks B and C, respectively. Species IRE-31 was detected only in Flask A, increasing from 5% of the total rDNA at two weeks to 23% by the end of the incubation. Species KW-40 was detected only in Flask B and contributed a relatively low percent of the rDNA (2% or less) at both the second transfer and final time point. Since all three species were isolated from each of the enrichment flasks at the final time point, their contribution to the enrichment rDNA is assumed to be <1% (below the limit of detection using dot-blot hybridizations) when a hybridization signal was not detected at the six-week sample (i.e. KW-40 in Flask A, IRE-31 in Flask B, and KW-40 and IRE-31 in Flask C; Figure 3). Although population levels of only three representative species were followed in this study, the fact that the sum of their contributions varied widely (from < 2 to 55%) indicates that the relative abundance of the bacterial species that were not tracked with 16S rRNA-based probes must also have varied considerably.

Substantial differences among the population levels in the replicate enrichments were already evident early in the study, at the time of the two-week transfer. Thus variability in relative abundance of these three species could not be traced solely to the points where the enrichments were reinoculated into fresh medium. Population levels continued to diverge after transfers were made at the two-week and four-week time points, and were greatest at the end of the experiment. The large differences found among replicates in this study may not be characteristic of less disturbed laboratory systems in which there is reduced opportunity for growth of the microbial community, for example in microcosms established with intact soil or sediment cores. Nonetheless, even relatively small disturbances associated with sample collection and microcosm establishment in the laboratory will likely

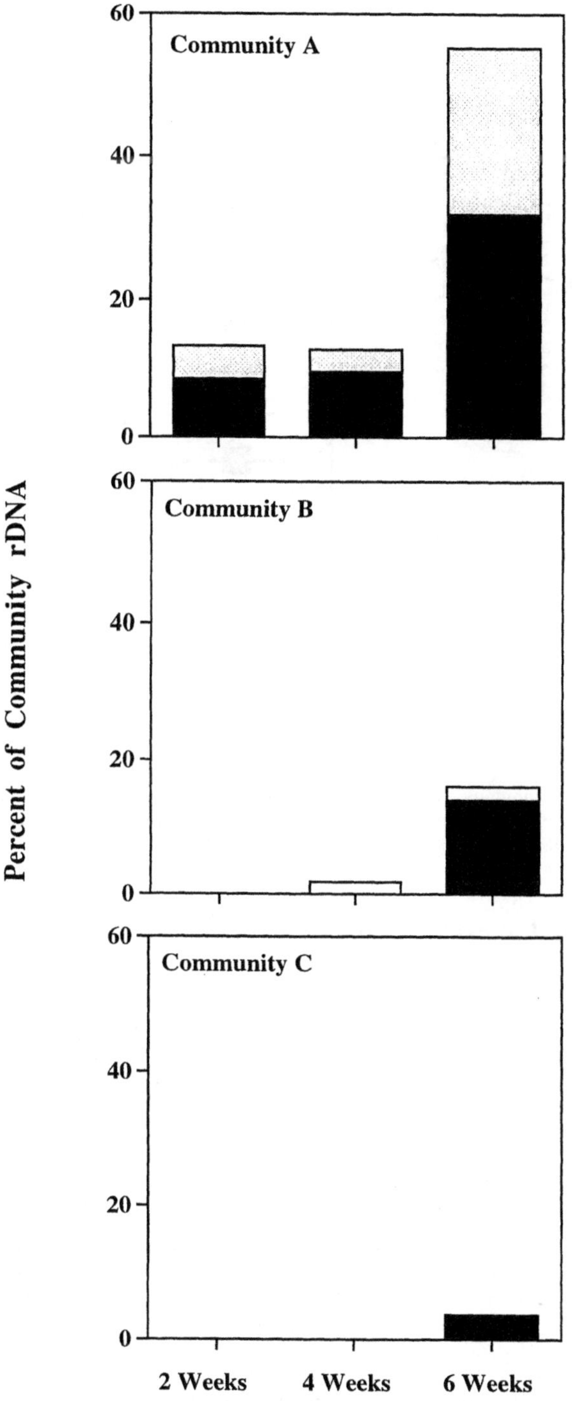

Figure 3. Contribution of three bacterial species to the total community DNA in replicate enrichment communities (A, B and C) after two, four and six weeks of incubation. Percent contribution was determined based on duplicate hybridizations of species-specific 16S rRNA oligonucleotide probes with community DNA. Black, gray and white bars are percent contributions of E-37, IRE-31 and KW-40, respectively.

stimulate growth of the microbial community, and results of this study suggest that differences in population levels of individual species can readily develop even under (apparently) equivalent external conditions.

Given the substantial differences in relative abundance of these three species among replicate enrichments, we determined whether or not significant functional differences were also evident by comparing rates of degradation of synthetic lignin among the enrichments (Figure 4). The final percent mineralization of the radiolabeled synthetic lignin (after 25 days of incubation) was lower in Flask A than in the other two enrichments by about 30% ($P > 0.002$, ANOVA; Figure 4). A comparison of mineralization rates (based on 95% confidence limits of a fit of the time series data to an exponential model) indicated similar results. Although it is difficult to determine the exact role of these three species in lignin degradation within the mixed community, rates of mineralization by two of the species (E-37 and KW-40) and by the three species mixed together were about one third those of the intact enrichment communities; isolate IRE-31 showed little ligninolytic activity (Figure 4). The rates of synthetic lignin degradation by the individual isolates in this study are similar to those found for other ligninolytic bacteria (Vicuña, 1988).

The development of very different population levels of these three bacterial species may be due to differences in the composition of the inoculum, or to differences in the outcome of ecological interactions within the flasks, or both. Evidence for microscale variability in the composition of bacterial communities in seawater at the scale of centimeters or less has been found previously (Azam & Hodson, 1981; Azam & Ammerman, 1984; Mitchell & Fuhrman, 1989). Such variability might give rise to differences in composition of the bacterial inoculum added to each enrichment, although pre-screening the inoculum through a 1 μm pore-size filter would be likely to temporarily disrupt microscale structure. Direct count data indicate that each enrichment was inoculated with 11 million bacterial cells, so that random differences in the composition of the inoculum would have to be at the level of fairly rare cells. However, the number of viable cells in the inoculum may have been considerably lower (Zweifel & Hagström, 1995).

The bacterial communities that became established in the enrichment flasks were functional subsets of the original seawater community, biased toward organisms able to utilize some component of the lignin-rich

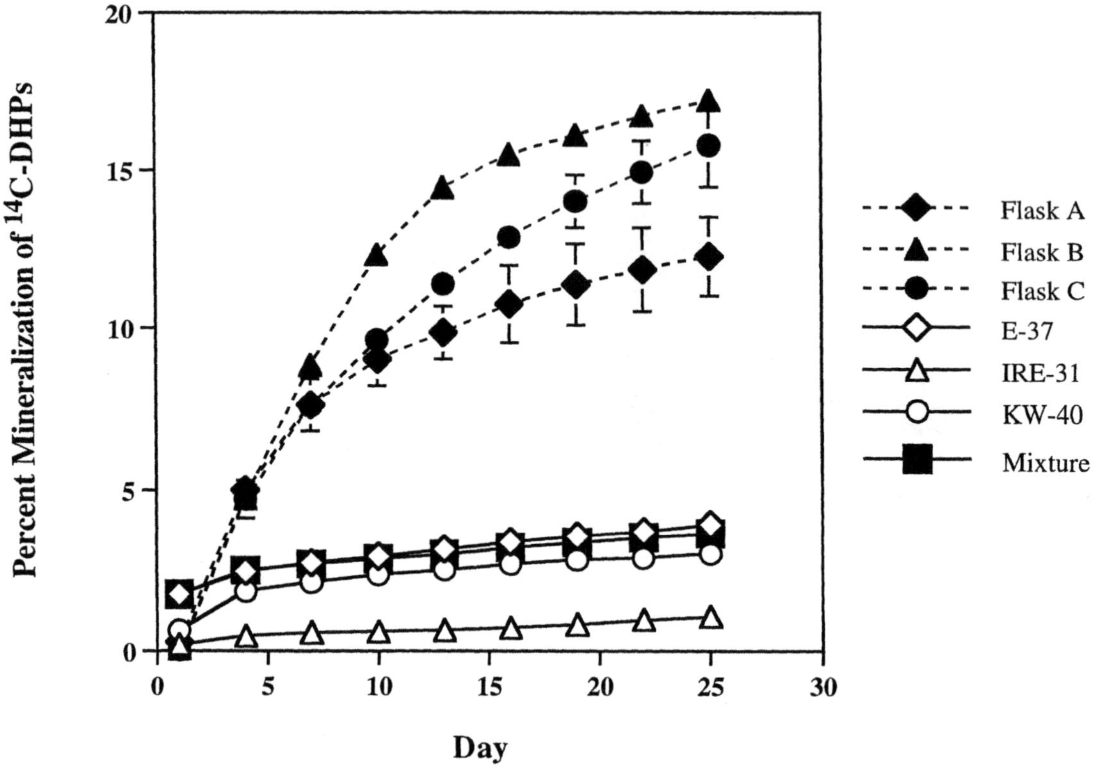

Figure 4. Rates of mineralization of [^{14}C-β]DHP by replicate enrichment communities (A, B and C), individual isolates (E-37, IRE-31 and KW-40), and a mixture of all three isolates, expressed as a percentage of the total radioactivity initially added to the medium. Uninoculated controls showed no mineralization. Error bars represent 1 S.D., n=3; where not shown, error bars are smaller than symbols.

pulp mill wastes. Lignin degradation in natural environments is known to be carried out by a number of bacterial and fungal organisms (Vicuña, 1988), with bacteria dominating in aquatic environments (Kaushik & Hynes, 1971; Mason, 1976; Benner et al., 1986). Although the diversity of bacterial lignin degraders is thought to be high (Vicuña, 1988; Zimmermann, 1990), relatively little is known about the identity or ecological interactions of this important group of microorganisms. Molecular approaches based on analysis of 16S rRNA genes provide valuable tools for tracking individual species (or groups of related species) within diverse microbial systems such as these, and for increasing our understanding of the relationships between structure and function in complex bacterial communities.

Acknowledgements

The authors thank W. M. Sheldon Jr. for expert assistance with statistical analyses. This work was supported by Office of Naval Research grant NOOO14-91-J-1826 and NOAA Office of Sea Grant NA66RG0282.

References

Azam, F. & J. W. Ammerman, 1984. Cycling of organic matter by bacterioplankton in pelagic marine ecosystems: microenvironmental considerations. In Fasham M. J. R. (ed.), Flows of Energy and Materials in Marine Ecosystems. Plenum Press, New York: 345–360.

Azam, F. & R. E. Hodson, 1981. Multiphasic kinetics for D-glucose uptake by assemblages of natural marine bacteria. Mar. Ecol. Prog. Ser. 6: 213–222.

Bauer, J. E. & D. G. Capone, 1988. Effects of co-occurring aromatic hydrocarbons on degradation of individual polycyclic aromatic hydrocarbons in marine sediment slurries. Appl. envir. Microbiol. 54: 1649–1655.

Benner, R., M. A. Moran & R. E. Hodson, 1986. Biogeochemical cycling of lignocellulosic carbon in marine and freshwater ecosystems: relative contributions of procaryotes and eucaryotes. Limnol. Oceanogr. 31: 89–100.

Brock, T. D., 1987. The study of microorganisms in situ: progress and problems. Symp. Soc. Gen. Microbiol. 41: 1–17.

Dwyer, D. F., F. Rojo, K. N. Timis, 1988. Bacteria with new pathways for the degradation of pollutants and their fate in model ecosystems. In Klingmüller W. (ed.), Risk Assessment for Deliberated Releases. Springer-Verlag, Berlin: 100–109.

Findlay, R. H., P. C. Pollard, D. J. W. Moriarty & D. C. White, 1985. Quantitative determination of microbial activity and community nutritional status in estuarine sediments: evidence for a disturbance artifact. Can. J. Microbiol. 31: 493–498.

Freudenberg, K. & A. C. Neish, 1968. Constitution and biosynthesis of lignin. Springer-Verlag, Berlin, 129 pp.

Giovannoni, S. J., T. B. Britschgi, C. L. Moyer & K. G. Field, 1990. Genetic diversity in Sargasso Sea bacterioplankton. Nature (London) 345: 60–63.

González, J. M., F. Mayer, M. A. Moran, R. E. Hodson & W. B. Whitman, 1997a. *Microbulbifer hydrolyticus* gen. nov., sp. nov. and *Marinobacterium georgiense* gen. nov., sp. nov., two marine bacteria from a lignin-rich pulp mill waste enrichment community. Int. J. syst. Bacteriol. 47: 369–376.

González, J. M., F. Mayer, M. A. Moran, R. E. Hodson and W. B. Whitman, 1997b. *Sagittula stellata* gen. nov., sp. nov., a lignin-transforming bacterium from a coastal environment. Int. J. syst. Bacteriol. 47: 773–780.

González, J. M., W. B. Whitman, R. E. Hodson & M. A. Moran, 1996. Identifying numerically abundant culturable bacteria from complex communities: An example from a lignin enrichment culture. Appl. envir. Microbiol. 61: 4433–4440.

Haider K. & J. Trojanowski, 1975. Decomposition of specifically ^{14}C-labeled phenols and dehydropolymers of coniferyl alcohol as models for lignin degradation by soft and white rot fungi. Arch. Microbiol. 105: 33–41.

Hobbie, J. E., R. J. Daley & S. Jasper, 1977. Use of Nuclepore filters for counting bacteria by fluorescence microscopy. Appl. envir. Microbiol. 33: 1225–1228.

Jørgensen, B. B. & T. Fenchel, 1974. The sulfur cycle of a marine sediment model system. Mar. Biol. 24: 189–201.

Kaushik, N. K. & H. B. Hynes, 1971. The fate of the dead leaves that fall into streams. Arch. Hydrobiol. 68: 465–515.

Lane, D. J., B. Pace, G. J. Olsen, D. A. Stahl, M. L. Sogin & N. R. Pace, 1985. Rapid determination of 16S ribosomal RNA sequences for phylogenetic analyses. Proc. natl. Acad. Sci. U.S.A. 82: 6955–6959.

Mason, C. F., 1976. Relative importance of fungi and bacteria in the decomposition of *Phragmites* leaves. Hydrobiologia 51: 65–69.

Mitchell, J. G. & J. A. Fuhrman, 1989. Centimeter scale vertical heterogeneity in bacteria and chlorophyll *a*. Mar. Ecol. Prog. Ser. 54: 141–148.

Pratt, J. R., N. J. Bowers, B. R. Niederlehner & J. Cairns, Jr., 1988. Effects of chlorine on microbial communities in naturally derived microcosms. Envir. Toxicol. Chem. 7: 679–687.

Søndergaard, M. & M. Middelboe, 1995. A cross-system analysis of labile dissolved organic carbon. Mar. Ecol. Prog. Ser. 118: 283–294.

Spain, J. C., P. H. Pritchard & A. W. Bourquin, 1980. Effects of adaptation on biodegradation rates in sediment/water cores from estuarine and freshwater environments. Appl. envir. Microbiol. 40: 726–734.

Vicuña, R., 1988. Bacterial degradation of lignin. Enz. Microb. Technol. 10: 646–655.

Wagner-Döbler, I., R. Pipke, K. N. Timmis & D. F. Dwyer, 1992. Evaluation of aquatic sediment microcosms and their use in assessing possible effects of introduced microorganisms on ecosystem parameters. Appl. envir. Microbiol. 58: 1249–1258.

Zimmermann, W., 1990. Degradation of lignin by bacteria. J. Biotechnol. 13: 119–130.

Zweifel, U. & A. Hagström, 1995. Total counts of marine bacteria include a large fraction of non-nucleoid-containing bacteria (ghosts). Appl. envir. Microbiol. 61: 2180–2185.

Hydrobiologia **401:** 77–96, 1999.
J.P. Zehr & M.A. Voytek (eds), Molecular Ecology of Aquatic Communities.

Diversity of bacterial communities in Adirondack lakes: do species assemblages reflect lake water chemistry?

B. A. Methé & J. P. Zehr*
Department of Biology, MRC 303, Rensselaer Polytechnic Institute, Troy, NY 12180-3590, U.S.A.
Tel: [+1]518-276-8386; Fax: [+1]518-276-2162; E-mail: zehrj@rpi.edu
Current address for J. P. Zehr: Ocean Sciences Department, Earth and Marine Sciences Bldg., University of California, Santa Cruz, CA 95064, U.S.A.
Current address for B. A. Methé: Department of Microbiology, 203N Morrill IVN, University of Massachusetts, Amherst, MA 01003, U.S.A.

Key words: bacterial diversity, acidification, Adirondack lakes, aluminum

Abstract

Bacterial diversity was examined in six lakes located in the Adirondack Mountains of upstate New York. This region receives significant anthropogenic atmospheric inputs of strong mineral acids resulting in decreases in pH and acid neutralizing capacity (ANC) in many of the lakes. Statistical methods including correlation and cluster analysis were used to determine if there were significant associations between phylotypes and water chemistry variables. Direct effects of acidification, pH and ANC, were not significantly correlated with any of the broadest level taxonomic classifications (equivalent to class or order), but may be correlated with subgroups within these classifications. Indirect influences of acidification were suggested by significant correlations of phylotypes with aluminum chemistry. There were positive correlations between the relative abundance of the γ subdivision of the *Proteobacteria* and total aluminum (r^2= 0.70, p= 0.04), monomeric aluminum (r^2= 0.78, p= 0.02) and non-labile aluminum (r^2= 0.92, p= 0.002). The ACK1 clade of the β-*Proteobacteria* (Adirondack clade 1) was correlated with monomeric aluminum (r^2= 0.71, p=0.03) and non-labile aluminum (r^2= 0.73, p= 0.03). Significant negative correlations were found between the relative abundance of the *Cytophaga-Flexibacter-Bacteroides* phylum and total aluminum (r^2= 0.74, p= 0.03), and the High G+C subdivision of the Gram Positive phylum with total aluminum (r^2= 0.70, p= 0.04). Dissolved organic carbon (DOC) concentrations may also influence bacteria through amelioration of aluminum toxicity and as a carbon source. There were significant positive correlations between DOC and the relative abundance of the γ (r^2= 0.66, p= 0.05) and β (r^2= 0.78, p= 0.02) subdivisions and the ACK1 clade (r^2= 0.84, p= 0.01). Additional significant correlations were also noted between specific phylotypes and certain macro- and micro-nutrients. The results of this study indicate that water chemistry can have a direct influence on bacterial lake assemblages and that in acid stressed lakes aluminum chemistry and DOC concentrations may play a particularly important role.

Introduction

Acid deposition, the input of strong mineral acids to terrestrial and aquatic ecosystems, has been a concern in a number of aquatic environments sensitive to perturbations in acid-base chemistry. The Adirondack Mountains, located in upstate New York, receives large inputs of nitrogen and sulfur oxides as a result of emissions from anthropogenic, largely industrial sources, which are located hundreds of kilometers away. Prevailing winds carry these emissions to the Adirondacks, where the acidic compounds are precipitated in the mountainous terrain. Adirondack surface waters and watershed soils lack buffering capacity (acid neutralizing capacity, or ANC) to resist changes in pH resulting from these acidic inputs. Many of the lakes of this region are considered acid stressed and are characterized by decreased pH and ANC.

* Author for correspondence.

The water chemistry of these lakes is also affected by acidification in other ways, for instance, by coincident increases in aluminum concentrations (Baker, 1990a,b).

A variety of effects of acidification on the aquatic biota of Adirondack lakes have been documented, primarily in community structure shifts and reduced biodiversity. Decreases in lakewater pH are known to have direct effects on survivorship and abundance of species of fish, zooplankton and phytoplankton (extensive review in Baker & Schofield, 1982; Sadler & Lynam, 1987; Havens & DeCosta, 1987a, b; Baker et al., 1990b). Although shifts in taxonomic structure due to changes in hydrogen ion or aluminum concentrations have been documented at all trophic levels, effects due to changes in pH have not been documented for bacterioplankton populations. Additionally, bacterial diversity in Adirondack lakes, or any freshwater system, has received little attention (Bahr et al., 1996; Hiorns et al., 1997; Methé et al., 1998). Therefore, it is not known whether acidification causes shifts in bacterial assemblages, as it does at other trophic levels. With the advent of molecular biological techniques, powerful new approaches have been developed for the examination of bacterial diversity in the environment independent of culture based methods.

The characterization of prokaryotic 16S rRNA sequences obtained from a variety of environments has identified a multitude of organisms currently not maintained in culture (Pace, 1997). Novel 16S rRNA sequences have been reported from marine (Giovannoni et al., 1990; DeLong, 1992; Fuhrman et al. 1993; Mullins et al., 1995; Gray & Herwig, 1996; Giovannoni et al., 1996; Bowman et al., 1997), soil (Liesack & Stackebrandt, 1992; Stackebrandt, 1993; Borneman et al., 1996) and extreme (Ekendahl et al., 1994) environments. One of the limitations of interpretation of ribosomal RNA sequence information obtained from the environment, is the lack of understanding of the relationship between the structure and function of specific phylogenetic clades. The identification of the roles and functions of specific phylotypes is key to understanding the importance of phylogenetic diversity to ecosystem function. Very few studies have attempted to document the relationships between phylogenetic composition and environmental conditions. Links between phylogenetic shifts and chemistry changes have been suggested in artificially maintained environments (e.g. bioreactors) (Raskin et al., 1996). At least one study has used an oligonucleotide approach to examine links between phylogenetic clades and marine water chemistry parameters (Ramsing et al., 1996). Such studies that identify the relationships of the composition of microbial assemblages to environmental characteristics will provide a basis for understanding the roles of microbial phylotypes in ecosystem structure and function.

This work attempts to examine the nature of patterns between bacterial diversity, determined by phylogenetic analysis of 16S rRNA gene sequences, and water chemistry in acid stressed lakes in the Adirondacks. The lakes in this study all receive inputs of strong mineral acids from atmospheric deposition (Baker, 1990a). Thus, examining relationships between lake bacterial assemblages and acidification is particularly important in this study.

The most direct measure of acidification in a lake is pH, or the concentration of free hydrogen ions. Changes in pH have been demonstrated to cause direct effects on the viability of aquatic biota (Baker, 1990b). Another important measurement of acidification is ANC, which determines the net strong base in solution (or the net strong mineral acid if the value is negative). As such, this parameter determines the buffering capacity, or ability of water to resist acid inputs (Drever, 1988). As a result, pH and ANC values are frequently significantly, positively correlated.

ANC does not produce direct effects on biota, but the individual components of ANC can (Baker, 1990a). ANC values are equated to the sum of base cation concentrations, minus the sum of mineral acid anion concentrations with all values measured in equivalents. Therefore, these components can also be correlated to ANC. The concentration of base cations such as calcium, often correlate significantly with ANC (Drever, 1988). Individual components of ANC, such as calcium can have direct, and important, effects on aquatic biota.

Perturbations in pH can translate into additional perturbations in other water chemistry parameters. Metals are more soluble in low pH solutions often causing their increase as lake water pH decreases. In acid stressed waters, increases in aluminum concentration are of particular importance.

Dissolved aluminum exists in several different states, which vary in biological activity and toxicity. These fractions are operationally defined by analytical methods. Monomeric aluminum (mono Al) is the fraction which represents the inorganic dissolved aluminum (including free ions and those associated with hydroxides, fluoride, sulfate and silica) and is believed to produce the greatest toxic effects in biota. Non-labile aluminum (NLAl) is that fraction of aluminum which is complexed with organic compounds (e.g. DOC) and therefore, less likely to be biologically active (Baker, 1990a, b).

Other water chemistry variables essential to bacterial metabolism include carbon, nitrogen, phosphorous, and sulfur and micro-nutrients including iron. Other water chemistry characteristics, including the availability of these macronutrients (C, H, N, P, S) and micronutrients (e.g. Fe) are also likely to be important in determining the abundance and diversity of microbial populations. Biotic factors such as predation by protozoa and zooplankton are also likely factors influencing bacterial assemblages, however, their effects were not measured in this study. Nitrate and sulfate are important parameters to consider in this analysis for two reasons. First, they are strong mineral acid anions associated with atmospheric deposition. Second, they can be assimilated by at least some bacteria as nitrogen and sulfur sources and/or used as electron acceptors in dissimilatory reductive processes.

The rapidly expanding use of molecular techniques to study bacteria in the environment has provided a wealth of new information on genetic diversity. In this study, we attempt to discern the patterns of phylogenetic composition of aquatic bacterial communities that are related to differences in water chemistry, with emphasis on acidification.

Materials and methods

Water samples

Water samples were collected from the epilimnion (depth-integrated) at the location of maximum lake depth, from six Adirondack lakes (Table 1) during mid-summer stratification in 1994. All samples were maintained on ice until they were filtered (approximately 0.5 l per filter) through 0.2 μm low protein binding, hydrophilic filters (Gelman Supor, or Whatman), which were subsequently preserved in sterile 10 mM Tris–HCl, 100 mM EDTA, 30 mM NaCl, pH 8 buffer in Whirl Pak bags and frozen at a temperature of no less than −20 °C. Samples were analyzed for pH, ANC (acid neutralizing capacity), chlorophyll *a*,total P, soluble reactive P, total soluble P, nitrate, ammonium, silica, iron, dissolved inorganic carbon, dissolved organic carbon, total aluminum, total monomeric aluminum, nonlabile aluminum, sulfate and major ions (calcium, magnesium, sodium, potassium, chloride). The analytical techniques and Quality Assurance/Quality Control plan were approved by the Environmental Protection Agency (Momen & Zehr, 1998).

DNA extractions

Genomic DNA was extracted with the method of Giovannoni et al. (1990). Briefly, 2 ml ENST buffer (400 mM NaCl, 0.75 mM sucrose, and 50 mM Tris–HCl, pH 9.0) was added to the filter samples in Whirl Pak bags and the filters were subsequently treated to physical manipulation and lysozyme (100μl from a 50 mg/ml stock solution), Proteinase K (50 μl of a 20 mg/ml stock solution) and 100 μl of a 10% w/v stock solution of SDS during a 37 °C incubation of two hours. Genomic DNA was then removed through extractions with phenol:chloroform (1:1) and chloroform, respectively. Genomic DNA was then concentrated by ethanol precipitation.

Cloning and sequencing

The 16S rRNA gene was amplified with the polymerase chain reaction (PCR) using the bacterial domain universal primer sets fd1 and rd2 (Weisburg et al., 1991). Appropriate positive and negative controls were run for each set of reactions. The thermal cycling program was 95 °C for 1 min, 54 °C for 1 min and 72 °C for 2 min for 30–35 cycles, followed by a final 7 min extension step. The resulting product was agarose gel purified using a commercial purification kit (QIAEX II, Qiagen Inc.). Purified product was then ligated into a plasmid (pCR II or pCR 2.1, Invitrogen), and subsequently transformed into a strain of competent *E. coli*. All positive transformants were selected at random from solid medium for further evaluation. Plasmid DNA was purified from positive, randomly selected transformants and the correct size insert verified by a restriction enzyme digest. Cloned 16S inserts were sequenced using a cycle sequencing protocol that is a modification of the Sanger dideoxynucleotide chain terminating reaction (Sanger et al., 1977). A primer complementary to a variable internal region of the gene (*E. coli* positions 522–536) was then used to produce partial sequences of approximately 350–400 nucleotides in length.

Table 1. Characteristics of study lakes. Mean (averaged epilimnetic) water chemistry values for the summers of 1994 and 1995

Lake	Max Depth (m)	Secchi Depth (m)	pH	ANC (μeq l^{-1})	Chl*a* (μg l^{-1})	Total P (μg P l^{-1})	DOC (mg l^{-1})	Nitrate (μg N l^{-1})	Total Al (μg Al l^{-1})
Carry Pond	4.3	2.3–Zmax	4.9	−18.05	2.5	13.0	2.6	18	66
Dart's Lake	16.5	4.3–5.7	5.7	22.1	1.3	6.6	3.5	263	255
Grass Pond	6	4.2–Zmax	5.9	34.9	1.8	6.5	4.3	99	166
Moss Lake	14.1	4.5–8.7	6.8	135.7	1.1	4.8	3.9	227	129
Sagamore Lake	22.5	2.4–4.7	6.1	50.8	1.2	6.1	6.5	293	279
Windfall Pond	5.7	3–4.8	6.9	140.5	6.7	10.0	4.4	271	80

Phylogenetic analysis

Nucleotide sequences were phylogenetically analyzed as described previously (Hiorns et al., 1997; Methé et al., 1998). Briefly, sequences were initially sorted into the broadest taxonomic groups and nearest neighbors determined with the SIMRANK and ALIGN SEQUENCE algorithms of the Ribosomal Database Project (RDP, release 6.1) (Maidak et al., 1997). All taxonomic listings used in this study are based on those of the RDP. Sequences were manually aligned and edited through the use of the Genetic Data Environment (GDE) sequence editor (Smith et al., 1994). Analysis was performed by the construction of phylogenetic trees from the unambiguous nucleotide alignments through the use of tree building software programs such as PHYLIP 3.5 (Felsenstein, 1993) and Treecon for Windows (version 1.3b) (Van de Peer et al., 1994). Representatives of this dataset have been assigned GenBank accession numbers U85098 to U85191 (Hiorns et al., 1997).

Statistical analysis

The percentage represented by each of the broadest taxonomic groups in each of the six lakes was then computed, creating a percentage by lake composition for use in the correlation matrix. The analysis was repeated using narrower, well-defined clades as appropriate. Only clades containing representatives in at least three of the six lakes and whose phylogenetic placement was supported by well-defined branch lengths and/or high bootstrap values were used for further statistical analysis. The water chemistry values used in the matrix correspond to water samples taken concurrently (at the same location and time) with the water samples used for the genomic extractions. Therefore, each lake sampled for water chemistry and phylogeny functions as a replicate in the subsequent statistical analysis.

Pearson Correlation Coefficients were calculated among phylogenetic groups and water chemistry values from the appropriate data sets. The SAS procedure, CORR, was used to compute coefficients of correlation (r) from which subsequent coefficients of determination (r^2) were derived. Probability values (p values) were also computed using the SAS, CORR program. From this initial screening, only r^2 values ≥ 0.60 and p-values ≤ 0.05 were considered significant and chosen for further evaluation. Multiple linear regression including residual analysis was then performed using selected phylogenetic components and water chemistry variables using the SAS procedure, GLM. Important phylogenetic and water chemistry associations were also examined visually by graphing.

Pattern recognition was pursued further by the use of an agglomerative hierarchical cluster analysis. Cluster analysis was performed using the SAS CLUSTER procedure with the Ward's minimum-variance method of distance calculation. This methodology seeks to group (or cluster) similar variables by starting with each variable acting as its own set (or cluster). Subsequently, these individual clusters are combined so as to minimize the ANOVA sum of squares within a cluster in each successive step. Distances between clusters are based on ANOVA sum of squares added over all variables. The clustering patterns are represented visually in the form of a tree graph. In this manner, the lakes can be grouped based on various combinations of water chemistry and phylogenetic composition. Matching clustering patterns based on the phylogenetic information and other variables can be determined by comparison of the resulting tree graphs. The STD option was used to standardize all variables to mean 0 and standard deviation of 1.

Finally, specific clusters of sequences observed within the broad taxonomic classifications were also analyzed with the water chemistry data. Those containing significant numbers and distributions across lakes were treated to the statistical analyses as previously described. If such conditions did not exist, then conclusions based only on the observation of these sequences and their association with water chemistry variables could be drawn.

Results

The objective of this study was to determine whether the composition of bacterial assemblages reflects water chemistry characteristics. Six lakes were chosen to reflect a range of pH and other water chemistry variables. Relationships between sequence assemblages and water chemistry were investigated using statistical methods including correlation and cluster analysis.

Analysis of water chemistry trends

Some water chemistry variables were strongly correlated. Significant positive correlations were calculated between pH and ANC (r^2= 0.88, p= 0.005), and DOC and non-labile aluminum (NLAl), (r^2= 0.71, p= 0.03). DOC was positively correlated with total aluminum (r^2= 0.51, p= 0.11) and monomeric aluminum concentrations (r^2= 0.57, p= 0.08), however these correlations were not considered significant. A significant negative correlation was calculated between nitrate and TSP (r^2= 0.78, p= 0.02). Total iron and chlorophyll *a* concentrations were significantly, positively correlated (r^2= 0.68, p=0.04).

Cluster analysis was used to group the study lakes based on a set of water chemistry variables (pH, ANC, Total Al, mono Al, NLAl, DOC, NO_3^-, TN,TSP, chl*a*, $SO_4^{2-}$and Fe). Two major clusters were obtained. One cluster contained Dart's Lake, Windfall Pond and Moss Lake, whereas the second cluster contained Carry Pond, Grass Pond and Sagamore Lake (Figure 1).

Different clustering patterns were observed when other variables, or combinations of variables, were used for the cluster analysis. The combination of pH and ANC produced two major clusters consisting of Windfall Pond and Moss Lake separate from the remaining lakes, and this same pattern was obtained using total aluminum concentrations (Figure 2). The use of mono aluminum or non-labile aluminum variables either separately, or in combination, produced a different clustering pattern, with Grass Pond and Sagamore Lake forming a cluster distinct from the remaining lakes (Figure 3). Cluster analysis based on DOC produced two major clusters, with Sagamore Lake by itself. This pattern was repeated when DOC and total aluminum were used for the analysis.

Analysis of total phylogenetic composition

Cluster analysis was performed using phylogenetic data to determine whether lakes clustered similarly based on bacterial assemblage data and water chemistry data. Lakes were clustered using phylogenetic information at broad classifications (phyla or subdivision) for each lake. Two major clusters were obtained. The first cluster consisted of Sagamore Lake, Dart's Lake and Grass Pond. The second cluster consisted of Carry Pond, Windfall Pond and Moss Lake (Figure 4).

The clustering pattern based on water chemistry does not match that observed for phylogeny. However, upon closer examination it can be noted that both Windfall Pond and Moss Lake cluster together in each analysis as do Grass Pond and Sagamore Lake. The two lakes which do not cluster consistently between the two analyses are Dart's Lake and Carry Pond.

Verrucomicrobium subdivision

All sequences belonging to the Planctomyces and Relatives phylum were members of the Verrucomicrobium subdivision. This group was not strongly correlated to any of the water chemistry variables analyzed. The sequences obtained formed three clades, designated Verruco1, Verruco2 and Verruco3 (Figure 5a). Verruco1, with representatives from four lakes (Dart's Lake, Moss Lake, Grass Pond and Windfall Pond), as well as sequences obtained from a diversity study of acidified Australian soil (Liesack & Stackebrandt, 1992), was not strongly correlated to any water chemistry variables. Verruco2 consists entirely of sequences from Dart's Lake, while Verruco3 contains two representatives from Carry Pond and one from Grass Pond.

Cytophaga-Flexibacter-Bacteroides (CFB Phylum)

The relative abundance of members of the CFB Phylum was negatively correlated with total Al concentrations (r^2= 0.74, p= 0.03) (Figure 6a). Negative

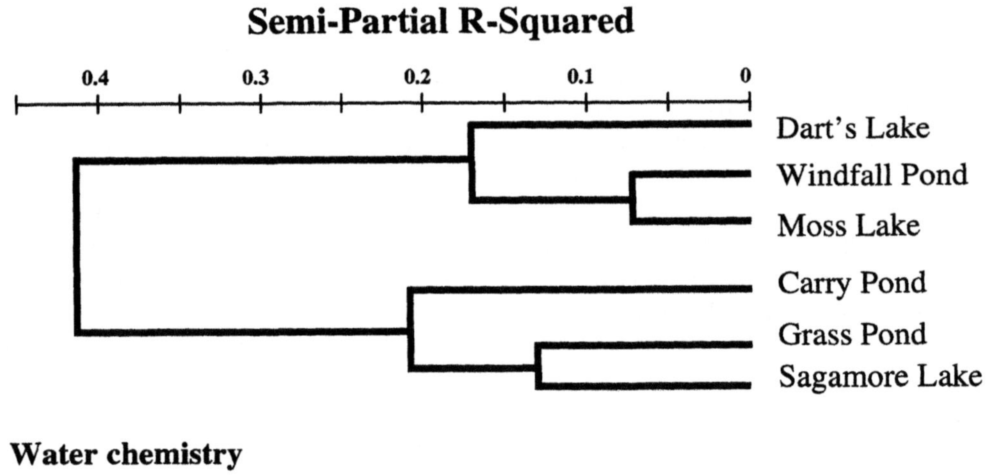

Figure 1. Cluster analysis of study lakes based on the water chemistry data (pH, ANC, Total Al, mono Al, NLAl, DOC, NO_3^-, TN,TSP, chl *a*, SO_4^{2-} and Fe).

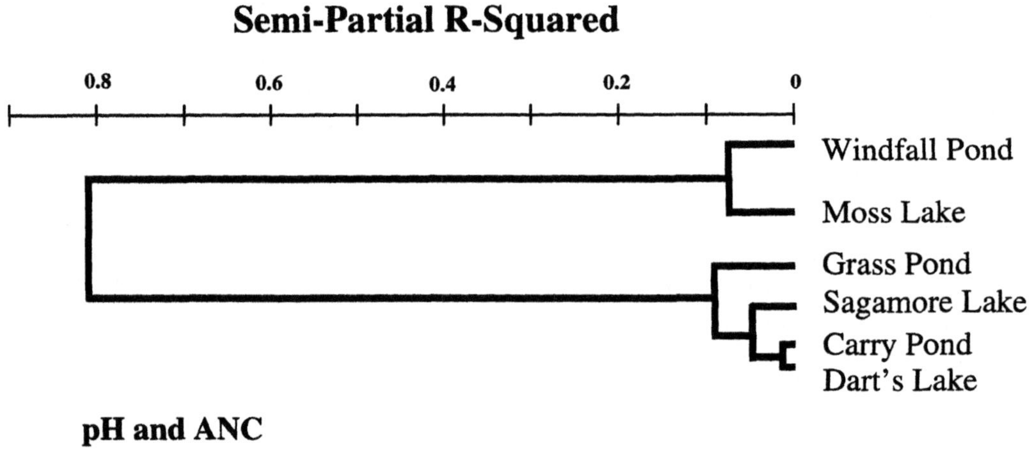

Figure 2. Cluster analysis of study lakes based on pH and ANC. A treegram with the same two major clusters was also obtained using total aluminum concentrations.

correlations with mono Al (r^2= 0.38, p= 0.19) (Figure 6b), non-labile Al (r^2= 0.44, p= 0.15) and DOC (r^2= 0.39, p= 0.19) were not significant.

With the exception of one sequence from Windfall Pond, all CFB sequences clustered in subdivision II. Sequences in the CFB formed two clades (Figure 5b). One clade contained representatives from four lakes, and was designated CFB1. The relative abundance of this clade was not correlated with any of the water chemistry variables. The second clade consisted of two sequences from Moss Lake and one from Windfall Pond. An additional sequence most closely related (of bacteria in culture) to *Prevotella buccalis* of Subdivision I was obtained from Windfall Pond.

α-Proteobacteria (subdivision of the class Proteobacteria)

The relative abundance of the *α-Proteobacteria* was positively correlated with nitrate concentration (r^2= 0.73, p= 0.03) (Figure 7a). Although nitrate is negatively correlated to total soluble phosphorus (TSP) (r^2= 0.79, p= 0.019) (Figure 7b), the relative abundance of the *α-proteobacteria* was not significantly correlated to this variable (r^2= 0.50, p= 0.12). However, the *α-proteobacteria* were positively correlated with sulfate concentrations (r^2= 0.65, p= 0.05).

Six distinct clades of *α-Proteobacteria* sequences, with a minimum of two sequences per clade, were ob-

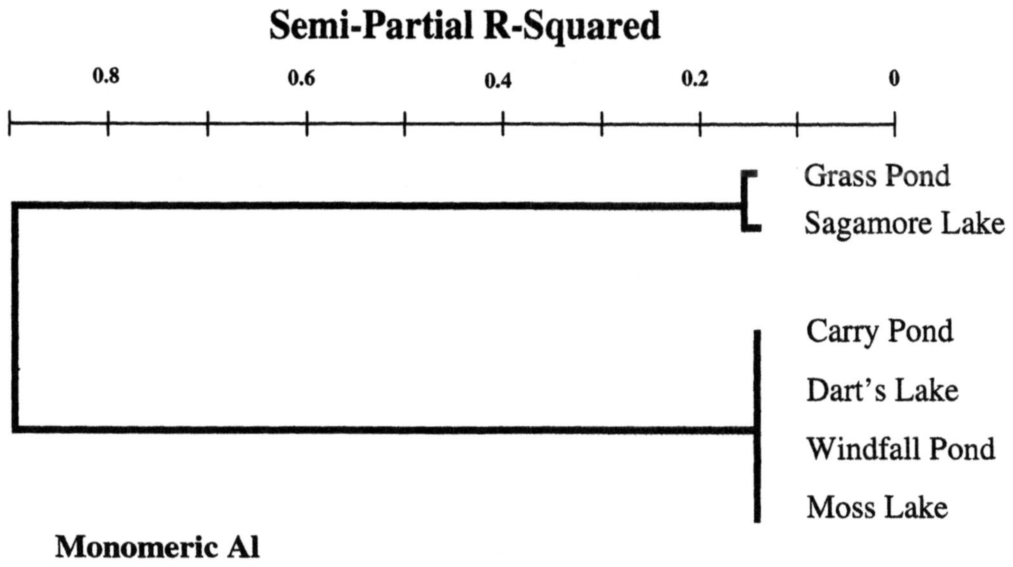

Figure 3. Cluster analysis of study lakes based on monomeric and non-labile aluminum concentrations. A treegram with the same two major clusters was obtained when the relative abundance of ACK1 and *γ-Proteobacteria* were used.

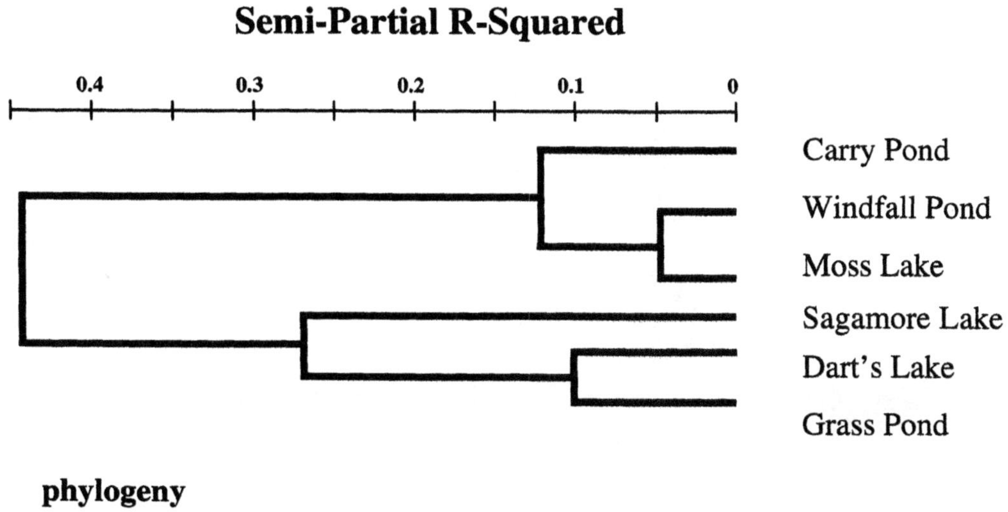

Figure 4. Cluster analysis of study lakes based on the relative abundances of major taxonomic classifications.

tained from the study lakes. Ten sequences were not assigned to a specific clade (Figure 5c). The most numerically abundant clade consists of a clade related to the marine SAR 11 group (Mullins et al., 1995). The freshwater relative clade (designated SAR 11 relatives in this paper) was the only one examined which correlated to ANC (r^2= 0.78, p= 0.02), which is directly related to acidification. Despite the strong correlation between pH and ANC in this data set, only a moderate negative correlation to pH in comparison was observed for this clade (r^2= 0.50, p= 0.12). The SAR 11 relatives are negatively correlated with total aluminum (r^2= 0.75, p= 0.02) and total iron values (r^2= 0.79 p= 0.02), but not with DOC concentration (r^2= 0.36, p= 0.21). In addition, cluster analysis using the relative abundance of SAR11 relatives produced similar patterns as in Figure 1, where the cluster analysis was based on a set of water chemistry variables (pH, ANC, total Al, mono Al, NLAl, DOC, NO_3^-, TN, TSP, chl*a*, $SO_4^{2-}$and Fe).

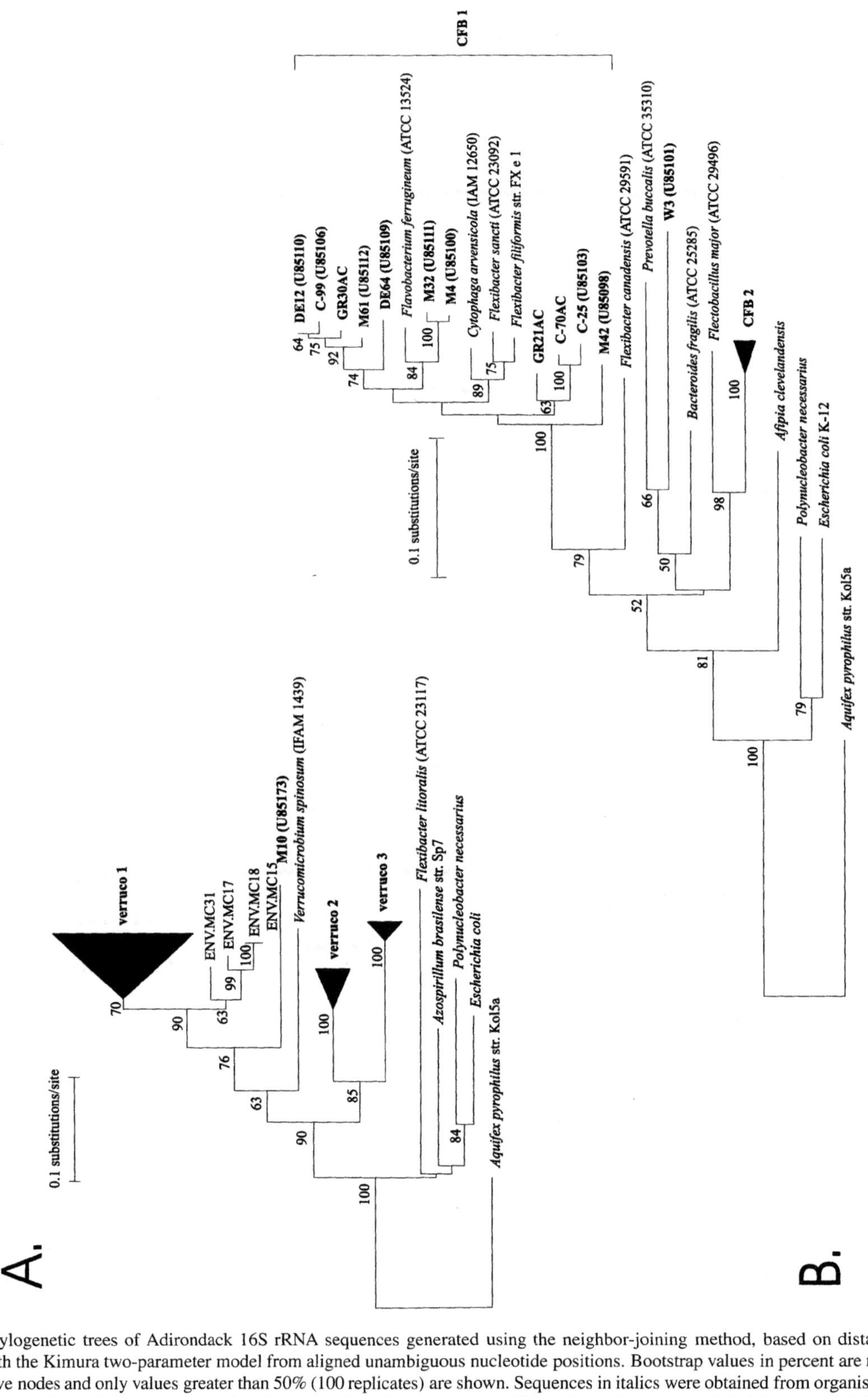

Figure 5. Phylogenetic trees of Adirondack 16S rRNA sequences generated using the neighbor-joining method, based on distance matrices computed with the Kimura two-parameter model from aligned unambiguous nucleotide positions. Bootstrap values in percent are represented at their respective nodes and only values greater than 50% (100 replicates) are shown. Sequences in italics were obtained from organisms in culture. Sequences in type face print represent uncultured bacteria from the environment. Triangles and sequences in bold represent Adirondack lake sequences. All sequences other than those from Adirondack lakes were obtained from the RDP release 6.1 or GenBank. (A) Verrucomicrobium subdivision. (B) CFB phylum.

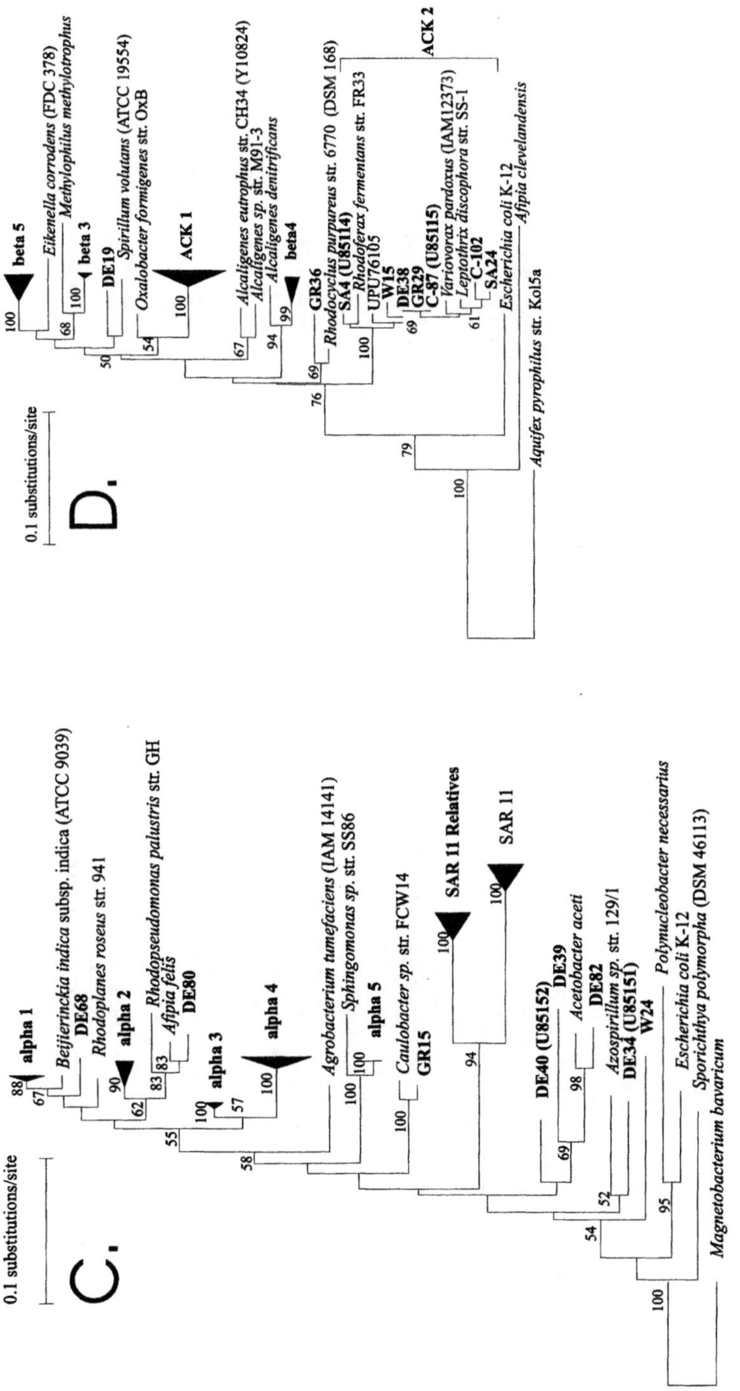

Figure 5. continued (C) α subdivision of the *Proteobacteria*. (D) β subdivision of the *Proteobacteria*.

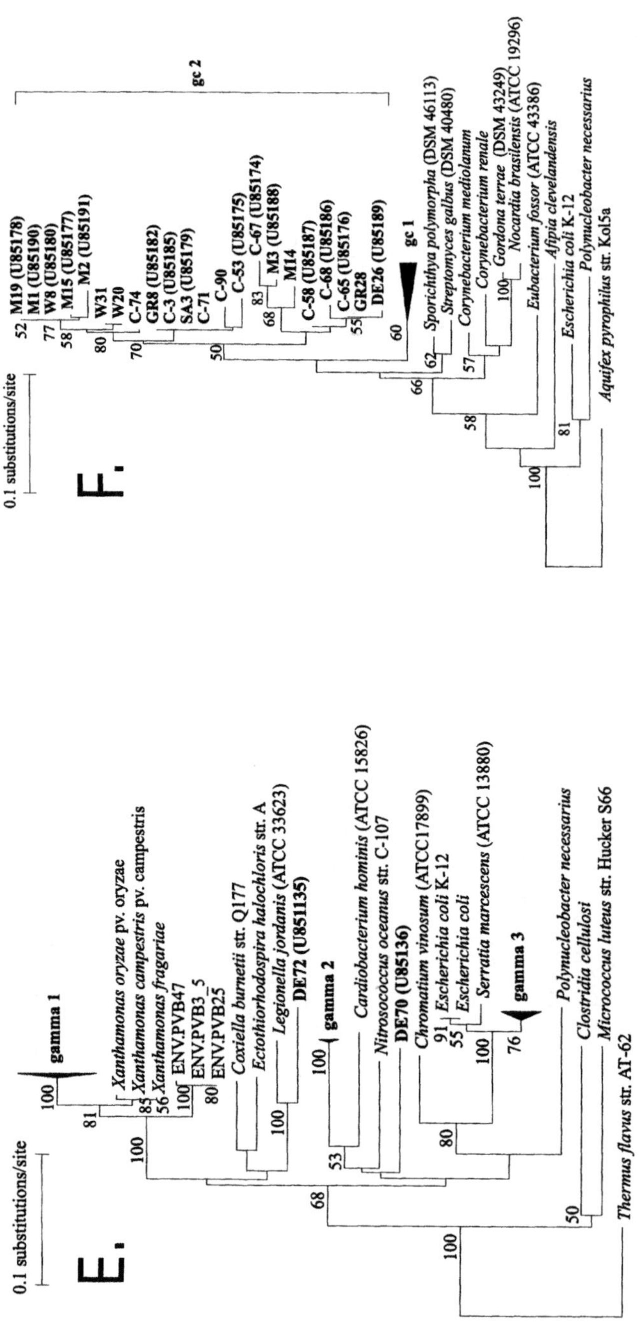

Figure 5. continued (E) γ subdivision of the *Proteobacteria.* (F) High G+C subdivision of the Gram Positive phylum.

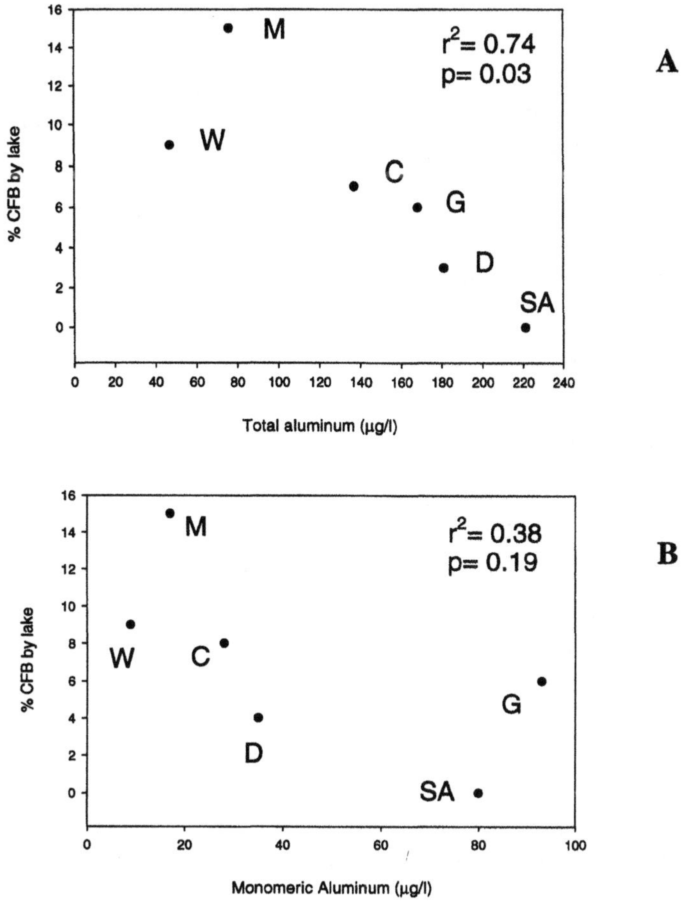

Figure 6. Plot of percent of total number of sequences that are members of the CFB phylum versus (a) total aluminum concentrations (μg/l), and (b) monomeric aluminum concentrations (μg/l).

β-Proteobacteria (subdivision of the class Proteobacteria)

Sequences in the β subdivision of the *Proteobacteria* were the most numerically abundant taxonomic group in the data set (~46% of the total). A further striking result is that the majority of these sequences (~92%) can be placed in one of two phylogenetic clades designated ACK1 and ACK2, respectively (Methé et al., 1998; Hiorns et al., 1997) (Figure 5d). Positive correlations were obtained between the relative abundance of the *β-Proteobacteria* and DOC (r^2= 0.78, p= 0.02), total iron (r^2= 0.79, p= 0.02) (Figure 8a) and chl*a* (r^2= 0.80, p= 0.01) (Figure 8b) concentrations. Cluster analysis of this subdivision produced two distinct groups, with Sagamore Lake separate from the remaining lakes. A similar pattern was produced when the lakes were clustered using DOC concentrations.

Multiple regression analysis, using both DOC and total iron as independent variables and the *β-Proteobacteria* as the dependent variable, produced an adjusted r^2 value of 0.94. Both independent variables were significant to the model with p-values of 0.03, respectively. Results of analysis of the two clades produced different results.

The more diverse of the two clades, ACK2, did not correlate significantly with any of the water chemistry variables. In contrast, the relative abundance of the tightly clustered (sequence similarity > 97% over 350 nt) and numerically dominant ACK1, was positively correlated with several variables. The strongest correlation was between ACK1 and DOC (r^2= 0.84, p= 0.01) (Figure 9a). Positive correlations were also noted between ACK1 and monomeric (r^2= 0.71, p= 0.03) (Figure 9b) and non-labile aluminum fractions (r^2= 0.73, p= 0.03). These aluminum chemistry variables were not significantly correlated with each other in this data set (r^2= 0.57, p= 0.084) in comparison.

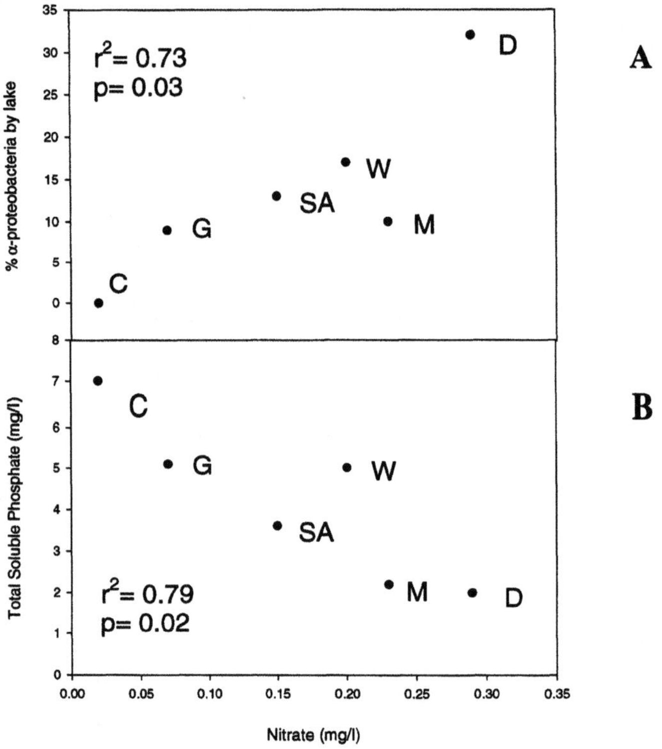

Figure 7. Plot of percent of total number of sequences that are members of the α subdivision of the *Proteobacteria* versus (a) nitrate concentrations (mg/l), (b) total soluble phosphate concentrations (mg/l) versus nitrate concentrations (mg/l).

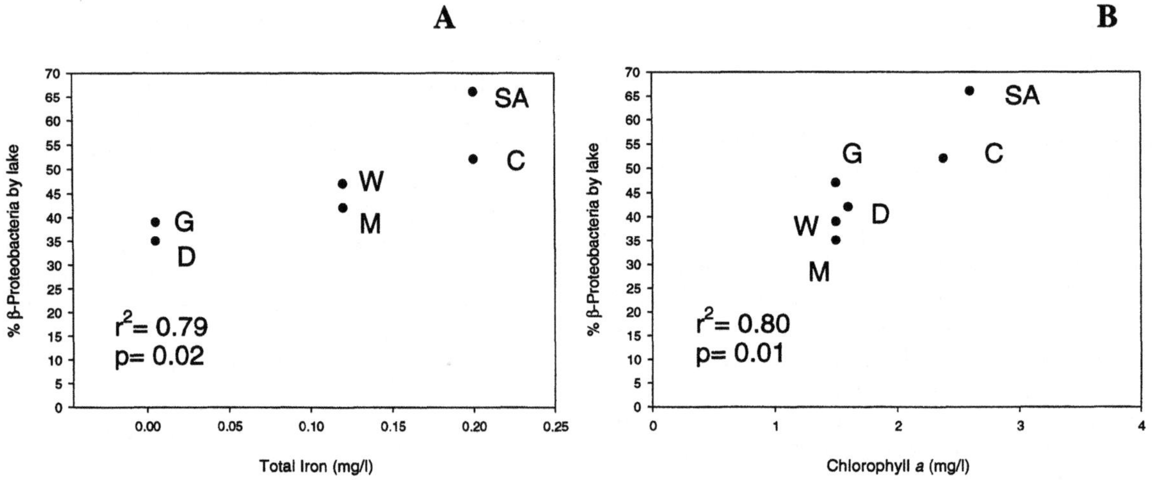

Figure 8. Plot of percent of total number of sequences that are members of the β subdivision of the *Proteobacteria* versus (a) total iron concentrations (mg/l), (b) chlorophyll *a* concentrations (mg/l).

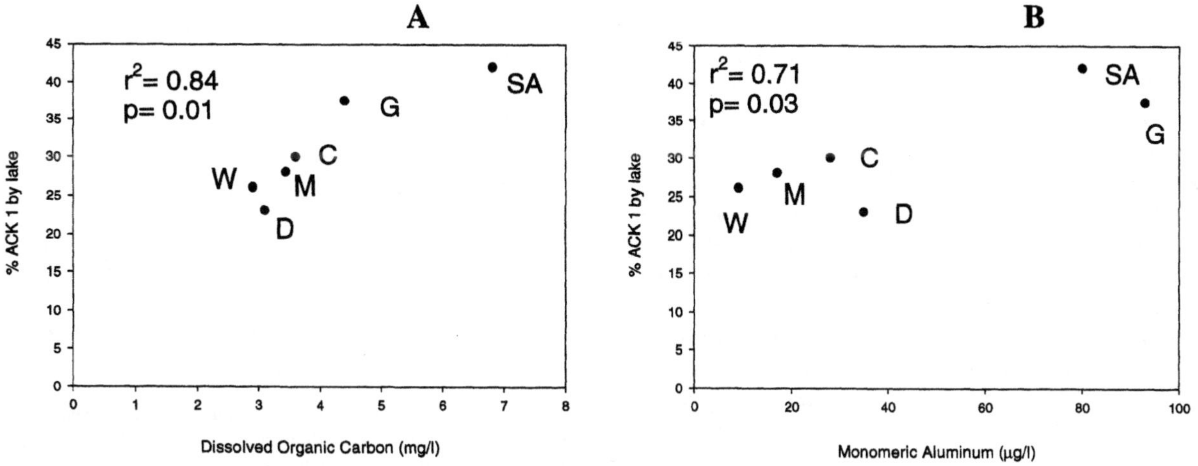

Figure 9. Plot of percent of total number of sequences that are members of the ACK1 clade of the β subdivision of the *Proteobacteria* versus (a) dissolved organic carbon concentrations (mg/l), (b) monomeric aluminum concentrations (μg/l).

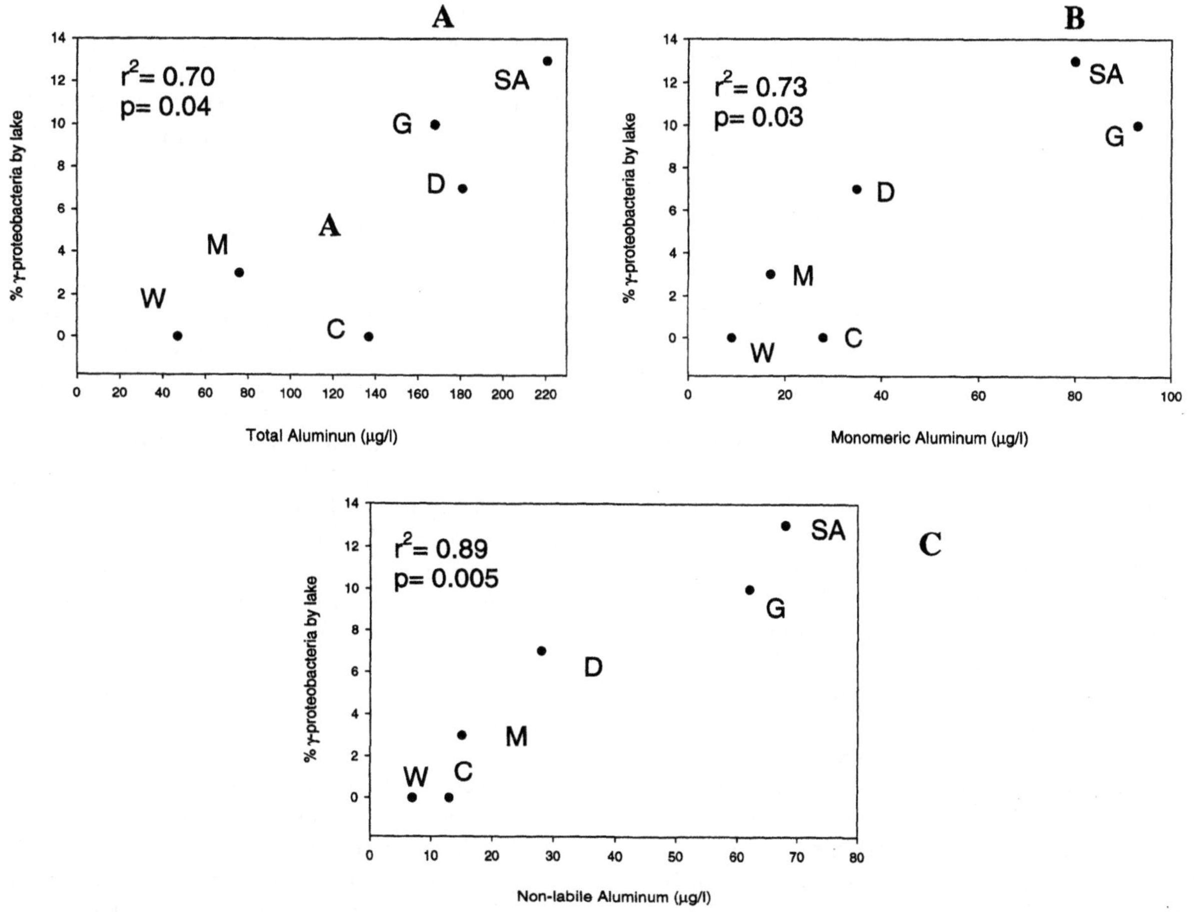

Figure 10. Plot of percent of total number of sequences that are members of the γ subdivision of the *Proteobacteria* versus (a) total aluminum concentrations (μg/l), (b) monomeric aluminum concentrations (μg/l), (c) non-labile aluminum concentrations (μg/l).

The abundance of ACK1 was not correlated with total aluminum (r^2= 0.35, p= 0.22). Cluster analysis of the lakes based on ACK1 relative abundance produced two major clusters, with Grass and Sagamore separated from the other lakes. This pattern was repeated when cluster analysis was performed using the relative abundance of the ACK1 clade and the γ subdivision of the *Proteobacteria.* A matching treegram was produced when the lakes were clustered based on monomeric and non-labile aluminum values together, as well as when each was analyzed separately. Although significant positive correlations were noted between the β-*Proteobacteria* and total iron and chl*a* values, they were not present when these water chemistry values were compared with either the ACK1, or ACK2 clades.

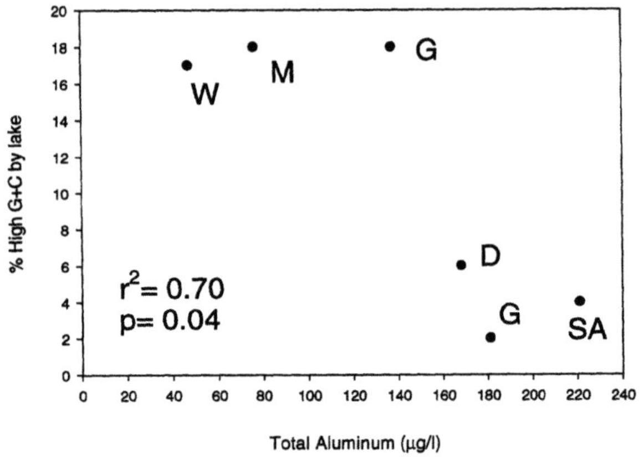

Figure 11. Plot of percent of total number of sequences that are members of the High G+C subdivision of the Gram Positive Phylum versus total aluminum concentrations (μg/l).

γ-Proteobacteria (subdivision of the class Proteobacteria)

The relative abundance of γ-*Proteobacteria* sequences was positively correlated with DOC (r^2 = 0.66, p= 0.05). The relative abundance of this group was also correlated with total aluminum concentration (r^2 = 0.70, p= 0.04) (Figure 10a), as well as monomeric, aluminum (r^2= 0.73, p= 0.03) (Figure 10b) and non-labile aluminum (r^2= 0.89, p= 0.005) (Figure 10c).

Sagamore Lake contained the highest percentage of γ-*Proteobacteria* sequences (3/24, 13%) of all the study lakes. Each of the γ-*Proteobacteria* sequences analyzed from this lake consisted exclusively of members that are close relatives of the enteric bacteria, *E. coli.* These enteric relatives were only found in Sagamore Lake in this analysis (although they were obtained from Moss Lake in June of 1995, which was not a part of this study).

The γ-*Proteobacteria* were next most abundant on a percentage basis in Grass Pond (9%, 3/32). The phylogenetic placement of these γ-*Proteobacteria* sequences stands in contrast to the γ-*Proteobacteria* from Sagamore Lake. Two of the γ-*Proteobacteria* sequences from Grass Pond belong in a cluster in the *Xanthomonas* group with two sequences from Dart's Lake, the lake in which the γ-*Proteobacteria* were the third most abundant based on percentage of the lake total (7%, 4/57). All members of this cluster are similar to each other (>98% over 350 nt). Interestingly, this clade shares a common ancestor from which branched a close, but distinct clade of organisms of marine origin (from Pele's Vents in the Pacific Ocean) (Moyer et al., 1995), and appear to have most recently branched from a group of *Xanthamonads* found in culture. The third sequence from Grass Pond is located in the *Ectothiorhodospira* assemblage and is 99% similar to the lone sequence found in Moss Lake. The remaining two sequences from Dart's Lake are nearest neighbors of the Legionella and Sulfur Oxidizing Symbiont groups, respectively (Figure 5e).

High G+C subdivision (of the Gram Positive Phylum)

Sequences from the Gram Positive Phylum were placed in two sub-classifications, Clostridia and Relatives and the High G+C subdivision. Clostridial relatives were found in only two lakes and were very low in abundance therefore, they were not considered for further evaluation. Abundance of sequences from the High G+C subdivision (Figure 5f) were negatively correlated with total aluminum concentration (r^2= 0.70, p= 0.04) (Figure 11). They were not correlated to monomeric aluminum (r^2= 0.50, p= 0.12), non-labile aluminum (r^2= 0.59, p= 0.08) or DOC concentration (r^2= 0.24, p= 0.32).

Discussion

The objective of this study was to relate the phylogenetic composition of freshwater bacterial assemblages to the chemical characteristics of a set of lakes that range in pH and other water chemistry variables. Statistical techniques were used to identify patterns among 16S rRNA phylotype abundances and a comprehensive water chemistry data set. The identified patterns suggest potential relationships between water chemistry and the composition of lake water bacterial populations. A primary goal of this study was to determine whether anthropogenic acidification, which is an important influence in the study region, has affected the composition of freshwater bacterial assemblages.

Bacterial phylogenetic classification, based on 16S rRNA sequences, define clusters of evolutionarily related organisms. The classification scheme defines major phylogenetic clades, such as the subdivisions of the *Proteobacteria*, containing sequences that are not necessarily closely related, but do share a common ancestor. The species represented by these higher order clades, equivalent to Class or Order, are diverse in terms of metabolic capabilities, and thus do not generally form obvious physiological groups. Despite this, several correlations between these higher order groups and specific water chemistry characteristics were observed. Correlations were also calculated between water chemistry variables and less diverse clades located within these higher taxonomic classifications.

pH and Acid Neutralizing Capacity (ANC)

In this study, no significant linear correlation between pH and bacterial class, subdivision, or order was found, nor for any of the discrete subgroups of sequences found in Adirondack lakes. The only phylotypes that were correlated with ANC, were the α-*Proteobacteria* relatives of the SAR 11 cluster, a distinct group with member sequences possessing >97% (over 350 nt) similarity to one another.

ANC in and of itself does not produce direct effects on biota, however, its concentration (or the lack thereof) is indicative of acidification. Although ANC is often highly correlated with pH, only a modest positive correlation was observed between this variable and the relatives of the SAR11 clade. This finding suggests that any effects on this group due to acidification may be due in larger part to changes in base cation concentrations such as calcium which contribute to ANC, rather than the free hydrogen ion content of the lake water. Changes in base cation composition, especially calcium, have been determined to provide an ameliorative effect on aluminum toxicity, thereby enhancing the survival of some species at other trophic levels (Baker, 1990b).

Relatives of the SAR11 cluster have been found to exist in other freshwater environments. Sequences bearing a percent similarity of >97% to those from Adirondack lakes have been found in Lake George, NY, an oligotrophic Adirondack lake not receiving acid deposition (Methé et al., 1998), Toolik Lake, Alaska, an arctic oligotrophic lake (Bahr et al. 1996) and Lake Loosdrecht, a eutrophic lake in the Netherlands (Zwart et al. 1998). Each of these aquatic environments, while chemically different, is characterized as 'softwater' and circumneutral in pH, and therefore, possess high ANC and low aluminum concentrations.

DOC and aluminum chemistry

Acidification affects other water chemistry variables, which in turn could influence bacterial assemblage composition. One important variable affected by acidification in Adirondack waters is aluminum, which is solubilized from clay minerals from watershed soils as pH is depressed (Cronan & Schofield, 1979). DOC concentration not only reflects the biologically available carbon pool, but also the presence of a potential mitigating factor in aluminum toxicity. This amelioration is due to functional groups from DOC that are capable of reacting with aluminum, thereby rendering it less biologically active. Reductions in aluminum toxicity to biota at other trophic levels have been observed in previous investigations (Baker & Schofield, 1980; Peterson et al., 1989).

The relative abundances of the γ- and β-*Proteobacteria* were positively correlated with DOC concentrations. A significant positive correlation was also observed between the relative abundance of the ACK1 clade and DOC. It is not clear what the reason is for this correlation, but many β-*Proteobacteria* (e.g. *Sphaerotilus natans*) have been found to be associated with high DOC environments, including activated sludge wastewater treatment systems (Wagner et al., 1994).

The significant positive correlations between the γ-*Proteobacteria* and the ACK 1 clade and non-labile aluminum, raises the possibility that DOC may reduce toxic effects of aluminum. It cannot be concluded if the correlations between DOC and non-labile Al with the γ-*Proteobacteria* and the ACK 1 clade are independent correlations, or are merely due to the colinearity between DOC and non-labile Al. However, the association of non-labile aluminum with each of these groups is stronger than that observed between non-labile aluminum and DOC.

The positive correlations between the γ-*Proteobacteria* and each of the aluminum chemistry variables, as well as those between the ACK 1 clade and monomeric and non-labile aluminum, suggest several possibilities. First, the effect of aluminum on this bacterial population could be neutral. This result, however, would more likely be expected to occur in those phylogenetic classifications lacking significant correlations either negatively or positively with these variables.

Second, there is no known biological requirement for aluminum (Ehrlich, 1990) therefore, a significant positive correlation especially with monomeric aluminum suggests that these bacterial groups may possess a mechanism for detoxification of this metal conferring a selective advantage on them. Less information has been published concerning the reactions of aluminum with bacteria, in comparison to other metals. However, various detoxification methods have been postulated including intracellular and extracellular chelation, expulsion from the cytoplasm, and chemical transformations (Pina & Cervantes, 1996). A microscopy based investigation of the ultrastructure of natural biofilm microcolonies from a riverine system revealed that the matrix surrounding cells in the colonies formed a selective barrier to the intrusion of metals including aluminum into the cytoplasm (Lunsdorf et al., 1997). The water samples obtained in our study were not pre-screened prior to bacterial cell collection by filtration. Therefore, they would contain larger particles capable of supporting attached growth of bacteria. Some members of the *β-Proteobacteria* are known to form extracellular matrices (a glycocalyx). Examples of sheath formers include *Sphaerotilus natans* and *Leptothrix* sp. that are capable of depositing metal ions in their sheaths and participate in biofilm formations. Therefore, if the ability to produce a glycocalyx and form aggregates are metabolic and physiological capabilities possessed by members of the ACK1 clade and *γ-Proteobacteria*, then they may potentially provide a protection mechanism against aluminum toxicity. An observation of abundance of *β-Proteobacteria* with lake snow aggregates was noted in a study using fluorescently-labeled oligonucleotide probes in a European freshwater lake (Weiss et al., 1996).

Recently, a member of the *β-Proteobacteria*, *Alcaligenes eutrophus* CH34, a strain closely related to the ACK 1 clade, was discovered to possess a plasmid encoding resistance genes for Co^{2+}, Zn^{2+} and Cd^{2+}. Organisms possessing this plasmid are capable of actively expelling these metal cations from their cytoplasm (Diels et al., 1995; Van der Lelie et al., 1997). Metal resistance genes have also been ascribed to members of the *γ-Proteobacteria* including copper resistance genes in *Pseudomonas syringae* (Mills et al., 1993) and*Xanthamonas campestris* (Lee et al., 1994). Therefore, it is possible that the positive correlations between these groups and aluminum chemistry reflect their possession of gene(s) conferring resistance to aluminum.

The correlations of DOC with the *γ-* and *β-Proteobacteria* and the ACK 1 clade suggest that these organisms may benefit from increased DOC levels due to the greater availability of carbon for metabolism. The increased relative abundance of the *γ-Proteobacteria* and ACK 1 clade with both DOC and non-labile aluminum suggest that these groups additionally benefit from increased DOC due to its mitigation of aluminum toxicity.

A significant negative correlation between total aluminum levels and the CFB phylum, High G+C subdivision and SAR 11 relatives raises the possibility of adverse toxicity effects on these populations. The correlations for each of these groups with DOC was negative and not significant in contrast to the DOC correlations calculated for the ACK1 and *β-* and *γ-Proteobacteria.* These comparisons suggest that the CFB phylum, High G+C subdivision and SAR 11 relatives may be more sensitive to toxic effects due to increased aluminum; even increasing DOC levels are not capable of ameliorating this toxicity.

The diverse phylogenetic placement of sequences within the *γ-Proteobacteria* and CFB raises the possibility that relationships encountered between these bacteria are not exclusive to a narrowly defined clade. Instead, they may be widespread and indicative of these higher level taxonomic classifications. The possible relationships encountered between the less diverse, High G+C, ACK1 and SAR 11 relatives suggest a more limited range of these associations.

*Total iron and chl*a

The strength of the positive correlations between the ACK1 and ACK2 clades with iron and chl*a* levels were greatly reduced when compared to those calculated using the *β* subdivision as a whole. Total iron and chlorophyll *a* concentrations were positively correlated, therefore, it is not possible to determine if one of these correlations is merely the result of this colinearity, or if each is independent in its occurrence.

However, it is interesting to note that a number of bacteria phylogenetically placed within the *β-Proteobacteria* are capable of producing an extracellular glycocalyx in which iron (and manganese whose concentration was not measured in this study) can be deposited. It has been determined that the β-*Proteobacteria*, *Leptothrix discophora* SS-1 possess an iron oxidizing protein (Corstjens et al., 1992). This metabolic ability may be relatively widespread among the β-*Proteobacteria* in this study, accounting for this positive correlation. Interestingly, previous reports have suggested that aluminum uptake in bacteria may involve iron transport systems (Davis et al., 1971). Two studies of bacteria in culture strengthened this idea, one suggesting that aluminum toxicity may be exacerbated by the absence of iron from the growth media (Guida et al., 1992) and that porphyrin synthesis and excretion increased with elevated aluminum concentration (Scharf et al., 1994).

A positive correlation with chloropyll *a* concentration could be indicative of an association between members of the β-*Proteobacteria* and phytoplankton. Phytoplankton are known to produce extracellular products including dissolved organic carbon compounds which can in turn be utilized by heterotrophic bacteria (Chen & Wangersky, 1996; Freirenordi et al., 1996).

Nitrate and sulfate

The strong positive correlation between the α-*Proteobacteria* and nitrate represents the only association with a nutrient which was significant. As a nutrient, nitrate can be assimilated directly by some bacteria, providing a nitrogen source. This compound can also serve as a terminal electron acceptor in respiration. Although this process generally occurs under anaerobic conditions, some bacteria are capable of denitrification at low oxygen tensions (Knowles, 1982) which could occur in particle-associated microenvironments. In this data set, nitrate is significantly negatively correlated with TSP. By comparison, a moderate negative correlation was calculated between TSP and the relative abundance of the α-*Proteobacteria*, lending credence to the importance of the nitrate correlation.

The α-*Proteobacteria* also correlated positively with sulfate. Bacteria do possess a biological requirement for sulfur, with sulfate serving as the principal source. Like nitrate, sulfate can be used by certain bacteria as a terminal electron acceptor in respiration. However, this process occurs in anaerobic environments and is not likely to occur in an oxygenated water column.

Both nitrate and sulfate are major components of the strong mineral acids associated with anthropogenic acid deposition. Collectively, the correlations of the α-*Proteobacteria* with these variables suggest an indirect effect of acidification on this subdivision, although the reason for this effect is unknown. Despite correlations with these strong mineral acids, this group was not significantly correlated with pH and in fact was completely absent from the lake with the lowest pH in this data set (Carry Pond, pH= 4.85).

Observations of individual sequences and small clusters with water chemistry

Examination of Adirondack lake phylotypes revealed an overall polyphyletic placement of sequences. These placements were not random and therefore suggest the importance of factors such as water chemistry in the determination of a lake's bacterial assemblage. At a finer level of phylogenetic detail (at the "tips" of the branches), it was observed that for each lake in this study, there were sequences which were only obtained in that lake. The Verruco2 clade is an example consisting entirely of sequences from Dart's Lake. It was also observed that some clades exist which are numerically rare and consist of representatives from one to three of the lakes in this study. The ACK1 and ACK2 clades were the only ones in which representatives from all six lakes could be found. Observations such as these do raise the possibility that there may be certain members of a lake that are present only in that lake, or a limited set of lakes. Due to their lack of numerical representation and distribution, it is difficult to apply statistical analyses to these small clusters, however it is possible to make observations concerning their associations with their lake water chemistry.

In the CFB phylum, the group designated CFB2 consists of two sequences from Moss Lake and one from Windfall Pond. These two lakes are higher in pH and ANC and lower in aluminum concentrations compared to the others. Therefore, this cluster could represent a group of organisms sensitive to acidification. A similar situation is observed in the High G+C subdivision. Within the cluster designated gc2, there are four sequences (approximately 99% similar) consisting of three sequences from Moss Lake and one from Windfall Pond, which could again represent bacteria sensitive to acidification.

In contrast, the β-*Proteobacteria* contain a group of three sequences similar to one another which cluster together consistently (bootstrap values > 95%), and are distant from other members of this subdivision (C30, C51 and C86). These sequences could represent organisms adapted to decreased pH levels. Alternatively, Carry Pond is consistently low in nitrate and these sequences could reflect this relationship.

Within the *α-Proteobacteria*, the clade designated Alpha1 contains three similar sequences from two different lakes (DE48, DE 81 and M37), with this placement supported by consistently high bootstrap values (>95%). These bacteria could reflect ones adapted to increased nutrient levels as each lake is elevated in nitrate and they share the same high total soluble phosphate values. Moss Lake drains into Dart's Lake, and therefore, this cluster of sequences could reflect the close proximity of the two lakes. Interestingly, based on cluster analysis of the overall phylogenetic composition, these two lakes are found in opposite clusters suggesting that the bacterial assemblages contained in these two lakes are different despite the shared drainage system.

Within the order, *Verrucomicrobiales*, the clusters Verruco2 and Verruco3 represent an example suggesting the development of different bacterial lake populations over geologic time scales. Verruco2 consists entirely of sequences (3 total) from Dart's Lake, while Verruco3 is represented by two Carry Pond sequences and one Grass Pond sequence. Phylogenetic analysis reveals that these two clades are sister taxa that diverged from a common ancestor. The relationship of these two clades might also be related to differences in lake water chemistry. Based on cluster analysis of lakes using nitrate values, Carry Pond and Grass Pond form a distinct major cluster based on low nitrate values. In contrast, Dart's Lake contains the highest nitrate lake value in the data set and is consistently elevated in this parameter throughout the summer.

Other clusters possess less obvious potential explanations. The clusters designated Gamma1 and Gamma2 are such examples. Gamma1 contains two Dart's Lake and two Grass Pond sequences all highly similar to one another (~99% over 350 nt). However, chemically each of these lakes is dissimilar when a number of parameters such as pH, nitrate and TSP are compared individually or in combination using cluster analysis. These two lakes appear in opposite groups when cluster analysis was performed using the comprehensive water chemistry set. They are more similar in terms of aluminum chemistry and DOC, variables to which positive statistical correlations have been calculated at the subdivision level. A similar situation to that just described exists when examining Gamma2, which consists of one representative each from Grass Pond and Moss Lake. No obvious additional explanations due to water chemistry can be offered at this level of analysis.

Conclusions

This study examined associations between phylogenetically determined bacterial groups from freshwater environments based on 16S rDNA sequences with water chemistry characteristics. Central to the study was elucidating the nature of the relationships of variables indicative of acidification to the phylogenetic composition. The results reveal that while causal relationships cannot be inferred from such a study, it is possible to determine where associations between variables exist, providing a framework on which to base future studies examining causality.

The direct influence of acidification on bacterial phylogenetic composition could not be detected, especially when higher taxonomic levels were examined. At narrower levels of sequence classification, clusters of sequences which could represent associations to changes in pH were observed.

The results of this study indicate that a direct linear effect of water chemistry on the phylogenetic composition of lake bactrerioplankton is not likely to be observed. Instead, acidification effects on phylogenetic composition are more likely to be indirect, and involve complex associations with water chemistry influenced by perturbations in acid-base water chemistry, or through effects on other trophic levels. This result is suggested by significant correlations, both negatively and positively, with aluminum chemistry. DOC concentrations also appear to be an important influence on bacterial populations and the interaction with aluminum may be significant. Elevations in nitrate and sulfate concentrations which can be indicative of acidification, were associated with only one *Proteobacteria* subdivision. These associations were the only examples where nutrient variables were significantly associated with a phylogenetic component.

The results of this study suggest that the water chemistry of a lake can have a significant influence on bacterial diversity and abundance. The influence of acidification appears to be an indirect one resulting from the host of changes it subsequently produces in lake water chemistry characteristics.

The findings of this study represent an initial foray into the understanding of potential relationships between bacterial phylogenetic composition and lake-water chemistry. This work reveals both the benefits and pitfalls of such a study. The use of sequence information to determine the phylogenetic composition of lake bacterioplankton provides the greatest detail and is necessary when examining an environment whose genetic diversity is unknown. However, the use of such an approach necessarily limits the amount of replication, specifically the number of lakes studied, in turn limiting the statistical power and methodologies that can be employed. The results at this basic level of statistical examination provide a framework for determining relationships that bear further, more rigorous evaluation. The sequence database obtained provides the necessary information to design oligonucleotide probes to increase sample throughput and allow for greater statistical replication. The analysis of larger numbers of samples will then allow the statistical analysis of the potential relationships of phylogenetic composition to water chemistry that have been suggested by the analysis presented here.

Acknowledgements

This study is part of the Adirondack Effects Assessment Program which is supported by the U.S. Environmental Protection Agency (contract number 68D20171). Although this work has been funded by the U.S. Environmental Protection Agency, it has not been subjected to the Agency's review and, therefore, does not necessarily reflect the views of the Agency, and no official endorsement should be inferred. We are greatly indebted to our colleagues in the Adirondack Project, particularly James Sutherland, Larry Eichler and Jennifer Slater for their dedication and expertise in the field and laboratory water chemistry program, and Bahram Momen for his advice and encouragement. We thank S. Nold and M. Frischer for their critical reviews.

References

Bahr, M., J. E. Hobbie & M. L. Sogin, 1996. Bacterial diversity in an arctic lake: a freshwater SAR 11 cluster. Aquat. microb. Ecol. 11: 271–277.

Baker, J. P. & C. L. Schofield, 1982. Aluminum toxicity to fish in acidic waters. Wat. Air Soil Pollut. 18: 289–309.

Baker, J. P., S. A. Gherini, S. W. Christensen, J. Gallagher, R. K. Munson, R. M. Newton, K. H. Reckhow & C. L. Schofield, 1990a. Adirondack Lakes Survey: an interpretive analysis of fish communities and water chemistry, 1984–87. Adirondack Lakes Survey Corporation, Ray Brook, NY, U.S.A.

Baker, J. P., D. P. Bernard, S. W. Christensen, M. J. Sale, J. Freda, K. Heltcher, D. Marmorek, L. Rowe, P. Scanlon, G. Suter, W. Warren-Hicks & P. Welbourn, 1990b. Biological Effects of Changes in Surface Water Acid-Base Chemistry. NAPAP Report 13. In National Acid Precipitation Assessment Program, Acidic Deposition: State of Science and Technology. Volume II.

Borneman, J., P. W. Skroch, K. M. O'Sullivan, J. A. Palus, N. G. Rumjanek, J. L. Jansen, J. Nienhuis & E. W. Triplett, 1996. Molecular microbial diversity of an agricultural soil in Wisconsin. Appl. envir. Microbiol. 62: 1935–1943.

Chen, W. H. & P. J. Wangersky, 1996. Rates of microbial degradation of dissolved organic carbon from phytoplankton cultures. J. Plankton Res. 18: 1521–1533.

Corstjens, P. L. A. M., J. P. M. De Vrind, P. Westbroek & E. W. De Vrind-De Jong, 1992. Enzymatic Iron Oxidation by *Leptothrix discophora*: identification of an iron-oxidizing bacteria. Appl. envir. Microbiol. 58: 450–454.

Cronan, S. S. & C. L. Schofield, 1979. Aluminum leaching response to acid precipitation: effects on high-elevation watersheds in the northeast. Science 204: 304–306.

Davis, W. B., M. J. McCauley & B. R. Byers, 1971. Iron requirements and aluminum sensitivity of hydroxamic requiring strain of *Bacillus megaterium*. J. Bact. 105: 589–594.

DeLong, E. F., 1992. Archaea in Coastal Marine Environments. Proc. natn. Acad. Sci. U.S.A. 89: 5685–5689.

Diels, L., Q. Dong, D. van der Lelie, W. Baeyens & M. Mergeay, 1995. J. Ind. Microbiol. 14: 142–153.

Ehrlich, H. L., 1990. Geomicrobiology. 2nd edn. New York City: Marcel Dekker, Inc. 646 pp.

Ekendahl, S., J. Arlinger, F. Stahl & K. Pedersen, 1994. Characterization of attached bacterial populations in deep granitic groundwater from the Stripa research mine by 16S rRNA gene sequencing and scanning electron microscopy. Microbiology 140: 1575–1583.

Felsenstein, J., 1993. PHYLIP-phylogeny inference package (version 3.5c). Department of Genetics, University of Washington, Seattle. Distributed by the author.

Freirenordi, C. S. & A. A. H. Vieira, 1996. Utilization of extracellular polysaccharides from Ankistrodesmus Densus (Chlorophyceae) by heterotrophic bacteria. Revista De Microbiologia 27: 27–32.

Fuhrman, J. A., K. McCallum & A. A. Davis, 1993 Phylogenetic diversity of subsurface marine microbial communities from the Atlantic and Pacific oceans. Appl. envir. Microbiol. 59: 1294–1302.

Giovannoni, S. J., E. J. DeLong, T. M. Schmidt & N. R. Pace, 1990. Tangential flow filtration and preliminary phylogenetic analysis of marine picoplankton. Appl. envir. Microbiol. 56: 2572–2575.

Giovannoni, S. J., M. S. Rappe, K. L. Vergin & N. L. Adair, 1996. 16S rRNA genes reveal stratified open-ocean bacterioplankton populations related to the green non-sulfur bacteria. Proc. natn. Acad. Sci. U.S.A. 93: 7979–7984.

Gray, J. P. & R. P. Herwig, 1996. Phylogenetic analysis of the bacterial communities in marine sediments. Appl. envir. Microbiol. 62: 4049–4059.

Guida, L., Z. Saidi, M. N. Hughes & R. K. Poole, 1991. Aluminum toxicity and binding to *Escherichia coli*. Arch. Mikrobiol. 156: 507–512.

Havens, K. E. & J. Decosta, 1987a. The role of aluminum contamination in determining phytoplankton and zooplankton responses to acidification. Wat. Air Soil Pollut. 33: 277–293.

Havens, K. E. & J. DeCosta, 1987b. Freshwater plankton community succession during experimental acidification. Arch. Hydrobiol. 111: 37–65.

Hiorns, W. D., B. A. Methé, S. A. Nierzwicki-Bauer & J. P. Zehr, 1997 Bacterial diversity in Adirondack mountain lakes as revealed by 16S rRNA gene sequences. Appl. envir. Microbiol. 63: 2957–2960.

Knowles, R., 1982. Denitrification. Microbiol. Rev. 46: 43–70.

Lee, Y. A., M. Hendson, N. J. Panopoulos & M. N. Schroth, 1994. Molecular cloning, chromosomal mapping and sequence analysis of copper resistance genes from *Xanthamonas campestris* pv. *juglandis*: homology with small blue copper proteins and multicopper oxidase. J. Bact. 176: 173–188.

Lunsdorf, H., K. Brummer, K. N. Timmis & I. Wagner-Dobler, 1997. Metal selectivity of *in situ* microcolonies in biofilms of the Elbe River. J. Bact. 179: 31–40.

Maidak, B. L., G. J. Olsen, N. Larsen, R. Overbeek, M. J. McCaughey & C. R. Woese, 1996. The Ribosomal Database Project (RDP). Nucleic Acids Res. 24: 82–85.

Methé B. A., W. D. Hiorns & J. P. Zehr, 1998. Contrasts between marine and freshwater bacterial community composition: analyses of communities in Lake George, NY and six other Adirondack lakes. Limnol. Oceanogr. 43: 368–374.

Mills, S. D., C. A. Jasalavich & D. A. Cooksey, 1993. A two-component regulatory system required for copper-inducible expression of the copper resistance operon of *Pseudomonas syringae*. J. Bact. 175: 1656–1664.

Momen, B. & J. P. Zehr, 1998. Watershed classification by discriminant analyses of lakewater-chemistry and terrestrial characteristics. Ecol. Appl. 8: 497–507.

Moyer, C. L., F. C. Dobbs & D. M. Karl, 1995. Phylogenetic diversity of the bacterial community from a microbial mat at an active, hydrothermal vent system, Loihi Seamount, Hawaii. Appl. envir. Microbiol. 1995. 61: 1555–1562.

Mullins, T. D., T. B. Britschgi, R. L. Krest & S. J. Giovannoni, 1995 Genetic comparisons reveal the same unknown bacterial lineages in Atlantic and Pacific bacterioplankton communities. Limnol. Oceanogr. 40: 148–158.

Pace, N. R., 1997. A molecular view of microbial diversity and the biosphere. Science 276: 734–740.

Peterson, R. H., R. A. Bourbonniere, G. L. Lacroix, D. J. Martin-Robichaud, P. Takats & G. Brun, 1989. Responses of Atlantic salmon (*Salmo salar*) alevins to dissolved organic carbon and dissolved aluminum at low pH . Wat. Air Soil Pollut. 46: 399–413.

Pina, R. G. & C. Cervantes, 1996. Microbial interactions with aluminum. BioMetals 9: 311–316.

Ramsing, N. B., H. Fossing, T. G. Ferdelman, F. Andersen & B. Thamdrup, 1996. Distribution of bacterial populations in a stratified fjord (Mariager Fjord, Denmark) quantified by *in situ* hybridization and related to chemical gradients in the water column. Appl. envir. Microbiol. 62: 1391–1404.

Raskin, L., B. E. Rittmann & D. A. Stahl, 1996. Competition and coexistence of sulfate-reducing and methanogenic populations in anaerobic biofilms. Appl. envir. Microbiol. 62:3847–3857.

Sadler, K. & S. Lynam, 1987. Some effects on the growth of brown trout from exposure to aluminum at different pH levels. J. Fish. Biol. 31: 209–219.

Sanger, F., S. Nicklen & A. R. Coulsen, 1977. DNA sequencing with chain termination inhibitors. Proc. natn. Acad. Sci. U. S. A. 74: 5463–5467.

Scharf, R., R. Mamet, Y. Zimmels, S. Kimchie & N. Schoenfeld, 1994. Evidence for the interference of aluminum with bacterial porphyrin biosynthesis. BioMetals 7: 135–141.

Smith, S. W., R. Overbeek, C. R. Woese, W. Gilbert & P. M. Gillevet, 1994. The genetic data environment an expandable GUI for multiple sequence analysis. CABIOS 10: 671–675.

Stackebrandt, E., W. Liesack & B. M. Goebel, 1993. Bacterial diversity in a soil sample from a subtropical Australian environment as determined by 16S rDNA analysis. FASEB Journal 7: 232–236.

Van de Peer, Y. & R. De Wachter, 1994. TREECON for Windows: a software package for the construction and drawing of evolutionary trees for the Microsoft Windows environment. Comput. Applic. Biosci. 10: 569–570.

Van der Lelie, D., T. Schwuchow, U. Schwidetzky & S. Wuertz, 1997. Two-component regulatory system involved in transcriptional control of heavy-metal homeostasis in *Alcaligenes eutrophus*. Molecular Microbiol. 23: 493–503.

Wagner, M., R. Erhart, W. Manz, R. Amann, H. Lemmer, D. Wedi & K.-H. Schleifer, 1994. Development of an rRNA-targeted oligonucleotide probe specific for the genus *Acinetobacter* and its application for *in situ* monitoring in activated sludge. Appl. envir. Microbiol. 60: 792–800.

Weisburg, W. G., S. M. Barns, D. A. Pelletier & D. J. Lane, 1991. 16S ribosomal DNA amplification for phylogenetic study. J. Bact. 173: 697–703.

Weiss, P., B. Schweitzer, R. Amann & M. Simon, 1996. Identification *in situ* and dynamics of bacteia on limnetic organic aggregates (lake snow). Appl. envir. Microbiol. 62: 1998–2005.

Zwart, G., W. D. Hiorns, B. A. Methé, M. P. Agterveld, R. Huismans, S. L. Nold, J. P. Zehr & H. J. Laanbroek, 1998. Near-identical 165 rRNA sequences recovered from lakes in North America and Europe indicate the existence of clades of freshwater bacteria with global distribution. Syst. appl. Microbiol. 21: 546–556.

Hydrobiologia **401:** 97–112, 1999.
J.P. Zehr & M.A. Voytek (eds), Molecular Ecology of Aquatic Communities.

New insights on old bacteria: diversity and function of morphologically conspicuous sulfur bacteria in aquatic systems

N.D. Gray & I.M. Head*
Fossil Fuels & Environmental Geochemistry, Postgraduate Institute, University of Newcastle, Newcastle upon Tyne, NE1 7RU, U.K. Tel: [+44] (0) 191 2226605; Fax: [+44] (0) 191 222 5431; E-mail: I.M.Head@newcastle.ac.uk

Key words: Sulfur-oxidation, phylogenetic diversity, function, niche adaptation

Abstract

Colourless sulfur bacteria were among the first groups of biogeochemically important bacteria to be studied by microbial ecologists. This was in part due to the fact that several species are large and often produce macroscopically visible structures such as mats. Although some of these sulfur bacteria have been studied for over a century, few have been obtained in axenic culture and thus, until recently, relatively little was known regarding their relationship to other bacteria or their role in the environment. The introduction of molecular biological methods to microbial ecology has allowed many of these distinctive uncultured bacteria to be characterized phylogenetically and *in situ* measurements have revealed aspects of their behaviour and their involvement in specific biogeochemical processes. Culture based studies of some morphologically distinctive sulfur bacteria, principally isolates of different *Beggiatoa* species, have revealed that they exhibit a wide range of metabolic activities that span heterotrophy to obligate chemolithautotrophy. This knowledge has informed studies of uncultured sulfur bacteria such as *Thioploca* and *Achromatium*. The combination of molecular biological analyses, chemical and process measurements, and culture based studies that have allowed the ecology of these bacteria to be interrogated are discussed.

Introduction

Diversity within the bacterial domain of life has become increasingly well documented in recent years, and the application of molecular biology to microbial ecology has greatly expanded our knowledge of the complexity of bacterial communities in natural ecosystems. Consequently, one of the important challenges now facing microbial ecologists is how to relate this diversity to bacterial function. While the influence of bacteria on biological and geochemical processes can be deduced from *in situ* process measurements, the activities of distinct bacterial groups requires information on their physiology, abundance and distribution. Some of these properties cannot be accessed by molecular characterisation approaches. Without the ability to culture the majority of organisms in natural ecosystems, understanding the link between bacterial diversity and function has, therefore, proved difficult and the investigation of organisms *in situ* has in the past been hampered by difficulties of identification and enumeration.

Bacteria that are involved in the oxidative side of the sulfur cycle are often gradient dwelling and couple the oxidation of reduced sulfur species to the reduction of dissolved oxygen, nitrate and possibly oxidised metals. Many of these sulfur bacteria are morphologically conspicuous (La Rivière & Schmidt, 1992) and have been known for many years. Although few have been grown in axenic culture and little was known of their physiology and hence biogeochemical role, their size, shape and unusual structural or behavioural features, have facilitated their study and yielded new insights into the relationship between bacterial diversity and function in aquatic environments. Recent studies have employed a combination of molecular biological methods, *in situ* process measurements, microscopy and, where it has been possible, culture-based investigations to gain an understanding of the ecology and evolution of these organisms.

* Author for correspondence.

Studies of the filamentous sulfur bacteria *Beggiatoa* and *Thioploca* and the single-celled bacterium *Achromatium oxaliferum* are the focus of this paper. All are morphologically conspicuous and have been known for approximately 100 years (Winogradsky, 1887; Schewiakoff 1893; Lauterborn, 1907). *Thioploca* spp. and *Beggiatoa* spp. are found in both freshwater and marine environments. They form long motile filaments and often inhabit the oxic/anoxic boundary of sediments where they mediate the oxidation of reduced sulfur compounds. Though easily identifiable, none of the three clearly distinguishable *Thioploca* spp. has been successfully cultured in the laboratory and, until recently, have been characterised on the basis of morphology alone. *Thioploca ingrica* has been identified in various freshwater lakes in Europe, Russia and North America (Maier, 1984), while *T. araucae* and *T. chileae* have been collected from benthic sediments off the Chilean coast (Maier & Gallardo, 1984a). It is the marine *Thioploca* spp. that have stimulated the most scientific interest in recent years, primarily as a result of the discovery in 1977 of extensive bacterial mats off the southern Peruvian and northern and central Chilean coast (Gallardo, 1977). These bacterial mats, the largest known, occur over a 3000 km stretch along the South American continental shelf (Fossing et al., 1995). In contrast to *Thioploca*, research on *Beggiatoa* has been pivotal in the development of microbial ecology and studies on this organism led to the concept of chemolithoautotrophic growth (Winogradsky, 1887). Strains of *Beggiatoa* can be found in freshwater and marine sediments, salt marshes and at sulfide-rich deep-sea hydrothermal vents (Nelson, 1992). A number of the freshwater and marine strains have been isolated and cultured and consequently there has been extensive research on *Beggiatoa*, though few strains have been comprehensively characterised (Strohl, 1989).

A. oxaliferum, a morphologically conspicuous but uncultured bacterium, is also found in some freshwater and marine environments (De Boer et al., 1971; La Rivière & Schmidt, 1992). Conspicuous for its size alone (5–40 μm in width and 15–100 μm in length), it is packed with regularly arranged crystals of calcium carbonate in the form of calcite (La Rivière & Schmidt, 1989, 1991; Head et al., 1996), a feature which makes it unique among bacteria. A number of authors have suggested that the presence of calcite and intracellular inclusions of elemental sulfur implicate the organism in the carbon and sulfur cycle of some sediments (De Boer et al., 1971; Babenzien, 1991; La Rivière & Schmidt, 1992). However, until recently little more was known about this fascinating bacterium. *A. oxaliferum* is distributed globally and is locally abundant. For instance, over 90% of the bacterial biovolume in some sediments in the English Lake District can be accounted for by *A. oxaliferum* cells (Head et al., 1995).

Morphological diversity within the sulfur oxidizing bacteria *A. oxaliferum*, *Beggiatoa* spp. and *Thioploca* spp.

An individual *Thioploca* trichome consists of a gliding filament made up of numerous cells which usually contain elemental sulfur inclusions (Figure 1) (Maier, 1986). Bundles of these filaments share common gelatinous polysaccharide sheaths (Huettel et al., 1996). *Beggiatoa* filaments do not form sheaths, but in other respects are similar to *Thioploca* spp. (Figure 2) (Nelson, 1992). Each species of *Thioploca*, has a defined filament diameter. This property has been used to differentiate individual species in the mixed communities of dense marine microbial mats found off the South American coast (Maier, 1986; Schulz et al., 1996). Filament diameter is also employed for the differentiation of *Beggiatoa* spp. (Strohl, 1989; Nelson et al., 1989b). In the case of the large autotrophic *Beggiatoa* spp. from the Guaymas Basin hydrothermal vents an order of magnitude difference in filament diameter between species was observed (Nelson et al., 1989b). The cell size classes of marine *Thioploca* spp., though still clearly defined, are less pronounced and identification is confused by the occurrence of more than one filament class in the same sheath (Schultz et al., 1996). One interesting morphological feature of large-diameter *Thioploca* and *Beggiatoa* spp. is the occurrence of large central vacuoles which have been implicated in the storage and transport of nitrate (Fossing et al., 1995; McHatton et al., 1996).

In the case of *A. oxaliferum*, electron microscopy studies have revealed morphological differences based on external appearance and the distribution of cellular elemental sulfur and calcium carbonate (Figure 3) (Head et al., 1996), but these morphotypes have not been clearly correlated with phylogenetic diversity within *A. oxaliferum* populations. Although cell size defines different species of *Thioploca* and *Beggiatoa* a similar analysis of *A. oxaliferum* cells has failed to reveal similar structure in *A. oxaliferum* populations. However, analysis of the volume of *A. oxaliferum* cells

Figure 1. Bundle of *Thioploca* trichomes protruding from a common sheath. The inset shows two large *T. araucae* trichomes and three smaller *T. chileae* trichomes. The balck dots at the cell membranes are sulfur globules (Schultz et al., 1996), (Reproduced with permission).

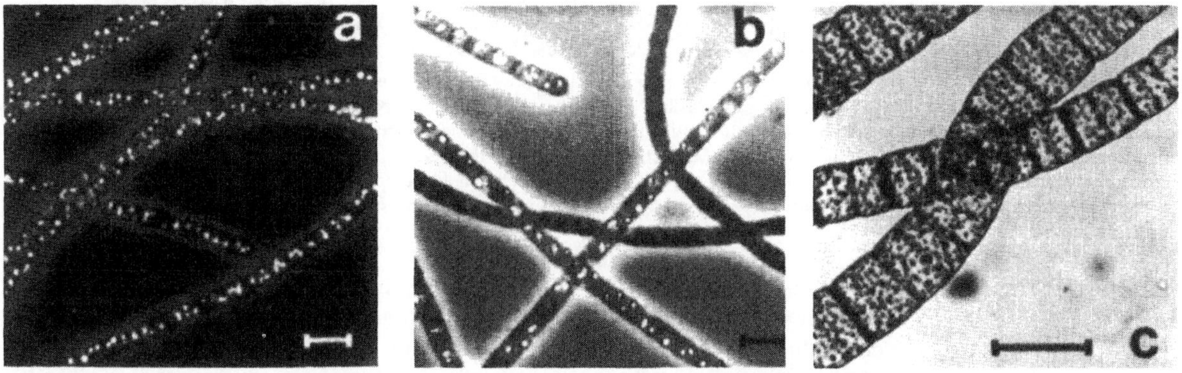

Figure 2. Light micrographs of pure cultures and natural samples of *Beggiatoa* spp. (a) and (b) phase contrast; (c) bright field illumination. (a) Obligate chemoautotrophic marine strain MS-81-1c, Scale bar represents 5 μm. (b) Freshwater strain OH-75-2a , Scale bar represents 5 μm. (c)*Beggiatoa* sp. collected at a deep sea hydrothermal vent in the Guaymas basin, Gulf of California. These chemoautotrophic filaments are representative of the wider material frequently found in natural samples. Scale bar represents 50 μm (adapted from Nelson, 1992), (Reproduced with permission).

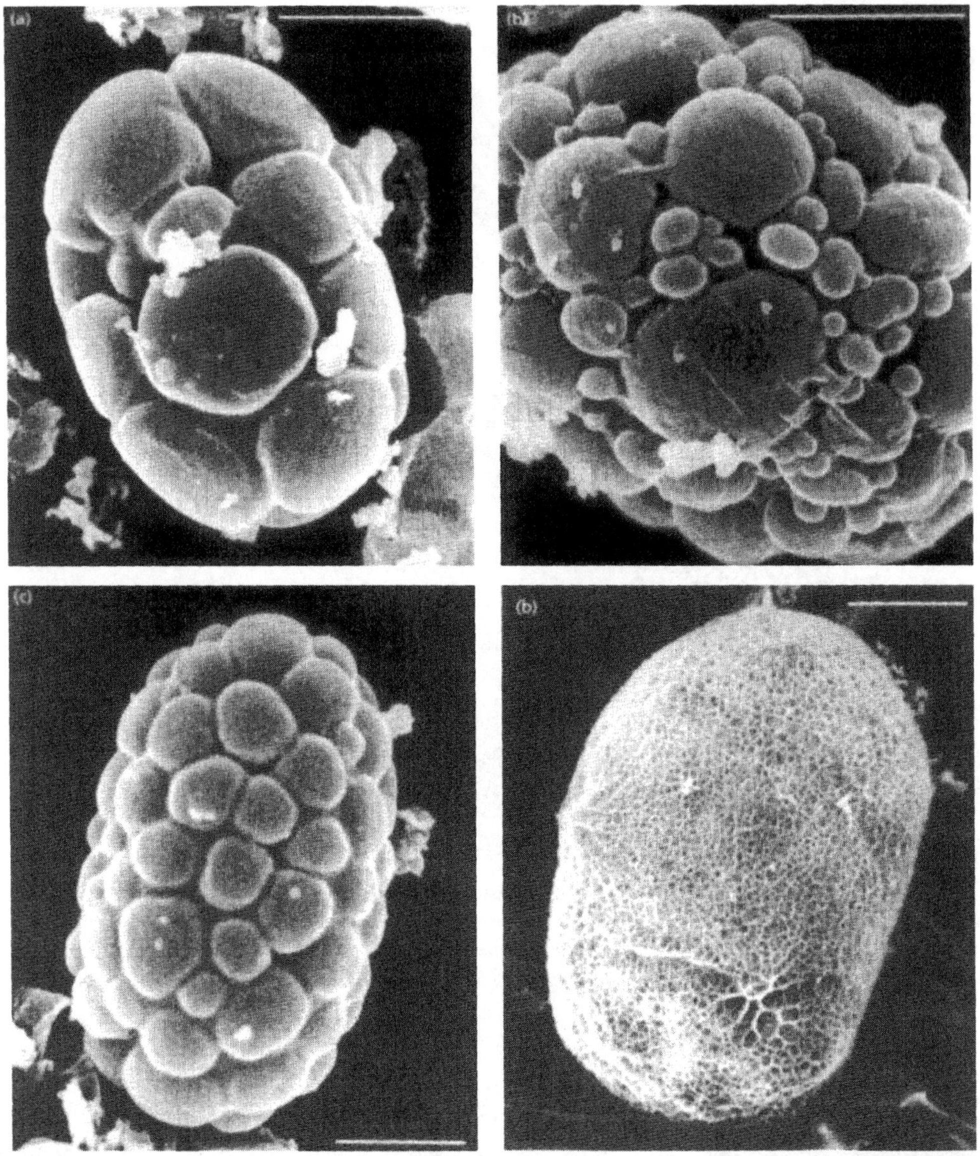

Figure 3. Scanning electron micrographs of spray frozen *A. oxaliferum* cells. Spray freezing results in the cell envelope forming a 'shrink wrap' around the cells revealing the structure of the internal inclusions. (a–c) Spray frozen morphotypes of cells with a 'smooth' surface; (a) Cell with large intracellular calcite inclusions. (b) Cell with small sulfur inclusions dispersed among large calcite inclusions. (c) Cell containing regularly sized calcite inclusions. (d) Cell of a 'rough' surface cell type, in this case the freeze spraying was less severe and the cell envelope is not contracted over the intracellular inclusions. The pitted appearance of the mucilaginous cell covering is due to removal of small ice crystals during preparation (Head et al., 1996). The scale bar represents 5 μm, (Reproduced with permission).

from a single sampling site revealed a wide range of cell sizes (1×10^3 to $3.5 \times 10^4 \mu m^3$). A frequency distribution of *A. oxaliferum* cell volumes (Figure 4) shows one dominant and several much smaller secondary maxima. However, unlike *Thioploca* and *Beggiatoa* populations there is no clear distinction between the maxima observed and they may simply reflect variability in the data.

Phylogenetic diversity and evolution of *A. oxaliferum*, *Beggiatoa* sp. and *Thioploca* sp.

Within the bacterial domain, the non-photosynthetic sulfur-oxidizing phenotype is widely distributed and is found in four of the five sub-divisions of the *Proteobacteria* (Figure 5) (Lane et al., 1992). The morphologically conspicuous sulfur bacteria *Beggia-*

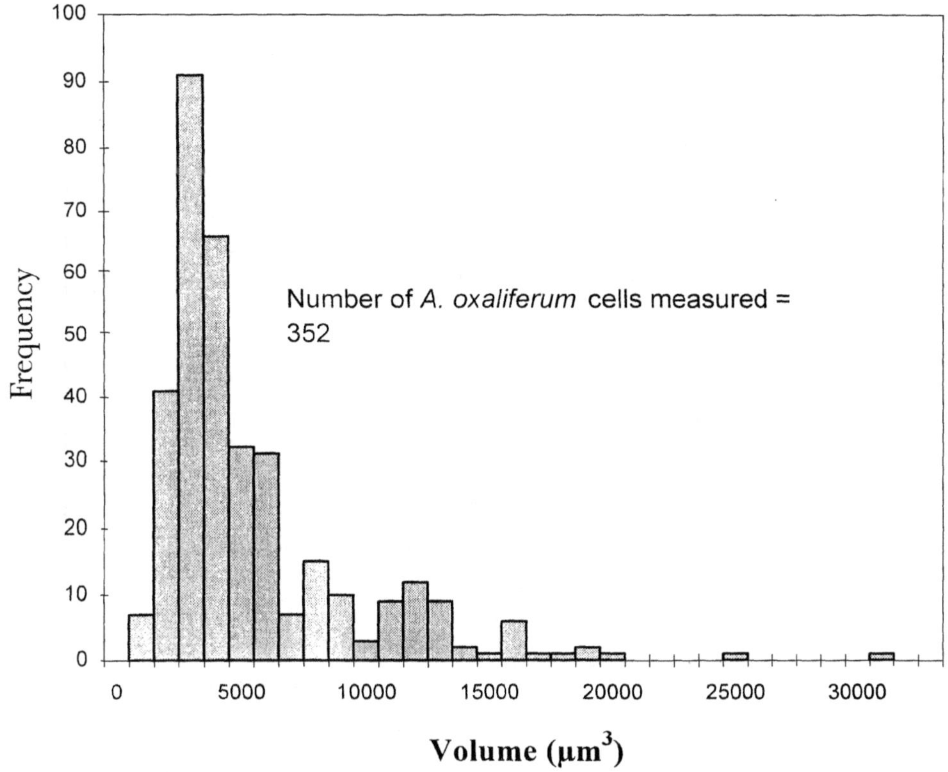

Figure 4. Size distribution of *A. oxaliferum* cells from a wetland area of Rydal Water, Cumbria U.K.

toa, *Thioploca* and *A. oxaliferum* all belong to the *γ-Proteobacteria* (Teske et al., 1995; Head et al., 1996). This radiation also contains endo- and ectosymbionts of marine invertebrates, *Thiothrix* spp., *Thiomicrospira* spp. and anoxygenic phototrophs related to *Chromatium* spp. (Muyzer et al., 1994; Polz et al., 1994; Teske et al., 1995; Polz et al., 1996). 16S rRNA sequence analysis demonstrated that the three validly described *Thioploca* species (*T. ingrica, T. chileae* and *T. araucae*) belong to a monophyletic group related to *Beggiatoa alba* and *Beggiatoa* sp. strain 1401–13 (Teske et al., 1995). The use of whole cell *in situ* hybridisation with fluorescently labelled oligonucleotide probes specific for different *Thioploca* 16S rRNA sequences confirmed the validity of using filament diameter as a differential character for species of *Thioploca* and has demonstrated that members of different species may sometimes inhabit the same sheath (Teske et al., 1995). As yet, comparatively few validly described *Beggiatoa* spp. have been analysed phylogenetically and we do not yet posses an accurate picture of the diversity within this group of organisms. For example, no information is available on the large vacuolate *Beggiatoa* strains found at a Monterey Canyon cold seep and at the Guaymas Basin hydrothermal vents (McHatton et al., 1996). These bacteria, although possessing wider filament widths, bear close resemblance to marine *Thioploca* spp.

Comparative rRNA sequence analysis of *A. oxaliferum* cells purified from a freshwater sediment indicated that *A. oxaliferum*-derived sequences also formed a deep branch within the *γ-Proteobacteria* (Head et al., 1996). A number of different *A. oxaliferum*-like 16S rRNA sequences were recovered from what appeared to be morphologically homogeneous purified preparations of cells. This monophyletic group was resolved into three distinct lineages, two of which possessed a unique structural motif in the V6 region of the 16S rRNA molecule (Head et al., 1996). The sequence differences between the lineages (approximately 10%) were similar to the diversity observed between rRNA sequences from different *Thioploca* and *Beggiatoa* species. This suggests that phylogenetically distinct sub-groups might exist within the *A. oxaliferum* population. However, heterogeneous 16S rRNA operons are known to occur at different loci in many bacteria (e.g. Nübel et al., 1996). It is therefore possible that heterogeneity in multiple

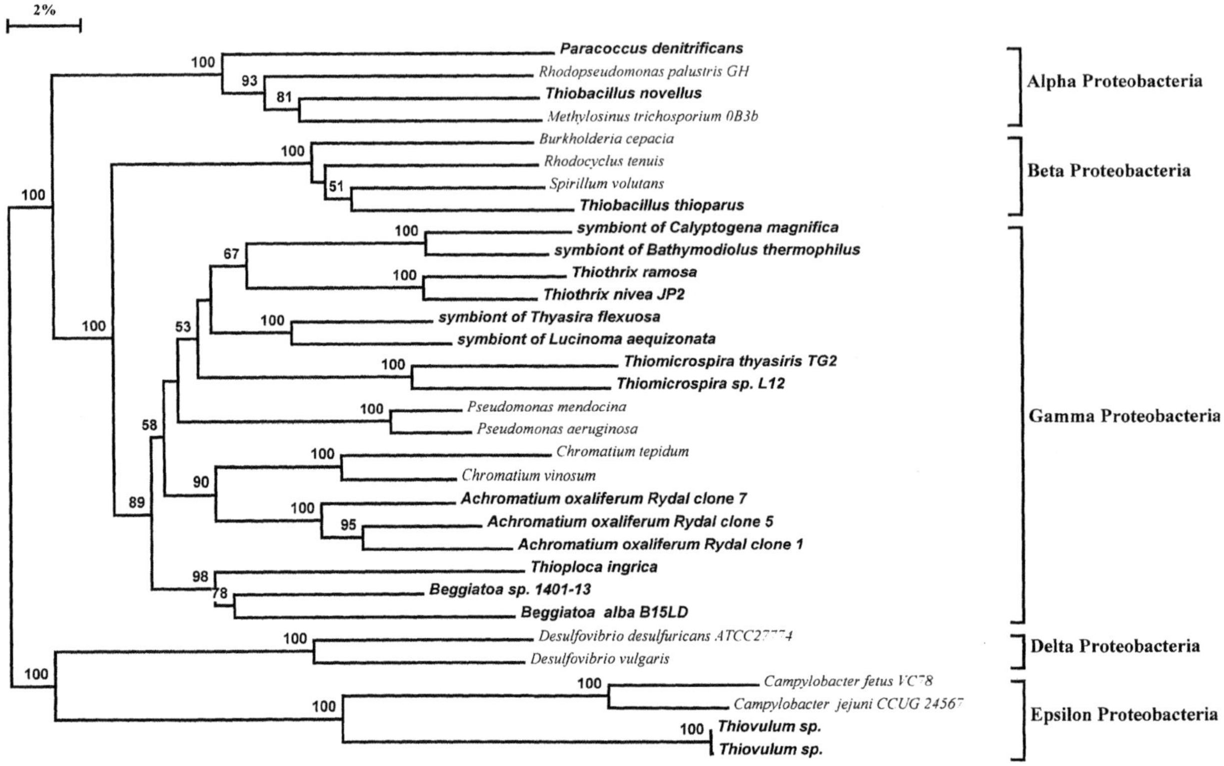

Figure 5. Phylogenetic distance tree of representatives of the *Proteobacteria*. Non-photosynthetic sulfur-oxidizing bacteria are highlighted in bold typeface. The tree was constructed from an alignment of partial 16S rRNA sequences (positions 145–203, 228–893 and 971–1367, *E. coli* numbering) using the Jukes and Cantor correction as implemented in TREECON (Van de Peer & de Wachter, 1994).

copies of 16S rRNA genes may account for the diversity observed in 16S rRNA gene clone libraries derived from purified *A. oxaliferum* cells. Recent results, using oligonucleotide probes specific for the divergent *A. oxaliferum*-derived 16S rRNA gene sequences, have confirmed that the three lineages identified, indeed represent genetically distinct sub-populations of cells morphologically identifiable as *A. oxaliferum* (Figure 6). The expression by *A. oxaliferum* cells of heterogeneous 16S rRNA operons is thought unlikely since the uptake of more than one probe by individual cells was never observed in cells simultaneously hybridized with oligonucleotides differentially labelled with fluorescein and rhodamine. Furthermore, like *Thioploca* and *Beggiatoa* spp., phylogenetically distinct *A. oxaliferum* cells fall into defined size classes. However, due to the considerable overlap in the size-class distributions, the different *A. oxaliferum* lineages could only be clearly discriminated using whole cell hybridisation techniques (unpublished data).

It seems likely that the sulfur-oxidizing bacteria from the γ-*Proteobacteria* have evolved from a sulfur-oxidizing progenitor (Canfield & Teske, 1996) and it has been proposed that the non-photosynthetic sulfur-oxidizing bacteria evolved in response to the oxygenation of the Earth's atmosphere 0.64–1.05 billion years ago (Teske et al., 1995; Canfield & Teske, 1996). They have since diversified to inhabit a large number of different environments in which reduced-sulfur oxidation occurs under different physicochemical conditions including in symbiotic relationships with many marine invertebrates. The evolution and diversification of the sulfur oxidizing phenotype has occurred both between and within habitats since, in some cases, genetically distinct populations of morphologically similar sulfur-oxidizing bacteria exist at a single site (Teske et al.,1995; Head et al., 1996). This may reflect the adaptation of closely related bacteria to slightly different ecological niches within the same environment. This observation may to some extent explain the clusters of closely related sequences often found in clone libraries of rRNA genes obtained from many environmental samples (Giovannoni et. al., 1990; Ward et al., 1992; Bond et al., 1995; Amann et al., 1996). For instance,

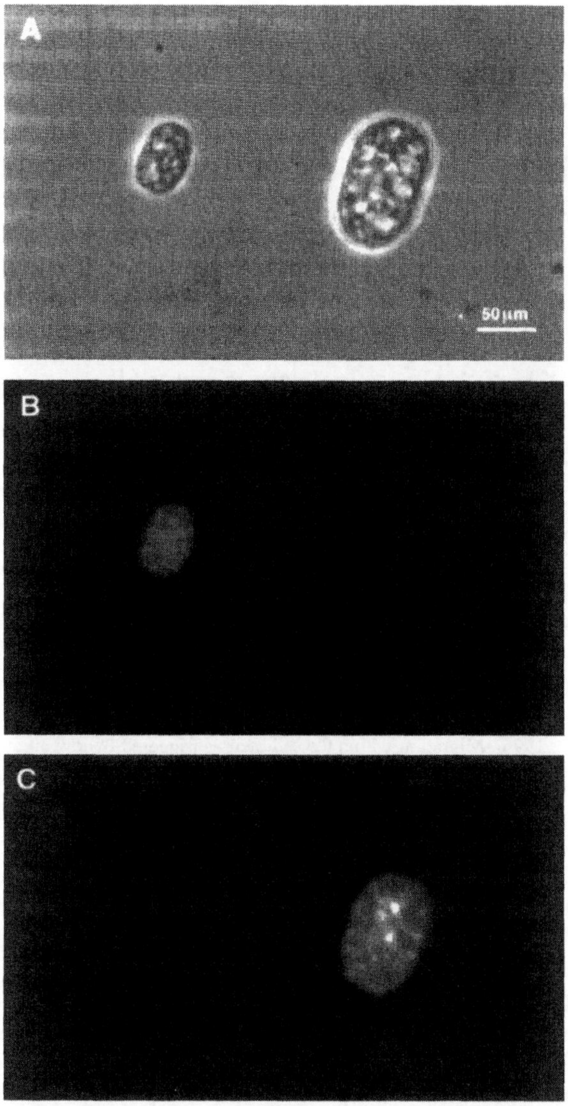

Figure 6. Whole cell *in situ* hybridization of *A. oxaliferum* cells purified from a freshwater sediment. The cells were simultaneously hybridized with oligonucleotide probes specific for the Rydal clone 1 sequence (rhodamine-labelled) and the Rydal clone 5 sequence (fluorescein-labelled). The *A. oxaliferum*-derived sequences are represented in Figure 5. (a) Phase contrast micrograph, (b and c) epifluorescence micrographs of (a) rhodamine fluorescence and (c) fluorescein fluorescence.

there is some evidence that niche specialisation may have occurred in marine planktonic bacteria first identified in the Sargasso Sea (SAR 11 cluster). Different representatives of the SAR 11 gene cluster appear to be more abundant at particular depths in the water column (Field et al., 1997). In addition, isolates of *Prochlorococcus* obtained from the same North Atlantic prochlorophyte community, though phylogenetically closely related (ca. 2% 16S rRNA sequence divergence), have been found to be adapted to different illumination conditions (Moore et al., 1998). It has been postulated that this is a consequence of adaptive radiation within the prochlorophyte community allowing exploitation of similar but distinct niches present within the oceanic water column (Moore et al., 1998). In these studies the differences in physicochemical and biological phenomena that define niches occur over tens or hundreds of meters. If this is also the case in populations of sediment-dwelling *A. oxaliferum* and *Thioploca* spp., then the variations in environment that engender niche specialization in these bacteria must exist on a much smaller scale (μm to cm). However, it remains to be demonstrated that niche adaptation accounts for the diversity found in *Thioploca* and *A. oxaliferum* clusters (Figure 5).

Ecology of *A. oxaliferum*, *Beggiatoa* spp. and *Thioploca* spp.

There are a number of different environments where sulfur-oxidizing bacteria flourish. The most widely studied of these are shallow, sulfide-rich sediments where bacteria form thin mats positioned in opposed sulfide and oxygen gradients. These mats dominate sulfur oxidation and prevent the abiotic reaction of reduced sulfur compounds with molecular oxygen. However, conspicuous communities of sulfur-oxidizing bacteria also occur under circumstances where the levels of dissolved sulfide and/or oxygen are rather low. This precludes the formation of thin mats that rely on the diffusion of electron donor and acceptor. In these environments, low dissolved sulfide concentrations are maintained either because of low rates of sulfate reduction (Gray et al., 1997) or precipitation of sulfide by reaction with reactive metal species (e.g. Thamdrup & Canfield, 1996; Gray et al., 1997). Given the different conditions under which sulfide-oxidizing bacteria flourish it is possible to interpret the physiological and behavioural adaptations of *A. oxaliferum*, *Beggiatoa* spp. and *Thioploca* spp. in relation to their predominance in specific environments. This type of investigation has highlighted a number of strategies employed by sulfur bacteria to exploit the oxidation of reduced sulfur.

Some studies have shed light on the ecology of *Thioploca* mats found off the west coast of South America (Fossing et al., 1995; Schultz et al., 1996; Huettel et al., 1996). A novel hypothesis has been proposed to explain the massive occurrence of sulfur-

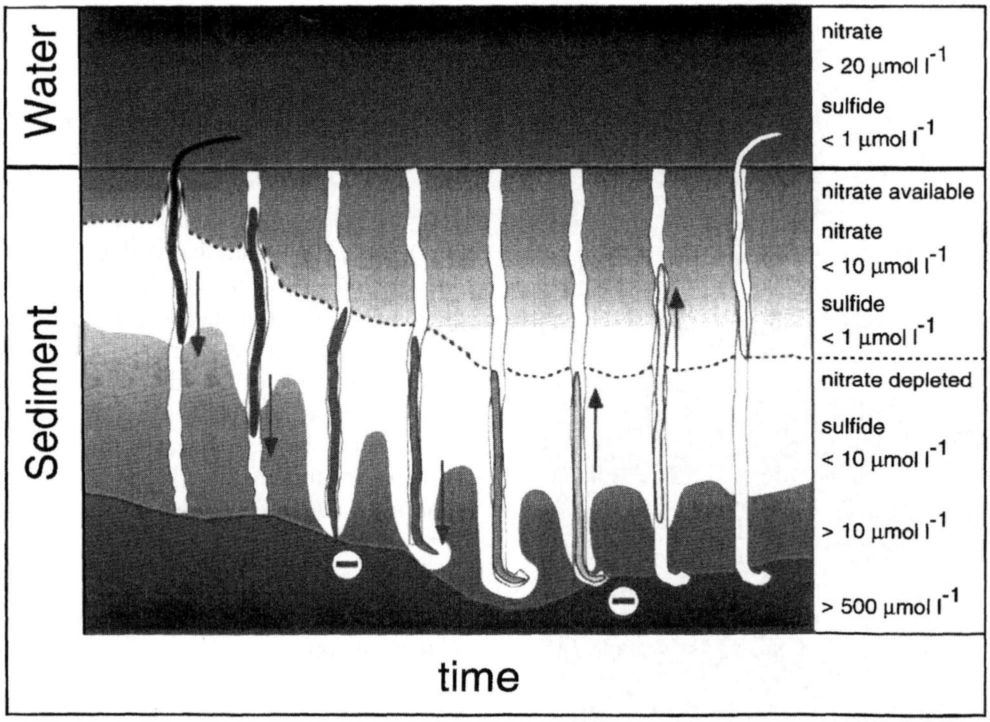

Figure 7. Vertical shuttling strategy adopted by *Thioploca* spp. (Huettel et al., 1996). Migration within the sediment is triggered by low nitrate concentrations (denoted by shading) within the filaments. The arrows indicate the chemotactic attraction towards sulfide and nitrate respectively, (Reproduced with permission).

oxidizing bacteria in this environment that is characterized by low levels of dissolved sulfide and oxygen. Some marine *Thioploca* inhabit sediments which are overlain by sea-water depleted in dissolved oxygen (< 5 μm) (Thamdrup & Canfield, 1996). In these sediments, nitrate is depleted within the top 2–3 cm of the sediment surface, however, nitrate concentrations in the overlying water are as high as 25 μM (Fossing et al., 1995). The oxidation of reduced sulfur compounds coupled to nitrate reduction is known among unicellular chemoautotrophic sulfur-oxidizing bacteria (Timmer-Ten Hoor, 1975; Schedel & Trüper 1980) and it has been proposed that *Thioploca* spp. may be denitrifiers (Fossing et al., 1995). In marine *Thioploca* spp. up to 80% of the cell volume is taken up by a central vacuole (Fossing et al., 1995). Presumably utilising these vacuoles as a storage reservoir, *Thioploca* accumulates nitrate to concentrations 20000 times greater than the surrounding sea water (Fossing et al., 1995). By continuous vertical migration between the nitrate rich overlying waters and the zone of sulfate reduction within the sediment *Thioploca* spp. may couple the oxidation of dissolved sulfide to the reduction of nitrate (Figure 7). This is achieved without the coexistence of these two substrates in the sediment pore-waters.

The ability of the *Thioploca* filaments to project above the sediment surface (Rosenberg et al., 1983; Huettel et al., 1996) facilitates uptake of nitrate from the overlying water and overcomes the diffusion limitation that maintains spatially separated pools of sulfide and nitrate in these sediments. *Thioploca* sheaths extend across sedimentary redox horizons allowing the free movement of filaments through different redox zones (Shultz et al., 1996). The strategy adopted by *Thioploca* allows it to flourish in sediments lacking nitrate and the activity of *Thioploca* spp. may actually promote the separation of pools of nitrate and dissolved sulfide, effectively allowing *Thioploca* spp. to out-compete other sulfur bacteria (Fossing et al. 1995).

Coupling the oxidation of dissolved sulfide to reduction of nitrate by filamentous sulfur-oxidizing bacteria has not only been postulated for *Thioploca* spp. Denitrification activity of *Beggiatoa* spp. has also been reported (e.g. Sweerts et al., 1990) but the validity

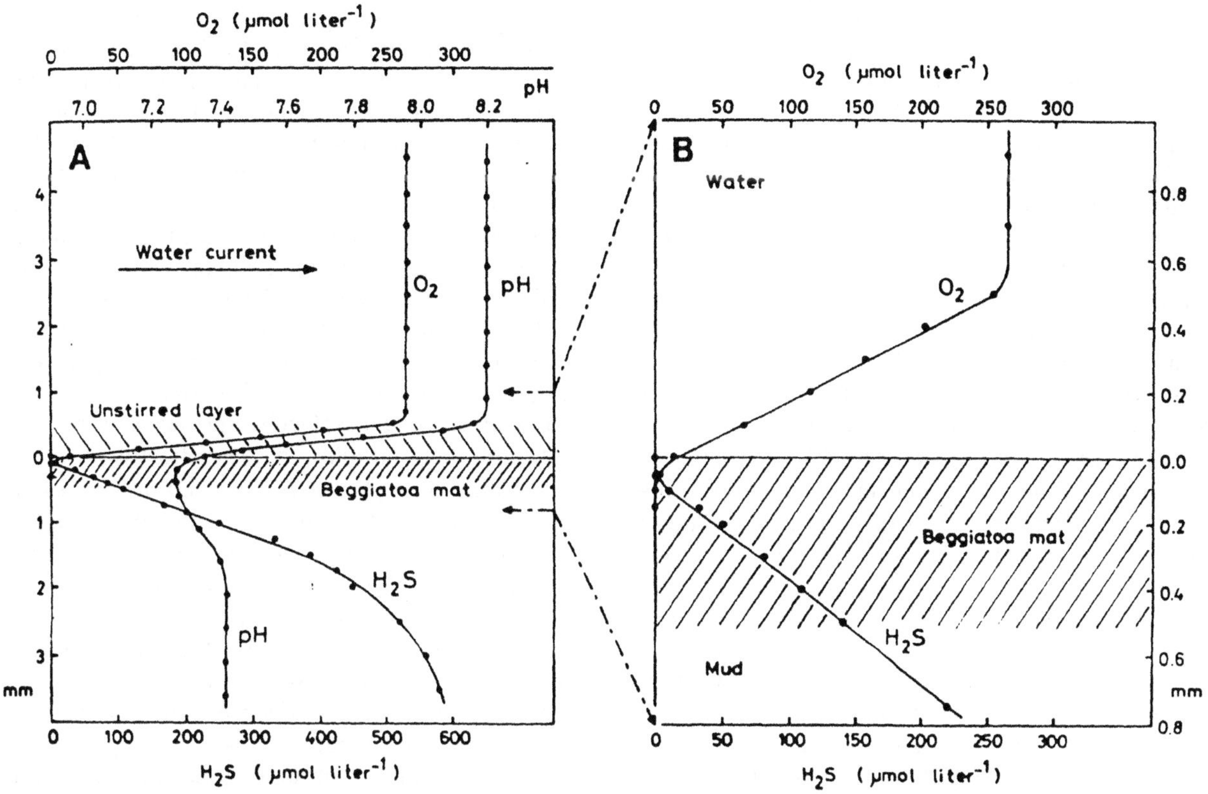

Figure 8. (a) Vertical distribution of O_2, dissolved sulfide and pH through a *Beggiatoa* mat growing at the sediment/water interface. (b) The interface shown between O_2 and dissolved sulfide is shown in expanded scale. An unstirred boundary layer in the water at 0.0–0.5 mm created a steep diffusion gradient of O_2 from the circulating water above (Jørgensen & Revsbech, 1983), (Reproduced with permission).

of this has been challenged recently (McHatton et al., 1996). However, studies of *Beggiatoa* from cold seeps and hydrothermal vents, that also possess large central vacuoles and hyperaccumulate nitrate, suggest that these bacteria may genuinely be chemoautotrophic nitrate reducers though not necessarily denitrifiers (McHatton et al., 1996).

A further interesting feature of the microbial mat-covered marine sediment environments dominated by *Thioploca* spp. is that despite sulfate reduction rates being very high and the process of carbon mineralization being almost entirely coupled to the sulfur cycle (Thamdrup & Canfield, 1996), high levels of reactive iron ensure that levels of dissolved sulfide in pore waters remain low as a consequence of the precipitation of iron sulfides. It was concluded that the *Thioploca* and *Beggiatoa* spp. within these microbial mat communities must be effective scavengers of dissolved sulfide to compete with this reactive iron. To explain this, it was suggested that there may be an intimate association between the sulfur-oxidizing bacteria and sulfide-producing, sulfate reducers (Thamdrup & Canfield, 1996).

Large vacuolate *Beggiatoa* spp., which form thick mats at hydrothermal vent and cold-seep sites, also appear to adopt a *Thioploca*-like shuttling strategy (McHatton et al., 1996). On the other hand, due to limitations imparted by substrate diffusion, narrow non-vacuolate *Beggiatoa* spp. are restricted to a narrow position within sediments often forming mats of a few hundred microns in depth (Jørgensen 1983; Nelson et al., 1986). The sediments on which such mats are formed are rich in dissolved sulfide and *Beggiatoa* are able to use dissolved oxygen for the complete oxidation of these sulfides to sulfate. Micro-electrode studies of these bacterial mats [Jørgensen & Revsbech, 1983) confirmed the coexistence of dissolved sulfide and oxygen over only a 50 μm depth interval in a mat 500–700 μm thick (Figure 8). The rate of sulfide oxidation within natural and cultured *Beggiatoa* mats was shown to be up to 4–5 orders of magnitude or more, than that attributable to the purely chemical oxidation of sulfide [Jørgensen & Revsbech, 1983). Furthermore, in a cultured marine chemolithotrophic *Beggiatoa* sp. the oxidation of dissolved sulfide occurred exclusively in the bacterial mat (Nelson et al., 1986). The formation of these thin mats requires conditions of high dissolved sulfide diffusing from below and high levels of oxygen diffusing from above. Under these environmental conditions, the mat bacteria are able to monopolize the use of both sulfide and oxidant to such an extent that little oxidation of reduced sulfur species occurs outside the mat and the amount of energy the bacteria obtain is maximised. In order to achieve optimal sulfur oxidation rates, *Beggiatoa* that form mats must maintain their position within a narrow redox zone where electron donor and acceptor coexist. It is, therefore, important that they are able to respond to movements in the redox gradient. There are possibly three mechanisms by which species of *Beggiatoa* in shallow aquatic environments maintain their position in sedimentary gradients. Some *Beggiatoa* spp. posses a phobic response to light (Nelson & Castenholz, 1982) coupled with a negative chemotactic response to high dissolved oxygen and possibly to high sulfide concentrations (Møller et al., 1985). In these shallow environments, *Beggiatoa* spp. migrate to the surface of sediments during periods of darkness when, due to the lack of oxygenic-photosynthesis, the oxic/anoxic boundary is positioned at the sediment surface or even in the water column. This initial migration of the bacteria is attributable to a chemotactic rather than phototactic mechanism, since movements below the sediment surface, where light does not penetrate, must be in response to gradients of oxygen and sulfide. However, it has been suggested that a response to light is still vital to these bacteria during diurnal cycles in shallow aquatic environments (Møller et al., 1985). In these sediments *Beggiatoa* employ both photic and chemotactic responses and react to the first or strongest stimulus they encounter. This dual response mechanism may be particularly important for *Beggiatoa* that inhabit photosynthetic mats, since throughout periods of darkness *Beggiatoa* spp. may be positioned above the mat surface. To return below the mat surface at the onset of daylight the bacteria must pass through a zone of high oxygen concentration engendered by the oxygenic-photosynthetic bacteria. A purely chemotactic response would leave the *Beggiatoa* trapped above the sediment surface (Møller et al., 1985).

A. oxaliferum inhabit sediments that are characterised by low levels of dissolved sulfide with high levels of reactive iron and oxidised overlying water (Gray et al., 1997). Here, the organism resides around the oxic-anoxic interface (Head et al., 1996) and demonstrates motility in response to movement of the oxycline, presumably in response to changes in the concentration of oxygen or other electron acceptors (Gray et al., 1997). The role of *A. oxaliferum* in the sulfur cycle in these sediments was inferred from experiments where the population of *A. oxaliferum* was artificially increased. The rate of sulfate accumulation in sediment cores in which sulfate reduction had been inhibited was directly correlated with the size of the *A. oxaliferum* population (Gray et al., 1997). For an organism to play a significant role in sulfur oxidation in these sediments, where dissolved sulfide levels were below detection limits, it must compete effectively with reactive iron for the available dissolved sulfide, perhaps by being intimately associated with sulfate-reducing bacteria as suggested for *Thioploca* spp. (Thamdrup & Canfield, 1996). Alternatively, it may be able to mobilise insoluble metal sulfides. To achieve either of these, the organism must be dispersed throughout the sulfate reduction zone where sulfides are being actively produced. *A. oxaliferum* cells have been observed from the sediment surface to a depth of 4 cm (Figure 9) and the majority of cells are located in regions where both oxygen and nitrate are absent. This suggests either, that in these regions *A. oxaliferum* is capable of using electron acceptors other than nitrate and oxygen or employs a shuttling strategy analogous to *Thioploca* spp. However, there is no evidence for the rapid vertical migration of *A. oxaliferum* cells and no large vacuoles to facilitate electron acceptor uptake and storage.

Given the widely dispersed nature of *A. oxaliferum* cells and the existence of genetically distinct populations in a single sediment, one can speculate that the genetic diversity observed reflects their exploitation of different niches. The relationship between the genetically distinct *A. oxaliferum* populations and their preference for different redox zones within a sediment would go some way to addressing the issue of niche-specialisation in phylogenetically related bacterial clusters derived from natural environments.

Nutrition and physiology of *A. oxaliferum*, *Beggiatoa* spp. and *Thioploca* spp.

A relatively large number of *Beggiatoa* spp. have been isolated in axenic culture and consequently a great deal more is known regarding their physiology and

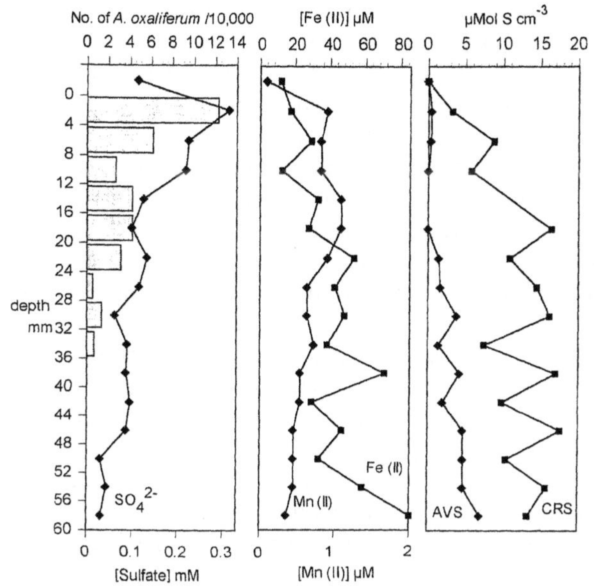

Figure 9. Distribution of *A. oxaliferum* cells in a freshwater sediment. *A. oxaliferum* cells are present across a number of redox boundaries defined by gradients of redox sensitive chemical species (Gray et al., 1997), (Reproduced with permission).

hence, biogeochemical function than is known about *Thioploca* spp. and *A. oxaliferum.*

While marine *Beggiatoa* spp. appear to display predominantly chemolithoautotrophic nutrition (Nelson, 1992), a recent study indicated that one marine strain (MS-81-6) exhibits facultative autotrophic growth (Hagen & Nelson, 1996). In the obligately chemolithotrophic strain MS-81-1c, grown in the presence of acetate, inorganic carbon fixation was not significantly reduced by the regulation of ribulose-1,5-biphosphate carboxylase/oxygenase activity (Hagen & Nelson, 1996). It was also demonstrated that this organism was unable to respire acetate and that 2-oxoglutarate reductase, an enzyme necessary for the respiration of organic substrates via the tricarboxylic cycle, was absent. Even in the presence of organic compounds at least 80% of cell carbon was obtained from CO_2. Additionally, this strain obtained more energy from the oxidation of reduced sulfur species than the facultative chemolithoautotrophic marine strain MS-81-6 (Hagen & Nelson, 1997). MS-81-6, although able to grow chemolithoautotrophically (Nelson & Jannasch, 1983), was found to be metabolically more versatile and could utilise acetate freely for energy generation or biosynthesis (Hagen & Nelson, 1996). This strain strongly regulated the expression of Calvin cycle enzymes in response to the presence of organic carbon (acetate). In samples of large marine *Beggiatoa* spp. collected from mats associated with deep sea hydrothermal vents, the presence of Calvin cycle enzymes and the incorporation of $^{14}CO_2$, indicated that they contributed to the lithoautotrophic primary production of the vent ecosystems (Nelson et al., 1989b).

In contrast to marine *Beggiatoa* there has been considerable debate as to whether freshwater strains of *Beggiatoa* are capable of genuine autotrophic or mixotrophic growth (Nelson, 1992; Hagen & Nelson, 1997). To date, no freshwater *Beggiatoa* has been grown lithoautotrophically, however mixotrophic nutrition has been claimed for a number of freshwater strains (Pringsheim, 1967; Güde et al., 1981). There appears to be considerable metabolic diversity amongst these bacteria but it is clear that most freshwater *Beggiatoa* isolates posses a functional TCA cycle (Nelson, 1992) and use the oxidation of simple organic compounds for energy conservation and biosynthesis (Nelson & Castenholz, 1981b; Strohl et al., 1981). Most freshwater *Beggiatoa* spp. also deposit intra-cellular sulfur under suitable growth conditions (Larkin & Strohl, 1983). However, unequivocal evidence that sulfide oxidation can be coupled to energy conservation has remained elusive. Several possible reasons why freshwater *Beggiatoa* may oxidise reduced sulfur compounds have been considered. These include: the reduction of the toxic effects of dissolved sulfide in microaerophillic environments; the removal of metabolically produced H_2O_2 without the need for catalase; or as a supplementary source of energy or reducing power for mixotrophic growth (Schmidt et al., 1987). In continuous culture experiments with *Beggiatoa alba* B18LD grown under heterotrophic and mixotrophic conditions, cell yield was found to be greater in media containing acetate and sulfide compared to media containing acetate alone (Güde et al., 1981). However later interpretation of these data (Kuenen & Beudeker, 1982; Nelson & Jannasch, 1983) indicated that the growth yield with respect to sulfide was unexpectedly high and it was suggested that the anomalous results were a consequence of the interference of sulfur compounds with the protein determination method used in this study (Kuenen & Beudeker, 1982). Mixotrophy has recently been more convincingly demonstrated in a freshwater *Beggiatoa* strain (AA5A) grown in gradient cultures (Ahmad et al., 1998). When supplied with sulfide under acetate-limited conditions a two to three fold increase in biomass yield, greater uptake of ^{14}C-labelled acetate and an increased rate of respiration was noted compared with control cultures provided with acetate alone.

For most freshwater *Beggiatoa* strains, only circumstantial evidence of mixotrophy has been provided. This evidence has been based on their potential to express and regulate enzymes concerned with biosynthesis and respiration. The ability of an organism to fix inorganic carbon is dependent on it possessing a mechanism for carbon fixation such as the Calvin cycle. Therefore, detection of sequences homologous to the gene encoding the large sub-unit of RuBisCO from a cyanobacterium in *Beggiatoa* strain OH-75-2a indicated the autotrophic potential of this freshwater isolate (Nelson et al., 1989a). The expression of the Calvin cycle enzymes (RuBisCO and Phosphoribulokinase) in strain OH-75-2a and *Beggiatoa alba* B18LD was also confirmed albeit at low activity. Nonetheless, it was shown that under conditions of organic substrate limitation the specific activity of these enzymes increased and in cultures of *Beggiatoa alba* B18LD supplemented with sulfide the specific activity of the Calvin cycle enzymes was approximately twice that obtained when sulfide was absent. Calvin cycle enzyme activities in the freshwater *Beggiatoa* strains were however, an order of magnitude lower than found in the facultative marine autotroph *Beggiatoa* sp. MS-81-6.

The ability of *Beggiatoa alba* strains to direct electrons from sulfide oxidation to electron carriers has also been demonstrated (Schmidt et al., 1987) and conversion of sulfide to sulfur was prevented by the addition of inhibitors of electron transport. In the same study, sulfide-dependent oxygen consumption was found to be consistent with the organism obtaining both energy and reducing power from sulfide.

A further feature of sulfur metabolism in *Beggiatoa alba* B18LD, that contrasts with autotrophic marine *Beggiatoa* spp., is that elemental sulfur deposited intracellularly is not oxidized to sulfate. However under anoxic conditions and in the presence of H_2 or endogenous carbon reserves the organism was able to reduce elemental sulfur to sulfide (Schmidt et al., 1987). Nonetheless, it has been demonstrated that the freshwater *Beggiatoa* strain OH-75-2a showed some activity of sulfite:acceptor oxidreductase an enzyme involved in the oxidation of sulfite to sulfate. The presence of this enzyme is consistent with lithotrophic energy conservation in this freshwater isolate (Hagen & Nelson, 1997) However, Strain OH-75-2a does not require a source of reduced sulfur for growth (Nelson & Castenholz, 1981a,b) and can gain carbon, reducing power and energy solely from organic compounds.

Thus, although there is evidence that most freshwater *Beggiatoa* strains may gain energy from oxidation of reduced sulfur compounds there is still debate as to the importance of this property in freshwater environments, where a low percentage of the organic carbon mineralisation is channelled through the sulfur cycle (Robertson & Kuenen, 1992). Here, sulfur oxidation may supplement the principally heterotrophic nutrition of the organism. However, an additional source of energy or reducing power (in the form of low levels of reduced sulfur) may impart some competitive advantage over both obligate chemoautotrophs or purely heterotrophic bacteria.

Thioploca spp. and *A. oxaliferum* have not yet been grown in axenic culture and consequently there is considerably less known about their metabolic capabilities. However, microautoradiographic techniques have been used to determine the metabolic activities of some marine and freshwater species of *Thioploca* (Maier & Gallardo, 1984b). It was reported that, in the presence of sulfide, the marine species *Thioploca araucae* and *Thioploca chileae* incorporated radiolabelled acetate, some amino acids, glucose, thymidine and carbonate in approximately equal quantities. In the absence of sulfide no assimilation of any of these radiolabelled substrates occurred. It was concluded that mixotrophic nutrition based on sulfide oxidation under reduced oxygen tension was indicated for these marine bacteria (Maier & Gallardo, 1984b). The freshwater *Thioploca ingrica*, however, incorporated methanol, acetate, dextrose and bicarbonate again in roughly equal measure with or without the presence of sulfide (Maier, 1986). On the basis of incorporation of bicarbonate by *Thioploca ingrica*, it was concluded that this organism was also mixotrophic and in the absence of an external inorganic energy source was able to oxidise internal elemental sulfur. While these experiments do show that *Thioploca* spp. have a potential for mixotrophy, it has been pointed out (Nelson, 1989) that the study on the marine *Thioploca* spp. took no account of the dilution effect of the high bicarbonate concentrations of sea-water (roughly 2.5 mmol) which would have reduced the specific activity of the added $H^{14}CO_3^-$. The degree of bicarbonate incorporation was therefore, perhaps 20 times higher than for any of the ^{14}C-labelled organic compounds. Given that bicarbonate concentrations in freshwater environments are typically in the millimolar range (Stumm & Morgan, 1981), the same conclusion may be drawn for

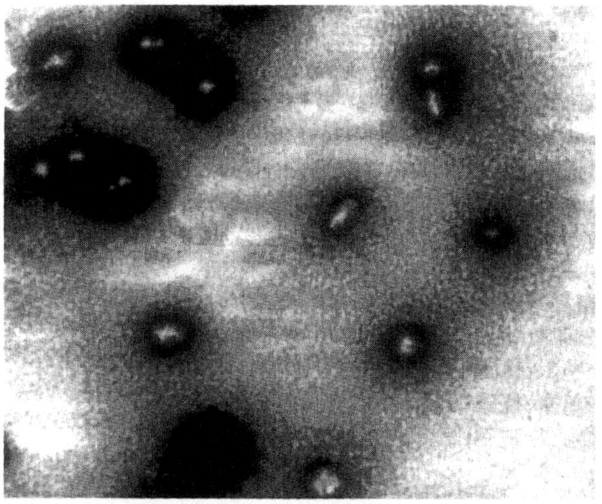

Figure 10. Microautoradiograph of *A. oxaliferum* cells incubated with radiolabelled bicarbonate. Sediment cores containing *A. oxaliferum* were incubated with ^{14}C-bicarbonate for 3 h. *A. oxaliferum* cells were purified from the sediment, placed on glass microscope slides and coated with photographic emulsion. The slides were developed after 14 days of exposure.

Thioploca ingrica, however, no experimental methods were given in this report (Maier, 1986). On the basis of re-examination of the original experiments it has therefore been suggested that some *Thioploca* spp. may also be capable of lithoautotrophic growth (Nelson, 1989).

A. oxaliferum has long been assumed to be an autotrophic sulfur-oxidizing bacterium (La Rivière & Schmidt, 1992) and recently it has been demonstrated that *A. oxaliferum* is capable of oxidizing reduced sulfur species to sulfate (Gray et al., 1997). Complete oxidation of reduced sulfur to sulfate is a feature of *Beggiatoa* strains for which there is strong evidence for energy conservation from sulfide oxidation (Nelson et al., 1986; Schmidt et al., 1987). Furthermore, microautoradiographic studies of *A. oxaliferum*, have indicated that individual cells can assimilate ^{14}C-labelled simple organic substrates and ^{14}C-labelled bicarbonate (Figure 10; Howarth, 1995) suggesting that *A. oxaliferum* may be mixotrophic and utilises both organic and inorganic carbon sources. However, the microautoradiographic studies of *A. oxaliferum* were not quantitative (Howarth, 1995) and thus, it was not possible to determine if the level of inorganic carbon fixation would be sufficient to sustain autotrophic growth.

The potential for the oxidation of reduced sulfur species by *A. oxaliferum* and *Thioploca* spp. as a source of energy or reducing power for biosynthesis still remains to be fully investigated, however, on balance, these initial studies on the carbon and sulfur metabolism of *A. oxaliferum* and *Thioploca* spp. indicate that, like some freshwater *Beggiatoa* they are most likely capable of utilizing reduced sources of inorganic sulfur for generating energy and/or reducing equivalents. They are also capable of using both organic and inorganic sources of carbon. The balance of the different energy generating mechanisms in the bacteria is probably dictated by the environment where the organisms occur. In freshwater environments where reduced sulfur levels are relatively low for example, there may be greater reliance on organic substrates as a source of energy and reducing power.

Conclusions

Morphologically conspicuous sulfur-oxidizing bacteria were among the first organisms to spark scientific interests in the ecology and ecophysiology of bacteria. These organisms, that so intrigued early microbial ecologists, now provide fruitful avenues of research for a new generation of scientists equipped with tools to study their activity, diversity and evolutionary relationships. Over a number of years, the morphologically conspicuous sulfur-bacteria have provided a useful insight into the link between the evolutionary diversity of bacteria and the functional adaptations they have adopted to exploit different environmental conditions. *Beggiatoa* spp., *Thioploca* spp. and *A. oxaliferum* all have the common characteristics of reduced sulfur oxidation, mixotrophic or autotrophic growth and share a common ancestry. However, it is clear that they have evolved distinct strategies such as mat formation, shuttling or sulfide scavenging to exploit the energy available from oxidation of reduced sulfur compounds in geochemically distinct environments. These morphologically conspicuous bacteria also exhibit phylogenetic heterogeneity within a single habitat. Genetic heterogeneity is consistently found in many populations of related bacteria and not just among the morphologically conspicuous sulfur bacteria (e.g. Amann et al., 1996). It is tempting to speculate that this evolutionary radiation corresponds to niche adaptation in response to subtle variations within an environment. Functional differences corresponding to this heterogeneity are easy to envisage, encompassing nutritional, physiological, behavioural and morphological characteristics, however they are difficult to demonstrate. The nutritional versatility of

Beggiatoa isolates from different habitats, highlights the effect on evolutionary and functional diversity that may be engendered by the environment. In freshwater environments with characteristically low levels of sulfide, lithotrophic energy generation from sulfide-oxidation may supplement the use of organic substrates, whereas in sulfidic marine sediments, organisms capable of lithoautotrophic growth may flourish.

Recent methodological advances may now allow further progress to be made on understanding the relationship between function and diversity of hitherto unculturable bacteria such as *Thioploca* and *A. oxaliferum*. Results of studies conducted to date, have already raised questions about the ecology of these bacteria. For instance: to what extent does the distribution of *A. oxaliferum* cells across distinct redox zones in the sedimentary environment reflect population diversity ? Do *Thioploca* trichomes inhabiting the same sheath compete for the same resources ? To what extent is diversity linked to physiological differences between genetically distinct cells within an *A. oxaliferum* or *Thioploca* population ? There is now the potential to link the use of molecular-biological tools such as whole cell fluorescent in situ hybridization to the measurement of single cell activity by techniques such as substrate specific microautoradiography (Andreasen & Nielsen, 1997), or other single cell activity measurements (Whiteley et al., 1996). Molecular probes have already been used to understand the abundance, relative spatial distribution and phylogenetic variation in bacterial communities (Amann et al., 1996) and a recent study has found that there was a link between diversity within closely rated organisms and their distribution in the environment (Field et al., 1997). Extending these approaches to the study of sulfur-oxidizing bacteria that can be clearly identified both morphologically and at the genetic level will now enable the relationship between genetic diversity and specific functions in natural microbial ecosystems to be more clearly defined.

Acknowledgements

Richard Howarth and Nick Davis are thanked for providing data, and Gwyn Jones and Roger Pickup are thanked for stimulating discussions on *A. oxaliferum*. We would also like to thank an anonymous reviewer for their helpful comments. Markus Huettel and Heidi Schultz provided photographs and figures from their work on *Thioploca* for which we are grateful. Work on *A. oxaliferum* was supported by the IFE and FBA. Financial support for work on *A. oxaliferum* was provided by the Natural Environmental Research Council (Grant GR3/9148).

References

Ahmad, A. D. & C. Nelson, 1998. A newly isolated strain of freshwater *Beggiatoa* sp. (AA5A) exhibits a mixotrophic mode of nutrition. Abstracts of the Eight International Symposium on Microbial Ecology, Halifax, Nova Scotia. August 9–14, 86pp.

Amann, R., J. Snaidr, M. Wagner, W. Ludwig & K-H. Schleifer, 1996. In situ visualization of high genetic diversity in a natural microbial community. J. Bact. 178: 3496–3500.

Andreasen, K. & P. H. Nielsen, 1997. Application of microautoradiography to the study of substrate uptake by filamentous microorganisms in activated sludge. Appl. envir. Microbiol. 63: 3662–3668.

Babenzien, H-D., 1991. *Achromatium oxaliferum* and its ecological niche. Zentralbl. Mikrobiol. 146: 41–49.

Bond, P. L., P. Hugenholtz, J. Keller & L. L. Blackall, 1995. Bacterial community structure of phosphate-removing and non-phosphate-removing activated sludges from sequencing batch reactors. Appl. envir. Microbiol. 61: 1910–1916.

Canfield, D. E. & A. Teske, 1996. Late proterozoic rise in atmospheric oxygen concentration inferred from phylogenetic and sulfur-isotope studies. Nature 382: 127–132.

De Boer, W. E., J. W. M. La Rivière & K. Schmidt, 1971. Some properties of *Achromatium oxaliferum*. Antonie van Leeuwenhoek 37: 533–563.

Field, K. G., D. Gordon, T. Wright, M. Rappé, E. Urbach, K. Vergin & S. J. Giovannoni, 1997. Diversity and depth distribution of SAR11 cluster rRNA genes from marine planktonic bacteria. Appl. envir. Microbiol. 63: 63–70.

Fossing, H., V. A. Gallardo, B. B. Jørgensen, M. Hüttel, L. P. Nielsen, H. Schultz, D. E. Canfield, S. Forster, R. N. Glud, J. K. Gunderson, J. Küver, N. B. Ramsing, A. Teske, B. Thamdrup & O. Ulloa, 1995. Concentration and transport of nitrate by the mat-forming sulphur bacterium *Thioploca*. Nature 374: 713–714.

Gallardo, V. A., 1977. Large benthic microbial communities in sulphide biota under Peru-Chile subsurface counter-current. Nature 268: 331–332

Gray, N. D., R. W. Pickup, J. G. Jones & I. M. Head, 1997. Ecophysiological evidence that *Achromatium oxaliferum* is responsible for the oxidation of reduced sulfur species to sulfate in a freshwater sediment. Appl. envir. Microbiol. 63: 1905–1910.

Giovannoni, S. J., T. B. Britschgi, C. L. Moyer & K. G. Field, 1990. Genetic diversity in Sargasso sea bacterioplankton. Nature 345: 60–63.

Güde, H., W. R. Strohl & J. M. Larkin, 1981. Mixotrophic and heterotrophic growth of *Beggiatoa alba* in continuous culture. Arch. Microbiol. 129: 357–360.

Hagen, K. D. & D. C. Nelson, 1996. Organic carbon utilisation by obligate and facultative autotrophic *Beggiatoa* strains in homogeneous and gradient cultures. Appl. envir. Microbiol. 62: 947–953.

Hagen, K. D. & D. C. Nelson, 1997. Use of reduced sulfur species by *Beggiatoa* spp.: enzymology and physiology of marine and freshwater strains in homogeneous and gradient cultures. Appl. envir. Microbiol. 63: 3957–3964.

Head, I. M., N. D. Gray, R. W. Pickup & J. G. Jones, 1995. The biogeochemical role of *Achromatium oxaliferum*. In Grimalt J. O. & C. Dorronsoro (eds), Organic Geochemistry: Developments and Applications to Energy, Climate, Environment and Human History. Selected papers from the 17th International Meeting on Organic Geochemistry, 4th–8th September 1995. Donostia-San Sebastián. The Basque Country Spain. AIGOA, Donostia-San Sebastián: 895–898.

Head, I. M., N. D. Gray, K. J. Clarke, R. W. Pickup & J. G. Jones, 1996. The phylogenetic position and ultrastructure of the uncultured bacterium *Achromatium oxaliferum*. Microbiology 142: 2341–2354.

Howarth, R., 1995. An investigation into the biogeochemistry of *Achromatium oxaliferum*. M.Sc. Thesis, University of Newcastle.

Huettel, M., S. Forster, S. Klöser & H. Fossing, 1996. Vertical migration in sediment-dwelling sulfur bacteria *Thioploca* spp. in overcoming diffusion limitations. Appl. envir. Microbiol. 62: 1863–1872.

Jørgensen, B. B., 1983. The Microbial Sulfur Cycle. In Krumbien W. E. (ed.), Microbial Geochemistry. Blackwell Scientific, Oxford: 91–124.

Jørgensen, B. B., & N. P. Revsbech, 1983. Colourless sulfur bacteria, *Beggiatoa* spp. and *Thiovulum* spp., in $O_2$and H_2S micro-gradients. Appl. envir. Microbiol. 45: 1261–1270.

Kuenen, J. G. & R. F. Beudecker, 1982. Microbiology of thiobacilli and other sulfur oxidising autotrophs, mixotrophs and heterotrophs. Phil. trans. R. Soc., Lond. B 298: 473–497.

Lane, D. J., A. P. Harrison Jr., D. Stahl, B. Pace, S. J. Giovannoni, G. J. Olsen & N. R. Pace, 1992. Evolutionary relationships amoung sulfur- and iron-oxidizing eubacteria. J. Bacteriol. 174: 269–278.

Larkin, J. M. & W. R. Strohl, 1983. *Beggiatoa*, *Thiothrix*, and *Thioploca*. Annu. Rev. Microbiol. 37: 341–367.

La Rivière, J. W. M. & K. Schmidt, 1989. The Genus *Achromatium*. In Staley J. T., M. P. Bryant, N. Pfennig & J. G. Holt (eds), Bergey's Manual of Systematic Bacteriology, Williams & Wilkins, Baltimore 3: 2131–2133.

La Rivière, J. W. M. & K. Schmidt, 1992. Morphologically conspicuous sulfur-oxidizing eubacteria. In Balows A., H. G. Trüper, M. Dworkin, W. Harder & K. H. Schleifer (eds), The Prokaryotes, 2nd edn, Springer Verlag, New York: 3934–3947.

Lauterborn, R., 1907. Eine neue Gattung der Schwefelbakterien (*Thioploca schmidlei* nov. gen. nov spec.). Ber. Dtsch. Bot. Ges. 25: 238–242.

Maier, S., 1984. Description of *Thioploca ingrica* sp. nov., nom. rev. Int. J. System. Bacteriol. 34: 344–355.

Maier, S., 1986. Diversity and ecology of the *Thioploca* group of aquatic bacteria. In Megusar F. & M. Gantar (eds), Perspectives in Microbial Ecology, Proceedings of the Fourth International Symposium on Microbial Ecology. Ljubljana 24-29 August 1986: 143–146.

Maier, S. & V. A Gallardo, 1984a. Description of *Thioploca araucae* sp. nov. and *Thioploca chileae* sp. nov.. Int. J. system. Bacteriol. 34: 414–418.

Maier, S. & V. A Gallardo, 1984b. Nutritional characteristics of two marine thioplocas determined by autoradiography. Arch. Microbiol. 139: 218–220.

McHatton, S. C., J. P. Barry, H. W. Jannasch & D. C. Nelson, 1996. High nitrate concentrations in vacuolate, autotrophic marine *Beggiatoa* spp.. Appl. envir. Microbiol. 62: 954–958

Møller, M. M., L. P. Nielsen & B. B. Jørgensen, 1985. Oxygen responses and mat formation by *Beggiatoa* spp. Appl. envir. Microbiol. 50: 373–382.

Moore, L. R., G. Rocap & S. W. Chisholm, 1998. Physiology and molecular phylogeny of coexisting *Prochlorococcus* ecotypes. Nature 393: 464–467.

Muyzer, G., A. Teske, C. O. Wirsen & H. Jannasch, 1995. Phylogenetic relationships of *Thiomicrospira* species and their identification in deep sea hydrothermal vent samples by denaturing gradient gel electrophoresis of 16S rDNA fragments. Arch. Microbiol. 164: 165–172.

Nelson, D. C., 1989. Physiology and biochemistry of filamentous sulfur bacteria. In Schlegel H. G. and B. Bowien (eds), Autotrophic Bacteria, Science Tech. Publishers; Madison, WI, U.S.A.

Nelson, D. C., 1992. The genus *Beggiatoa*. In A. Balows, H. G. Trüper, M. Dworkin, W. Harder & K-H. Schleifer (eds), The Prokaryotes, 2nd edn, Springer-Verlag, New York: 3171–3180.

Nelson, D. C. & R. W. Castenholz, 1981a. The use of reduced sulfur compounds by *Beggiatoa* sp. J. Bact. 147: 140–154.

Nelson, D. C. & R. W. Castenholz, 1981b. Organic nutrition of *Beggiatoa* sp. J. Bact. 147: 236–247.

Nelson, D. C. & R. W. Castenholz, 1982. Light responses of *Beggiatoa* spp.. Arch. Microbiol. 131: 146–155.

Nelson, D. C. & H. W. Jannasch, 1983. Chemoautotrophic growth of a marine *Beggiatoa* in sulfide gradient cultures. Arch. Microbiol. 136: 262–269.

Nelson, D. C., B. B. Jørgensen & N. P. Revsbech, 1986. Growth pattern and yield of a chemoautotrophic *Beggiatoa* sp. in oxygen-sulfide microgradients. Appl. envir. Microbiol. 52: 225–233.

Nelson, D. C., C. A. Williams, B. A. Farah, & J. M. Shively, 1989a. Occurrence and regulation of Calvin cycle enzymes in non-autotrophic *Beggiatoa* strains. Arch. Microbiol. 151: 15–19.

Nelson, D. C., C. O. Wirsen & H. W. Jannasch, 1989b. Characterization of large, autotrophic *Beggiatoa* spp. abundant at hydrothermal vents of the Guaymas Basin. Appl. envir. Microbiol. 55: 2909–2917.

Nübel, U., B. Engelen, A. Felske, J. Snaidr, A. Wieshuber, A. Amann, W. Ludwig & H. Backhaus, 1996. Sequence heterogeneities of genes encoding 16S ribosomal-RNAs in *Paenibacillus-polymyxa* detected by temperature-gradient gel-electrophoresis. J. Bact. 178: 5636–5643.

Polz, M. F., D. L. Distel, B. Zarda, R. Amann, H. Felbeck, J. A. Ott & C. M. Cavanaugh, 1994. Phylogenetic analysis of an highly specific association between ecto-symbiotic, sulfur-oxidizing bacteria and a marine nematode. Appl. envir. Microbiol. 60: 4461–4467.

Polz, M. F., E. V. Odintsova & C. M. Cavanaugh, 1996. Phylogenetic relationships of the filamentous sulfur bacterium *Thiothrix ramosa* based on 16S ribosomal RNA sequence analysis. Int. J. System. Bacteriol. 46: 94–97.

Pringsheim, E. G., 1967. Die mixotrophie von *Beggiatoa*. Arch. Mikrobiol. 59:247–254.

Robertson, L. A. & J. G. Kuenen. 1992. The colorless sulfur bacteria., In Balows A., H. G. Trüper, M. Dworkin, W. Harder & K. H. Schleifer (eds), The Prokaryotes, 2nd edn. Springer-Verlag, New York. 385–413.

Rosenberg, R., W. E. Arntz, E. Chumán des Flores, L. A. Flores, G. Carbajal, I. Finger & J. Tarazona, 1983. Benthos biomass and oxygen deficiency in the upwelling system of Peru. J. Mar. Res. 41: 263–279.

Schedel, M. & H. Trüper, 1980. Anaerobic oxidation of thiosulfate and elemental sulfur in *Thiobacillus denitrificans*. Arch. Microbiol. 124: 205–210.

Schewiakoff, W,. 1893. Über einen neuen bakterienähnlichen Organismus des Süßwassers. Heidelberg: Habilitationsschrift.

Schmidt, T. M., B. Arieli, Y. Cohen, E. Padan & W. R. Strohl, 1987. Sulfur metabolism in *Beggiatoa alba*. J. Bact. 169: 5466–5472.

Schultz, H. N., B. B. Jørgensen, H. A. Fossing & N. B. Ramsing, 1996. Community structure of filamentous, sheath building sulfur bacteria, *Thioploca* spp. off the coast of Chile. Appl. envir. Microbiol. 62: 1855–1862.

Strohl, W. R., 1989. Genus I. *Beggiatoa*. In Staley J.T., M. P. Bryant, N. Pfennig & J. G. Holt. (eds), Bergey's Manual of Systematic Bacteriology, Williams & Wilkins, Baltimore, 3: 2091–2097.

Strohl, W. R., G. C. Cannon, J. M. Shively, H. Güde, L. A. Hook, C. M. Lane & J. M. Larkin, 1981. Heterotrophic carbon metabolism by *Beggiatoa alba*. J. Bact. 148: 572–583.

Stumm, W. & J. J. Morgan, 1981. Aquatic Chemistry: An Introduction Emphasizing Chemical Equilibria in Natural Waters, Wiley-interscience, New York

Sweerts, J. P. R. A., D. De Beer, L. P. Nielsen, H. Verdouw, J. C. Van Den Heuvel, Y. Cohen & T. E. Cappenberg, 1990. Denitrification by sulfur oxidizing *Beggiatoa* spp. mats on freshwater sediments. Nature 344: 762–763.

Teske, A. P., N. B. Ramsing, J. Küver & H. Fossing, 1995. Phylogeny of *Thioploca* and related filamentous sulfide-oxidizing bacteria. Syst. Appl. Microbiol. 18: 517–526.

Thamdrup, B. & D. E. Canfield, 1996. Pathways of carbon oxidation in continental margin sediments off central Chile. Limnol. Oceanogr. 41: 1629–1650.

Timmer-TenHoor, A., 1975. A new type of thiosulfate oxidizing, nitrate reducing organism: *Thiomicrospira denitrificans* sp. nov. Neth. J. Sea Res. 9: 344–350.

Van de Peer, Y. & R. De Wachter, 1994. TREECON for Windows: a software package for the construction and drawing of evolutionary trees for the Microsoft Windows environment. Comput. Appl. Biosci. 10: 569–570.

Ward, D. M., M. M. Bateson, R. Weller & A. L. Ruff-Roberts, 1992. Ribosomal RNA analysis of microorganisms as they occur in nature. Adv. Microb. Ecol. 12: 219–286.

Whiteley, A. S., A. G. O'Donnell, S. J. MacNaughton & M. R. Barer, 1996. Cytochemical colocalization and quantitation of phenotypic and genotypic characteristics in individual bacterial-cells. Appl. envir. Microbiol. 62: 1873–1879.

Winogradsky, S., 1887. Über Schwefelbacterien. Bot. Zeitung. 45: 489–610.

Hydrobiologia **401:** 113–130, 1999.
J.P. Zehr & M.A. Voytek (eds), Molecular Ecology of Aquatic Communities.

The distribution and relative abundance of ammonia-oxidizing bacteria in lakes of the McMurdo Dry Valley, Antarctica

M. A. Voytek[1], J. C. Priscu[2] & B. B. Ward[3]
[1]*Water Resources Division, U.S. Geological Survey, Reston, VA 20192, U.S.A.*
Tel: [+1](703) 648-6894; E-mail: mavoytek@usgs.gov
[2]*Montana State University, Bozeman, MT, 59717, U.S.A.*
[3]*Dept. of Geosciences, Princeton University, Princeton, NJ 08540, U.S.A.*

Key words: nitrification, ammonia–oxidizing bacteria, amoA, Antarctic lakes, family and species specific PCR detection of nitrfying bacteria

Abstract

Marked differences in the concentrations of major ions and cations, macronutrient chemistry and general trophic status exist among the lakes of the McMurdo dry valleys in Antarctica. These differences have been attributed to both variations in stream inputs and *in situ* lake processes (Priscu, 1995; Lizotte et al., 1996, Spigel and Priscu, 1996). This study examines the role of nitrifying bacteria in nitrogen transformations in these lakes. Applying two polymerase chain reaction (PCR) assays targeting the 16S rRNA genes of ammonia-oxidizing bacteria and the active site of the ammonia monooxygenase gene (amoA), the distribution of ammonia-oxidizers was examined in six Antarctic lakes: Lake Bonney, Lake Hoare, Lake Fryxell and Lake Joyce in the Taylor Valley, Lake Miers in the the Miers Valley and Lake Vanda in the Wright Valley. Using a two stage amplification procedure, ammonia-oxidizers from both the beta and gamma- subclasses of the Proteobacteria were detected and their relative abundances were determined in samples collected from all sites. Ammonia-oxidizers were detected in all lakes sampled. Members of the gamma subclass were only present in the saline lakes. In general, nitrifiers were most abundant at depths above the pycnocline and were usually associated with lower concentrations of NH_4 and elevated concentrations of NO_3 or NO_2. The distribution of nitrifiers suggests that the primary N_2O peak observed in most of the lakes was produced via nitrification. Preliminary data on the rate of nitrification (Priscu et al., 1996) support the occurrence of nitrification and the presence of nitrifiers at the depth intervals where nitrifiers were detected. In all lakes, except Lake Miers, the data indicate that nitrifying bacteria have an important role in the vertical distribution of nitrogen compounds in these systems.

Introduction

Biogeochemical processes in the nitrogen cycle of aquatic systems are controlled primarily by microorganisms. One of these processes, nitrification, the two-step oxidation of NH_4^+ through NO_2^- to NO_3^-, is a link between oxidized and reduced forms of nitrogen and is, therefore, of fundamental importance in all ecosystems. Nitrification has been investigated for its potential importance as an oxygen sink, as a source of nitrite and nitrate, substrates for denitrification, which results in the subsequent loss of fixed nitrogen from the system, and in aquatic systems as the supply of nitrate fueling surface primary productivity (Ward, 1986). Under low oxygen conditions, nitrification can also result in the production and release of trace gases which are not available for phytoplankton uptake and may contribute significantly to the distribution and flux of nitrous oxide (N_2O) and nitric oxide (NO) in aquatic environments and the atmosphere (Goreau et al., 1980; Poth & Focht, 1985; Downes, 1988; Priscu et al., 1996). Nitrification is often tightly coupled to denitrification (dissimilatory nitrate reduction) in environments containing oxygen gradients. Nitrification and denitrification are performed by phylogenetically and metabolically distinct groups of bacteria; denitrification is an anaerobic heterotrophic process, but shares many of the same substrates and intermediates as nitrification. The spatial overlap of the distribution of nitrifying and denitrifying bacteria around environmental gradients and their metabolic overlap confound direct interpretation of rate measurements and features in nutrient profiles. As with nitrification, the significance of denitrification on global nitrogen budgets is still in dispute and varies considerably among systems.

Tracer methods have provided means to measure in-situ nitrification rates. Within marine aquatic environments, consistent patterns have emerged with respect to the magnitude and distribution of nitrifying activity. In general, higher nitrification rates are associated with more eutrophic regions or lower light environments (Ward, 1986). Minimum nitrification occurs in the surface photic zone where irradiance is high and substrates low; maximum rates are observed just below, in association with the nitrocline (Ward, 1986). Much less is known about water column nitrification in lakes and generalizations about the distribution of nitrification are not as straight forward (e.g. Hall, 1986). The environmental interpretation of these measurements often relies on laboratory studies examining the interaction of cultured nitrifying bacteria with their substrates and with environmental factors which may control their activity. A more appropriate analysis would employ studies of *in situ* populations using accurate methods for detecting and enumerating nitrifying bacteria. Immunofluorescent assays have been modified to study the serological diversity and distribution of nitrifying bacteria in soils (Schmidt, 1974), lake sediments (Smorczewski & Schmidt, 1991) and marine environments (Ward & Perry, 1982; Ward & Carlucci, 1985). However, this approach is limited by the difficulties associated with culturing representatives from the natural environment. Moreover, little is known about the diversity or genetic composition of lacustrine lithotrophic nitrifiers. The number of species that have been isolated and described from lakes is small even compared with the accepted low diversity of terrestrial and marine lithotrophic nitrifiers (Hall, 1986).

Studies based on the analyses of 16S rRNA and the use of the polymerase chain reaction have become common in the investigation of the spatial distribution of microbial taxa in the natural environment which have not been grown in culture (Giovanonni & Cary, 1993, Amann, 1995). Cultured ammonia-oxidizing bacteria belong to two phylogenetically distinct lineages within the Proteobacteria (Head et al., 1994; Teske et al., 1995); the gamma subclass line contains a single species and the beta subclass cluster contains several genera. Using a two stage amplification technique (Voytek & Ward 1995; Voytek, 1996), ammonia-oxidizing bacteria of the beta and gamma subclass can be detected and the relative abundance determined. There is some evidence that nitrifiers may be more phylogenetically diverse than we thought (i.e. not all nitrifying bacteria in natural systems belong to these two groups; Hovanec & DeLong, 1996). To try to capture the entire nitrifying population, techniques have been developed recently targetting a functional gene in the process of nitrification, ammonia monoxoygenase (AMO) (Holmes et al., 1995; Rotthauwe et al., 1997). AMO is the enzyme which catalyzes the oxidation of ammonia to hydroxylamine. The product of the amoA gene is the membrane bound active site polypeptide of AMO and primers designed to amplify this polypeptide have been developed as a means for analyzing indigenous ammonia-oxidizing bacteria (Rotthauwe et al., 1997).

In this study, we examined six permanently ice covered lakes in the McMurdo dry valleys, which is the largest relatively ice-free region on the Antarctic continent. The trophic status and chemistry of these lakes vary markedly and these differences are thought to be due to variations in meltwater input as well as in-situ biotic and abiotic lake processes (Priscu, 1995; Green et al., 1988). Using sensitive and specific molecular assays, we determined the distribution of ammonia-oxidizing bacteria to examine the role of nitrification in controlling the vertical distribution and magnitude of inorganic nitrogen compounds in these lakes.

Methods and materials

Chlorophyll *a*, ammonia, nitrite, nitrate and nitrous oxide concentrations were measured and the distribution of ammonia-oxidizers was examined in six Antarctic lakes: Lake Bonney (East and West Lobe), Lake Hoare, Lake Fryxell and Lake Joyce in the Taylor Valley, Lake Miers in the Miers Valley and from Lake Vanda in the Wright Valley.

Study sites

The McMurdo dry valleys ($\sim$ 4800 km^2, 77° 00′ S, 162° 52′ E) constitute the largest relatively ice-free region on the Antarctic continent and were formed by the advances and retreats of glaciers through the coastal ranges of the Transantarctic Mountains along

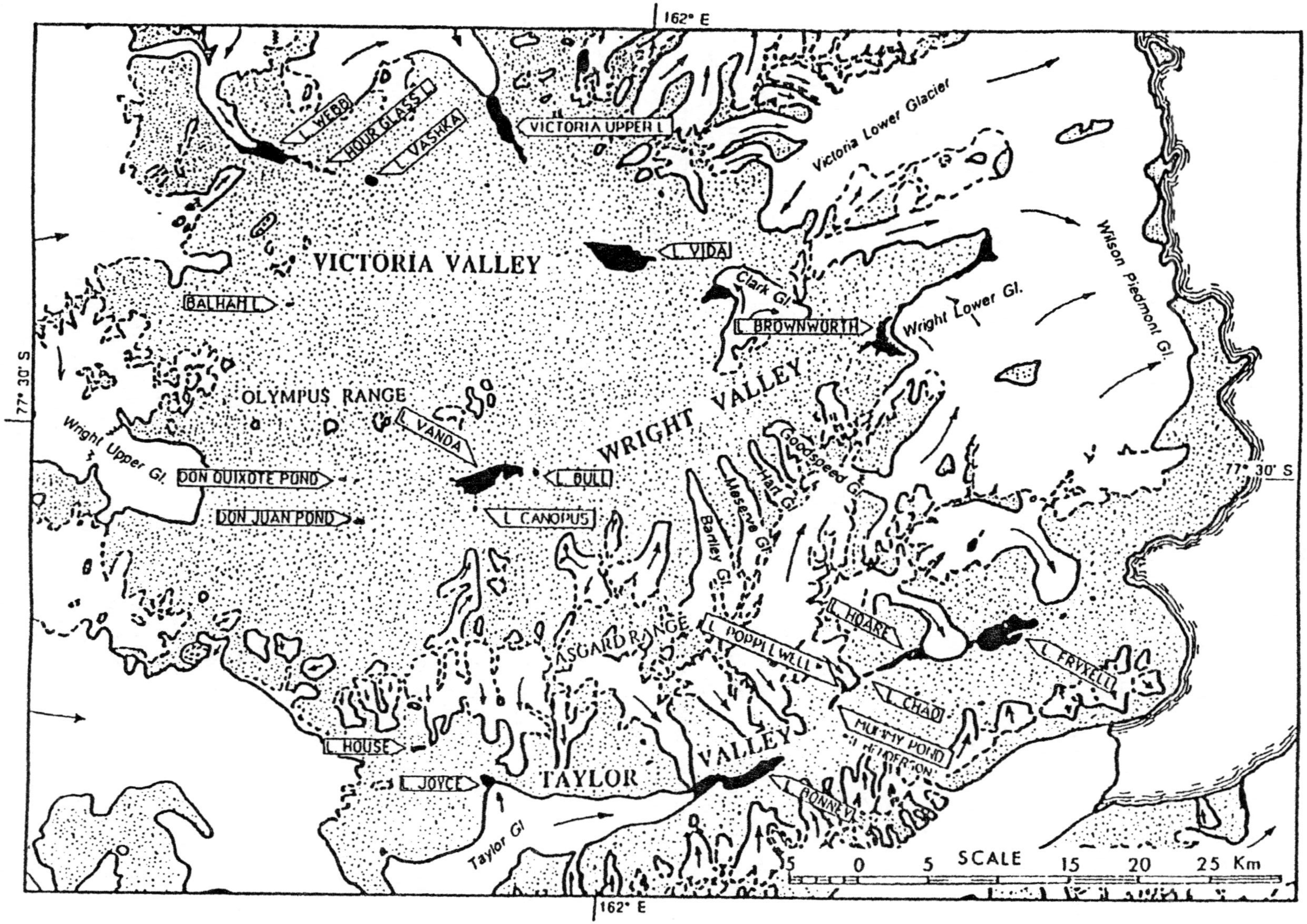

Figure 1. Map of the Antarctic Dry Valleys region (reprinted from Green et al., 1988).

the western coast of the Ross Sea. The valleys contain numerous closed basins (Figure 1) in which permanently ice-covered, chemically stratified lakes are found. The ice cover (3–6 m) has a profound effect on the chemical, biological and physical processes in the lakes. Specifically, it limits the transfer of mechanical energy from the wind and gas exchange with the atmosphere, modifies the quantity and spectral quality of solar radiation reaching the underlying water, and prevents the introduction of foreign species (Palmisano & Simmons, 1987; Lizotte & Priscu, 1992; Wharton et al., 1993; Priscu et al., 1996). Glacial meltwater and leaching from soils adjacent to streams provide freshwater and nutrients to the lakes.

Lake Fryxell is located near the mouth of the Taylor Valley proximate to McMurdo Sound (77° 37′ S, 163° 07′E, Figure 1). The total area of the lake is 7 km^2 and it is the shallowest lake in the study with a maximum depth of 18.5 m. Dissolved O_2 is supersaturated (>40 mg l^{-1}) above 9.5 m (Figure 2A). Water below 9.5 m is anoxic and the brackish, sulfidic bottom waters have salinities reaching 30% of seawater (Vincent, 1988). The temperature ranges from 0–4 °C. Inorganic nutrient concentrations are moderate, total organic carbon ranges from 1.4–29.1 (Matsumoto, 1989) and annual primary production is reported to be 7–50 g C m^{-2}yr^{-1} (Simmons et al., 1993).

Lake Hoare is a wet-based lake ponded at the western margin of the piedmont lobe of the Canada Glacier (77° 38′ S, 163° 07′ E, Figure 1). Its total area is 2.9 km^{2}and it has a maximum depth of 34 m. Lake Hoare is relatively fresh and is oxygenated throughout the water column (> 50 mg l^{-1}) with isolated anoxic zones near the bottom (Figure 3A). The temperature is nearly uniform at ~0 °C. Lake Hoare is oligotrophic with relatively low nutrient concentrations; annual production has been estimated to be 1.7 g C m^{-2} yr^{-1}(Simmons et al., 1993).

Lake Bonney (77° 43′ S, 162° 20′ E, Figure 1) is the largest lake in the Taylor Valley (area > 4 km^2). It is divided into two lobes by a sill (approximately 45 m wide and 13 m deep) which hydrographically isolates the deeper layers of each basin. Therefore, each lobe has been considered a separate lake system. The area of East Lobe Lake Bonney is 3.2 km^2 and the maximum depth is 40 m. Dissolved O_2 is supersaturated above 20 m (> 45 mg l^{-1}) and the water is fresh above 13 m (Figure 4A). Below the pycnocline, the water becomes anoxic and hypersaline reaching salinities of almost 10 times seawater. The temperature ranges from −2 to +6 °C (Priscu et al., 1993; Priscu, 1995). The West Lobe of Lake Bonney is a glacial margin lake at the advancing snout of the Taylor Glacier. It is smaller than the East Lobe with a total surface area of 0.8 km^2. The maximum depth is 40 m. Above 15 m the dissolved O_2 is supersaturated with values greater than 20 mg l^{-1}(Figure 5A). The deep water is anaerobic and salinities reach 6 times seawater (Priscu et al., 1993). The temperature ranges from −2 to +3 °C (Priscu et al., 1993; Priscu, unpublished).

Lake Joyce (77° 43′ S, 161° 33′ E, Figure 1) is an ice-dammed, proglacial lake against the marginal lobe of the Taylor Glacier in the Pearse Valley, an extension of the Taylor Valley. Its total surface area is 1 km^2 and the maximum depth is 35 m. The upper 20 m are well oxygenated and fresh (Figure 8A). Below 20 m, O_2 concentrations decrease and the salinity is 4‰. The temperature throughout the water column is nearly uniform and barely above freezing.

Lake Vanda (77° 32′ S, 161° 33′ E, Figure 1) is the largest lake in the Wright Valley with a total area of 5.2 km^2 and maximum depth of 80 m. The upper 65 m are oxic and relatively fresh (Figure 6A). The deep waters are warm (24 °C), hypersaline (9 times seawater) and sulfidic. Nutrient concentrations in the trophogenic zone of Lake Vanda are low and estimated annual primary production is low 0.34–2.5 g C m^{-2} yr^{-1} (Parker et al., 1982; Vincent & Vincent, 1982). Lake Vanda is relatively oligotrophic.

Lake Miers lies in the Miers Valley (78° 07′ S, 163° 54′ E, Figure 1). Its total area is 1.05 km^2 and maximum depth is 20 m. Glacial meltwater inflow from the Miers and Adams Glaciers is more significant at Lake Miers than at many of the other dry valley lakes (Andersen et al., 1993; Bell, 1967); it is the only lake we studied that has a surface outflow. The lake is relatively fresh and oxygen is present throughout the water column ranging from 4 mg l^{-1} to supersaturation (Figure 7A). The temperature range is 0–5 °C. Lake Miers is oligotrophic.

Sample collection

Each lake was sampled at a central site at least once during the 1993 austral summer. Sampling was done through a 10 inch diameter hole drilled through the ice cover. Depths are reported relative to the free water surface, i.e. the level to which water would rise in the sampling hole. Temperature and conductivity profiles were measured with a Seabird SBE 25 CTD system modified for use in these lakes (Spigel et al.,

1990). Oxygen was measured using a YSI oxygen probe. The YSI oxygen probe can measure a maximim oxygen concentrations of 20 mg l^{-1}, values higher than 20 mg l^{-1} were obtained by the Winkler method (see below). Water sampling was done with a 5 liter Niskin bottle at various depths based on the oxygen profile and CTD cast and encompassed significant hydrographic features. Samples were kept dark and on ice during transport back to the laboratory (<2 h). Samples for nutrient analysis were filtered (Whatman GF/F) and frozen until analysis. Bacterioplankton samples (approximately 4 liters) were concentrated 100 fold by ultrafiltration using a Filtron (Northborough, MA) open channel Ultrasette with a 300 kd nominal molecular weight cutoff membrane or a Pellicon tangential flow filtration system using GVLP, 0.22 μm ultrafilters (Millepore). The concentrate was filtered onto a 47 mm, 0.2 μm pore size Gelman Supor filter. Filters were stored frozen in EDTA (0.5 M, 0.5 ml) until total DNA was extracted.

Chlorophyll a *and nutrient analysis*

Chlorophyll *a* was measured on material collected on Whatman GF/F filters. Samples were extracted overnight in 90% acetone at −20 °C. Phaeophytin corrected chlorophyll *a* was quantified by comparing fluorescence of acidified and nonacidified samples measured on a Turner Designs AU10 fluorometer (Holm-Hansen et al., 1965). Ammonia, NO_2^- and NO_3^- were measured on 10 ml replicates of filtered (Whatman GF/F) lakewater samples following the protocols of Parsons et al. (1984). It was necessary to dilute samples exceeding seawater salinity with deionized water to yield salinities at or below seawater before analysis. Recovery of internal nutrient standards ranged between 90–110%. Samples for N_2O were collected following the procedure described in Priscu & Downes (1985) and Priscu et al. (1995) and was measured with a gas chromatograph fitted with a ^{63}Ni, electron capture detector (Priscu & Downes, 1985). Oxygen values obtained at Lakes Bonney, Hoare and Fryxell from the YSI oxygen probe were corroborated by Winkler titrations (Parson et al., 1984).

DNA extraction

High molecular weight DNA was extracted and purified from the frozen filters of concentrated bacteria following a standard protocol with slight modification (Ausubel et al., 1989; Kerkhoff & Ward, 1993). After ethanol precipitation the DNA was resuspended in TE2 (10 mM Tris and 0.1 mM EDTA).

Oligonucleotide primers

The universal eubacterial 16S rDNA primers EUB1 and EUB2 corresponding to sites at positions 9 through 27 and 1525 through 1542 respectively, of the Escherichia coli 16S rDNA (Liesack et al., 1991) were used to confirm the PCR quality of the DNA template. PCR primers corresponding to conserved sequences within the 5' and 3' regions of the 16S rDNAs of beta subclass (NITA and NITB; Voytek and Ward, 1995) and gamma-subclass (NOC1 and NOC2; Voytek, 1996) ammonia oxidizers were used to detect these groups. Amplification with these primers yields a 1080 bp NIT product and a 1130 bp NOC product, both internal to the EUB fragment. Primers designed to target the gene encoding the active site of the ammonia monoxygenase gene (amoA) were used to amplify an approximately 500 bp fragment (amoA-1F and amoA-2R; Rotthauwe et al., 1997). Primers were synthesized commercially (OPERON, Alameda, CA).

PCR amplification

PCR amplification was performed in a total volume of 100 μl using a DNA thermal cycler (Thermolyne Amplitron I, Barnstead) following the protocol described in Voytek & Ward (1995) and Voytek (1996) with slight modification. To reduce amplification of non-specific products the total number of cycles was reduced to 30. As discussed in Voytek & Ward (1995) a two stage PCR procedure is necessary for amplification of ammonia oxidizer template from the complex mixture of prokaryote and eukaryote DNA found in DNA extracts from natural samples. DNA templates were first amplified using the EUB primers. A 1 μl aliquot of the EUB amplification product mixture was added to a new reaction mixture and reamplified with either the NIT or the NOC primers. Amplification of the amoA product was performed following Rotthauwe et al. (1997). PCR-amplified products were resolved by electrophoresis.

Relative abundance of ammonia oxidizers

Direct quantification of the number of organisms in a natural sample based on specific PCR products amplified from DNA extracted from these samples is difficult. In all steps of the collection, extraction and amplification procedures error and uncertainty are introduced. Some cells are more efficiently collected by ultrafiltration and not all cells readily lyse with standard DNA extraction protocols. Additionally, the template (e.g. rRNA operons) copy number per cell may vary almost ten-fold and some templates amplify more efficiently than others (Suzuki & Giovannoni, 1996). Farrelly et al. (1995) have shown that without foreknowledge of the genome size and gene copy number in all members, accurate estimations of the abundance of a particular species is impossible. The two-step amplification protocol used here reinforces these errors. For these reasons, an absolute determination of the abundance of ammonia oxidizers was not possible. In order to estimate the relative abundance of ammonia oxidizers in each natural sample, the EUB amplification product was diluted before amplification with the NOC or NIT primers. The dilution series included 5 dilutions (undiluted, 1:2, 1:5, 1:10; and 1:50). A number (1–5) was assigned to a sample based on the highest dilution in the series (e.g. 1=undiluted; 5= 1:50) that yielded the correct amplification product. This number should reflect the relative abundance of nitrifiers. The rationale for this was based on the assumption that the greater the dilution factor the higher the number of the original template and thus the higher the number of ammonia oxidizers present in the original sample.

Results and discussion

Lake Fryxell

The temperature in Lake Fryxell was 0 °C just under the ice and increased to a maximum of 3 °C at 11 m. Oxygen was supersaturated in the surface (Figure 2A) and decreased steeply to zero between 9 and 10 m. A deep chlorophyll *a* maximum existed at 9 m (12 μg l^{-1}) with a secondary peak at 11 m (6 μg l^{-1}). All nitrogenous compounds were low or undetectable in the trophogenic zone (Figure 2B). Below the oxycline, NH_4^+ increased to 400 μM. N_2O was low at the surface (0.2 μg-at N l^{-1}) and decreased to below detection with depth. Nitrate was undetectable throughout the water column. Nitrite had two small peaks, one at 8 m and the other just above the bottom.

In general, ammonia-oxidizers of the beta-subclass were more abundant than those of the gamma subclass (Figure 2B). Beta nitrifiers were most abundant at depths above the oxycline. Although the distribution of the two groups overlaps, gamma subclass nitrifiers were found at deeper depths and into the oxygen transition zone. The presence of ammonia oxidizers is consistent with a significant contribution from nitrification to the shallow nitrite peak and the N_2O which accumulates above the oxycline. However, the sharp drop off of N_2O and the low concentrations of NO_2^- and NO_3^- below the oxycline are evidence of the anaerobic process, denitrification. No nitrifiers from either subclass were detected in the sulfidic bottom waters of Lake Fryxell. This is not surprising since sulfide has been reported as an inhibitor of nitrification (Snra & Baggaley, 1975; Capone & Kiene, 1988) and nitrifiers are obligate aerobes. Indeed, Joye & Hollibaugh (1995) suggested that H_2S levels may control the spatial and temporal differences in nitrification rates observed in certain marine environments. Ammonia oxidizers would not be expected to persist in the presence of sulfide since they derive the energy necessary for all cellular metabolism from nitrification, the oxidation of ammonia using oxygen.

Lake Hoare

The vertical temperature distribution in Lake Hoare varied little (range 0.2 – 0.7 °C) and the water column was well oxygenated throughout (Figure 3A). There was a deep chlorophyll *a* peak at 14 m (4 μg l^{-1}). The overall concentration of nitrogenous compounds in Lake Hoare were generally lower than the other lakes examined (Figure 3B). NH_4^+ was very low, less than 0.5 μM, throughout the water column. Nitrate was undetectable in the trophogenic zone and accumulated (up to 8 μM) below 15 m suggesting that nitrification is occurring and that it surpasses biotic utilization in deeper waters. None of the intermediates in nitrification accumulated in this lake: nitrite was below detection and N_2O was between 0.3- and 0.6 μg-at N l^{-1} throughout the water column. Although not very abundant, beta nitrifiers were detected at all depths sampled and support the assumption of nitrifying activity (Figure 3B). Gamma nitrifiers were not found at any depth in this lake. Although pockets of anoxic waters have been observed near the sediment water interface (D. Anderson, pers. comm.), there was no evidence suggesting that denitrification is an important process controlling nutrient concentrations in the overall well oxygenated waters of Lake Hoare.

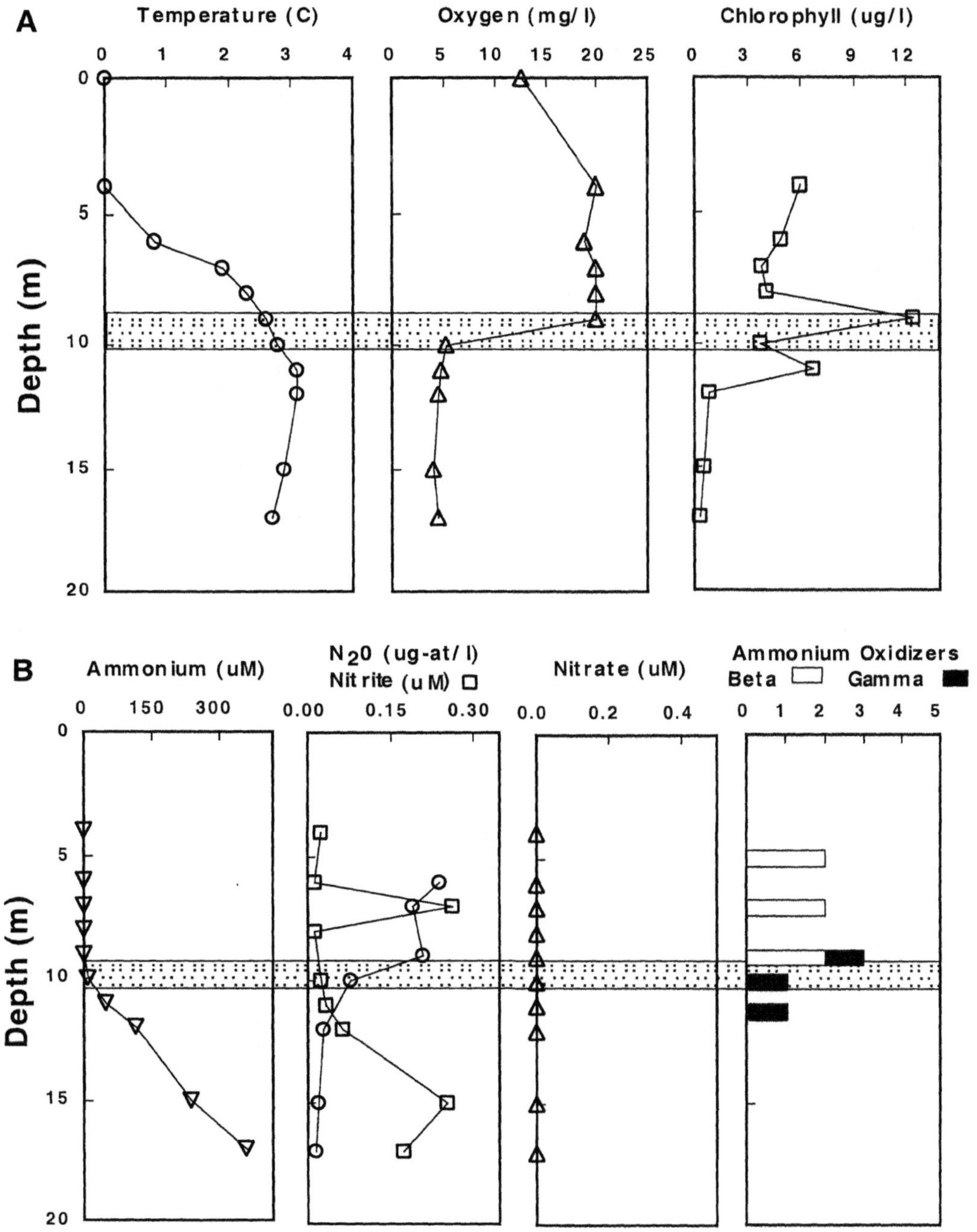

Figure 2. Temperature, oxygen, chlorophyll *a*, DIN, and beta and gamma ammonia oxidizer distributions in Lake Fryxell, November 1993. Stipled bar in panels A and B demarcates the oxic/anoxic transition.

Lake Bonney east lobe

Temperature increased in the east lobe of Lake Bonney from 0 °C just beneath the ice to 6.1 °C at 15 m and decreased to −2 °C near the bottom (Figure 4A). Chlorophyll *a* concentrations were highest just beneath the ice (2 $\mu g\ l^{-1}$) and decreased rapidly with depth. A strong chemocline occurred at about 17 m, above which conditions were aerobic and nutrient depleted. Below this region, O_2 dropped from supersaturated levels to nearly zero and dissolved inorganic N, including oxidized forms, increased rapidly, even though the deep waters (down to the bottom at

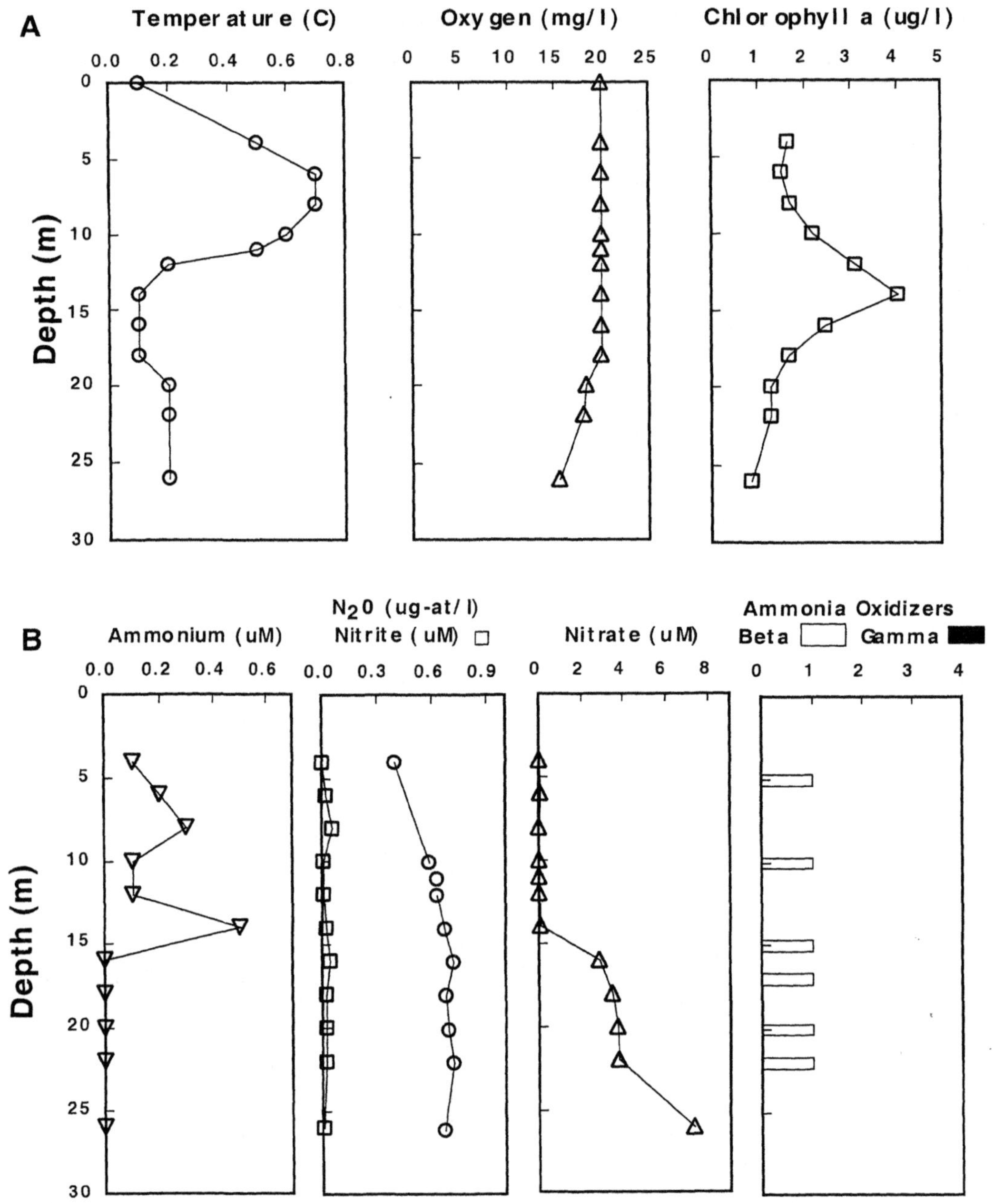

Figure 3. Temperature, oxygen, chlorophyll *a*, DIN, and beta and gamma ammonia oxidizer distributions in Lake Hoare, November 1993.

40 m) were virtually anoxic (Figure 4B). A broad N_2O peak occurred between 20–24 m reaching extremely high values of 40 μg-at N l^{-1}.

The presence of nitrifying bacteria was detected by the PCR assay from 7 depths above 25 m in the East Lobe of Lake Bonney (Figure 4B). Beta ammonia oxidizers were far more abundant than gamma ammonia oxidizers with relative abundances highest in the aerobic waters above the chemocline. Gamma nitrifiers had a small peak around 15 m. Despite the abundance of ammonia oxidizers in the surface waters, NO_2^- and NO_3^- do not accumulate presumably because these nutrients are readily used by phytoplankton and bacteria in the euphotic zone.

^{15}N natural abundance studies in Lake Bonney (Wada et al., 1984), indicate that the NO_3^- in the

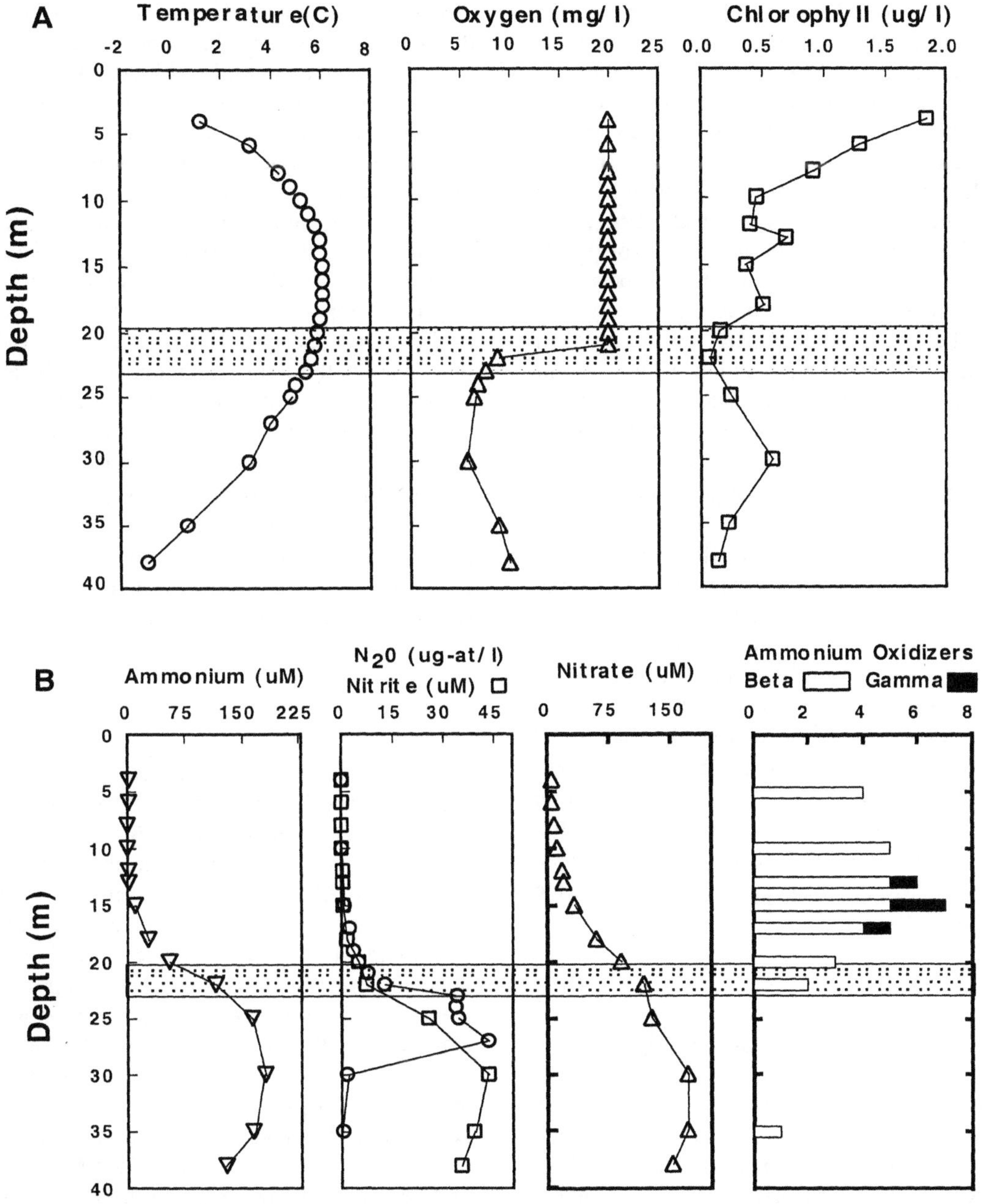

Figure 4. Temperature, oxygen, chlorophyll *a*, DIN, and beta and gamma ammonia oxidizer distributions in Lake Bonney (East Lobe), November 1993. Stipled bar in panels A and B demarcates the oxic/anoxic transition.

deep water of the east lobe is highly enriched in ^{15}N. Wada et al. (1984) proposed that microbial activity such as denitrification or nitrification may cause the observed variation in ^{15}N, as no other processes could produce the magnitude of enrichment observed. N_2O levels measured just below the chemocline exceeded 550 000% of air saturation for that depth and salinity, the highest recorded in any natural system (Priscu et al., 1996). Denitrification could be responsible for high levels of N_2O since it is an intermediate in the process; however, the NO_3^- and NO_2^- profiles show high levels (>250 μM and > 40 μM respectively) in the hypoxic deep water indicating bulk denitrification is not occurring. Moreover, although the denitrification substrates are available, oxygen is low and denitrifying bacteria have been detected and isolated from these waters (Ward & Priscu, 1996), denitrification based on acetylene block assays could not be detected in this lobe (Priscu et al., 1993; Priscu et al., 1996).

It has been shown that cultures of ammonia oxidizing bacteria can produce N_2O at low oxygen tension (Goreau et al., 1980; Anderson et al., 1993) and nitrification has been suggested as the source of N_2O in many natural systems (Vincent et al., 1981; Kaplan & Wofsy, 1985; Priscu et al., 1986; Tortoso & Hutchinson, 1990). Using ^{14}C-bicarbonate based inhibitor experiments, ammonia oxidizer activity was detected at the top of the N_2O peak but not at 24 m (Priscu et al., 1996). The distribution of nitrifiers in the east lobe suggested that nitrification could be at least partially responsible for the shallower N_2O peak but does not support nitrification as the source of the lower peak. Priscu et al. (1996) have suggested that the peak we observe today is a 'fossil record'. They proposed that even small transformations of N are relatively easy to detect as chemical gradients within the water column of these ice covered lakes. Lack of turbulence allows chemical gradients to persist over long periods of time and to develop even if reaction rates are extremely low. Moreover, chemical gradients can persist even when there is no present day biological activity. Hence, the broad N_2O maximum observed in the East Lobe may be the result of past nitrification activity. There is no evidence of consumption within the peak and although diffusion may have dispersed the peak, the rates are not great enough to dissipate the signal substantially.

Lake Bonney west lobe

The chemistry and *in situ* processes shaping nutrient profiles in the west lobe of Lake Bonney appear to be very different from the adjacent east lobe. The temperature maximum (3 °C) occurred at 10 m and temperature decreased below the pycnocline to −2 °C (Figure 5A). The surface standing stock of biomass was higher in the west lobe than in the east lobe: chlorophyll *a* concentrations were 1–1.5 μg l^{-1} under the ice and peaked (6 μg l^{-1}) near the base of the trophogenic zone (13 m). The chemocline and oxycline began at 13 m and 15 m respectively, both shallower than in the east lobe (Figure 5 A & B). The reduced N substrate, NH_4^+ was undetectable in the trophogenic zone and steadily increased to 300 μM at the bottom. Nitrite, nitrate and N_2O peaked between 13–17 m, concentrations reaching 0.6 μM, 25 μM and 1.1 μg at N l^{-1}, respectively, and then decreased rapidly. Ammonia-oxidizing bacteria were detected throughout the water column but were less abundant here than in the east lobe (Figure 5B). Beta nitrifiers were most abundant above the oxycline and the distribution of the less abundant gamma nitrifiers was limited to the oxygen transition zone. Nitrification by these bacteria was probably responsible for the peaks of N_2O, NO_2^- and NO_3^-. Nitrifiers detected in the anoxic bottom waters were probably inactive.

Wada et al. (1984) found the highest d^{15}N values for N_2 yet reported in an aquatic system (1.5–2.5‰) in the anoxic layers of the West Lobe of Lake Bonney. These authors suggest intense denitrification may be responsible for the high d^{15}N observed. The concentrations of inorganic oxidized nitrogen, both NO_2^- and NO_3^-, decreased steeply in the deep water below the oxycline in this lobe, which supports this contention. Furthermore, Priscu et al. (1996) have measured significant rates of denitrification in the West Lobe and Ward & Priscu, (1997) isolated and enumerated (by immunofluorescence) denitrifying bacteria in this region.

Lake Vanda

The upper layer of Lake Vanda extending to 55 m depth was essentially fresh, well oxygenated and the temperature ranged between 0–6 °C (Figure 6A). Below 55 m, the temperature steadily increased to a maximum of 23 °C and the oxygen steeply decreased, reaching zero below 67 m. A deep chlorophyll *a* maximum of 0.45 μg l^{-1} occurred at 67 m. The concentrations of all nitrogenous compounds measured were low or undetectable in the trophogenic zone (Figure 6B). Below the oxycline, NH_4 increased to 1500 μM. N_2O began to accumulate at 55 m, peaked at 67 m (7.9 μg-at N l^{-1}) and rapidly decreased to zero below the oxycline. Nitrate and nitrite peaked at 67 (268 μM) and 70 m (1.65 μM), respectively and decreased below. In general, ammonia-oxidizers of the beta-subclass were more abundant than those of the gamma subclass (Figure 6B). Beta nitrifiers were detected at all depths sampled above the oxycline and were most abundant just above the oxycline. The less abundant gamma subclass nitrifiers were not found in the shallowest (5 and 10 m) samples but were present at deeper depths. Neither group was present in the oxygen transition zone at oxygen concentrations below 4 mg l^{-1} or in the anoxic zone. The nitrification activity of these ammonia oxidizers may be responsible for the peaks in N_2O and nitrate. The disappearance

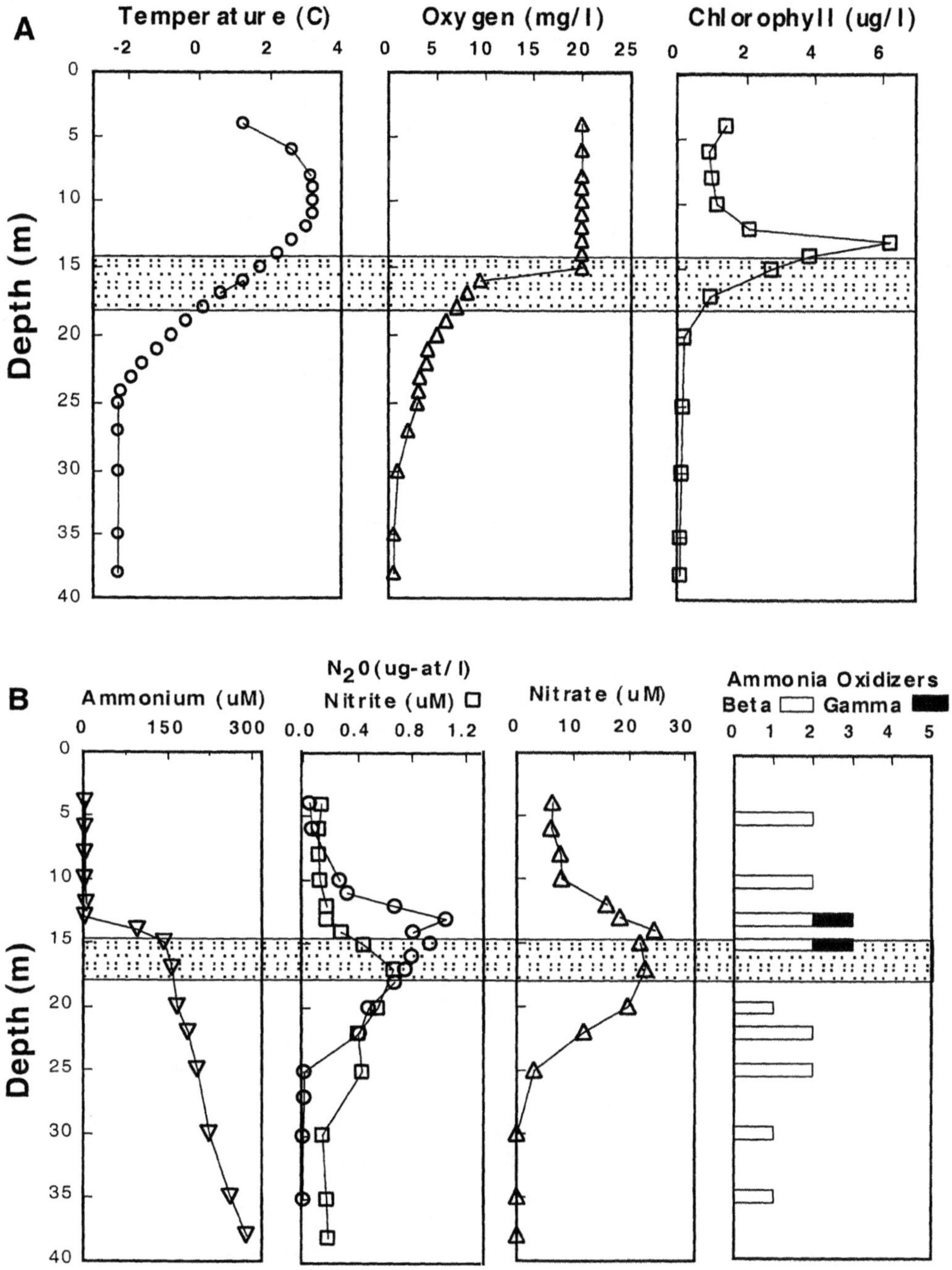

Figure 5. Temperature, oxygen, chlorophyll *a*, DIN, and beta and gamma ammonia oxidizer distributions in Lake Bonney (West Lobe), November 1993. Stipled bar in panels A and B demarcates the oxic/anoxic transition.

of N_2O, NO_2^- and NO_3^- below the oxycline are evidence of the anaerobic process denitrification. No nitrifiers were detected in the sulfidic bottom waters of Lake Vanda. As discussed above, it would be unlikely for nitrifiers to persist in the presence of sulfide which shuts down nitrification.

Vincent et al. (1981) examined N-fluxes in the Dry Valley Lakes. They found a region of intense N_2O cycling near the chemocline in Lake Vanda, mediated by nitrification, denitrification and diffusion. Maximum N_2O concentrations in Lake Vanda exceeded 20 000% of air saturation. They found the rate of

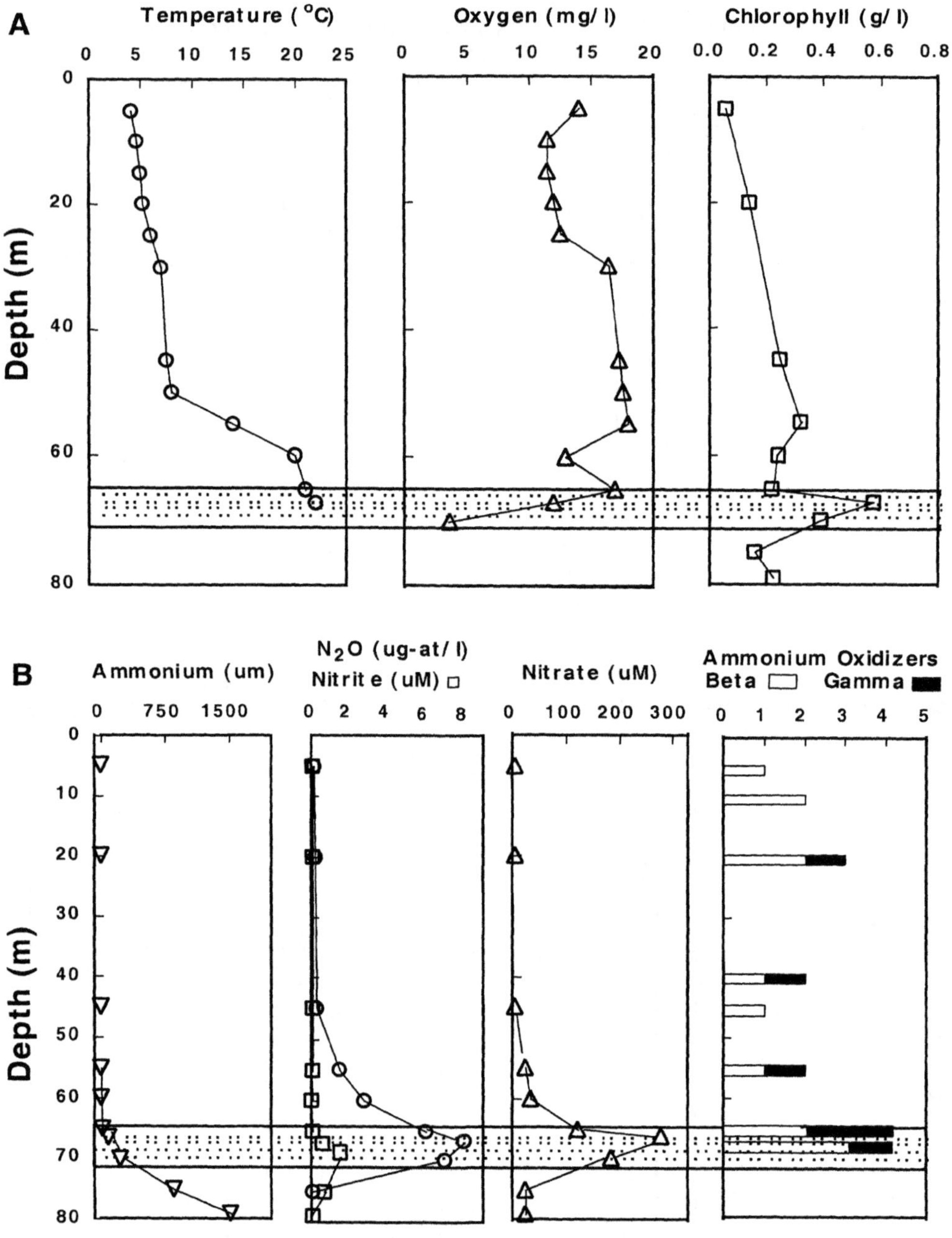

Figure 6. Temperature, oxygen, chlorophyll *a*, DIN, and beta and gamma ammonia oxidizer distributions in Lake Vanda, November 1993. Stipled bar in panels A and B demarcates the oxic/anoxic transition.

nitrification highest at a layer just above the deep chlorophyll maximum (50–55 m) and denitrification was poised between the DCM and the sulfate reducing community (61–65 m). Vincent (1987) suggested that the various forms of nitrogen in Lake Vanda regulated the discrete layering of organisms. The lake levels in the Dry Valleys have been rising, which should result in a transposition of the relative position of the populations, but the processes should continue in the same relative depth horizons. At present lake level, N_2O, NO_3^- and NO_2^- begin to accumulate and then peak at or just above the deep chlorophyll maximum, which occurs at 67 m. These chemical features and

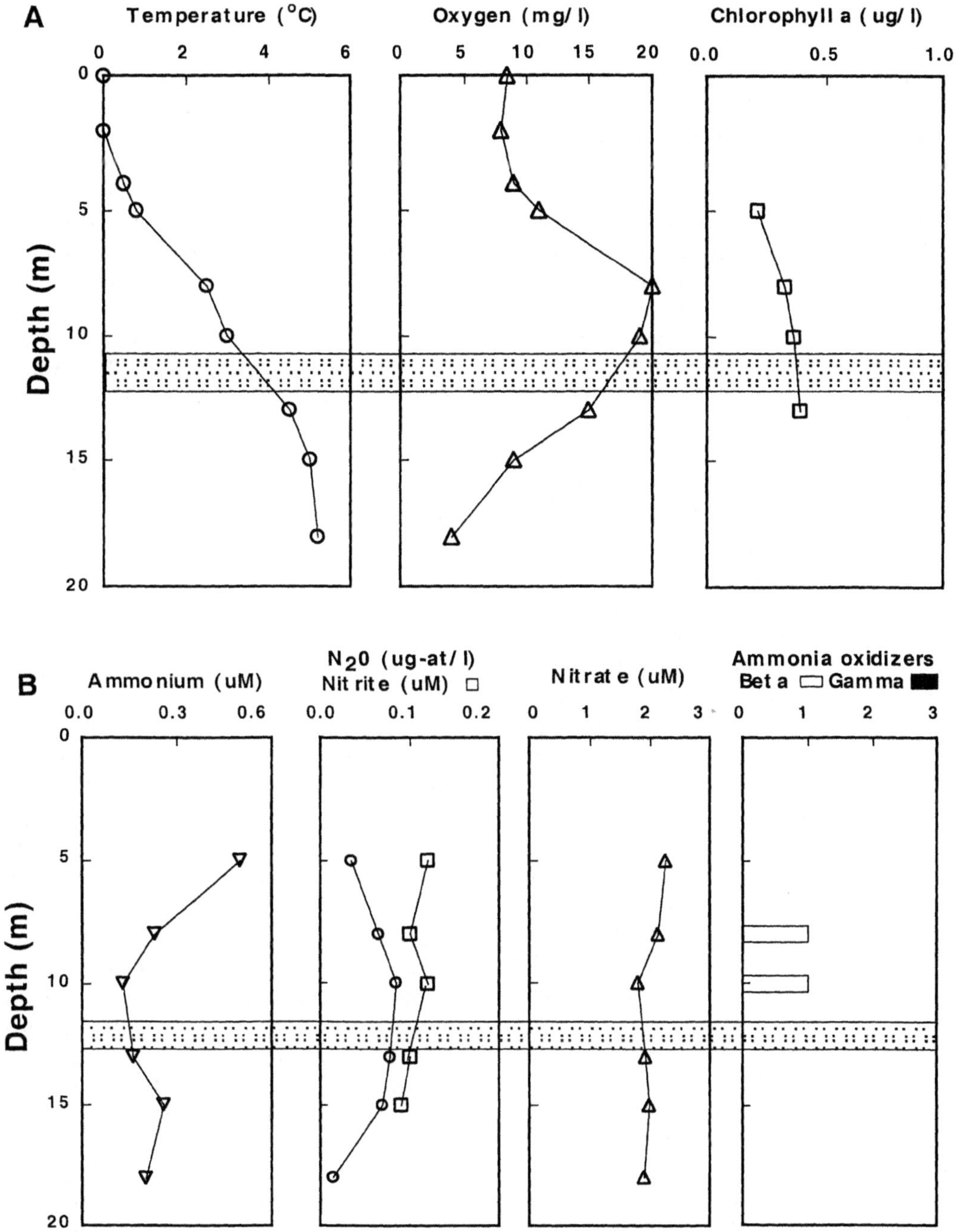

Figure 7. Temperature, oxygen, chlorophyll *a*, DIN, and beta and gamma ammonia oxidizer distributions in Lake Miers, November 1993. Stipled bar in panels A and B demarcates the oxic/anoxic transition.

the observed distribution of nitrifiers support Vincent's model.

Lake Miers

The temperature increased steadily from 0 °C at the surface to 5.2 °C at 18 m (Figure 7A). Oxygen concentrations were above 4 mg l^{-1} at all depths and oxygen was supersaturated at 15 m. Chlorophyll *a* was very low (less than 0.4 $\mu g\ l^{-1}$). Reported total heterotrophic bacterial abundances are very low (Simmons et al., 1993). The concentrations of all nitrogenous substrates, products and intermediates were low throughout the water column and the profiles were nearly uniform (Figure 7B). NH_4 never exceeded 0.5 μM, nitrite <0.12 μM, nitrate <2.3 μM and N_2O was less than 0.09 μg-at N l^{-1}. Beta subclass nitrifiers were detected at two of the depths sampled and were not very abundant (Figure 7B). The low substrate and product levels observed here suggest that nitrification rates would be low and transformations due to nitrification would be insignificant in this ultra-oligotrophic lake. Of all the lakes sampled, the chemistry of Lake Miers most closely resembles the chemistry of the glaciers and meltwater that feed into the lake (Green et al., 1988) suggesting that there has been little change due to in-situ processes. A possible explanation may be that Lake Miers is relatively young and biogeochemical cycles are still evolving. Green et al. (1988) estimated the chloride age of Lake Miers to be 300 years. In addition, Lake Miers is the only lake with hydraulic through flow with a large annual influx of meltwater and substantial outflow.

Lake Joyce–There was a narrow temperature range in Lake Joyce (0-0.6°C) with a slight warming at 10 m (Figure 8A). Oxygen was supersaturated in the surface and decreased to zero below 20 m. The chlorophyll *a* maximum (5 μg l^{-1}) occurred at 10 m. With the exception of nitrate, the concentrations of nitrogenous compounds at the surface were low (Figure 8B). Below the oxycline, NH_4^+ was regenerated and increased to 12 μM. N_2O was low at the surface (0.2μg-at N l^{-1}), increased to a maximum of 0.97 μg-at N l^{-1} at 20 m and disappeared below the oxycline. Nitrate was relatively high in the upper water column, peaking at 15 m (130 μM) and disappeared rapidly below 15 m. Nitrite was 0.6 μM at the surface and decreased with depth. Beta nitrifiers were detected at depths above the oxycline and at the top of the transition (Figure 8B). Gamma subclass nitrifiers are found at deeper depths and through the oxygen transition zone. Ammonia oxidizers are probably responsible for the nitrate peak and the N_2O above the oxycline since their distribution overlaps with the distritbution of N_2O. The sharp drop off of N_2O and the low concentrations of NO_2^- and NO_3^- below the oxycline are evidence of the anaerobic process denitrification. The distribution of nitrifiers in this lake was also restricted to depths above the region of sulfate reduction (sulfide was detected in samples collected from 25 to 30 m). Detection and distribution of the amoA gene.

AmoA was detected in all lakes and followed the same distribution pattern as the nitrifier 16S rRNA genes (data not shown). Surprisingly, no additional information on the distribution of ammonia oxidizers was obtained from the use of amoA. Amplification of the amoA gene yielded the correct product in all samples where either the gamma or beta or both subclasses were detected and was not amplified in samples where the 16S rRNA genes were absent. Since none of the products from any of the amplifications were sequenced, we cannot determine if any additional nitrifier subclasses were present in our samples. If they were present and if we had used the amoA product to determine the relative abundance of nitrifiers, we may have observed different distribution patterns and depth profiles. Moreover, abundance estimates calculated from a functional gene may be less variable and more reliable. In general, the copy number of functional genes is 1–3 copies per genome (Norton et al., 1996) whereas, the number of rRNA genes can vary from 2 to 13.

Conclusion

Factors controlling ammonia oxidizer distributions

The distributions of ammonia-oxidizing bacteria observed in these Antarctic lakes are generally consistent with expectations based on available information on the factors that control nitrification rates and the growth of nitrifiers in culture. In this study nitrifiers were always found in the surface waters and in most lakes they were detected just beneath the ice. Nitrifiers have been reported to be light sensitive and in general, nitrifying activity is higher in low light environments, e.g. deeper in the water column (Olson, 1981; Ward, 1986). All of these lakes are permanently covered by 2.8–6 m of ice which limits penetration of light to between 0.8 and 14.7% (10–190 μE m^{-2} sec^{-1} PAR) incident irradiation (Palmisano & Simmons, 1987; Wharton et al., 1993). Therefore, light is not likely to limit the distribution of nitrifiers in surface waters. The rates of nitrification and growth of nitrifiers are also controlled by the concentration and/or supply of their energy substrate NH_4^+ (Ward, 1986; Koops & Möller, 1992). Although some of the lakes have well developed benthic microbial mats that could fix nitrogen (Wharton et al., 1983), the new nitrogen is primarily supplied by meltwater, usually in the form of nitrate. Lithoautotrophic bacteria would have to compete with primary producers and heterotrophic bacteria for these nutrients in the trophogenic zone. Rapid utilization in excess of inputs

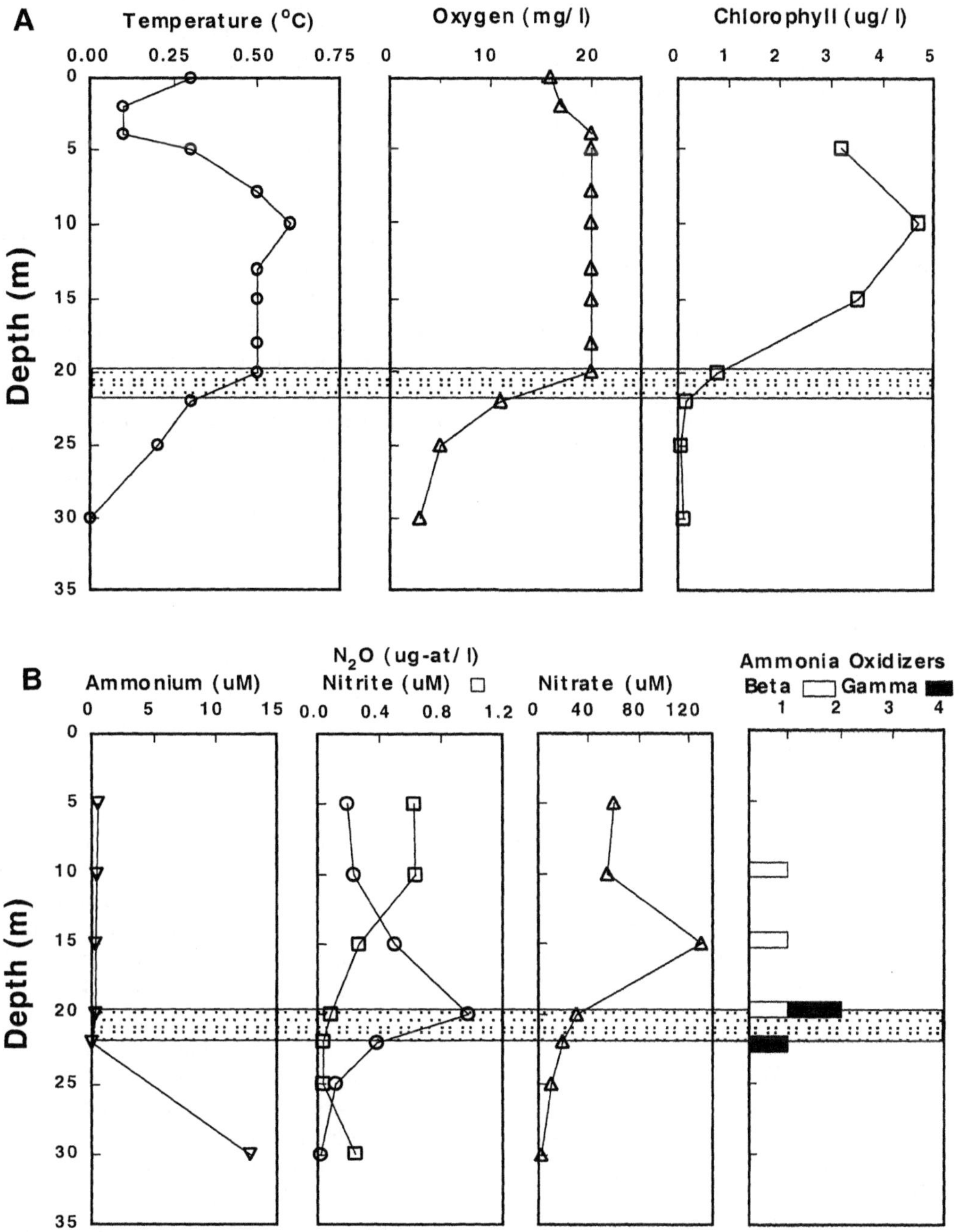

Figure 8. Temperature, oxygen, chlorophyll *a*, DIN, and beta and gamma ammonia oxidizer distributions in Lake Joyce, November 1993. Stipled bar in panels A and B demarcates the oxic/anoxic transition.

during the late austral summer when melt stream flow and primary production are highest, presumably results in the low concentrations observed there. Under these conditions the nutrient supply to nitrifiers may be low and we would expect low activity. In temperate marine systems, water column nitrification rates are usually lowest in the trophogenic zone (Ward, 1986). Although often present at surface depths in Antarctic lakes, the highest abundances of nitrifers were usually observed deeper in the water column.

As particulate organic matter containing fixed nitrogen settles, it undergoes microbial degradation, thus releasing nitrogen (NH_4^+ and R-NH_2) which is then available to organisms at those depths or the reduced compounds can diffuse upward to fuel more production. In the hypersaline lakes, Lakes Bonney and Vanda for example, the sinking of organic detritus is severely retarded by the steep density gradients and the high density of deep waters, thus most of the material is degraded before reaching the sediments. High concentrations of refractory organic carbon are often observed in the anoxic bottom layers of these lakes, supporting this contention (Matsumoto et al., 1993). The regeneration and small, but significant, upward supply of NH_4^+ should favor nitrification and indeed, the highest relative abundances of both subclasses were found at depths near the nitracline. Rates of nitrification have been difficult to measure in some of these lakes since nitrifiers are slow growing and rates may be retarded by low temperatures. However, Vincent et al. (1981) and Priscu et al. (1996) were able to measure nitrification rates directly in Lake Vanda and Bonney, respectively. They found the rate of nitrification highest at depths just below the nitracline. In addition, at deeper depths in Lake Bonney, Priscu et al. (1996) postulated the possibility of 'fossil' nitrification activity, i.e. product accumulation from nitrification occurring hundreds to thousands of years ago.

A final environmental factor controlling nitrification and nitrifiers is oxygen (Koops & Möller, 1992). Peaks in nitrifier abundance were almost always found near the oxygen transition zone. Ammonia oxidizers are obligate aerobes but are often found in microaerophilic environments (Koops & Möller, 1992). Under low concentrations of oxygen N_2O is produced during nitrification (Goreau et al., 1980). N_2O is also a product and substrate in the anaerobic microbial process of denitrification. Within the transition zone, nitrification and denitrification compete for substrates and the resultant chemical profiles reflect the activity of organisms responsible for both processes. In anaerobic waters, denitrification becomes the primary process controlling nitrogen; denitrifiers use oxidized forms of nitrogen as alternative respiratory substrates. Although nitrifiers were detected in these waters, nitrification should not be occurring in the complete absence of oxygen. In the more extreme case where sulfur reduction has occurred, denitrification has completely removed the oxidized forms of nitrogen. Sulfide has been shown to inhibit nitrification (Joye & Hollibaugh, 1995) and therefore, should limit the distribution of nitrifiers. Indeed, the distribution of nitrifiers in Lakes Fryxell, Vanda and Joyce, where sulfide was detected, appears to be restricted to depths above the region of sulfate reduction.

Origin and relative abundances of beta and gamma subclass ammonia oxidizers

In general, there were more beta subclass than gamma subclass ammonia oxidizers present in these lakes and the vertical distribution pattern of the beta subclass nitrifiers was broader. One reason for this difference is simply that the beta subclass of ammonia oxidizers contains members from many genera which have been isolated from diverse environments: soils, marine and freshwater environments. The gamma subclass has one genetically confirmed member, thus far, and it is the marine species Nitrosococcus oceanus. It is not surprising that one species from a limited environmental niche may be less abundant and have a restricted distribution in a different environment compared to a group with environmentally diverse members. Alternatively, the distribution we observed for the gamma nitrifier may be a result of a difference in its response to environmental factors. The gamma nitrifier was found in deeper waters, through the transition zones and often below in anaerobic waters. It may be more light sensitive and more tolerant of sulfides and low oxygen than members of the beta group. Finally, the original mechanism of species seeding of these lakes when they were original formed may also account for some of the differences. The origin and precise mechanism of formation and development of these lakes is still controversial. Lake Fyxell, Bonney, Joyce and Vanda have ion signatures that indicate a marine source (Green et al., 1988; Chin, 1993). Chinn (1993) suggested that many of the lakes in the Dry Valleys were originally formed when the area lifted isostatically from the sea and sea water was trapped in basins formed by large outlet glaciers. However, Green et al., (1988) note that the major ionic components in Lakes Fryxell, Hoare, Joyce and Miers can be largely understood in terms of their present inflows without invoking trapped sea water. Others have suggested that Lakes Vanda, Bonney, Joyce and Fryxell have been influenced by trapped seawater in glacier or marine aerosols or that Lakes Fryxell and Vanda have had direct contact with McMurdo Sound in their recent past (Hendy et al., 1977; Wilson, 1981; Chinn, 1993). Thus the source of the gamma subclass nitrifier may be ultimately marine. No physical or chemical evidence of a marine origin is found in Lake Miers or Hoare where only members of the beta subclass are found.

References

Amann, R. I, W. Ludwig & K-H. Schleifer, 1995. Phylogentic identification and *in situ* detedction of individual microbial cells without cultivation. Microbiol. Rev. 59: 143–169.

Anderson, I. C., M. Poth, J. Homstead & D. Burdige, 1993. A comparison of NO and N_2O production in the autotrophic nitrifier Nitrosomonas europaea and the heterotrophic nitrifier Alcaligenes faecalis. Appl. envir. Microbiol. 59: 3525–3533.

Ausubel, F. M., R. Brent, R. E. Kingston, D. D. Moore, J. G. Seidman, J. A. Smith & K. Struhl (eds), 1989. Current Protocols in Molecular Biology. Green Publishing Associates and Wiley-Interscience, New York.

Bell, R. A. I., 1967. Lake Miers, South Victoria Land Antarctica. J. Geol. Geophys. 10: 540–556.

Canfield, D. E. & W. J. Greene, 1985. The cycling of nutrients in a closed-basin Antarctic lake: Lake Vanda. Biogeochem. 1: 233–256.

Capone, D. G. & R. P. Kiene, 1988. Comparison of microbial dynamics in marine and freshwater sediments: Contrasts in anaerobic carbon catabolism. Limnol. Oceanogr. 33: 725–749.

Chinn, T. J., 1993. Physical hydrology of the dry valley lakes. In W. J. Green & E. I. Friedmann (eds), Antarctic Research Series: Physical and Biogeochemical Processes in Antarctic Lakes. A.G.U., Washington, D.C. 59: 1–52.

Downes, M. T., 1988. Aquatic nitrogen transformations at low oxygen concetrations. Appl. envir. Microbiol. 54: 172–175.

Farrelly, V., F. A. Rainey & E. Stackebrandt, 1995. Effect of genome size and rrn gene copy number of PCR amplification of 16S rRNA genes from a mixture of bacterial species. Appl. envir. Microbiol. 61: 2798–2801.

Giovannoni, S. J. & S. C. Craig, 1993. Probing marine systems with ribosomal RNAs. Oceanogr. 6: 95–104.

Goreau, T. J., W. A. Kaplan, S. C. Wofsy, M. B. McElroy, F. W. Valois & S. W. Watson, 1980. Production of NO_2 and N_2O by nitrifying bacteria at reduced concentrations of oxygen. Appl. envir. Microbiol. 40: 526–532.

Green, W. J., M. P. Angle & K. E. Chave, 1988. The geochemistry of antarctic streams and their role in the evolution of four lakes of the McMurdo Dry Valleys. Geochim. Cosmochim. Acta 52: 1265–1274.

Green, W. J., T. J. Garner, T. G. Ferdelman, M. P. Angle, L. C. Varner & P. Nixon, 1989. Geochemical processes in the Lake Fryxell basin (Victoria Land Antarctica). Hydrobiologia 172: 129–148.

Hall, G. H., 1986. Nitrification in Lakes. In J. I. Prosser (ed.), Nitrification IRL Press, Washington, D.C. 127–156.

Head, I. M, W. D. Hiorns, T. Martin, A. J. McCarthy & J. R. Saunders, 1993. The phylogeny of autotrophic ammonia-oxidizing bacteria as determined by analysis of 16S ribosomal RNA gene sequences. J. gen. Microbiol. 139: 1147–1153.

Hendy, C. H., A. T. Wilson, K. B. Popplewell & D. A. House, 1977. Dating of geochemical events in Lake Bonney, Antarctica and their relation to glacial and climate changes. N.Z. J. Geol. Geophys. 20: 1103–1122.

Holmes, A. B., A. Costello, M. E. Lidstrom & J. C. Murell, 1995. Evidence that particulate methane monooxygenase and ammonia monooxygenase may be evolutionarily related. FEMS Microbiol. Lett. 132: 203–208.

Holm-Hansen, O., C. J. Lorezen, R. W. Holmes & J. D. H. Strickland, 1965. Fluorometric determination of chlorophyll. J. Conc. int. Explor. Mer. 30: 169–204.

Horrigan, S. G., A. F. Carlucci & P. M. Williams, 1981. Light inhibition of nitrification in sea surface films. J. mar. Res. 39: 557–565.

Hovanec, T. A. & E. F. DeLong, 1996. Comparative analysis of nitrfying bacteria associated with freshwater and marine aquaria. Appl. envir. Microbiol. 62: 2888–2896.

Joye, S. B. & J. T. Hollibaugh, 1995. Influence of sulfide inhibition of nitrification on nitrogen regeneration in sediments. Science 270: 623–625.

Kaplan, W. A. & S. C. Wofsy, 1985. The biogeochemistry of nitrous oxide: a review. In Advances in Aquatic Microbiology. Academic Press Inc. London. Vol. 3: 181–206.

Kerkhof, L. & B. B. Ward, 1993. Comparison of nucleic acid hybridization and fluorometry for measurement of the relationship between RNA/DNA ratio and growth rate in a marine bacterium. Appl. envir. Microbiol. 59: 1303–1309.

Koops, H.-P. & U. C. Möller, 1992. The lithotrophic ammonia-oxidizing bacteria. In Balows A., H. G. Truper, M. Dworkin, W. Harder & K.-H. Schleifer, (eds), The Prokaryotes: a Handbook on the Biology of Bacteria: Ecophysiology, Isolation, Identification, Applications. 2nd edn. Springer-Verlag, New York. 2625–2637.

Liesack, W., H. Weyland & E. Stackenbrandt, 1991. Potential risks of gene amplification by PCR as determined by 16S rDNA analysis of a mixed-culture of strict barophilic bacteria. Microb. Ecol. 21: 191–198.

Lizotte M. P. & J. C. Priscu, 1992. Spectral irradiance and bio–optical properties in perennially ice-covered lakes of the dry valleys (McMurdo Sound, Antarctica). In Elliot D. (ed.), Contribution to Antarctic Research III, Antarctic Research Series. 57: 1–14. Washington, D.C. AGU.

Matsomuto, G. I., 1989. Biogeochemical study of organic substances in Antarctic lakes. Hydrobiologia 172: 265–289.

Norton, J. M., J. M. Low & M. G. Klotz, 1996. The gene encoding ammonia monooxygenase subunit A exists in three nearly identical copies in Nitrosospira sp. NpAV. FEMS Microbiol. Lett. 139: 181–188.

Palmisano, A. A. C. & G. M. Simmons, 1987. Spectral downwelling irradiance in an Antarctic lake. Polar Biology 7: 145–151.

Parker, B. C., G. M. Simmons, K. G. Seaburg, D. D. Cathey & F. C. T. Allnut, 1982. Comparative ecology of plankton communities in seven Antarctic oasis lakes. J. Plankton Res. 4: 271–286.

Parsons, T. R., Y. Maita & C. M. Lalli, 1984. A Manual of Chemical and Biological Methods for Seawater Analysis. Pergamon Press, 173 pp.

Poth, M. & D. D. Focht, 1985. ^{15}N kinetic analysis of N_2O production by Nitrosomonas europaea: an examination of nitrifier dentrification. Appl. envir. Microbiol. 49: 1134–1141.

Priscu, J. C. & M. T. Downes, 1985. Relationship among nitrogen uptake, ammonia oxidation and nitrous oxide concentration in the coastal waters of western Cook Strait, New Zealand. Estuar. Coast. Shelf Sci. 20: 529–542.

Priscu, J. C., B. B. Ward & M. T. Downes, 1993. Water column transformations of nitrogen in Lake Bonney, a perennially ice-covered Antarctic lake. Ant. J. U.S. 28: 237–239.

Priscu, J. C., M. T. Downes & C. P. McKay, 1996. Extreme supersaturation of nitrous oxide in a non-ventilated Antarctic lake. Limnol. Oceanogr. 41: 1544–1551.

Rotthauwe, J. H., K.-P. Witzel & W. Liesack, 1995. The ammonia monooxygenae structural gene amoA as a functional marker: molecular fine scale analysis of natural ammonia-oxidizng populations. Appl. envir. Microbiol. 63: 4704–4712.

Schmidt, E. L., 1974. Quantitative autecological study of microorganisms in soil by immunofluorescence. Soil Sci. 118: 141–149.

Simmons, G. M., J. R Vestal & R. A. Wharton, 1993. Environmental regulators of microbial activity in continental Antarctic lakes. In Vincent W. (ed.), Antarctic Microbiology, Wiley-Liss, Inc.: 491–541.

Smorczewski, W. T. & E. L. Schmidt, 1991. Numbers, activities, and diversity of autotrophic ammonia-oxidizing bacteria in a freshwater, eutrophic lake sediment. Can. J. Microbiol. 37: 828–833.

Snra, R. F. & A. Baggaley, 1975. Kinetic response to perturbed marine nitrification systems. J. Wat. Poll. Cont Fed. 47: 472–486.

Spigel, R. H., I. V. Sheppard & J. C. Priscu, 1990. Temperature and conductivity fine structure from Lake Bonney. Ant. J. U.S. 1990. Rev. Vol.

Teske, A., E. Alm, J. M. Regan, S. Toze, B. E. Rittmann & D. A. Stahl, 1994. Evolutionary relationships among ammonia- and nitrite oxidizing–bacteria. J. Bacteriol. 176: 6623–6630.

Tortoso, A. C. & G. L. Hutchinson, 1990. Contributions of autotrophic and heterotrophic nitrifiers to soil NO and N_2O emissions. Appl. envir. Microbiol. 56: 1799–1805.

Vincent, W., 1988. Microbial Ecosystems of Antarctica. Cambridge University Press, Cambridge, England. 304pp.

Vincent, W. & C. Howard-Williams, 1985. Ecosystem properties of Dry Valley lakes. N.Z. Ant. Rec. (Suppl) 6: 11–20.

Vincent, W., M. T. Downes & C. L. Vincent, 1981. Nitrous oxide cycling in Lake Vanda, Antarctica. Nature 292: 618–620.

Vincent, W. F., R. W. Castenholz, M. T. Downes & C. Howard Williams, 1993. Antarctic cyanobacteria – light, nutrients and photosynthesis in the microbial mat environment. J. Phycol. 29: 745–755.

Vincent, W. F., M. T. Downes, R. W. Castenholz & C. Howard Williams, 1993. Community structure and pigment organisation of cyanobacteria-dominated microbial mats in Antarctica. Eur. J. Phycol. 28: 213–221.

Vincent, W. F. & C. L. Vincent, 1982. Factors controlling phytoplankton production in Lake Vanda (77 °S). Can. J. Fish. aquat. Sci. 39: 1602–1609.

Voytek, M. A., Relative abundance and species diversity of autotrohpic ammonia-oxidizing bacteria in aquatic systems, Ph.D. Dissertation, University of California, Santa Cruz, 1996.

Voytek, M. A. & B. B. Ward, 1995. Detection of ammonia-oxidizing bacteria of the beta-subclass proteobacteria in aquatic samples with the PCR. Appl. envir. Microbiol. 56: 2430–2435.

Wada, E., R. Imaizuma, S. Nakaya & T. Torii, 1984. ^{15}N abundance in the dry valley area, south Victoria Land, Antarctica: Ecophysiological implications of microorganisms. Mem. natl. Inst. Polar Res. Spec. Issue Jpn. 32: 130–139.

Ward, B. B., 1986. Nitrification in marine environments. In J. I. Prosser (ed.), Nitrification IRL Press, Washington, D.C. 157–184.

Ward, B. B. & A. F. Carlucci, 1985. Marine ammonia- and nitrite-oxidizing bacteria: Serological diversity determined by immunofluorescence in culture and in the environment. Appl. envir. Microbiol. 50: 194–201.

Ward, B. B. & M. J. Perry, 1980. Immunofluorescent assay for the marine ammonia-oxidizing bacterium, Nitrosococcus oceanus. Appl. envir. Microbiol. 39: 913–918.

Ward, B. B., & J. C. Priscu, 1997. Detection and characterization of denitrifying bacteria from a permanently ice-covered Antarctic lake. Hydrobiologia

Wharton, R. A., C. P. McKay, G. D. Clow & D. Andersen, 1993. Perrenial ice covers and their influence on Antarctic lake ecosystems. In W. J. Green & E. I. Friedmann (eds), Antarctic Research Series: Physical and Biogeochemical Processes in Antarctic Lakes, A.G.U., Washington, D.C. 59: 53–70.

Wilson, A. T., 1981. A review of the geochemistry and lake physics of the antarctic dry areas. In L. McGinnis (ed.), Dry Valley Drilling Program, Antarctic Res. Ser. 33: 185–192. Washington, D.C.

Hydrobiologia **401:** 131–138, 1999.
J.P. Zehr & M.A. Voytek (eds), Molecular Ecology of Aquatic Communities.

Microscopic detection of the toluene dioxygenase gene and its expression inside bacterial cells in seawater using prokaryotic *in situ* PCR

F. Chen[1], W. A. Dustman[2] & R. E. Hodson[1,2,3]
[1]*Department of Marine Sciences,*
Tel: [+1]706 542-6625. Fax: [+1]706 542-5888. E-mail: fchen2@arches.uga.edu
[2]*Department of Microbiology, and*
[3]*Institute of Ecology, University of Georgia, Athens, GA 30602-2206, U.S.A.*

Key words: in situ PCR, *in situ* hybridization, *todC1* gene

Abstract

The toluene dioxygenase (*todC1*) gene and its mRNA transcripts were amplified by *in situ* PCR and *in situ* RT-PCR, respectively, in intact cells of the bacterium *Pseudomonas putida* F1. *In situ* amplicons of DNA and mRNA were then detected by hybridizing to a fluorescently labeled oligonucleotide. *In situ* PCR protocols were developed to distinguish between cells of *P. putida* F1 (possessing the *todC1* gene) and *P. putida* AC10R (lacking the *todC1* gene); the method was sensitive enough to detect amplified products from a single copy of the *todC1* gene. *P. putida* F1 cells were also introduced into seawater with toluene addition. Cells expressing *todC1* and total cells were detectable by *in situ* RT-PCR and Yo-Pro 1 counterstaining, respectively. Nearly 90% of cells expressing the *todC1* gene were detected in seawater amended with toluene at day 3, but no cells expressing *todC1* were detected in seawater not exposed to toluene. Our results suggest that *in situ* PCR amplification can be a useful technique for studying presence or absence of a specific gene and gene expression of bioremediative bacteria at the individual cell level following release into natural environments.

Introduction

Many aromatic chemicals, including toluene, benzene and *p*-xylene, are discharged into natural environments as the result of chemical and related industries. A number of bacterial species can degrade these compounds, with members of the genus *Pseudomonas* prominent among them. In many cases, the catabolic genes involved in degradation of such compounds encode a multi-enzyme system and are commonly carried on bacterial plasmids (Pemberton, 1983). The presence of degradative genes for aromatic compounds has been demonstrated in a number of bacterial isolates from natural environments, using both DNA-DNA hybridization and polymerase chain reaction (PCR) techniques (Sayler et al., 1985; Foght & Westlake et al., 1991; Sobecky et al., 1992). Neither of these approaches, however, can provide information about the micro-scale distribution of individual cells which possess the ability to degrade aromatic compounds or about the conditions necessary for the genes to be expressed *in situ*.

In situ hybridization using rRNA-targeted oligonucleotide probes has been developed for taxonomic identification of individual microbial cells (Giovannoni, 1988; DeLong et al., 1989; Amann et al., 1990, 1995), and this approach has been used to investigate bacterial populations and communities in a variety of environments. The *in situ* hybridization method relies on the presence of multiple target sequences within an individual cell, which in this application is all the rRNA associated with prokaryotic ribosomes. However, bacterial cells in natural aquatic communities, unlike cultured cells or cells in nutrient-rich environments, often do not contain sufficient rRNA to produce detectable signals (i.e. fluorescence or color precipitation) with *in situ* hybridization (Lee & Kemp, 1994). Methods that increase signal strength by using multiple rRNA-targeted fluorescent probes (Lee & Kemp, 1994), polyribonucleotide probes (Trebesius et al.,

1994), or more intensely fluorescent dyes (Yamaguchi et al., 1996; Alfreider et al., 1996) have increased the applicability of *in situ* hybridization for characterization of microbial community structure in natural environments. However, the existing standard *in situ* hybridization protocols are still limited to taxonomic identification, and cannot be used to detect the presence of single-copy functional genes or their expression (mRNA) at the individual-cell level.

Recently, prokaryotic *in situ* PCR (PI-PCR) and *in situ* reverse transcription PCR (RT-PI-PCR) methods have been developed to amplify and visualize the microscale distribution of specific genes and their expression, respectively, inside individual bacterial cells (Hodson et al., 1995; Porter et al., 1995; Tolker-Nielsen et al., 1997; Chen et al., 1997, 1998). PI-PCR is used to detect a specific gene target at the DNA level, while RT-PI-PCR can detect mRNA transcripts or 16S rRNA target sequences inside prokaryotic cells (Hodson et al., 1995; Chen et al., 1998). So far, these protocols have been used to detect the presence and expression of the *nahA* gene (nathalene degrading gene) and the *todC1* gene (toluene degrading gene) in *Pseudomonas* cells, the *nifH* gene (nitrogen fixation gene) in cells of *Azotobacter vinelandii* (Hodson et al., 1995), and the 16S rRNA sequences of lignin degrading bacteria in enrichment communities (Chen et al., 1997). The current existing PI-PCR and RT-PI-PCR protocols use either direct *in situ* PCR, in which fluorescently labeled-dUTP is directly incorporated into the PCR amplicon, or a two-stage amplification in which non-labeled monomers are incorporated into a primary unlabeled amplicon, followed by binding and extension of a primer internal to the amplicon using labeled monomers, thus yielding a lableled secondary amplicon. Direct *in situ* PCR sometimes results in a high background signal in negative controls probably due to nonspecific amplification. Therefore, it requires additional optimization of *in situ* PCR conditions which can be expensive and time consuming. Likewise, two-stage PI-PCR is time consuming and somewhat labor intensive. When applied to eukaryotes, *in situ* PCR, in which the unlabeled PCR amplicon is detected by hybridization to a fluorescently-labeled internal probe, has been reported to be more specific for gene target detection than direct *in situ* PCR (Zehbe et al., 1994; Long & Komminoth, 1995).

In this study, PI-PCR (including prokaryotic *in situ* PCR and *in situ* RT-PCR) coupled with *in situ* hybridization detection was used to detect the toluene dioxygenase gene (*todC1*) and its expression inside *Pseudomonas putida* F1 cells. Filtered seawater with added *P. putida* F1 cells were used as a model system to study the survival and expression of the *todC1* genes of a nonindigenous bioremediative species introduced into a marine system.

Materials and methods

Bacterial samples and pretreatment

Bacterial strain *Pseudomonas putida* F1 (Deutsche Sammlung von Mikroorganismen, DSM6899) has been described by Gibson et al. (1968). The *P. putida* F1 chromosome includes the toluene dioxygenase gene system, which encodes proteins that degrade toluene via a dihydrodiol pathway to a tricarboxylic acid cycle intermediate (Gibson et al., 1970). In this study, *in situ* PCR/RT-PCR was first used to detect gene and gene expression of *P. putida* F1 in the cultured media. Secondly, *P. putida* F1 added to seawater amended with toluene was monitored for the *todC1* gene expression using *in situ* RT-PCR.

P. putida F1 was grown at 30 °C in basal salts medium (BSM) (Schell, 1985) with toluene vapor as the sole carbon source. Three ml of toluene were placed into a glass tube with a hole, and the tube suspended in a 250 ml Erlenmeyer flask with 50 ml of BSM. *P. putida* (NAH7) AC10R-7 (which degrades naphthalene, but not toluene) grown in Luria-Bertani (LB) broth (Bertani, 1951) was used as the negative control. Samples were then subjected to fixation, permeabilization, and DNase or RNase digestion, following methods described in detail in Hodson et al. (1995) and Chen et al. (1998).

P. putida F1 cells added to the filtered seawater system were prepared as follows. *P. putida* F1 was grown in LB medium until mid-logarithmic phase. One ml of culture was centrifuged at 3000 rpm for 5 min in a microcentrifuge tube. The supernatant was discarded and the pellet was resuspended in 1 ml of 0.2 μm-filtered seawater. The cell concentration was immediately determined by direct microscopic counting (DAPI) (Porter & Feig, 1980) and adjusted to 10^9cells ml^{-1} using 0.2 μm-filtered seawater. An aliquot of 10 μl of the washed cells was added to each of two flasks containing 100 ml of 0.2 μm-filtered seawater collected from the pier at the Univ. of Georgia Marine Education Center on Skidway Island, GA, U.S.A. One flask was incubated with toluene addition and the other without toluene addition. Both flasks were incubated at 25 °C with a shaking speed of 50 rpm. Three ml was taken from each flask at day 3 and quickly concentrated to a 30 μl final volume in PBS by centrifugation at 10 000 rpm for 10 min. Two 5 μl subsamples from each subsample were spotted onto a silane-coated slide (Perkin Elmer) and dried in a 50°C oven for 5 min. Cells on slides were fixed in 4% paraformaldehyde in phosphate-buffered saline (PBS: 120 mm NaCl and 2.7 mm KCl in 10 mm phosphate buffer, pH 7.6) for 2 h in ice, after which the cells were washed twice with PBS. Prior to cell wall permeabilization, fixed cells were dehydrated sequentially in 50%, 80% and 98% ethanol in PBS. Dehydrated cells were treated with lysozyme (Boehringer Mannheim) at a final concentration of 1 mg/ml for 15 min at room temperature. If the amplification target was DNA, RNA was removed from the cells by adding DNase-free RNase I (0.01U μl^{-1}, final concentration, Boehringer Mannheim) along with the lysozyme. However, if the amplification target was RNA, permeabilized cells were treated with RNase-free DNase (0.1 U μl^{-1}, final concentration, Boehringer Mannheim) at room temperature for 2 h. Messenger RNA is very sensitive to RNase activity, and sample treatment for *in situ* RT-PCR was, therefore, conducted on ice throughout the experiment. Water, reagents (e.g. PBS and 0.5X SSC), reagent bottles, microcentrifuge tubes and glass slides were pretreated with DEPC. RNase inhibitor (0.8 U/μl, final concentration, Boehringer Mannheim) was added to the RT-PCR mixture (Chen et al., 1998).

Primer design

The primers were designed to specifically amplify a nucleotide segment from the large subunit of terminal dioxygenase (*todC1*) in the toluene dioxygenase system located on the chromosome of *P. putida* F1 (Zylstra & Gibson, 1989). Sequence alignment of the most similar dioxygenase genes from nine bacterial strains was performed using Genetic Computer Group programs, and three unique sites were selected for the *todC1* gene in *P. putida* F1. The relative positions of the upstream primer (TOL1), downstream primer (TOL2) and the internal primer/probe (TOL3) on the toluene dioxygenase gene system are shown in Figure 1. Sequences of TOL1, TOL2 and TOL3 primers are 5′ TATCAGGCTGCGCAGGAGC 3′, 5′ GACGTTGTTGCTGATCCGC 3′, 5′ AGCAAGGTCGGCCATTTCAAGG 3′, respectively.

PCR optimization

Extracted DNA or RNA from both positive and negative control strains was first used to optimize conditions for PCR amplification of the *todC1* gene fragment (DNA-PCR) or the mRNA from this gene (RT-PCR). Bacterial DNA was purified using QIAamp® Blood and Tissue kit (QIAGEN Inc.), while bacterial RNA was extracted using the RNeasy kit (QIAGEN Inc.). PCR was optimized by varying the annealing/extension temperatures and the concentration of primers, DNA or RNA template, $MgCl_2$ or dNTP. A two-temperature PCR approach, which combined annealing and extension temperatures into one temperature, was developed based on extracted DNA and used subsequently for *in situ* PCR.

The optimized protocol for DNA amplification was as follows. Extracted DNA (10 ng) was added to 25 μl of a PCR reaction mixture containing Thermalase Tbr buffer (10 mm Tris-HCl, 1.5 mm $MgCl_2$, 50 mm KCl and 0.1% Triton X-100), 200 μm of deoxyribonucleoside triphosphate (dNTP), 1.0 μm of TOL1 and TOL2 primers, and 2 U of Thermalase Tbr DNA polymerase (AMRESCO). Mineral oil (30 μl) was added to each reaction tube. Negative controls contained all reagents except the DNA template. PCR was conducted using a DNA thermal cycler 480 (Perkin-Elmer Corp., Norwalk, Conn.) with the following cycle parameters: an initial denaturation at 94 °C for 3 min, followed by 30 cycles of annealing/extension at 60 °C for 1 min and denaturation at 94 °C for 30 s.

The optimized protocol for mRNA amplification was as follows. Reverse transcription and PCR amplification were carried out using the GeneAmp® EZ r*Tth* PCR kit (Perkin Elmer), in which both reactions may be accomplished in a single solution mixture. A 25 μl of reaction mixture contained 1X EZ r*Tth* buffer, 200 μm each of dATP, dCTP, dGTP and dTTP, 1.0 μm of TOL1 and TOL2 primers, 2.5 mm of $MnCl_2$, 20 U of RNase inhibitor, 10 ng of RNA target, and 2.5 U of r*Tth* DNA polymerase. mRNA was reverse transcribed by incubation at 60 °C for 30 min, followed by 30 cycles of a two-temperature PCR protocol (denaturation at 94 °C for 30 s and annealing/extension at 60 °C for 1 min), ending with 60 °C for 7 min.

In situ PCR/RT-PCR

In situ amplification of the DNA or mRNA target inside *P. putida* F1 cells generally follows the optimized PCR conditions established above, except that extracted DNA or RNA was replaced by permeabilized cells,

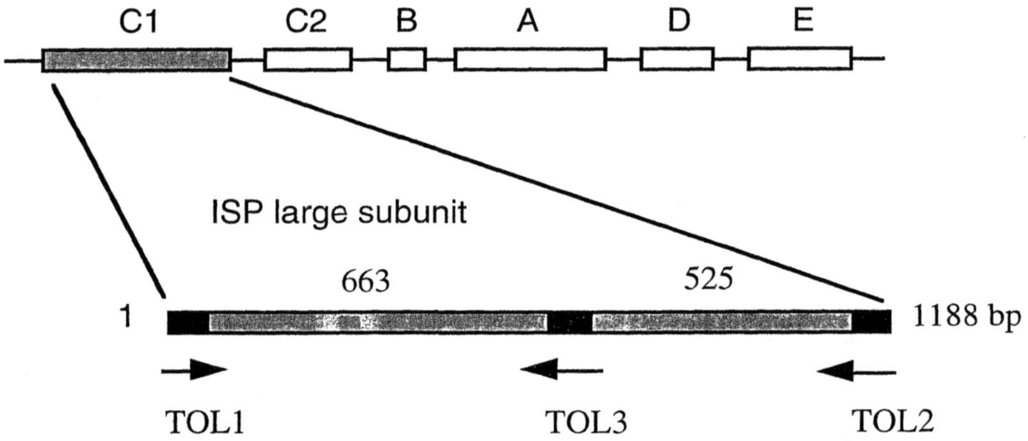

Figure 1. Physical map of the toluene dioxygenase system and location of three specific oligonucleotide primers for *P. putida* F1. The toluene dioxygenase system includes genes for reductase (A), ferredoxin (B), terminal dioxygenase (two subunits C1 and C2), dehydrogenase (D) and methylcatechol dioxygenase (E) (Zylstra & Gibson, 1989). ISP is iron-sulfur protein. TOL1, TOL2 and TOL3 are the three oligonucleotide primers designed based on the nucleotide sequence of the large subunit of terminal dioxygenase (*tod*C1).

and a higher $MgCl_2$ concentration (2.5 mm, final concentration) was used for *in situ* DNA amplification. A 40 μl aliquot of reaction mixture was added to each sample spot on the *in situ* PCR glass slide (Perkin-Elmer) after the slide was heated to 70 °C to achieve a hot start, and covered with an AmpliCoverþ disc (Perkin-Elmer Corp., Norwalk, Conn.). *In situ* PCR or RT-PCR was carried out using an *in situ* PCR 1000 thermal cycler (Perkin Elmer). After *in situ* amplification, slides were washed once with 0.5X SSC (750 mm NaCl, 75 mm trisodium citrate, pH 7.0) at 45 °C for 10 min. The *in situ* amplified gene sequences were then detected by hybridization to the CY3-labeled probe, TOL3.

In situ hybridization for detection of amplified gene

Forty μl of hybridization solution (900 mM NaCl, 20 mm Tris-HCl, 0.01% sodium dodecyl sulfate; pH 7.2) containing 5 ng/μl (final concentration) of CY3-labeled TOL3 probe (Life Technologies) was added to each sample on the slide. Hybridization solution on the slide was covered with plastic Amplicover® and metal Ampliclip® using the assembly tool (Perkin Elmer). Slides were incubated at 94 °C for 5 min and then 45 °C overnight in an *in situ* PCR 1000 thermal cycler (Perkin Elmer). After incubation, slides were washed with 0.5X SSC for 10 min at 45 °C.

Digital image analysis

Slides were viewed under an epifluorescence microscope (Olympus BX-40) with a 100 X high resolution U Plan Oil Objective lens. Images were acquired with a SenSys:1400 cooled, charge-coupled device (CCD) (Photometrics, Tucson, Ariz.) and processed using the Oncor Image software package version 2.02 (Oncor, Inc., Gaithersburg, Maryland) on a Power Macintosh 9500.

Results

Amplification of the *tod*C1 gene from extracted DNA by the oligonucleotide primer set TOL1/TOL2 resulted in a DNA fragment of approximately 1.2 kb (Figure 2, lane 2). Likewise, amplification with this primer set produced a 1.2 kb fragment from cDNA reverse-transcribed from total RNA (Figure 2, lane 5). When the 1.2 kb amplicons in lanes 2 and 5 were excised from a low melting point agarose gel and reamplifed by the internal primer set TOL1/TOL3, 660 bp amplicons resulted (Figure 2, lanes 3 and 6), indicating that the correct gene target was being amplified by the TOL1/TOL2 primer set. No product resulted when DNA was left out of the reaction mixture (Figure 2, lane 4).

The *tod*C1 gene and its mRNA were successfully detected inside intact *Pseudomonas* cells by *in situ* PCR/RT-PCR followed by hybridization with CY3-labeled TOL3 probe. Positive cells (*P. putida* F1) with

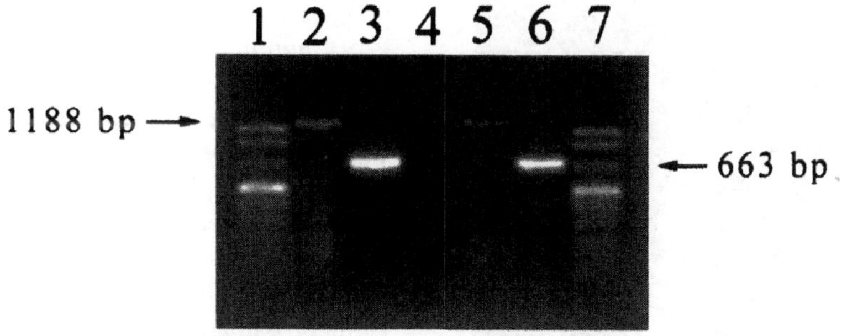

Figure 2. Gel electrophoresis of the *tod*C1 gene products amplified from extracted DNA of *P. putida* by PCR and RT-PCR. Lanes 1 and 7 are molecular marker VIII (Boehringer Mannheim), with bands corresponding to 1114, 900, 692, 501, 489, 404, 302 bp fragments (smaller bands were not visible). Lane 2, a 1.2 kb *tod*C1 gene fragment amplified by primer TOL1/TOL2; lane 3, PCR product (ca. 660 bp) reamplified from the 1.2 kb amplicon in lane 2 by nested primers TOL1/TOL3; lane 4, negative control with no addition of target DNA; lane 5, RT-PCR product (ca. 1.2 kb) amplified by primers TOL1/TOL2; lane 6, PCR product (ca. 660 bp) reamplified from 1.2 kb amplicon in lane 5 by nested primers TOL1/TOL3. PCR products were electrophoresed in 2% NuSieve® GTG® agarose gel (1X TBE buffer), and visualized with ethidium bromide staining.

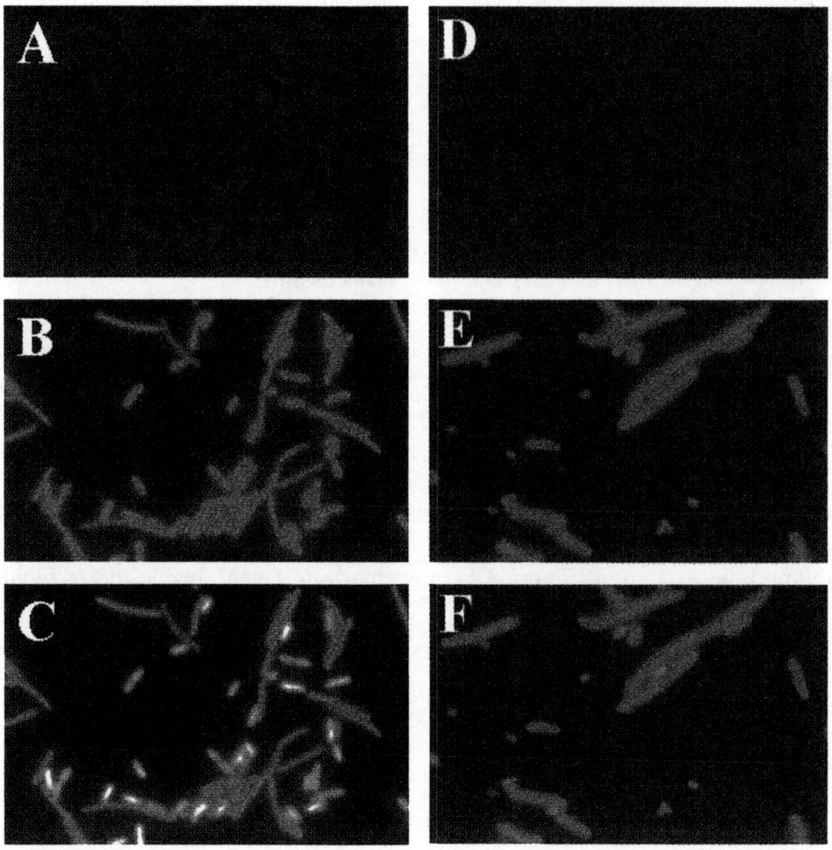

Figure 3. Gene and gene expression of *todC1* in a mixture of *P. putida* F1 cells and negative control cells were detected by *in situ* PCR (A, B, C) and *in situ* RT-PCR (D, E, F). (A) positive cells (orange red) with the *todC1* gene; (B) both positive and negative control cells counterstained with DAPI; (C) superimposed images from (A) and (B) showing different colors for positive and negative control cells; (D) positive cells (orange red) which express the *todC1* gene; (E) both positive and negative cells counterstained with Yo-Pro 1; (F) superimposed images from (D) and (E) showing different colors for positive and negative cells. Images (A) and (B) were from the same viewfield. Image (A) was acquired under yellow light excitation while image (B) was taken under UV light. Images (D) and (E) were from the same viewfield. Image (D) was acquired under yellow light excitation while image (E) under blue light excitation.

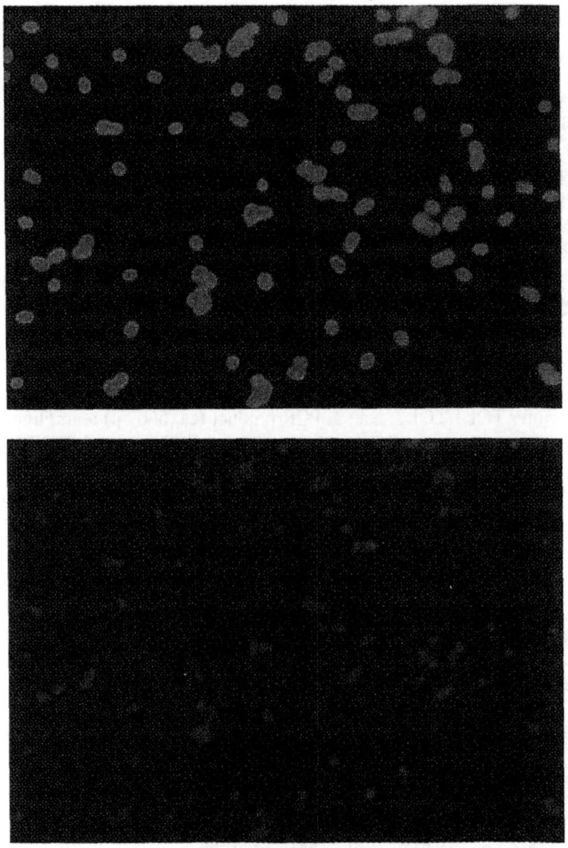

Figure 4. *P. putida* F1 cells expressing the *todC1* gene were detected by *in situ* RT-PCR at day 3 after they were introduced into seawater exposed to toluene. Cells expressing *todC1* showed orange-red color (CY3-dye) under yellow light excitation (upper photo). Under blue light excitation, total cells (counterstained with Yo-Pro 1) showed green color (lower photo). Both images were taken from the same viewfield with different filters for light excitation.

the *todC1* gene emitted orange-red fluorescence while negative cells (*P. putida* AC10R-7) were almost undetectable under yellow light excitation (Figure 3A). Both positive and negative cells in the same viewfield of Figure 3A were visible under UV excitation when counterstained with DAPI (Figure 3B). The superimposed images of Figure 3A and 3B allow the visualization of both positive and negative cells in the same image but with different colors (Figure 3C). For detection of *todC1* gene expression, the positive cells (*P. putida* F1) emitted orange-red fluorescence while negative cells (*P. putida* AC10R-7) were almost undetectable under yellow light (Figure 3D). Both positive and negative cells in the same viewfield of Figure 3D were visible under blue light excitation when counterstained with Yo-Pro 1 (Figure 3E). The superimposed images of Figure 3D and 3E allow visualization of both positive and negative cells in the same image but with different colors (Figure 3F).

P. putida F1 cells were introduced into filtered seawater with or without toluene, and expression of the *todC1* gene was detectable using *in situ* RT-PCR coupled with *in situ* hybridization detection. *P. putida* F1 cells from a LB culture were used to inoculate the flasks incubated with and without toluene. No cells expressing *todC1* were detected in both flasks at time 0. By day 3, cells expressing the *todC1* gene accounted for approximately 90% in the flask with toluene, and no cells expressing *todC1* were detected in the flask without toluene. According to Yo-Pro 1 counts, total cell number in the flask with toluene increased from 5.0×10^5 cells ml^{-1} at time 0 to 3.4×10^7 cells ml^{-1} at day 3, while cell number did not change in the flask without toluene. As seen in Figure 4, a majority of *P. putida* F1 cells in the flask with toluene were detected by CY3-labeled probe after *in situ* RT-PCR amplification *todC1* (orange-red cells in the upper photo of Figure 4). Because the sample was counterstained with Yo-Pro 1, total cells could also be visualized in the same viewfield under blue light excitation (green cells in the lower photo of Figure 4).

Discussion

Success in amplifying *todC1* genes from extracted DNA or mRNA of *P. putida* F1 using three newly-designed oligonucleotide primers allowed us to further develop *in situ* PCR and *in situ* RT-PCR protocols for detection of toluene degrading capability and activity in intact cells of *P. putida* F1. *In situ* PCR coupled with *in situ* hybridization was successfully used to detect the *todC1* gene inside the *P. putida* F1 cells. *In situ* PCR previously has been demonstrated to be sensitive enough to amplify a single copy of a gene target sequence in prokaryotes (Hodson et al., 1995). In *P. putida* F1, the toluene degrading gene is encoded on the bacterial chromosome (Zylstra & Gibson, 1989), although many aromatic compound degradation genes studied to date have been found to be located on bacterial plasmid DNA. When a genetic engineered microorganisms is released into a natural environment, its survival and environmental impact depends on its genetic stability and potential for growth (Sobecky et al., 1992). *In situ* amplification of a nucleic acid target will be an ideal approach to compare the genetic stability of chromosome-encoded genes and plasmid-borne genes in bioremediative bacteria under a wide range of environmental conditions.

In situ RT-PCR also was successfully used to detect expression of the *todC1* gene in *P. putida* F1. CY3 yields a bright red fluorescence and provides a good contrast with counterstaining dyes like DAPI (blue fluorescence) and Yo-Pro 1 (green fluorescence). DAPI is more specific for DNA and was used to counterstain cells when DNA was the target molecule. When RNA (gene expression) was the target, Yo-Pro 1 was used for counterstaining because cellular DNA had been removed by DNase treatment prior to *in situ* PCR amplification, and Yo-Pro 1 binds to either DNA or RNA (Hirons et al., 1994). Detection of target cells with CY3-labeled TOL3 probe coupled with counterstaining allowed us to simultaneously visualize positive cells and enumerate total cells in the same viewfield.

We demonstrated the use of *in situ* RT-PCR to detect cells expressing the *todC1* gene among the total bacteria in a microcosm. A high proportion of cells exposed to toluene vapor were found to contain mRNA transcripts of the *todC1* gene, while no *todC1* expression was observed in seawater without toluene. There are a number of possible explanations why *todC1* was expressed in less than 100% of cells. Some of the *P. putida* F1 cells may have utilized other organic carbon sources occurring naturally in the seawater rather than toluene. Perhaps utilization of one or more of such compounds may have inhibited expression of the *todC1* gene. Alternatively, cells that grew initially on toluene as carbon source may have become inactive secondarily due to the lack of other required substances, such as nitrogen or phosphorus, which were present only in the trace natural concentrations in the seawater used. And finally, it is to be expected that the level of gene expression may vary significantly over the cell cycle, and the detection limits of methods we used in this study have not yet been determined.

In situ PCR followed by *in situ* hybridization detection (referred as indirect *in situ* PCR) greatly improves detection specificity. Direct single-stage incorporation of fluorescently labeled nucleotides (e.g. CY3-dUTP and FITC-dUTP) into amplified products inside cells (direct *in situ* PCR) often yields a high background and can result in false positives. These nonspecific results are probably due to:

1. incorporation of labeled nucleotides into cellular DNA by the repair mechanism of DNA polymerase;
2. mispriming, in which added primers bind to nontarget sequence regions;
3. endogenous priming, in which endogenous DNA or RNA fragments act as primers for PCR amplification;
4. unstable binding between fluorochome and dUTP during heating and cooling cycles of PCR which releases fluorochrome that might act like a general stain for all cells; or
5. binding of fluorecently-labeled dUTP to the cellular components inside cells due to high temperatures reached during PCR.

Some of these problems cannot be eliminated even if a second nested *in situ* PCR is used. Use of *in situ* hybridization for *in situ* PCR product detection greatly reduces factors which cause false positive signals and hybridization stringency can be controlled by varying concentrations of formamide and salt. Recent studies of eukaryotic*in situ* PCR suggest that *in situ* PCR coupled with *in situ* hybridization provides much higher specificity compared to direct *in situ* PCR (Zehbe et al., 1994; Long & Komminoth, 1995).

Indirect *in situ* PCR also has the potential to amplify multiple gene targets (DNA , mRNA or rRNA) simultaneously inside the cell using more than one set of primers which have similar annealing temperatures and will not anneal with one other. Signal detection with *in situ* hybridization permits simultaneous utilization of multiple probes labeled with different fluorochromes. For example, the *todC1* gene can be detected with CY3-labeled probe while members of the genus *Pseudomonas* can be detected with a 16S rRNA probe labeled with a distinguishable fluorochrome such as Quantam Dye (UV excitable, Research Organics). Even though rRNA molecules are present in higher numbers in bacterial cells than mRNA transcripts, *in situ* hybridization alone is still often hindered by a low copy number of rRNA molecules in slow-growing or dormant bacteria in natural environments (Lee & Kemp, 1994). Jin et al. (1995) reported that *in situ* RT-PCR detects 24 times more number of cells expressing the target gene compared to *in situ* hybridization alone, suggesting that *in situ* RT-PCR provides a much higher sensitivity than *in situ* hybridization. Using the techniques outlined in this paper, manipulation of one sample on a single slide will not only provide information on activity of specific functional genes, but indicate which taxonomic groups of bacteria possess that gene and respond to environmental perturbation or stimuli by gene expression. Future studies will be directed toward applying this approach to investigate *in situ* microbial responses to specific aromatic pollutants (e.g. toluene and benzene) in coastal marine environments. The ability to examine *in situ* microbial activities in terms of population distribution and genetic constitution and capabilities will provide new insight into microbial communities and the important biogeochemical functions they mediate.

Acknowledgements

We thank Dr D. T. Gibson and B. Parales from the University of Iowa for providing information on *P. putida* F1. This research was funded by grants from the Office of Naval Research, Marine Environmental Quality Program (N00041-93-0850); Department of Energy (DE-FG02-97ER62451), and the NSF LMER program (DEB-9412089).

References

Alfreider, A., J. Pernthaler, R. Amann, B. Sattler, F-O. Glockner, A. Wille & R. Psenner, 1996. Community analysis of the bacterial assemblages in the winter cover and pelagic layers of a high mountain lake by *in situ* hybridization. Appl. envir. Microbiol. 62: 2138–2144.

Amann, R. I., B. J. Binder, R. J. Olson, S. W. Chisolm, R. Devereux & D. A. Stahl, 1990. Combination of 16S rRNA-targeted oligonucleotide probes with flow cytometry for analyzing mixed microbial population. Appl. envir. Microbiol. 56: 1919–1925.

Amann, R. I., W. Ludwig & K.-H. Schleifer, 1995. Phylogenetic identification and *in situ* detection of individual microbial cells without cultivation. Microbiol. Rev. 59: 143–169.

Bertani, G., 1951. Studies on lysogenesis. I. The mode of phage liberation by lysogenic *Escherichia coli*. J. Bact. 62: 293–300.

Chen, F., W. Dustman, M. A. Moran & R. E. Hodson, 1998. *In situ* PCR methodologies for visualization of microscale genetic and taxonomic diversities of prokaryotic communities. Ch. 3.3.9. In Akkermans A. D. L., J. D. van Elsas & F. J. DeBruijn (eds), Molecular Microbial Ecology Manual, suppl. 3. Kluwer Academic Publishers, The Netherlands: 1–17.

Chen, F., J. M. Gonzalez, W. Dustman, M. A. Moran & R. E. Hodson, 1997. *In situ* reverse transcription: an approach to characterize genetic diversity and activities of prokaryotes. Appl. envir. Microbiol. 63: 4907–4913.

DeLong, E. F., G. S. Wickham & N. R. Pace, 1989. Phylogenetic stains: ribosomal RNA-based probes for the identification of single cells. Science 243: 1360–1362.

Foght, J. M. & D. W. S. Westlake, 1991. Cross hybridization of plasmid and genomic DNA from aromatic and polycyclic aromatic hydrocarbon degrading bacteria. Can. J. Microbiol. 37: 924–932.

Gibson, D. T., M. Hensley, H. Yoshioka & T. J. Mabry, 1970. Formation of (+)-*cis*-2,3,-dihydroxy-1-methylcyclohexa-4,6-diene from toluene by *Pseudomonas putida*. Biochem. 9: 1626–1630.

Gibson, D. T., J. R. Koch & R. E. Kallio, 1968. Oxidative degradation of aromatic hydrocarbons by microorganisms. I. Enzymatic formation of catechol from benzene. Biochem. 7: 2653–2662.

Giovannoni, S. J., E. F. DeLong, G. J. Olsen & N. R. Pace, 1988. Phylogenetic group specific oligonucleotide probes for identification of single microbial cells. J. Bact. 170: 720–726.

Hirons, G. T., J. J. Fawcett & H. A. Crissman, 1994. TOTO and YOYO: New very bright fluorochromes for DNA content analyses by flow cytometry. Cytometry 15: 129–140.

Hodson R. E., W. A. Dustman, R. P. Garg & M. A. Moran, 1995. *In situ* PCR for visualization of microscale distribution of specific genes and gene products in prokaryotic communities. Appl. envir. Microbiol. 61: 4074–4082.

Jin, L., X. Qian & R. V. Lloyd, 1995. Comparison of mRNA expression detected by *in situ* PCR and *in situ* hybridization in Endorcrine cells. Cell Vision 4: 314–321.

Lee, S. & P. F. Kemp, 1994. Single-cell RNA content of natural marine planktonic bacteria measured by hybridization with multiple 16S rRNA-targeted fluorescent probes. Limnol. Oceanogr. 39: 869–879.

Long, A. A. & P. Komminoth, 1995. *In situ* PCR: General methodology and recent advances. In Gu J. (ed.), *In situ* Polymerase Chain Reaction and Related Technology. Eaton Publishing Co., Natick, Massachusetts: 23–34.

Pemberton, J. M., 1983. Degradative plasmids. Internatl. Rev. Cytol. 84: 155–183.

Porter, J., R. Pickup & C. Edwards, 1995. Flow cytometric detection of specific genes in genetically modified bacteria using in situ polymerase chain reaction. FEM Microbiology 134: 51–56.

Porter, K. & Y. S. Feig, 1980. The use of DAPI for identifying and counting aquatic microflora. Limnol. Oceanogr. 25: 943–948.

Sayler, G. S., M. S. Shields, E. T. Tedford, A. Breen, S. W. Hooper, K. M. Sirotkin & J. W. Davis, 1985. Application of DNA-DNA colony hybridization to the detection of catabolic genotypes in environmental samples. Appl. envir. Microbiol. 49: 1295–1303.

Schell, M. A., 1985. Transcriptional control of the *nah* and *sal* hydrocarbon degradation operons by the *nah*R product. Gene 36: 301–309.

Sobecky, P. A., M. A. Schell, M. A. Moran & R. E. Hodson, 1992. Adaptation of model genetically engineered microorganisms to lake water: growth rate enhancements and plasmid loss. Appl. envir. Microbiol. 58: 3630–3637.

Tolker-Nielson, T., K. Holmstrøm & S. Molin, 1997. Visualization of specific gene expression in individual *Salmonella typhimurium* Cells by in situ PCR. Appl. envir. Microbiol. 63: 4196–4203.

Trebesius, K., R. Amann, W. Lugwig, K. Mühlegger & K-H. Schleifer, 1994. Identification of whole fixed bacterial cells with nonradioactive 23S rRNA-targeted polynucleotide probes. Appl. envir. Microbiol. 60: 3228–3235.

Yamaguchi, N., S. Inaoka, K. Tani, T. Kenzaka & M. Nasu, 1996. Detection of specific bacterial cells with 2-hydroxy-3-naphthoic acid-2′-phenylanilide phosphate and Fast Red TR *in situ* hybridization. Appl. envir. Microbiol. 62: 275–278.

Zehbe, I., J. F. Sällström, G. W. Hacker, C. Hauser-Kronberger, E. Rylander & E. Wilander, 1994. Indirect and direct *in situ* PCR for the detection of human papillomavirus. An evaluation of two methods and a double staining technique. Cell Vision 163: 163–167.

Zylstra, G. J. & D. T. Gibson, 1989. Toluene degradation by *Pseudomonas putida* F1. J. Biol. Chem. 264: 14940–14946.

Hydrobiologia **401:** 139–148, 1999.
J.P. Zehr & M.A. Voytek (eds), Molecular Ecology of Aquatic Communities.

Variability in bacterial community structure during upwelling in the coastal ocean

L. J. Kerkhof[1], M. A. Voytek[2], R. M. Sherrell[1], D. Millie[3,*] & O. Schofield[1]
[1]*Institute of Marine and Coastal Sciences, 71 Dudley Rd. Cook College, Rutgers University, New Brunswick, NJ 08901-8521, U.S.A. E-mail: kerkhof@ahab.rutgers.edu*
[2]*United States Geological Survey, MS430, 12201 Sunrise Valley Drive, Reston, VA 20192, U.S.A.*
[3]*Southern Regional Research Center, Agricultural Research Service, New Orleans, LA 70179, U.S.A.*

Key words: marine bacteria, phytoplankton, upwelling, rRNA clonal library, specific co-occurrence

Abstract

Over the last 30 years, investigations at the community level of marine bacteria and phytoplankton populations suggest they are tightly coupled. However, traditional oceanographic approaches cannot assess whether associations between specific bacteria and phytoplankton exist. Recently, molecular based approaches have been implemented to characterize specific members of different marine bacterial communities. Yet, few molecular-based studies have examined coastal upwelling situations. This is important since upwelling systems provide a unique opportunity for analyzing the association between specific bacteria and specific phytoplankton in the ocean. It is widely believed that upwelling can lead to changes in phytoplankton populations (blooms). Thus, if specific associations exist, we would expect to observe changes in the bacterial population triggered by the bloom. In this paper, we present preliminary data from coastal waters off New Jersey that confirm a shift in bacterial communities during a 1995 upwelling event recorded at a long-term earth observatory (LEO-15) in the Mid-Atlantic Bight. Using PCR amplification and cloning, specific bacterial 16S ribosomal RNA sequences were found which were present in upwelling samples during a phytoplankton bloom, but were not detected in non-bloom samples (surface seawater, offshore sites or sediment samples) collected at the same time or in the same area. These findings are consistent with the notion of specific associations between bacteria and phytoplankton in the ocean. However, further examination of episodic events, such as coastal upwelling, are needed to confirm the existence of specific associations. Additionally, experiments need to be performed to elucidate the mechanisms leading to the specific linkages between a group of bacteria and a group of phytoplankton.

Introduction

Bacteria and phytoplankton have a profound impact on the marine ecosystem since their combined activity accounts for nearly all of the biogeochemical cycling that occurs in the ocean (Goldhaber & Kaplan, 1974; Van Es & Meyer-Reil, 1982; Carpenter & Capone, 1983). For example, the importance of bacteria as consumers of organic carbon in marine food webs has long been recognized (Pomeroy, 1974; Fuhrman & Azam, 1980; Cho & Azam; 1988). Additionally, coupling between total bacteria and phytoplankton communities has been hypothesized and most often related to the extracellular release of photosynthetically produced dissolved organic carbon (for reviews see, Cole et al., 1988; Baines & Pace, 1991). These studies have shown that 30–50% of the bacterial carbon demand can be met from the direct excretion of photosynthetically-fixed carbon *in situ* in many oceanographic situations. The coupling between bacterial activity and phytoplankton is particularly important in the waters of the Middle Atlantic Bight (MAB). Only 5–10% of the particulate matter produced on the continental shelf is exported off the shelf, and

* Author for correspondence.

the remainder is remineralized and/or consumed by heterotrophic processes *in situ* (Kemp et al., 1994; Falkowski et al., 1994; Biscaye et al., 1994).

Although many of these traditional studies suggest a coupling between bacterial and phytoplankton populations overall, the whole-population methodology can not address whether specific associations exist between these two groups. If specific associations between bacteria and phytoplankton exist, changes in both these populations should result from physical forcing, such as upwelling events, on short time scales. For example, upwelling can very often lead to mono-specific phytoplankton blooms in the ocean. In a bloom, a diverse community of phytoplankton shifts to a community dominated by a few members of the phytoplankton community, i.e. phytoplankton diversity plummets. It is anticipated that bacterial diversity will likewise plummet during phytoplankton blooms if specific associations exist, since a few members of the bacterial community will also come to dominate during the bloom. Furthermore, these specific bacterial groups should be readily detectable in waters experiencing the bloom and otherwise absent or present at reduced numbers comparable to the 'host' phytoplankton species.

The MAB provides a unique opportunity to apply new methodologies and approaches to determine whether specific bacteria are associated with specific phytoplankton. Specifically, the region experiences episodic upwelling of offshore bottom waters into surface layers (Hicks & Miller, 1980). This upwelling is initiated by strong alongshore winds and can last for days or weeks (Glenn et al., 1996), leading to phytoplankton blooms and accumulation of organic matter (Warsh, 1987). An extensive array of oceanographic sensors have recently been put in place to monitor these upwelling events in real time (von Alt et al., 1997).

In this paper, we describe changes in a coastal planktonic community during an upwelling event in 1995 using 16S rRNA analysis (for review, Amann et al., 1995; Torsvik et al., 1996; Head et al., 1998; and Hugenholtz et al., 1998). On July 15, an upwelling began and lasted approximately 3 weeks. Initially, a small band of cold water was detectable near shore on the sea surface by satellite. By July 27, the upwelling feature was fully developed with colder surface water extending 80 km up and down the coast and 20 km offshore from the Rutgers Marine Field Station. Within a few days (August 4), the winds began to shift and a relaxation of upwelling became detectable. The entire upwelling feature was destroyed when Hurricane Felix began impacting the New Jersey coast on August 15,1995.

On July 17 and August 9, 1995, we fortuitously collected bacteria from upwelling and non-upwelling sites during a routine sampling cruise along a 34 km transect line. Differences in microbial diversity, as well as changes in individual bacterial species composition were observed during this upwelling event. Specifically, microbial diversity was found to decrease during the onset of upwelling. However, ten 16S rRNA genes were found in two upwelling/phytoplankton bloom gene libraries (9 m depth) that were not observed in six different gene libraries from samples where the phytoplankton concentrations were low (surface seawater, surface and bottom at an offshore site, and 2 sediment samples). This study suggests that distinct populations of bacteria are present during coastal upwelling and implies that physical events can drive phytoplankton community composition and bacterial diversity in the coastal ocean. Furthermore, our data provide important molecular markers for use in the future to ascertain whether the associations observed between specific bacteria and phytoplankton in this study are robust or circumstantial.

Methods

Field samples

Our routine, monthly sampling in 1995 covered a 34 km transect (10 stations) off the coast of New Jersey. Water was collected using Niskin bottles at two depths (1 m below the surface and 2 m above the bottom; see Figure 1). The depth of the water column at our study area is between 5 and 35 m. Seawater samples (500–1000 ml) were filtered onto 0.2 μm polysulfone filters to obtain bacterial and phytoplankton biomass. All filtered samples were placed at <0 °C at sea and stored at −80 °C in the laboratory until processing.

For the upwelling study described in this paper, biomass samples were characterized at two stations collected 4 and 19 km offshore (Station 9 and A4; see Figure 1). The samples were collected early (July 17) and late (August 9) during upwelling, and the stations were selected to represent samples within and outside the upwelling feature. Physical and chemical data for this area are described in Glenn et al. (1996) and Boehme et al. (1998).

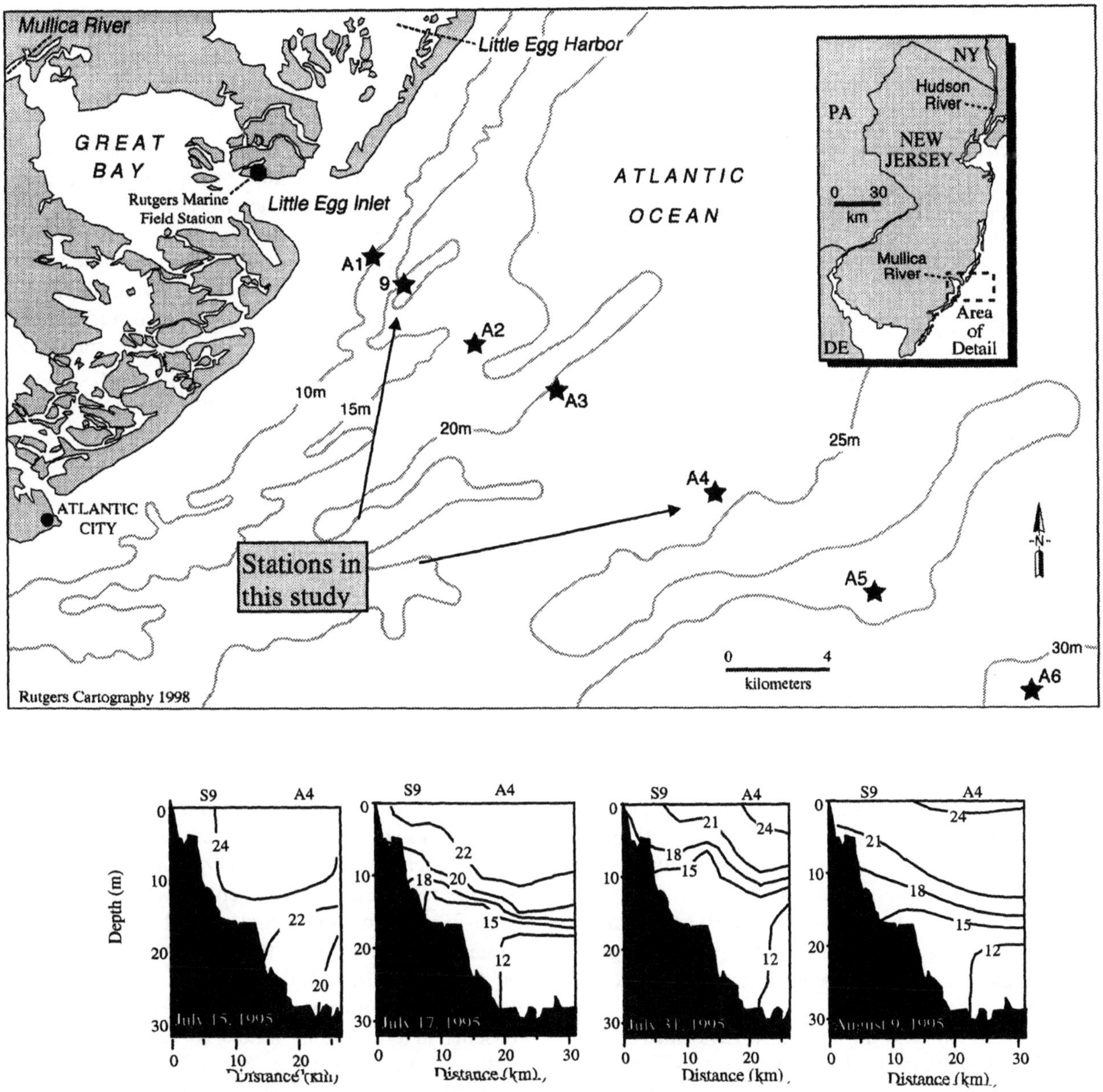

Temperature profiles along transect shown above.

Figure 1. Map of study area off the New Jersey Coast indicating stations along transect line. Stations 9 and A4 are indicated. Below are CTD profiles obtained during the upwelling event. Samples were collected on July 17 and August 9, 1995.

DNA extractions and PCR amplification of 16S rRNA genes

Bacterial biomass samples were subjected to a modified phenol/chloroform extraction as previously described (Kerkhof & Ward, 1993). Amplification of 16S rDNA was done using standard eubacterial primers [27 Forward (5′ cua cua cua cua AGA GTT TGA TCC TGG CTC AG 3′) and 1525 Reverse (5′ cau cau cau cau AAG GAG GTG WTC CAR CC 3′)] (Lane et al., 1991) which have a uracil-rich cloning site. The following amplification parameters were used: initial denaturation at 95 °C for 5 min, then 94 °C for 0.5 min, 55 °C for 0.5 min and 72 °C for 1.5 min for 22 cycles, and a final extension step at 72 °C for 10 min. The amplifications were performed in a Perkin-Elmer Gene Amp PCR system 2400 thermal cycler (Perkin-Elmer, Foster City, CA).

Cloning, screening and sequencing

The cloning of 16S rDNA was done using the Cloneamp System (Gibco BRL, Gaithersberg, MD) as per the manufacturer's instruction using 22 cycle amplifications. Colonies were screened by restriction enzyme analysis (Hae III, Boehringer-Mannheim (BMB), Indianapolis IN) on extracted plasmids or by PCR amplification of insert 16S rDNA and digestion with Rsa I and Msp I (BMB). The restriction digests were analyzed using Methaphor agarose (FMC Corp., Rockland, ME). Estimates of microbial diversity in these samples were made using rarefaction curves by tallying the number of unique 16S rRNA restriction patterns vs. the total number of clones screened within a 16S rRNA clonal library (Sakano & Kerkhof, 1998). The shapes of these curves reflect the underlying diversity in the bacterial population.

Plasmids bearing a unique restriction banding pattern were re-purified using a Flexiprep kit (Pharmacia, Piscataway, NJ). DNA sequence was determined by automated techniques (Perkin Elmer-ABI, Foster City, CA) with the following primers: 27 Forward (5′ AGA GTT TGA TCC TGG CTC AG3′), and 519 Reverse (5′ GWA TTA CCG CGG CKG CTG 3′), (Lane, 1991). Determinination of the phylogenetic affiliation of all RFLP unique clones in each library was performed by BLAST (Altschul et al., 1990) search and phylogenetic tree re-construction sequences and alignments in the Ribosomal Database Project (Maidak et al., 1994).

Initial screening of all unique RFLP clones from the six small ribosomal subunit libraries was done by phylogenetic analysis of unambiguously aligned sequence (321–375 bp) from a total of 470 bp sequence. The initial phylogenetic trees were re-constructed by the Neighbor joining method to identify all 16S rRNA clones common to both upwelling and non-upwelling samples and those clones bearing >99% identity to other clones in the library. This initial screening made it possible to identify the 16S rRNA genes unique to the bloom conditions. Final phylogenetic tree re-construction on the bloom-specific 16S rRNA genes was based on 294 bp of aligned sequence using fastDNAml (Olsen et al., 1994) and 100 bootstrap replications. The sequences have been submitted to Genbank and have the accession numbers: AF111859–AF111867.

Phytoplankton Pigment Analysis.

Pigments were extracted in acetone and concentrated before HPLC analysis. Chlorophylls and carotenoids were separated on a Vydac reverse-phase, C18 column using a binary gradient system (after Van Heukelem et al., 1992) optimizing for separation of the carotenoids, zeaxanthin and lutein. Pigment peaks were identified by comparing retention times and absorption spectra (380–670 nm) with those of authentic standards and of known pigments from purified algal cultures maintained at the Southern Regional Research Center (SRRC, Millie et al., 1993).

Results

Sea surface temperature from AVHRR images and CTD data indicated a coastal upwelling event occurred off New Jersey between July and August, 1995. A large difference in surface temperature was observed with the coldest surface water (between 18.7 and 25.5 °C) detectable along the coast. Station 9 is centered within this upwelling feature and station 4 is along the offshore edge during the greatest extent of upwelling (data not shown). CTD profiles (Figure 1) confirmed a shoaling of isotherms towards the surface during our sampling periods. Chemical analysis on July 17 detected minimal differences between nutrients at Station 9 for surface and bottom samples (0.11–0.13 μmol/kg nitrate + nitrite, () 0.12–0.13 μmol/kg ammonium, and 0.39–0.53 μmol/kg phosphate). However, Station A4 values during this same period had higher nutrients in the bottom waters (0.12–1.64 μmol/kg nitrate + nitrite), 0.14–2.78 μmol/kg ammonium,; and 0.12–0.85 μmol/kg phosphate. In August, Station 9 had higher values of some nutrients in bottom samples (0.11–0.13 μmol/kg nitrate + nitrite, 0.23–0.72 μmol/kg ammonium, and 0.46–0.56 μmol/kg phosphate). Station A4 values during this same period were 0.03–0.53 μmol/kg nitrate + nitrite, 0.15–2.07 μmol/kg ammonium, and 0.12–0.45 μmol/kg phosphate. These data suggest that the sources of upwelled waters in our study area are from offshore bottom sites, such as Station A4, and that nutrient were brought to the bottom waters at Station 9. A more detailed report of nutrient dynamics for a 2-year time series in this area is presented in Boehme et al. (1998).

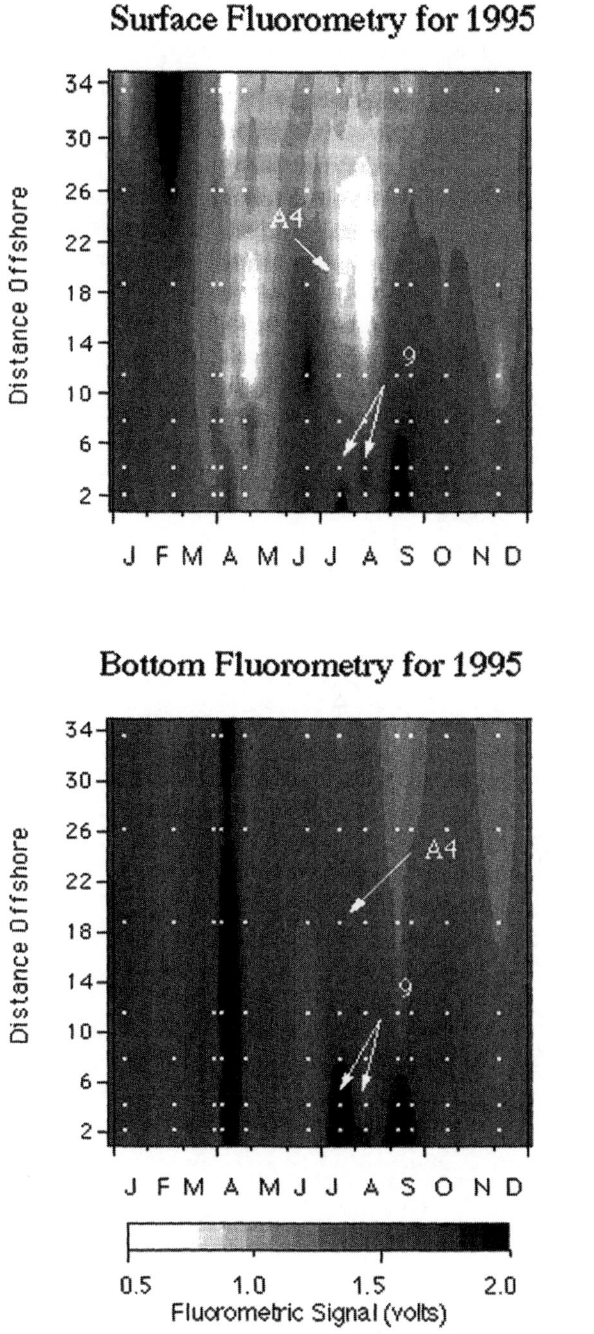

Figure 2. Annual survey of fluorescence along a 40 km transect offshore from New Jersey during 1995. Stations 9 and A4 sampled for this study are indicated.

Fluorometry data collected during 1995 indicated high fluorescence in bottom waters during the upwelling event (Figure 2). Surface water exhibited a reduced fluorescence throughout much of the year and during our sample period (Figure 2).

The greatest concentrations of chlorophyll *a* and other algal accessory pigments were observed in the summer (July 1995 and August 1995) in the bottom samples at our study site (Table 1). During the initial stages of upwelling, the chlorophyll *a* concentrations were 7-fold higher in bottom waters than in surface waters. The surface samples at these times contained extremely low concentrations of all algal pigments. By the end of the upwelling event, bottom waters had 100 times higher chlorophyll *a* concentrations than surface samples. The significant peridinin concentration in the bottom waters during July and August of 1995 indicated dinoflagellates were a major component of the phytoplankton population. The other major marker pigment present during 1995 was fucoxanthin, suggesting that diatoms were a dominant member of the phytoplankton community. Surface samples taken simultaneously, or offshore samples, had much lower pigment signals indicating lower concentrations of all phytoplankton in these waters during this time.

To determine the impact of the July–August 1995 upwelling event on microbial diversity, six clonal libraries of 16S rRNA genes were created for Station 9 and Station A4. Four libraries were created at Station 9 (surface and bottom in July 1995 and August 1995) and 2 libraries at Station A4 (surface and bottom in July 1995). Rarefaction curves were constructed from all libraries to estimate overall microbial diversity. A large drop in diversity is apparent in the Station 9 bottom July 1995 sample when compared with all other samples (Figure 3) concurrent with the initial stages of upwelling at our study site. However, an analysis of the phylogenetic affiliation of all clones in each library showed only minor changes in the relative percentages of the contributing members of the microbial community during upwelling suggesting minimal bias in the clonal libraries (Figure 4). At all sites/times, the small subunit (SSU) clones were comprised mainly of Proteobacteria sequences (80–90%). Representatives of the α and γ subdivisions accounted for the bulk of the clones in each library. A small number of clones from the β subdivision were also present in most of the marine libraries. Interestingly, a significant fraction of the SSU clones in the Station 9 bottom samples contained sequences from the ϵ-subdivision Proteobacteria. SSU clones from this subdivision were not observed in the Station 9 surface or Station A4 samples (surface and bottom) containing lower algal pigment concentrations. Clones from the Cytophaga phyla were present in all libraries at lower abundance (<20%).

Table 1. Concentrations of pigments from station 9 time series samples Units in μg/l

Date	Depth	Chl *a*	Chl *b*	Chl *c*1–*c*2	Peridinin	Fuco	Diadino	Anth	Allo
Jan-95	surface	-	-	-	-	0.116	-	-	-
	bottom	-	-	-	-	0.190	-	-	-
Feb-95	surface	-	-	-	-	0.049	-	-	-
	bottom	-	-	-	-	0.050	-	-	-
Mar-95	surface	0.116	-	-	-	0.119	-	-	-
	bottom	-	-	-	-	0.073	-	-	-
Apr-95	surface	0.088	-	-	-	0.073	-	-	-
	bottom	-	-	-	-	-	-	-	-
May-95	surface	0.178	-	0.173	-	0.351	0.033	-	0.039
	bottom	0.279	-	0.203	-	0.294	-	0.048	-
Jun-95	surface	0.119	-	0.094	-	0.112	-	-	0.099
	bottom	-	-	-	-	0.036	-	-	
Jul-95	surface	0.049	-	-	-	0.133	-	-	0.036
	bottom	0.339	-	2.628	2.269	0.501	0.179	0.389	-
Aug-95	surface	-	-	0.114	0.044	0.326	-	-	-
	bottom	4.853	0.294	6.42	8.151	1.338	0.437	-	-

Surface=1 m depth, bottom =15 m depth; Chl *a*=Chlorophyll *a*, Chl *b*=Chlorophyll *b*, Chl *c*1-*c*2=Chlorophyll *c* 1 and 2, fuco=fucoxanthin, anth=antherxanthin, allo=alloxanthin. - = below detection limit.

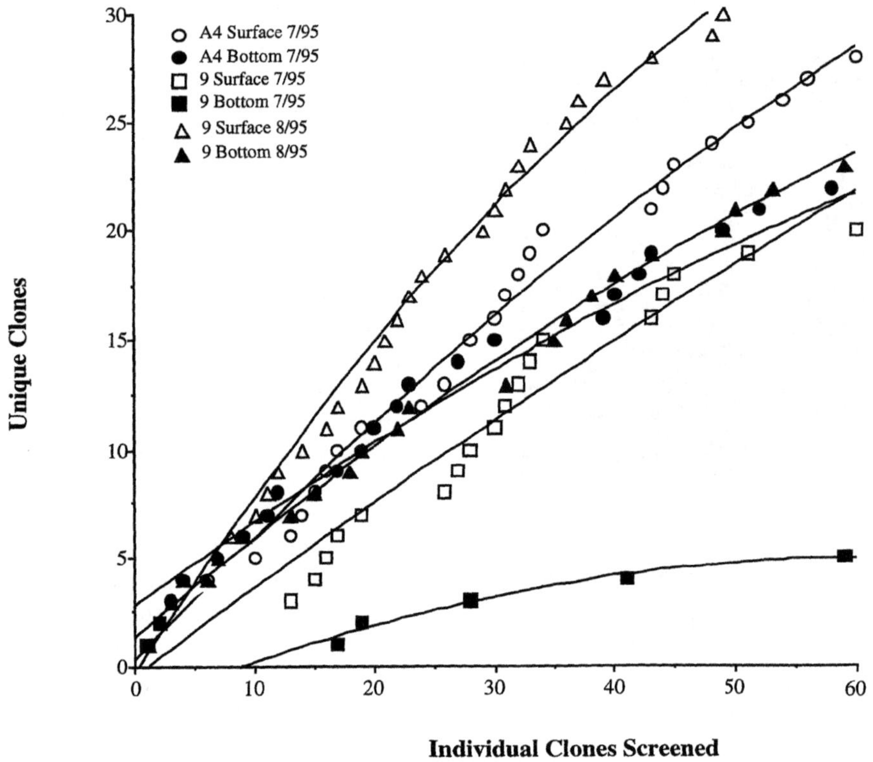

Figure 3. Rarefaction curve of surface and bottom samples during summer 1995. The samples are from: Station A4 Surface July 1995 (○), Station A4 Bottom July 1995 (●), Station 9 Surface July 1995 (□), Station 9 Bottom July 1995 (■), Station 9 Surface August 1995 (▲), and Station 9 Bottom August 1995.

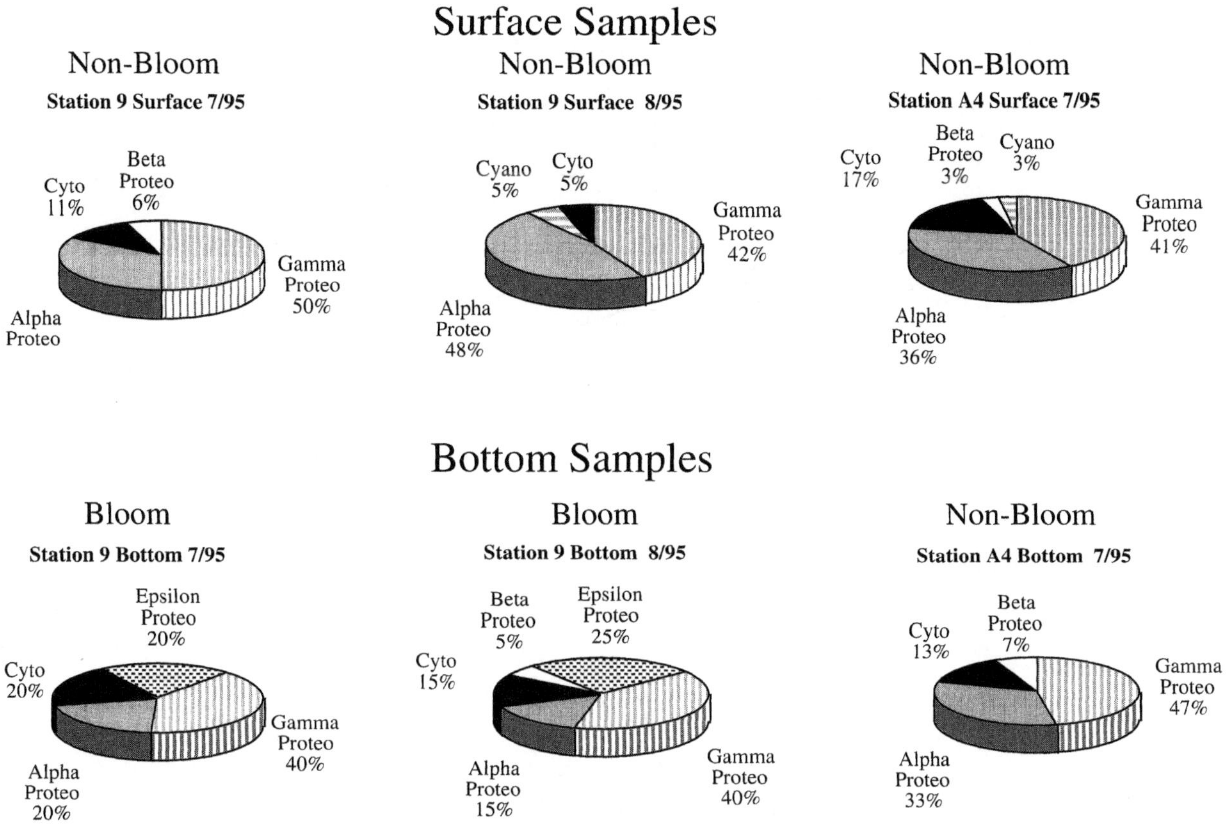

Figure 4. Proportion of phylogentic affiliation of the various clones observed in all 16S rRNA gene libraries collected in the study. Sites, times, and depths are indicated. Legend: Cyto= *Cytophaga*, Proteo=*Proteobacteria*, Cyano=*Cyanobacteria*.

To ascertain whether there were specific bacteria associated with the phytoplankton at Station 9 in July 1995 and August 1995, phylogenetic trees were reconstructed using sequences from the six 16S rRNA libraries. A total of 8 α-subdivision, 12 γ-subdivision, and 6 ϵ-subdivision SSU clones were observed in all six 16S rRNA libraries. Of the 26 Proteobacteria SSU clones, 9 were found to be unique to the Station 9 upwelling samples with 5 clones coming from the ϵ-subdivision. None of these upwelling SSU clones have been observed in 16S rRNA genes libraries from sediments within a kilometer of Station 9 in November 1995 and May 1996 (Kerkhof, unpubl. data), suggesting that bottom re-suspension was not a significant source of the unique rDNA clones obtained from upwelling samples (data not shown).

The phylogenetic affiliations of the bloom-specific clones (highlighted in gray) are presented with most of their closest matches from Genbank in Figure 5. The designation of UPA or B refers to SSU clones present in July 1995 or August 1995, respectively. Many of the clones exhibited identity with rRNA genes obtained from the marine environment. For most UPA or UPB clones, the nearest matches in Genbank were <93% similar. The clones UPB A1 and UPA D3 were related to gene sequences from other marine 16S rRNA gene libraries (OM 27, <88% similar) or uncultured microorganisms (Lucinid endosymbionts, <90% similar). One 16S rRNA gene sequence, UPB C7, discovered in our study was related (<92% similar) to an algicidal bacterial strain capable of lysing *Gymnodinium* sp. (Yoshinaga et al., 1995). The clone UPB G7 was found to be between 96 and 98% similar to the SSU gene sequences OM 42, GAC 2, and BAL 28 from libraries generated from Atlantic and Pacific samples (Rappe et al., 1997; Gonzalez & Moran, 1997; and Pinhassi et al., 1997). Finally, 5 of the SSU clones from this study were related (<92% similarity) to the epsilon subdivision of the Proteobacteria. The closest isolates are indicated in Figure 5. Because this group of microorganisms have diverse metabolic capabilities, it is difficult to extrapolate the metabolic capabilities of the epsilon bacteria detected in this study.

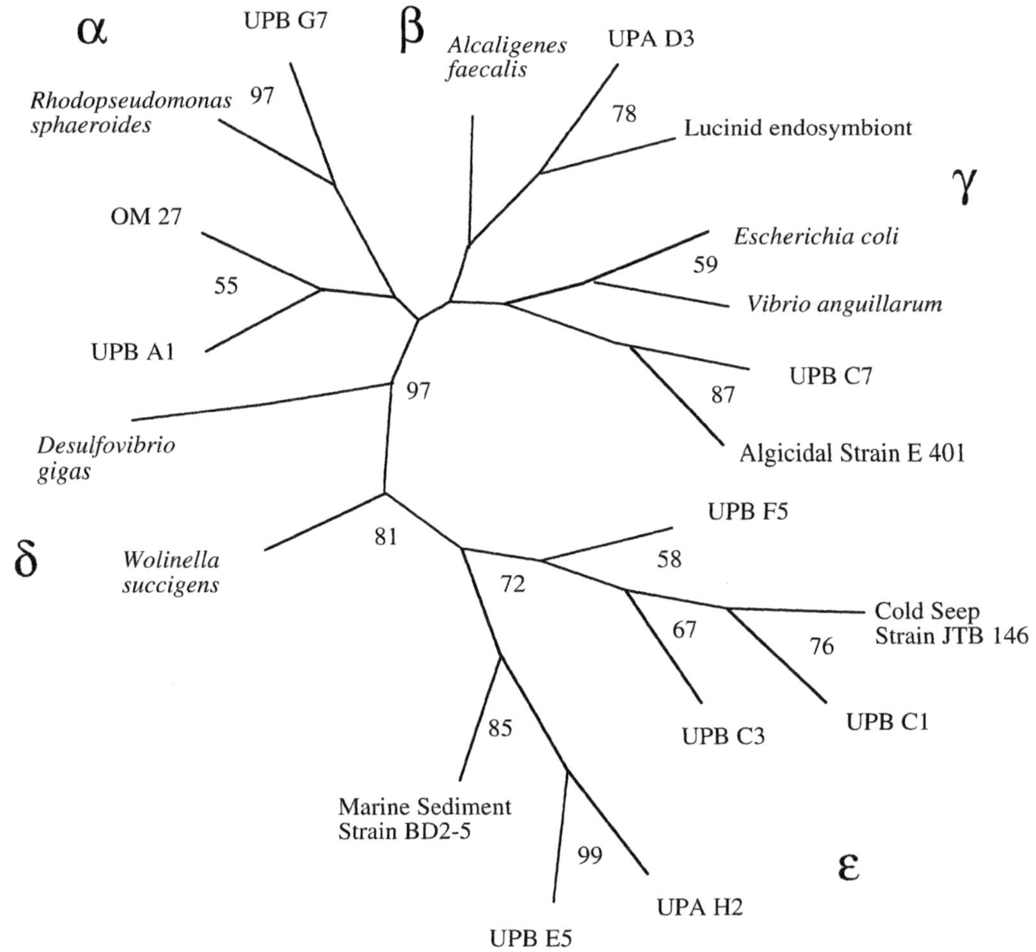

Figure 5. Maximum likelihood phylogenetic tree of Proteobacterial 16S rRNA clones re-constructed using 295 bp alignment. Clones unique to the upwelling condition are boxed in gray. The diatom (Diatom) and dinoflagellate (Dino) bloom associated clones are indicated.

Discussion

Other studies using a species approach to document the response of the microbial community to environmental changes have applied techniques including immunoassays, fatty acid analysis, or 5S rRNA analysis (Frostegard et al., 1993; Höfle & Brettar, 1995; Tuomi et al., 1997). Studies have also been performed to compare microbial communities from different aquatic habitats (Field et al., 1997; Rappe et al., 1997; Gonzalez & Moran, 1997; Pinhassi et al., 1997 and Methe' et al., 1998). The common thread in all of these reports is a switch from bulk measurements of bacteria to techniques that allow for detection of specific 'types' in response to environmental factors (e.g. depth, salinity, DOC, nutrients, heavy metals, virus abundance, protozoan grazing).

In this study, we used a 16S rRNA-based approach to address the response of specific bacteria and phytoplankton to an episodic event, such as upwelling, and assess the effect of physical factors on microbial community structure. Although 16S rRNA gene characterization is now routine in many labs, potential biases in the clone and sequence approach still exist. Primarily, DNA extraction procedures can miss entire groups that are difficult to lyse such as Gram-positive bacteria. Additionally, large amounts of template DNA, high cycle numbers in the PCR amplification, and large amounts of transforming DNA are used to maximize the amount of colonies obtained during cloning. However, this strategy will confound any attempts at quantitative compositional analysis due to PCR (Alard et al., 1993; Suzuki & Giovannoni, 1996) or cloning biases resulting from asymptotic transformation of *E. coli* (Hanahan, 1983; Dower et al. 1988). We have taken steps to minimize these biases inherent in the traditional approach. For example, our extraction procedure can successfully isolate Gram-positive 16S rRNA genes from marine sediments (Phelps et al., 1998). Furthermore, we used minimal template concentration (<10 ng of genomic DNA), low cycle numbers (20–25), and low transforming DNA concentrations(<6 ng) to create our clonal libraries for determining specific bacterial 16S rRNA genes associated with upwelling.

Previous research has demonstrated the bacterial community is linked to the phytoplankton community at a whole population level (Cole et al. 1988; Baines & Pace, 1991). Our data from an upwelling event off New Jersey is consistent with the concept of linkages between bacterial and phytoplankton assemblages. Primarily, shifts in bacterial populations were observed over the course of several weeks during upwelling in the MAB. Secondly, there appeared to be specific 16S rRNA phylotypes associated with phytoplankton bloom conditions that were not detected in samples where phytoplankton concentrations were low.

Although these differences in bacterial community assemblages co-occur with differences in algal pigment distributions, at present it is unclear what mechanism(s) control the bacterial and phytoplankton populations. Given the limitations of this study, we cannot yet determine how robust these associations are, or whether they represent physical structuring of microbial communities (advection) and/or biotic interactions between the bacteria and the phytoplankton. In order to firmly establish if specific bacterial groups are stimulated by various phytoplankton groups (e.g. diatoms, dinoflagellates), future studies during upwelling events in the MAB system should collect samples on ecologically relevant spatial and temporal scales and identify specific bacterial 16S genes associated with specific groups of phytoplankton. Furthermore, quantitative analysis on specific groups of bacteria and phytoplankton (abundance and growth rate analysis) are needed to verify any potential causal linkage. Ultimately, research such as described in this study will help to provide an understanding of the mechanisms connecting bacteria and phytoplankton in the ocean.

Given the specific linkages between bacterial and phytoplankton groups that were identified in this study, specific hypotheses can be tested in future studies: 1. Specific groups of bacteria and phytoplankton will be tightly coupled in time and space. That is, specific bacterial 16S rRNA genes will only become dominant during specific phytoplankton blooms and otherwise be minor constituents of the community. 2. 16S rRNA gene sequences identified within a specific phytoplankton bloom will undergo growth and senescence in phase with the bloom itself. 3. Specific associations are repeatable and can be observed in different bloom events either in the same year or from year to year. These hypotheses can be tested during episodic events (such as upwelling) which provide a physical/biological gradient over short spatial and temporal scales.

Acknowledgments

This research was supported in part by funds from Rutgers University, (NOAA/NURP NA76RU0165 to L. K.) and ONR CoMOP Program (N0014-97-1-0797 to O.S.). The authors wish to especially thank Mike Crowley and Clare Reimers for supplying physical and chemical data and Gary Taghon for constructive comments on the manuscript. We are indebted to David Scala, Larry Biohem, Mark Moline, Joe Grymski, Sue Boehme, Robert DeKorsey and Rose Petrecca for help in sample collection or processing.

References

Alard, P, O. Lantz, M. Sebagh, C. Clavo, D. Weill, G. Chavanel, A. Senik & B Charpentier, 1993. A versatile ELISA-PCR assay for mRNA quantitation from a few cells. Biotechniques 15: 730–737.

Altschul, S. F., W. Gish, W. Miller, E. W. Myers & D. J. Lipman, 1990. Basic local alignment search tool. J. mol. Biol. 215: 403–410.

Amann, R. I., W. Ludwig, & K. H. Schleifer, 1995. Phylogenetic identification and *in situ* detection of individual microbial cells without cultivation. Microbiol. Rev. 59: 143–169.

Baines, S. B. & M. L. Pace, 1991. The production of dissolved organic matter by phytoplankton and its importance to bacteria: Patterns across marine and freshwater systems. Limnol. Oceanogr. 36: 1078–1090.

Biscaye, P. E., C. N. Flagg & P. G. Falkowski, 1994. The shelf edge exchange processes experiment, SEEP-II: an introduction to hypotheses, results and conclusions. Deep Sea Res. Part II, 41: 231–255.

Boehme, S., C. Sabine & C. Reimers, 1998. CO_2 fluxes from a coastal transect: A time series approach. Mar. Chem. In Press.

Carpenter, E. J. & D. G. Capone, 1983. Nitrogen in the Marine Environment. Academic Press. London. 900 pp.

Cho, B. C. & F. Azam, 1988. Major role of bacteria in biochemical fluxes in the ocean's interior. Nature 332: 441–443.

Cole, J. J., S. Findlay & M. Pace, 1988. Bacterial production in fresh and saltwater ecosystems: a cross-system overview. Mar. Ecol. Prog. Ser. 43: 1–10.

Dower, W., J. Miller & C. Ragsdale, 1988. High efficiency transformation of *E. coli* by high voltage electroporation. Nucl. Acids. Res. 16: 6127–6144.

Falkowski, P. G., P. E. Biscaye & C. Sancetta, 1994. The lateral flux of biogenic particles from the eastern North American continental margin to the North Atlantic Ocean. Deep Sea Res. Part II, 41: 583–601.

Field, K. G., D. Gordon, T. Wright, M. Rappe, E. Urbach, K. Vergin & S. J. Giovanonni, 1997. Diversity and depth-specific distribution of SAR 11 cluster rRNA genes from marine planktonic bacteria. Appl. envir. Microbiol. 63: 63–70.

Frostegard, A., A. Tunlid & E. Baath, 1993. Phospholipid fatty acid composition, biomass, and activity of microbial communities from two soil types experimentally exposed to different heavy metals. Appl. envir. Microbiol. 59: 3605–3617.

Fuhrman, J. & F. Azam, 1980. Bacterioplankton secondary production estimates for coastal waters of British Columbia, Antarctica, and California. Appl. envir. Microbiol. 39: 1085–1095.

Goldhaber, M. & I. Kaplan, 1974. The Sulfur Cycle. In The Sea. Goldberg E., (ed.), John Wiley and Sons, New York: 569–655.

Glenn, S. M., M. F. Crowley, D. B. Haidvogel & Y. H. Song, 1996. Underwater observatory captures coastal upwelling events off New Jersey. EOS 77: 233–236.

Gonzalez, J. & M. A. Moran, 1997. Numerical dominance of a group of bacteria in the alpha-subclass of the class Proteobacteria in coastal seawater. Appl. envir. Microbiol. 63: 4237–4242.

Hanahan, D, 1983. Studies on transformation of *E. coli* with plasmids. J. mol. Biol. 166: 557–580.

Head, I. M., J. R. Saunders & R. W. Pickup, 1998. Microbial evolution, diversity, and ecology: A decade of ribosomal RNA analysis of uncultivated microorganisms. Micr. Ecol. 35: 1–21.

Hicks, D. C. & J. R. Miller, 1980. Meteorological forcing and bottom water movement off the northern New Jersey coast, Estuar. coast. Sci. II. 563–571.

Hofle, M. G. & I. Brettar, 1995. Taxonomic diversity and metabolic activity of microbial communities in the water column of the central Baltic Sea. Limnol. Oceanogr. 40: 865–874.

Hugenholtz, P., B. M. Goebel & N. R. Pace, 1998. Impact of culture-independent studies on the emerging phylogenetic view of bacterial diversity. J. Bact. 180: 4765–4774.

Kemp, P. F., P. G. Falkowski & C. D. Wirick, 1994. Modeling vertical oxygen and carbon flux during stratified spring and summer conditions on the continental shelf, Middle Atlantic Bight, eastern U.S.A. Deep Sea Res. Part II, 41: 629.

Kerkhof, L. & B. B Ward, 1993. Comparison of nucleic acid hybridization and fluorometry for measurement of the relationship between RNA/DNA ratio and growth rate in a marine bacterium. Appl. envir. Microbiol. 59: 1303–1309.

Lane, D. J., 1991. 16S/23S rRNA sequencing, In Stackebrandt E. & M. Goodfellow (eds), Nucleic Acid Techniques in Bacterial Systematics. John Wiley & Sons: 115–175.

Maidak B. L., G. J. Olsen, N. Larsen, R. Overbeek, M. J. McCaughey & C. R. Woese, 1997. The RDP (Ribosomal Database Project). Nucl. Acids Res. 25: 109–111.

Methe, B. A., W. D. Hiorns & J. P. Zehr, 1998. Contrasts between marine and freshwater bacterial community composition: Analyses of communities in Lake George and six other Adirondack lakes. Limnol. Oceanogr. 43: 368–374.

Millie, D., H. Paerl, J. Hurley & G. Kirkpatrick, 1993. Algal pigment determinations in aquatic ecosystems: Analytical evaluations, applications, and recommendations. Current Topic in Bot. Res. 1: 1–13.

Olsen, G. J., H. Matsuda, R. Hagstrom & R. Overbeek, 1994. FastDNAml – a tool for construction of phylogenetic trees of DNA sequences using maximum likelihood. Computer Applications in the Biosciences 10: 41–48.

Phelps, C., L. Kerkhof & L. Young, 1998. Molecular characterization of a sulfate-reducing consortium which mineralizes benzene. FEMS Micr. Ecol. 27: 269–279.

Pinhassi, J., U. Zweifel & A. Hagstrom, 1997. Dominant marine bacterioplankton species found among colony-forming bacteria. Appl. envir. Microbiol. 63: 3359–3366.

Pomeroy, L. R., 1974. The ocean's food web, a changing paradigm. Bioscience 24: 499.

Rappe, M. S., P. F. Kemp, & S. J. Giovannoni, 1997. Phylogenetic diversity of marine coastal picoplankton 16S rRNA genes cloned from the continental shelf off Cape Hatteras, North Carolina. Limnol. Oceanogr. 42: 811–826.

Reynolds, C. S., 1980. Phytoplankton assemblages and their periodicity in stratifying lake systems. Holarctic Ecol. 3: 141–159.

Sakano, Y. & L. Kerkhof, 1998. Assessing changes in microbial community structure during operation of an ammonia biofilter using molecular tools. Appl. envir. Microbiol. 64: 4877–4882.

Suzuki, M. T. & S. J. Giovannoni, 1996. Bias caused by template annealing in the amplification of mixtures of 16S rRNA genes by PCR. Appl. envir. Microbiol. 62: 625–30.

Tuomi, P., T. Torsvik, M. Heldal & G. Bratbak, 1997. Bacterial population dynamics in a meromictic lake. Appl. envir. Microbiol. 63: 2181–2188.

Torsvik, V., R. Sorheim & J. Goksoyr, 1996. Total bacterial diversity in soil and sediment communities – A review. J. industrial Microbiol. 17: 170–178.

Van Es, F. & L. Meyer-Reil, 1982. Biomass and Metabolic Activity of Heterotrophic Marine Bacteria. In: K. Marshall, (eds), Advances in Microbial Ecology Plenum, New York: 111–169.

Van Heukelem, L., A. J. Lewitus, T. M. Kana & N. E. Craft, 1992. High-performance liquid chromatography of phytoplankton pigments using a polymeric reversed-phase C^{18} column. J. Phycol. 28: 867–872.

Von Alt, C., M. DeLuca, S. Glenn, J. Grassle & D. Haidvogel, 1997. LEO-15: Monitoring and managing coastal resources. Sea Technology 38: 10–16.

Warsh, C., 1987. NOAA's Northeast monitoring program (NEMP): A report on progress of the first five years (1979–84) and a plan for the future. NOAA Technical Memorandum NMFS-F/NEC-44. Northeast Fisheries Center, Woods Hole.

Yoshinaga, I., T. Kawai & Y. Ishida, 1995. Lysis of *Gymnodinium nagasakiense* by marine bacteria. Proceedings of the International Conference On Harmful Algal Blooms: 687–692.

Hydrobiologia **401:** 149–175, 1999.
J.P. Zehr & M.A. Voytek (eds), Molecular Ecology of Aquatic Communities.

Application of molecular techniques to addressing the role of P as a key effector in marine ecosystems

D. J. Scanlan[1] & W. H. Wilson[2]
[1]*Department of Biological Sciences, University of Warwick, Gibbet Hill Road, Coventry, CV4 7AL, U.K. Tel: [+44] 1203 528363; Fax: [+44] 1203 523701; E-mail: dp@dna.bio.warwick.ac.uk*
[2]*Marine Biology Association of the U.K., The Laboratory, Citadel Hill, Plymouth, PL1 2PB, U.K. Tel: [+44] 1752 633356; Fax: [+44] 1752 633102; E-mail: whw@wpo.nerc.ac.uk*

Key words: bacterioplankton, denaturing gradient gel electrophoresis (DGGE), marine viruses, molecular markers, phosphate limitation, phytoplankton, *Synechococcus*

Abstract

One of the fundamental questions of marine biology is what limits the growth of marine organisms? Although there are many environmental variables that may impact on thsis problem, limitation for inorganic nutrients is a key component. Recent examples in the literature provide good evidence for phosphorus (P) limitation of phytoplankton and bacterioplankton growth in several ocean regions. This review seeks to address new molecular methods that may be used to evaluate the P status of bacterial and algal groups *in situ*. Physiological parameters of the photosynthetic apparatus and cell cycle variables are considered, but attention is focused on the development of protein biomarkers that can be used to derive antibody probes for the immunological interrogation of natural microbial populations. The need to identify suitable protein markers in key genera, and for a good knowledge of the regulation of the marker in the target organism to be obtained before assessing expression in natural populations, is emphasised. We also examine the role of viruses, which are now widely accepted as being abundant in marine environments, as a source of dissolved DNA, and hence an intermediate in P regeneration. Using molecular techniques, we present evidence that the availability of P can influence virus-host interactions, with particular reference to viral lysogeny in marine *Synechococcus* spp.

Introduction

Over the last 20 years, there have been several major methodological advances in the field of oceanography which are changing our way of investigating and understanding the biological functioning of marine ecosystems. Both satellite imagery and the new routine use of moored instrumentation *in situ* allow for rapid analysis of biological processes over large spatial scales, and can be directly linked with physical features of the water column. Moreover, the introduction of molecular biological techniques, as are being described in this set of papers (and see Falkowski & LaRoche, 1991; Joint, 1995), allow an analysis of the fundamental mechanisms controlling the diversity, adaptive responses and survival of the marine biota. Taken together, these techniques are giving new insights into the biological 'potential' of marine organisms, and will undoubtedly lead to a better understanding of the fate and rate fluxes of the key elements cycling within the marine environment.

However, it is becoming increasingly apparent that the wealth of genetic diversity within marine microorganisms will take much more exhaustive sequence analysis of individual genomes before we can begin to understand the biochemical and metabolic capacity of the whole community. Recent microcosm experiments have highlighted the role of species diversity in maintaining the predictability of microbial communities and thus ecosystem function (Naeem & Li, 1997; McGrady-Steed et al., 1997), but to understand the effects of changes in environmental conditions, and the subsequent extinction of 'redundant' species, will rely on exhaustive basal knowledge of the community as it is now. It is also clear that there are specific factors within the environment that strongly dictate the distribution and diversity of these marine organisms, for example, sedimentation, predation, temperature, light, nutrient availability or viral abundance. Since one of the potentially key changing environmental variables is nutrient input and cycling, and certainly there is considerable discussion of the precise nutrient limiting the productivity of marine phytoplankton (Is there P limitation in the marine environment section), and with the acknowledgement of viruses as an important factor in controlling phytoplankton numbers, it is opportune to identify how molecular techniques might begin to help elucidate the links between these particular environmental variables. Thus, we have focused here on the role of one nutrient, phosphorus (P), in potentially controlling the productivity and diversity of marine phyto- and bacterio-plankton, together with the interactions of this nutrient with marine viruses thus linking two trophic levels within the marine microbial food web. We assess the use of modern molecular techniques in addressing these issues and particularly emphasise the molecular biological and biochemical features of marine microorganisms that may be useful in developing 'molecular markers' of their P status.

P in the marine environment

P (in the form of phosphate) is an essential mineral macronutrient for biological growth and development. Chemical weathering of minerals on continents represents a significant source of bioavailable P, with its entry into the marine environment being via river and aeolian sources. However, the dominant allochthonous source of P to phytoplankton communities in oceanic regions is likely the subsurface nutrient stocks re-introduced into the photic zone by upwelling processes from deep water, since atmospheric deposition rates for P, certainly in the North Atlantic, are small (Prospero et al., 1996). A small proportion of the reactive phosphate is ultimately lost from the system and is deposited on the sea floor as organic phosphate, in organic matter and fish debris, or as inorganic phosphate through adsorption to iron and manganese (hydr)oxides and as phosphate minerals (Berner et al., 1993). Recycling of this particulate phosphate deposited on the sea floor seems to depend on the oxygen concentration of the bottom waters (Van Cappellen & Ingall, 1996). As yet, there seems to be a poor understanding of the microbiological processes responsible for the initial precipitation of phosphate at the sediment-water interface, and only recently some clues as to how bacteria might solubilise exogenous mineral phosphates (Goldstein, 1994). So, marine biological productivity, certainly over long time scales, is a function both of the supply rate of nutrient P from continental weathering and how efficiently P is recycled in the oceans. Föllmi (1996) has proposed that it is this response to changes in continental weathering rates that may well determine the long-term changes in marine P levels, and so ultimately any changes in marine ecology, productivity rates and thus, in turn, carbon dioxide fixation and the cycling of carbon.

In euphotic marine ecosystems, the focus of this review, inorganic orthophosphate (Pi) is probably the sole P source taken up directly by bacterial and phytoplankton cells (see Björkman & Karl, 1994). Numerous methods are now available for measuring not only Pi concentrations in seawater (Solorzano & Sharp, 1980; Parsons et al., 1984; Karl & Tien, 1992), but also for assessing the rates of microbial cycling of inorganic and organic phosphate compounds (Ammerman, 1993). Such radioactive and fluorescence methods can give measurements of the uptake and regeneration of Pi, enzymatic hydrolysis of components of the dissolved organic phosphate (DOP) pool, and uptake of the released Pi-coupled uptake (Ammerman, 1993). However, these are bulk measurements. What we ultimately require are biochemical molecular markers which are capable of assessing the P status of individual phytoplankton cells (see cyanobacteria section). This would then allow us to analyse and dissect the biological response of individual organisms to a specific environmental variable, the P supply. This is important because phytoplankton and bacterial communities are highly complex mixtures of species, with species-specific characteristics, especially with regard to meeting their nutrient requirements (Kilham & Hecky, 1988).

Is there P limitation in the marine environment?

Both (photo-) autotrophic and heterotrophic components of the microbial community encounter the same external environment, but by having different cellular phosphate quotas and nutrient acquisition capacities, (due to differences in cell surface area to volume ratio, and the presence or absence of high affinity Pi uptake systems) there may be considerable heterogeneity in the actual 'stress' these organisms experience. Since both primary production by phytoplankton and the metabolic activity of heterotrophic bacteria together control planktonic productivity, nutrient controls on both primary producers and heterotrophs can clearly affect global biogeochemical cycles.

Although there is little argument over the potential of nutrient availability to control productivity, the specific nutrient that ultimately controls this has been the subject of much debate. Even over geological timescales were P is deemed the ultimately controlling nutrient (Redfield, 1958; Codispoti, 1989), there appear to be well-reasoned arguments to doubt this (Falkowski, 1997). Martin (1990) suggested that the higher primary productivity of the glacial ocean resulted from enhanced aeolian supply of the micronutrient iron (Fe). Indeed, the considerable heterogeneity in the flux of aeolian Fe in the contemporary ocean (Barber & Chavez, 1991; Duce & Tindale, 1991) and the recent Fe fertilisation experiments in the high-nitrate low-chlorophyll (HNLC) regions of the equatorial Pacific ocean (Martin et al., 1994; Coale et al., 1996) demonstrate that in this and other HNLC regions (North Pacific and Southern Ocean) Fe may control phytoplankton photosynthesis. Much of the marine open-ocean environment is, however, typified by low to undetectable levels of the major inorganic nutrients nitrogen (N) and P. Where available, dissolved nutrient concentrations have been used to determine the N:P ratio with depth and geographical location (Fanning, 1992; Tyrell & Law, 1997). Such studies infer a predominance of N over P limitation in surface waters. Nutrient enrichment bioassays also provide evidence for nitrogen limitation of phytoplankton abundance in some coastal areas and subtropical gyres (Granéli, 1987; Fisher et al., 1992; DiTullio et al., 1993). Moreover, examination of the physiological response of a natural phytoplankton community in the North Atlantic, using pump and probe measurements of photochemical efficiency (F_v/F_m, i.e. the ratio of variable to maximum fluorescence), examined independently of cell abundance in nutrient addition experiments, also indicated N limitation of the taxa investigated (Graziano et al., 1996a). So, what of P limitation? Although, P *versus* N limitation of the marine environment, and a comparison of nutrient limitation of phytoplankton in freshwater and marine environments have been previously reviewed (Smith, 1984; Hecky & Kilham, 1988) – with the general consensus of P limitation of freshwater environments and N limitation of marine environments – there have been several recent reports of P limitation in marine systems.

Cotner et al., (1997) suggest that P limitation in the ocean may be more common than previously thought. Certainly, there are now well-documented examples in particular ocean provinces, e.g. of both phytoplankton and bacterioplankton limitation by P in the Mediterranean Sea (Berland et al., 1980; Krom et al., 1991; Thingstad & Rassoulzadegan, 1995) and the Gulf of Mexico (Smith & Hitchcock, 1994; Pomeroy et al., 1995), and of P-limited bacterioplankton growth in the Sargasso Sea (Cotner et al., 1997; Rivkin & Anderson, 1997). Noteworthy here is that observed high N:P ratios in the south-eastern Mediterranean Sea (25–30:1) (Krom et al., 1991) and in the upper 1000 m of the western Sargasso Sea (20–50:1) (Michaels et al., 1996) are consistent with observed P limitation of bacterioplankton or primary production from incubation experiments. These observations fit with the suggestion of Thingstad & Rassoulzadegan (1995), that the microbial food web in Mediterranean surface waters transforms imported phosphate to DOP of limited biological availability, although removal of phosphate by adsorption to Fe-rich dust particles may also control the P concentration in the water column (Krom et al., 1991). Other evidence, such as high activities of alkaline phosphatase, short turnover times for Pi, luxury consumption of Pi (a physiological response where P-starved cells fill their internal cell quotas with Pi) and stimulation of DNA synthesis in *Synechococcus* populations, support the hypothesis of P limitation of both bacterial and phytoplankton size fractions in the Mediterranean Sea (Thingstad & Rassoulzadegan, 1995; Vaulot et al., 1996).

There are also several reports of P deficit of both phytoplankton and bacterioplankton in Norwegian fjords (Sakshaug et al., 1983; Paasche & Erga, 1988; Thingstad et al., 1993), and of P appearing to have 'limiting' functions in certain shallow marine areas such as the Mississippi river delta (Macrae et al., 1994), Florida Bay (Fourqurean et al., 1993), Chesapeake Bay (Fisher et al., 1992), Chinese coastal waters (Harrison et al., 1990) and Shark Bay, Western Australia (Smith & Atkinson, 1984). It is important to clarify though, that most of the studies referred to above, address nutrient limitation of growth rate of individual organisms as distinguished from those aimed at addressing the factors limiting the net production rate of entire ecosystems (Smith & Atkinson, 1984).

In the surface waters of the oligotrophic Pacific Ocean, uptake rates of P greatly exceed the combined rates of atmospheric input or input from below the thermocline, suggesting P is rapidly recycled within the euphotic zone (Björkman & Karl, 1994). Since SRP levels and total dissolved phosphorus (TDP) levels are known to be very low in the upper levels of the tropical oceans (SRP<50 nM, TDP<350 nM) (Björkman & Karl, 1994), it has been suggested that P regeneration and recycling (see Marine microbial P cycle section) may ultimately control primary production in certain oceanic regions (Perry & Eppley, 1981; Orrett & Karl, 1987). Over slightly longer time scales, it has been proposed that P limits new and export production in the North Pacific subtropical gyre, alternating with N-limitation over ten-year periods (Karl et al., 1995; Karl, 1996).

In contrast to the studies of Fanning (1992) and Tyrell & Law (1997), Downing (1997) analysed the collected published data on total N:P inventories (dissolved inorganic and organic plus particulate forms) at 88 marine sampling stations. This study showed the total N concentration to vary by about 30-fold (6–200 μM), whereas total P varied 650-fold (0.03–20 μM), giving rise to total N:P ratios between 5–310 (average N:P=37). Of the open ocean euphotic zone sites under analysis, 78% had a N:P ratio indicating a deficiency of P or some other element other than N. Such work suggests there is a need for clarification and further study of the total N and P pools found in marine ecosystems. Michaels et al. (1996) and Ammerman et al. (1994) have provided data on DOP and particulate organic phosphate (POP) pools from the Sargasso Sea near Bermuda. They showed that DOP pools, typically of the order of 0.2 μM, to exceed the Pi concentration in surface waters, whilst POP pools were present at low concentrations, typically 10–20 nM, although these pools are 1–2 orders of magnitude higher in productive zones. The authors acknowledge however, that both DOP and POP are under-sampled and more poorly understood compared to particulate N. This is important since the dynamics and bioavailabilty of these pools will clearly influence P cycling and primary productivity in many ocean regions. This is where the application of molecular biological data can help. Though knowledge of the complexities of the potential bioavailable P is much needed, the use of molecular and biochemical techniques as a means of directly assessing the nutrient status of individual phytoplankton cells using molecular markers (Molecular markers section) will circumvent the need for such indirect analyses.

Marine microbial P cycle

The major pathways of the marine P cycle include the uptake of Pi by phyto- and bacterioplankton (for review see Cembella et al., 1984a,b), phagotrophy of bacterial and algal biomass (and thus P) by protozoan zooplankton (Caron, 1990), and regeneration of Pi from DOP (see Ammerman, 1991). In many aquatic ecosystems, >50% of the primary production is recycled through the microbial food web (i.e. the marine protozoan-bacterioplankton chain (Azam et al., 1983), and this rapid regeneration of nutrients is important in marine P cycling since no P input process exists analogous to N_2-fixation. The efficiency of P regeneration through the microbial loop may be complicated by several factors. In heterotrophic nanoflagellates for example, P regeneration is determined both by the nutritional status of the bacterial prey, and by interspecies variation among the grazers, with fast-growing species releasing significantly more P than slow-growing ones (Eccleston-Parry & Leadbeater, 1995).

P is also regenerated directly by phytoplankton, other zooplankton and viruses. Estimates of bacterial cell lysis by viral infection give very high release rates for dissolved organic compounds (Bratbak et al., 1992), and the role of viruses in this process is further discussed in Viruses as a source of P section. An idealised marine P cycle incorporating the role of viruses in releasing DOP for subsequent Pi regeneration is illustrated in Figure 1. That phage-induced cell lysis is a major mortality factor for bacteria is still open to argument, however, whereas there is now good evidence that the majority of the bacterial biomass is consumed by protozoa (e.g. see Wikner et al., 1990). DOP release, specifically in the form of DNA, during nanoflagellate predation on bacteria has already been demonstrated (Turk et al., 1992), and is likely to be an important intermediate step in P regeneration.

Since nucleotides and most other phosphorylated compounds do not cross cytoplasmic membranes (Lugtenberg, 1987), Pi regeneration mainly occurs outside the cell and involves enzymatic activities that release Pi from the particulate and DOP pools (Overbeck, 1991). Cell-surface localised alkaline phosphatase and 5′-nucleotidase are the enzyme activities which contribute most significantly to Pi regeneration in aquatic systems. These enzymes differ in their substrate specificities, with 5′-nucleotidase capable of hydrolysing nucleotides only, whereas alkaline phosphatase can hydrolyse a wide range of organic P substrates (Chrost, 1991; Ammerman & Azam, 1985,

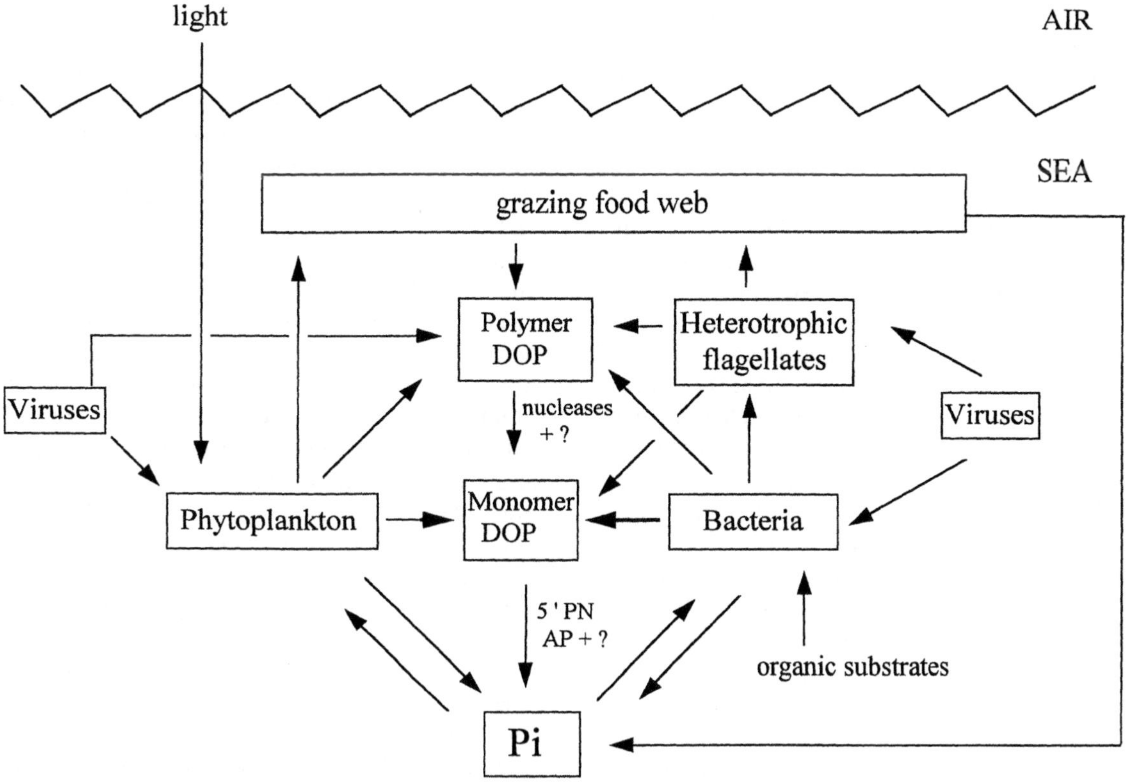

Figure 1. Idealised flow diagram of the marine microbial P-cycle, incorporating the role of viruses into this scheme. Hydrolytic enzymes capable of cleaving polymer DOP or phosphate monoesters (monomer DOP) may be produced by both bacteria and phytoplankton (after Thingstad et al., 1993).

1991a,b). The regeneration of Pi from particulate and DOP, in which the hydrolysed Pi is released outside the cell, suggests that without mechanisms that simultaneously connect the released Pi physically to the cell, or some form of coupled hydrolysis and uptake, then bacterioplankton and phytoplankton will be competing directly for a single Pi pool, irrespective of their ability to produce hydrolytic enzymes. Although both 5′-nucleotidase and alkaline phosphatase activities have been found in phytoplankton (e.g. Flynn et al., 1986; Hernández et al., 1995) and bacteria (e.g. Bengis-Garber & Kushner, 1981; Martinez & Azam, 1993a; Rath et al., 1993), there is evidence to suggest that 5′-nucleotidase may mainly be associated with the bacterial size fraction (Ammerman & Azam, 1985, 1991a,b). In addition, at least part of the Pi liberated by the membrane-bound 5′-nucleotidase is taken up directly without intermediate mixing into the free Pi pool (Ammerman & Azam, 1985, 1991a,b; Tamminen, 1989). Even so, liberation of enzymes directly into the water by active processes, sloppy feeding, cell lysis or osmotic stress may occur and indeed drive the system towards competition for a single Pi source (Thingstad et al., 1993). Thus, the functioning of the microbial P cycle should not compartmentalise phytoplankton and bacterioplankton simply as consumers and remineralizers of P, respectively.

Effects of P limitation on photosynthesis

Since the biogeochemistry of the ocean is largely driven by the downward flux of organic and inorganic carbon, the efficient functioning of the 'biological pump' – i.e. the photosynthetic phytoplankton, the responsible component, is clearly important. This 'pump' not only controls the partitioning of dissolved inorganic carbon between the surface and deep ocean, but also reduces the partial pressure of carbon dioxide in surface waters, thus ultimately determining atmospheric carbon dioxide levels over long timescales. Nutrient limitation is one way in which the pump's functioning may be perturbed. Nutrients modify photosynthetic responses, and either by limiting the specific growth rates of phytoplankton or by limiting their standing crop, can control marine photosynthesis (see Cullen et al., 1992). In order to accurately predict marine primary productivity then, it is important that we have a sound knowledge of the response of the photosynthetic apparatus to nutrient depletion, and that this is studied in a variety of phytoplankton genera. This latter objective is important, since although the physiological response of a phytoplankton clone in culture may provide us with a general correlation of the response of the photosynthetic apparatus to nutrient deficiency, there may be considerable inter- and intra-species variability, and this might not conform to previously proposed guidelines of how phytoplankton adjust their photosynthetic apparatus to variations in the environment. So, there is a need to address the question of how the genetic diversity of phytoplankton species might affect the response of the photosynthetic apparatus to nutrient limitation, so that this information can be incorporated into light-chlorophyll models for primary production prediction. As pointed out by Cullen et al. (1992), certainly for the long term temporal patterns of phytoplankton physiological responses to be predicted, knowledge of species succession or the exchange in dominance of physiologically distinct genotypes (Gallagher, 1982; Wood, 1988) is required.

As well as the direct role of P limitation of photosynthesis in controlling primary productivity, indirectly the consequent variability in the functioning and structure of the photosynthetic apparatus may potentially be useful as a diagnostic marker of nutrient limitation in the sea (Falkowski et al., 1992). For photosynthetic organisms, the ultimate result of nutrient limitation is a reduction in photosynthesis and growth. However, in addition to general effects on the photosynthetic apparatus (Kolber et al., 1988; Collier & Grossman, 1992; McKay et al., 1997), there are likely nutrient-specific responses which reflect the specific function of each nutrient within the cell. P is a component of nucleic acids and membranes, acts as a carrier of substrate or chemical energy and is a cytosolic signalling molecule. In addition, post-translational modification of proteins by phosphorylation is central to many adaptive responses, and the photosynthetic thylakoid membranes may be the site of metabolic integration for a range of dispartate signals via this process (Mann, 1994; Escoubas et al., 1995). So, what of some of these more specific effects of P limitation on photosynthesis, and are any of the responses potentially useful as a diagnostic marker?

Chlorosis, which increases chlorophyll *a* (chl *a*)-specific absorption, and although common to other nutrient stresses, is much less pronounced in P *versus* N and Fe-limited *Phaeodactylum tricornutum* with absorption increasing by about 15, 57 and 62% for P, N and Fe-starved cells, respectively (Geider et al., 1993). In *Dunaliella tertiolecta*, P limitation resulted in a 20% decrease in chl *a* per cell in P-starved batch cultures and P-limited chemostats (Graziano et al., 1996b). However, these observations of a linear decrease in chl *a*:C ratio with increasing limitation are also observed under nutrient sufficient growth conditions under variable light irradiance, so they are not a unique indicator of nutrient stress (Geider, 1987). Changes in the ratio of accessory pigments to chl *a* have also been observed in various phytoplankton species. Increases in the ratio of the photoprotective xanthophyll-cycle pigments diatoxanthin and diadinoxanthin to chl *a* have been observed in N, Fe and P-starved *P. tricornutum* cells, whilst lutein:chl *a* and neoxanthin:chl *a* both increased in P-limited *D. tertiolecta* under steady-state conditions (Geider et al., 1993, Graziano et al., 1996b). Of more use is that only in P-starved *P. tricornutum* cells did the ratio of diatoxanthin to diadinoxanthin increase dramatically (Geider et al., 1993). It remains to be determined whether this is true of other phytoplankton species containing these accessory pigments.

In freshwater cyanobacteria, the bleaching process (chlorosis) differs among nutrient starvation conditions (Collier & Grossman, 1992). N or sulphur starvation cause a rapid degradation of the phycobilisome, whereas in P-deprived cells extensive proteolysis of the phycobilisome does not occur. Since P is most likely the limiting nutrient in freshwater systems, this may reflect a specific strategy of the cell to survive P-limiting conditions, such as the use of polyphosphate reserves to survive short periods of P starvation. Whether these results, and thus the use of pigment-specific responses of the photosynthetic apparatus to particular nutrient stress as diagnostic markers of nutrient status, are also true of marine *Synechococcus* spp. remains to be elucidated. In these organisms, molecular markers have focused on nutrient-specific pathways such as those that control the expression of nutrient acquisition genes (Cyanobacteria section; Scanlan et al., 1997b).

The use of chl *a* as a measurement of photosynthetic biomass or use of ^{14}C methodology to estimate production (e.g. DiTullio et al., 1993) may also have their drawbacks in nutrient addition experiments. These variables respond slowly to stimulation, sometimes only after several days, by which time the phytoplankton community may have radically altered its composition (e.g. from picoplankton to diatoms, Chavez et al., 1991), whilst also reflecting only the net response of the community (the sum of gains and losses) and not the physiological response of individual primary producers (Vaulot et al., 1996). Thus, biological parameters that responding more rapidly to changes in the environment are required, such as biophysical fluorescence measurements of phytoplankton photosynthesis.

In batch cultures of *P. tricornutum* and *D. tertiolecta* the ratio F_v/F_m, describing the quantum efficiency of photosynthesis, declines markedly under N and Fe starvation, but only moderately under P starvation. F_v/F_m is also a non-linear function of growth rate in *D. tertiolecta* under P-limited steady state conditions (Geider et al., 1993; Graziano et al., 1996b). The parameter F_v/F_m provides an absolute index of photochemical efficiency which allows a comparison of measurements to be made, both spatially and temporally. Although the analysis is specific for phytoplankton and can be determined rapidly *in situ* avoiding sample manipulation, unfortunately this fluorescence measurement does not distinguish different types of nutrient limitation. Even so, recent advances in technology now allow analysis of photosynthetic parameters in individual cells (Olson & Zettler, 1995; Olson et al., 1996), rather than measuring a bulk water sample that provides an average value for the whole community, and it is realistic to expect that this will be extended for use in natural populations which will enhance its utility for correlating variations in phytoplankton photosynthetic parameters to the physics and chemistry of the ocean environments.

The use of the concentration of the photosynthetic enzyme ribulose-1,5-bisphosphate carboxylase (Rubisco) also has potential for use as a diagnostic marker of the cell's P status. Orellana & Perry (1992, 1995) have developed an immunofluorescent assay for quantitating the enzyme by flow cytometry in single phytoplankton cells. Based on the correlation of Rubisco concentration and maximal rates of photosynthesis in higher plants (Björkman, 1981), there is the potential for resolving patterns of relative growth and photosynthetic rates among individual cells. However, the response of Rubisco content to nutrient limitation seems to vary with phytoplankton species, the nutrient under investigation and growth rate (Graziano et al., 1996b). No change in Rubisco content was observed in *P. lutheri* under N or P limitation, whilst in *D. tertiolecta* Rubisco declined only by 30% in cells grown under P-starved conditions compared to P-replete conditions. A five-fold loss in Rubisco content occurred in N-limited *Isochrysis galbana*, but only when the growth rate was below half- maximal (Falkowski et al., 1989). Significantly, the relative abundance of Rubisco decreased in response to N and Fe-starvation in *P. tricornutum*, but did not change in P-starved cells. In these cells, Rubisco may have been present in an inactive form, or light-saturated photosynthesis may have been limited by factors other then Rubisco content. These results suggest that it may be unwise to use Rubisco content to infer P_{max} in natural phytoplankton populations that may be P-limited (Geider et al., 1993), and warrant further investigation of the interactions of nutrient limitation and irradiance in other phytoplankton.

The responses described above focus on the effects of a particular limiting nutrient, in isolation, affecting phytoplankton photosynthetic parameters and ultimately growth. It is important to remember that nutrient interactions will often occur and complicate the situation. For example, nutrient addition experiments have shown that the dual addition of phosphate with nitrate or ammonium has a synergistic effect upon the rate of photosynthesis or the production of chl *a*, in both the marine diatom *Skeletonema costatum* and in natural phytoplankton assemblages (Davies, 1988). This response has been explained by the implication of N in chlorophyll biosynthesis, and that carbon dioxide fixation is then controlled by both the P and N cell quotas when P is limiting. This again has ramifications for determining the suitability of a particular photosynthetic response as a suitable indicator of specific nutrient deficiency (see also J. LaRoche chapter this issue).

The input of molecular techniques for assessing the P status of marine organisms – can cell physiology be used to determine nutrient status?

Discussion of the processes regulating numbers of phytoplankton and bacterioplankton in time and space in the marine environment are not new (Yentsch, 1980; Barber, 1992). Indeed, an acknowledgment of the need to bring together the subjects of physical oceanography (advection diffusion, mixing), biological oceanography (biochemistry, microbiology, physiology – and we can now add molecular biology) and geochemistry (global cycles and deposition) (Yentsch, 1980) is clearly important to understanding the complex questions regarding marine phytoplankton and bacterioplankton growth. Although physiological experimental research is only one of the means to answering the general problem of growth in the sea, what is described efficiently is the organism's ability to sense and respond to various environmental changes in the water column (Yentsch, 1980). The advent of molecular phylogenetic techniques (Ward et al., 1990) has demonstrated the diversity of open ocean bacterioplankton populations (Giovannoni et al., 1990), and the same appears to be true of phytoplankton species (Rappé et al., 1998). This makes it difficult to choose a 'relevant' organism for physiological study. It will ultimately require culture of these organisms before we can begin to elucidate their biochemical roles. Since satellite imagery has enlightened us to the magnitude or scale of some 'bloom-forming' phytoplankton species (Holligan et al., 1993; Capone et al., 1996), for example *Emiliania huxleyi, Phaeocystis* spp., dinoflagellate spp. and *Trichodesmium* spp., and that we now know that oligotrophic oceans are dominated by photosynthetic picoplankton of the genera *Prochlorococcus* and *Synechococcus*, an alternative route is to direct physiological and molecular studies towards those species that are already known to play a major role in marine processes, and for which laboratory culture is not a problem. For these organisms at least, their biochemical and molecular responses can provide extremely useful indicators of nutrient status. The fact that diverse phytoplankton groups possess remarkably similar physiological and compositional responses to nutrient limitation (Healey, 1973), though see Effects of P limitation on photosynthesis section regarding the response of the photosynthetic apparatus, suggests the responses of these organisms may have more general applications. Moreover, since analysis can be performed on natural assemblages, prolonged incubations and nutrient enrichment bioassays can be avoided (Hecky & Kilham, 1988), whilst assessment can also be made of nutrient status at a particular time or with repetitive sampling over a longer time period to characterise the general state of the ecosystem.

Molecular markers

Traditional methods for determining limiting factors, including long-term incubations, chemical analysis and mathematical modelling do not permit us to ask questions regarding the nutrient status of specific components of natural assemblages, and hence new techniques are required.

Phytoplankton cells can survive in conditions varying from high-light, nutrient depleted surface waters to nutrient-replete waters at depths of 100–200 m receiving as little as 0.1% of the irradiance incident in surface waters. Similar ranges in distribution of heterotrophic bacterioplankton may occur, but it is less clear what the precise environmental parameters delimiting their distribution actually are. These observations suggest a high degree of genetic plasticity, or the existence of several genetically distinct strains adapted to specific niches, with the ability to rapidly modify gene expression when environmental conditions dictate. Whichever situation is correct, the fact that these organisms are able to acquire nutrients at submicromolar concentrations suggests that they possess particularly efficient nutrient transport systems. Components of these nutrient acquisition systems can provide readily useable markers for assessing the physiological state of the organism in time and space. It is already known that several cyanobacterial strains (Reddy et al., 1988; Scanlan et al., 1989, 1993; Laudenbach et al., 1991), as well as eukaryotic taxa (La Roche et al., 1993, 1995), induce or de-repress synthesis of unique proteins during environmental stress. Thus, by understanding the mechanisms for nutrient acquisition at the molecular level, it is possible to generate immunological markers or probes for uniquely induced stress proteins, so-called molecular markers, which can potentially be used to interrogate natural assemblages with regard to their nutrient status (see J. LaRoche this issue; Vrieling & Anderson, 1996; Palenik & Wood, 1997). Methodologies currently available allow for these molecular markers to be identified in single phytoplankton cells, thus extending this technology to allow dissection of an individual cell response from those of the whole population and ultimately the whole phytoplankton-bacterial assemblage (Lin & Carpenter, 1996; Scanlan et al., 1997b). Examples of these molecular markers, in particular functional groups, are discussed below, with reference to alternative approaches where known.

Phytoplankton

The coccolithophorid *Emiliania huxleyi* is an important component of the pelagic phytoplankton, and through its production of calcium carbonate plates (coccoliths) and dimethyl sulphide, may act to directly affect the Earth's climate and certainly influence the biogeochemical cycles of C and S (Westbroek et al., 1993). *E. huxleyi* has been shown to outcompete other phytoplankton at low Pi concentrations (Riegman et al., 1992; Van Bleijswijk et al., 1994; van Bleijswijk & Veldhuis, 1995), whilst P limitation causes a significant increase in the ratio of calcium carbonate to organic carbon (Paasche & Brubak, 1994), which may cause populations of this haptophyte to act as net exporters of carbon dioxide. Since P metabolism and calcification seem to be inextricably linked in this organism, molecular markers of the organism's nutrient status are highly desirable. Palenik & Koke (1995) using the biotinylating reagent succinimidyl 6-(biotinamido) hexanoate identified a N-regulated protein, Nrp1, that is the major membrane protein in N-limited cells. Recently, cell surface biotinylation has identified three cell surface proteins, designated Prp1, Prp2 and Prp3, induced by P stress. One of these is likely to be an alkaline phosphatase since enzyme activity correlates with induction of the three proteins. Synthesis of these proteins appears tightly controlled by external P levels; re-addition of Pi to the growth medium causes active degradation of the proteins and loss of alkaline phosphatase activity (Dyhrman & Palenik, 1998). Further characterisation of one or more of these proteins, and subsequent development of an antibody probe, will be extremely useful in elucidating the N *versus* P status of these organisms in the natural environment.

The use of cell surface biotinylation has also been applied to identify a cell surface alkaline phosphatase in the dinoflagellate *Prorocentrum minimum* (Dyhrman & Palenik, 1997). This phosphatase has a denatured molecular weight of ca. 200 000 daltons and is induced during the late log phase of growth. In contrast to *E. huxleyi* above, synthesis of the enzyme is repressed upon Pi re-addition, but is not actively degraded (Dyhrman & Palenik, 1998). Together with a protein of 130 000 daltons that is also induced by P stress in this organism (during mid-log phase), these proteins have considerable potential as molecular markers in this and perhaps other dinoflagellates. Reports that P stress increase toxin production in some dinoflagellates (e.g. Boyer et al., 1987), regulate toxin production in others (Flynn et al., 1994), and be linked with encystment, which is important for the dispersal and survival of toxic species (Anderson & Lindquist, 1985), underline the need for markers of P status in these organisms. The relationship between P stress and toxin production has additionally been demonstrated in the diatom *Pseudo-nitzschia multiseries* producing the neurotoxin domoic acid, increased toxin production correlating also with high levels of alkaline phosphatase activity (Pan et al., 1996), and in the haptophyte *Chrysochromulina polylepis* (Edvardsen et al., 1990).

^{14}C-bicarbonate labelling of cell proteins in the marine algae *Dunaliella tertiolecta* and *Phaeodactylum tricornutum* identified membrane-associated proteins of >200 kDa and 56 kDa, respectively, which were specifically induced under P limitation (LaRoche et al., 1993). Further characterisation of the induction of the 200 kDa polypeptide in *D. tertiolecta* revealed a good correlation between levels of protein and increasing limitation (i.e. decreasing growth rate in steady-state chemostat culture; Graziano et al., 1996b), though the protein declined in abundance in stationary-phase starved batch cultures. A knowledge of the regulation of the cell surface polypeptide relative to the cell's physiological state is particularly important since it is possible that the induction or de-repression of such proteins relative to internal or external P levels is species-specific (Palenik & Wood, 1997). Other requirements of putative molecular markers, including those that are necessary for immunological analysis in single cells, as well as their associated attributes, are summarised in Table 1. The 200 kDa polypetide satisfies several of these requirements since expression is unique to P limitation and synthesis is switched on shortly after P becomes limiting. The function of the polypeptide remains cryptic, though it has been proposed that its localisation in the internal membranes may allow the protein to play a role as an internal phosphatase for the hydrolysis and mobilization of stored polyphosphates (Graziano et al., 1996b). Unfortunately, antibodies to this protein are not yet available.

In addition to the 200 kDa polypeptide, *D. tertiolecta* also produces an alkaline phosphatase whose activity increases only after extreme P deficiency, and is thus insensitive to onset of, and changes in, P-limited growth rate (Graziano et al., 1996b). This contrasts with the situation in *P. minimum* where alkaline phosphatase activity was induced in the late-logarithmic phase of growth (Palenik & Wood, 1997). Since bulk alkaline phosphatase activity is widely used as a means of diagnosing P deficiency, even though the majority of activity is probably bacterial, it is important to understand that bulk seawater activities may include both inducible activities describing subtly different responses to P depletion. In addition, since

Table 1. Attributes and requirements of cell surface proteins and antibodies as indicators of nutrient limitation

Attributes	Requirements
(i) high sensitivity	(i) knowledge of the precise biochemical (and genetic) regulation of the marker protein in the target species*
(ii) nutrient specificity	(ii) protein purification
iii) potential quantitation per cell e.g. by flow cytometry or by western blotting and densitometry	(iii) antibody development
(iv) possible taxon specificity, hence a big advantage over bulk measurements	(iv) for single cell analysis, optimisation of permeabolisation conditions

*Indicators of nutrient deficiency in phytoplankton can be categorised as those that: (i) change in abundance over all nutrient-limited growth rates and which indicate the extent of limitation; (ii) are rapidly induced at the onset of limitation but there is no correlation with nutrient-limited growth rate over a range of growth rates – these are sensitive to the onset of limitation; and (iii) induction only under extreme starvation with little response to limitation – these are not sensitive to the onset of limitation (Healey & Hendzel, 1975; Graziano et al., 1996b).

enzyme activity is also a function of repression by the external Pi concentration (Veldhius et al., 1987) or by an internal P compartment (Gage & Gorham, 1985; van Boekel, 1991), which may well reflect the ability to store P in the form of polyphosphate (see for example Miyata et al., 1986), measured rates may hide different degrees of limitation between species or phytoplankton groups. Specific antibody markers will resolve this problem, assuming that within the range of cross-reactivity of the antibody the control of enzyme activity or protein is comparable.

Although molecular markers for assessing phytoplankton P status are beginning to be developed, components of the high affinity Pi transporters have not been identified in eukaryotic marine phytoplankton as yet. Recently, amino acid sequences for these transporters have been described in yeast (Bun-Ya et al., 1991), fungi (Harrison & van Buuren, 1995) and higher plants (Muchhal et al., 1996). Degenerate PCR primers designed to conserved regions of these Pi transporters, may be useful in amplifying homologous genes from various phytoplankton species for subsequent characterisation of the protein products and ultimately antibody production.

As well as the specific molecular markers identified or suggested above, potential also exists for the use of changes in the total protein, DNA and RNA content of phytoplankton cells as a specific marker of P status. In the dinoflagellate *Heterocapsa* sp., a dramatic decrease in the RNA content of the cell characterised P-starved batch cultures, whilst protein synthesis continued. In contrast, under N starvation the protein: RNA ratio remained constant (Berdalet et al., 1994). Thus, in this organism at least, the protein:RNA ratio may be a good indicator of P starvation.

Cyanobacteria

Marine photoautotrophs of the genera *Synechococcus* and *Prochlorococcus* are cyanobacteria that represent ubiquitous and abundant members of the picoplankton community (see Partensky et al., 1998). The contribution of *Synechococcus* to primary production has been estimated to be between 5 and 30% (Waterbury et al., 1986); *Prochlorococcus* often dominates over its *Synechococcus* relative, for example contributing 40% of the chlorophyll and 30% of living carbon in the central Pacific off Hawaii (Campbell et al., 1994). Hence, these organisms represent a substantial proportion of the photosynthetic biomass in many oligotrophic areas. Although their small cell size and subsequent large surface area/volume ratio allows for efficient nutrient uptake, recent observations have demonstrated a potential for limitation by inorganic nutrients in natural populations (Vaulot & Partensky, 1992; Vaulot et al., 1996; Scanlan et al., 1997b). Thus, there is obvious applicability for techniques capable of assessing the nutrient status of these organisms.

Knowledge of the nutrient acquisition systems of *Prochlorococcus* are relatively poor, though the recent availability of an axenic culture will surely bring rapid progress. Marine *Synechococcus* spp. have received more attention. With regard to P acquisition, both biochemical and genetic data have demonstrated the presence of a high affinity Pi uptake system comparable to that found in heterotrophic bacteria. Analysis of the kinetics of Pi uptake in *Synechococcus* sp. WH7803 and *Synechococcus* sp. NIBB1071 have shown this high affinity system to be energy-dependent (Donald et al., 1997; Ikeya et al., 1997). The latter authors also demonstrated a susceptibility of the uptake system to osmotic shock, a feature typical of high affinity transport systems containing a periplasmic-binding protein. Such a component has been identified in several marine *Synechococcus* spp. and *Prochlorococcus* spp. (Scanlan et al., 1993, 1997b), and has potential as a diagnostic molecular marker (see below). Other, membrane-bound components of this uptake system remain to be identified.

As alluded to in the previous section, an understanding of the regulation of a putative molecular marker is a major prerequisite for assessing its potential use. The periplasmic Pi-binding protein, PstS, is regulated via a two-component system pair PhoB/PhoR comprising part of the Pho regulon, certainly in various heterotrophic bacteria and freshwater cyanobacteria (for review see Bhaya et al., 1998). Homologues of these proteins have recently been identified in a marine *Synechococcus* sp. WH7803 (Watson et al., 1996), suggesting a similar regulatory network. However, in addition to this regulatory feature, a putative transcriptional activator of the Crp/Fnr family, designated PtrA, has also been identified, which lies immediately downstream (3′) of the genomic region encoding PstS (Scanlan et al., 1997a). This protein has been implicated in global P regulation in this organism, and since homologues exist in several other phycoerythrin-containing marine *Synechococcus* strains, PtrA may be a common feature of these organisms. This suggests a more complex mechanism of regulation of P acquisition than has so far been found in bacteria. The development of a transformation system for marine *Synechococcus* strains (Brahamsha, 1996) paves the way for the construction of mutants in this and other genes, which will give insights into the function of these proteins *in vivo*.

The PstS protein identified in the marine *Synechococcus* sp. WH7803 satisfies several of the requirements of a molecular marker, being nutrient specific, repressible by various organic phosphates and induced at external Pi concentrations of 50 nM or less, which are typical of oligotrophic marine environments. In addition, antibodies raised against the protein specifically recognise the marine genera *Synechococcus* and *Prochlorococcus*, but not marine heterotrophs. Use of PstS antibodies to interrogate *Synechococcus* populations in a natural mesocosm system using standard Western blotting techniques has indicated that the protein can be specifically detected in populations that have been artificially manipulated to force P depletion, but is repressed in enclosures where nutrients have been added in a Redfield ratio (Scanlan et al., 1997b). Moreover, differential PstS expression has been observed in a depth profile of natural picophytoplankton populations at the Bermuda Atlantic Time-Series (BATS) Station in the Sargasso Sea (Figure 2). Further use of this molecular marker over larger spatial and temporal areas in open ocean systems will be required to assess fully its usefulness and robustness, especially as in many oligotrophic regions, Pi levels may result in a constitutive expression of the protein in surface waters. The recent development of a single cell immunofluorescence assay for use with *Synechococcus* spp. extends the use of such a system to study heterogeneity of nutrient response within a single genera, though unfortunately the methodology cannot yet be linked for use with flow cytometry (Scanlan et al., 1997b).

The availability of PstS as a molecular marker of P status in both the *Synechococcus* and *Prochlorococcus* genera will be especially useful in comparing the expression of PstS in waters dominated by just one species. *Prochlorococcus* has been shown to become P-limited at Pi concentrations around 30 nM, suggesting that it has the ability to thrive at very low ambient P levels (Parpais et al., 1996). Recently, Cotner et al. (1998) examined uptake of inorganic (^{33}P-Pi) and organic P (d^{32}P-ATP) into picoplankton at the BATS Station using flow cytometry. Although there was no apparent selectivity for either inorganic or organic P by *Synechococcus* or *Prochlorococcus*, *Synechococcus* incorporated the majority of the labelled P sources (70% of total). Little difference, between genera was observed in cell-specific incorporation of ^{14}C-NaHCO$_3^-$, however, suggesting that *Synechococcus* cells may have a higher P requirement than *Prochlorococcus*. This suggests that *Synechococcus* spp. may be more susceptible to P limitation, and lends itself to careful interpretation of PstS expression data in waters containing both picophytoplankton genera.

A recent study has also used extracellular alkaline phosphatase activity as a means of elucidating the P status of planktonic communities in the northern Red Sea (Li & Post, 1998). These workers found the enzyme activity to be highly stable in seawater, whilst the presence and absence of freely dissolved and particulate alkaline phosphatase activity potentially could be used to reflect the recent P 'history' of the plankton community. Although the source of this 'bulk' alkaline phosphatase activity could not be directly attributed to a specific plankton group or groups, activities measured in the Gulf of Aqaba showed good correlation with the vertical depth profiles of marine *Synechococ-*

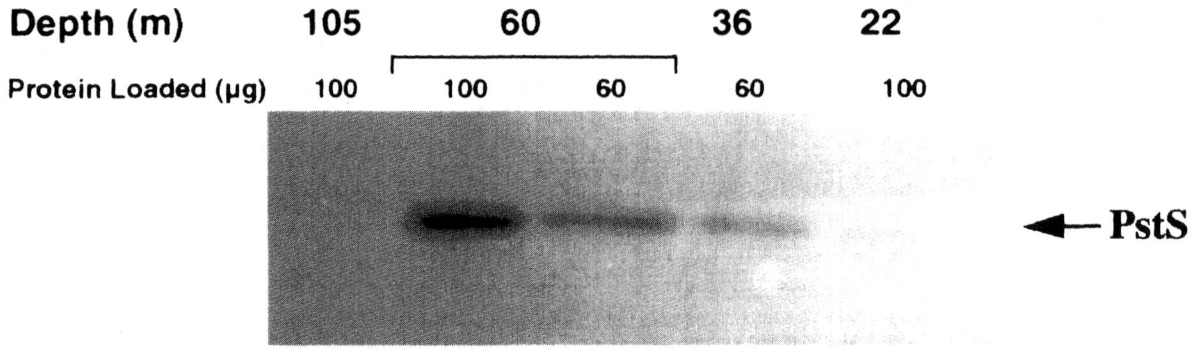

Figure 2. Expression of PstS in a depth profile of seawater samples at the BATS site during October 1996 by Western blotting of total cell extracts from phytoplankton-bacterioplankton populations (size, <20 μm) separated on a 10% SDS-polyacrylamide gel using PstS antiserum.

cus spp. Such data suggests these organisms encounter P stress in this environment, and the potential exists for the development of a *Synechococcus*-specific alkaline phosphatase antibody marker that could complement these activity measurements, together with parallel studies using the PstS protein to monitor the P status of the total picophytoplankton community.

In addition to the use of diagnostic molecular markers to assess the nutrient status of picophytoplankton, cell cycle variables have also been very useful in this respect, especially for rapidly monitoring nutrient bioassay experiments. This type of analysis has the added advantage that the cell cycle status of natural populations can be readily determined by flow cytometry, hence it is possible to discriminate between the autotrophic populations present (*Prochlorococcus*, *Synechococcus*, picoeukaryotes etc). The effect of manipulations in cell P status have been reported both in laboratory cultures and in natural seawater samples (Parpais et al., 1996; Vaulot et al., 1996). In the former, it was shown that P starvation of cultured *Prochlorococcus* isolates led to an arrest in the cell cycle at all phases with only minor differences between strains. The observed arrest during the DNA synthesis (S) phase of the cell cycle in these organisms appears specific to P starvation, and is an unusual response for prokaryotic and eukaryotic organisms generally, cells arrested in S being unable to continue their cycle upon P re-addition, and dying. Thus, the observation of large numbers of *Prochlorococcus* cells from natural samples in the S phase would suggest P is limiting and could be used as a diagnostic marker of nutrient status. The effect of P addition on the cell cycle of natural picophytoplankton has been investigated specifically only in *Synechococcus* populations from the Mediterranean Sea in summer (Vaulot et al., 1996). Here, P rather than N addition, advanced the later stages of the *Synechococcus* cell cycle, probably by shortening the length of the G_1 period. Additions of as low as 20 nM P induced a cell cycle response, illustrating the severity of P limitation of such populations. Thus, the use of cell cycle parameters as a means of assessing (pico-) phytoplankton, and perhaps heterotrophic bacterioplankton, cellular nutrient status have a bright future.

As well as the picophytoplankton genera *Synechococcus* and *Prochlorococcus* discussed above, there is another dominant planktonic marine cyanobacterium, the diazotroph *Trichodesmium* spp., that plays a major role in oceanic ecosystems (Capone et al., 1997), but for which markers of nutrient status are poorly developed. Because of its ability to fix atmospheric N, the growth of this organism is presumably limited by non-N nutrients. Although the micronutrients Fe and molybdenum are likely limiting factors, being required directly for the nitrogen-fixation process, the macronutrient P may also be important, and indeed models have been proposed to explain how *Trichodesmium* might accumulate P in the low-nutrient surface waters in which it blooms (Karl et al., 1992). Clearly, the development of specific antibody probes for one or more of these potentially limiting nutrients is desirable, particularly since this organism plays such a major role in global 'new' production.

Heterotrophic bacterioplankton

In contrast to the cyanobacterial and phytoplankton groupings mentioned above, which contain either dominant genera or at least a major bloom-forming species, the diversity and unculturability of the heterotrophic bacterioplankton component of marine ecosystems complicates the development of molecular probes for use in assessing their nutrient status. Thus, antibody probes might be restricted by an unusually narrow range of cross-reactivity, whilst cell cycle analysis may be compromised by the presence of slow-growing, starved or nongrowing bacteria. Nonetheless, with specific regard to how these organisms regulate and achieve an adequate cellular P status, the identification of potential diagnostic markers can be inferred from the wealth of information available for heterotrophic bacteria generally, although precise interpretation of data will probably require isolation and culture of those bacteria that are dominant and representative of marine ecosystems, so that specific analysis of the regulation of the desired protein or system can be undertaken.

Most of our knowledge of bacterial P acquisition comes from the enteric organism *Escherichia coli* (see Torriani-Gorini et al., 1994). Under P depletion, this bacterium induces various components of the Pho regulon which includes proteins required for the scavenging of Pi or the use of alternative P sources for growth (Figure 3). In addition to these adaptive responses, *E. coli*, like many heterotrophic bacteria, is able to survive periods without a P source, cells initiating a so-called 'starvation response'. Such a survival mechanism is also induced during starvation for other nutrients. The physiological and molecular mechanisms employed by marine bacteria during this 'general' response to nutrient starvation have been recently reviewed (Kjelleberg et al., 1993a). Studies on starvation of a marine *Vibrio* sp. have also shown that adaptation to carbon or multiple nutrient starvation, but not N or P starvation, results in the development of a stress resistant ultramicrocell which is capable of rapidly responding to nutrient input (Nyström et al., 1992; Kjelleberg et al., 1993b). Thus, these organisms appear to have a repertoire of responses to survive periods of nutrient deprivation. Since we are interested here only in deprivation of a single nutrient, we are essentially focusing on the adaptive responses of these organisms during the transition from growth to nongrowth at the critical threshold concentration of Pi, though it should be remembered that this nongrowing phase may be typical for marine bacteria.

The versatility of heterotrophic bacteria in adapting to a restricted P supply has been demonstrated using two-dimensional gel electrophoresis in both *E. coli* and *Bacillus subtilis*, and has identified several proteins that may be useful for the construction of antibody probes (Eymann et al., 1996; VanBogelen et al., 1996). For marine bacteria there is much less information. Some inference to mechanisms of P acquisition in marine bacterioplankton can be made from work performed on *Vibrio* spp. and *Pseudomonas* spp., known to be present in marine ecosystems, although the following cited studies didn't necessarily use marine isolates. Thus, specific physiological responses to P starvation and the development of an immunological P starvation marker have already been shown for *Pseudomonas* spp. (Givskov et al., 1994; Eberl et al., 1996; Leopold et al., 1997), whilst identification of proteins induced specifically under P starvation has been described in the marine *Vibrio* sp. strain S14 (Nyström et al., 1992). Moreover, Tanoue et al. (1995) have demonstrated that a major constituent of dissolved protein in marine ecosystems is a protein with an apparent molecular mass of 48 kDa, whose N-terminal amino acid sequence is identical to both a Pi-specific porin, porin P, and a polyphosphate-specific porin, porin O, from *Pseudomonas aeruginosa* (Hancock et al., 1982; Hancock et al., 1992). Synthesis of this Pi/polyphosphate-specific outer membrane protein channel occurs only under P starvation conditions in *P. aeruginosa*. Its presence in the dissolved protein pool of several ocean ecosystems (Tanoue et al., 1996), perhaps a result of virus-induced cell lysis, suggests that it is not only refractory to degradation but also indicates that P limitation of marine bacteria may be fairly common. Identification and characterisation of bacteria synthesising this protein *in situ*, perhaps using an antibody probe, should prove most interesting, since it is an easily accessible antigen and a diagnostic marker for the cell P status.

The continued use of bulk alkaline phosphatase activities as an indication of P starvation has not been mirrored by a thorough molecular characterisation of the enzyme from specific marine bacteria. Since the regulation and biochemical properties of the enzyme are likely species-specific, it is desirable that we have a knowledge of these in 'representative' marine bacterioplankton, albeit even if this is limited to those strains that have been cultured. The importance of understanding the regulatory network controlling alkaline phosphatase activity is reinforced by the recent modification of an *Achromobacter* strain isolated from coastal seawater, in which a plasmid containing a *phoA* hybrid gene that directed constitutive overproduction of alkaline phosphatase had a significant impact on the organism's ability to hydrolyse DOP

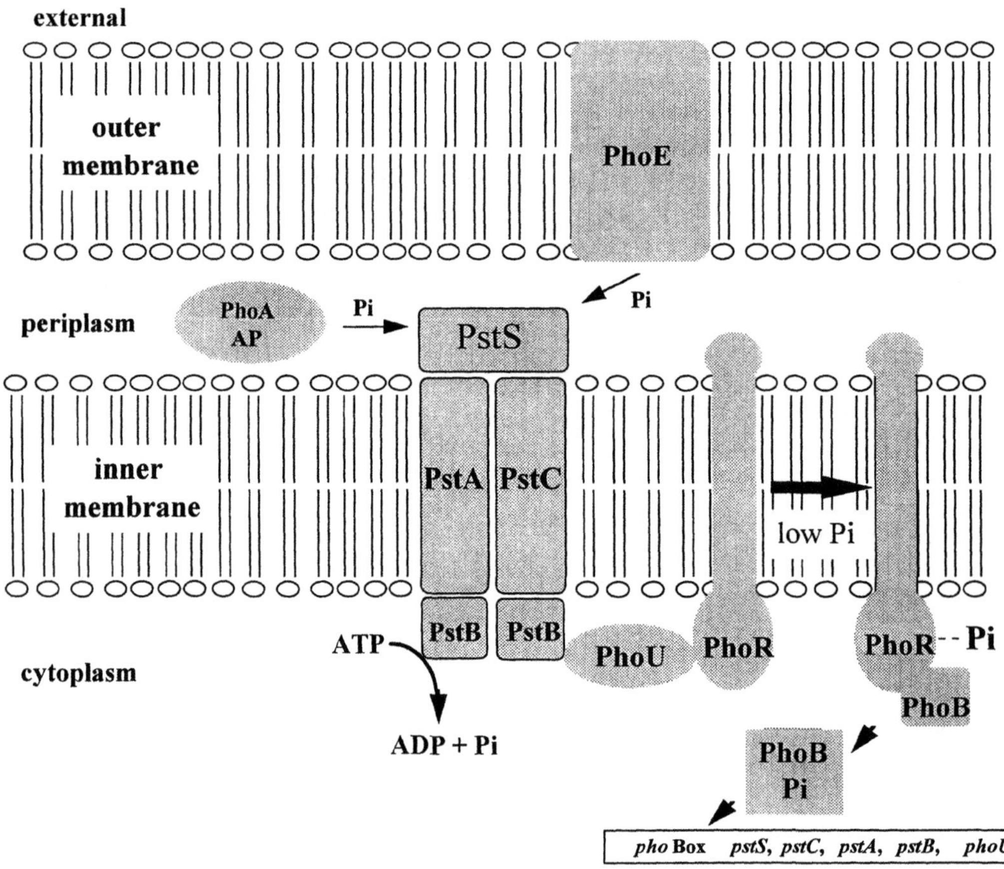

Figure 3. The high affinity Pst Pi transport system of *E. coli*. The pathway for Pi movement from the external environment into the cytoplasm begins with passage through an outer-membrane protein channel PhoE which is specific for Pi and polyphosphate. The Pi is transported across the periplasm by a binding protein (PstS) where it interacts with the inner membrane transport system composed of PstA, PstB and PstC polypeptides. The periplasmic enzyme alkaline phosphatase hydrolyses Pi from organic P sources, the released Pi also being transported via PstS. The two major regulatory proteins are PhoR and PhoB. PhoR is an integral membrane sensor protein and PhoB is a soluble cytoplasmic DNA-binding effector protein. The N-terminus of the PhoR protein is needed both for recognition of excess extracellular Pi and also for membrane anchoring of the protein. When cells sense low Pi levels, PhoR becomes autophosphorylated at a specific histidine residue and trans-phosphorylates PhoB. Phospho-PhoB activates transcription by binding to a consensus *pho* box. PhoU appears to act as a modulator of PhoR phosphatase/kinase functions.

and survive in a seawater microcosm (Sobecky et al., 1996). Similarly, the importance of 5′-nucleotidase activity in Pi regeneration, and the potential to couple nucleotide hydrolysis with uptake of the released Pi, which would appear to be particularly advantageous in an aqueous environment, belies the fact that the enzyme has been characterised in relatively few marine bacteria (Bengis-Garber, 1985). One group that is particularly under-represented are the Archaea, which have been shown to be widespread in several open-ocean regimes (Delong, 1997). There is thus a real need to determine subcellular location, kinetic properties and interactions if any, between these ectoenzymes and components of the high affinity Pi transport system in other marine heterotrophs.

In addition to the well-described induction of various proteins involved in maximising uptake and utilisation of the available P during P limitation, it has also been demonstrated that the same conditions can repress synthesis of the periplasmic binding proteins for both the oligopeptide and dipeptide permeases involved in peptide transport in *E. coli*, and potentially others involved in amino acid uptake (Smith & Payne, 1992). Given that hydrolytic aminopeptidase activity has been documented both for heterotrophic and autotrophic marine bacteria (Martinez & Azam, 1993a,b) it is possible that this and/or the peptide uptake systems described above may also be negatively regulated during P-deplete conditions in marine organisms. This has implications for the efficient processing of DOM by marine bacteria, and will be especially important in situations of a rapidly fluctuating nutrient supply, when both N and P sources are required, which may be typical of many marine environments.

Taken together, the above studies and the recent demonstrations of P limitation of bacterioplankton growth in the Sargasso Sea, Mediterranean Sea and Gulf of Mexico (see Is there P limitation in the marine environment? section), highlight the importance of inorganic nutrients, particularly P, in controlling carbon cycling in the sea. Coupling such studies to analysis of the nutrient status of marine bacteria *in situ* using molecular markers is highly desirable, especially if this can further discriminate whether only a component of the bacterial genera are experiencing limitation. This may further distinguish those rapidly growing (and dying) bacteria capable of providing a flux of inorganic nutrients to the phytoplankton, from the slow-growing ones which may limit the recycling of essential nutrients to the primary producers.

The role of viruses in marine ecosystems

Numerous studies within the last decade have established that viruses are ubiquitous throughout marine ecosystems. Knowledge of the factors regulating virus production in the ocean is limited, although they are being increasingly recognised and accepted as important contributors to element cycling within the microbial loop (for reviews see Fuhrman, 1992; Suttle, 1994; Bratbak et al., 1994). Recently, the effect of marine viruses on element flows has been incorporated into theoretical models for the control of bacterial growth rate, abundance, diversity and carbon demand (Thingstad & Lignell, 1997). In such simulated model situations, viruses appear to control only the diversity of bacterial communities. However, viral lysis of microbial populations has been shown experimentally to be important for carbon flow through microbial food webs. For example, throughout the course of a diel mesocosm study, using potassium cyanide to inhibit production of new viruses, Bratbak et al. (1992) estimated that viral lysis of bacteria exceeded bacterial production (measured by thymidine incorporation) by a factor of six. Despite this six-fold over-estimation of viral productivity, their results suggested that viral-mediated processes may be quantitatively significant in element cycling of microbial systems. Consequently, the microbial loop (after Azam et al., 1983) was formally extended to incorporate viruses by the proposal of the viral loop (Bratbak et al., 1992). Thus, a pool of phytoplankton-derived dissolved organic carbon (DOC) will fuel bacterial production. Bacteria are then removed by either predation or viral lysis, and bacterial-derived organic material, following lysis, is fed back into the DOC pool, hence completing the loop.

Viruses as a source of P

In addition to DOC, other organic nutrients, notably DON and DOP in the form of proteins and nucleic acids, are also released from bacteria or phytoplankton infected by lytic viruses and further utilised as a source of inorganic nutrients by the remaining planktonic organisms. The effects of such viral lysis on the growth and metabolism of non-infected organisms is poorly understood however. In a laboratory culture study, Middelboe et al. (1996) demonstrated that when a *Vibrio* sp. phage lysate was added to seawater containing both a natural marine bacterial community and a cultured marine *Vibrio* sp., there was an increase in metabolic activity and bacterial production of the non-infected bacterioplankton community. Interestingly, when viruses were added to P-limited cultures, both bacterial DOC uptake and alkaline phosphatase activity significantly increased, suggesting that the products of viral lysis are a valuable and readily available source of P to other organisms. In a more extensive study, Gobler et al. (1997) used radiotracer analysis to examine the biogeochemical pathways and bioavailability of elements following viral lysis of the brown tide chrysophyte *Aurecoccus anophagefferens*. Their experiments demonstrated that viral lysis of phytoplankton can significantly affect the partitioning and biological cycling of major nutrient elements and micronutrients. Specifically, it was found that P was released into the particulate phase following viral lysis of *A. anophagefferens*. This makes sense because P is relatively more concentrated in algal structural components (Reinfelder & Fisher, 1991). However, it was demonstrated that 80% of the P was released into the dissolved phase over a 4 week period, presumably through enzymatic cleavage by bacterial activity (Ammerman & Azam, 1991). Gobler et al. (1997) also revealed that P regenerated by viral lysis relieved diatom P-limitation, consistent with findings made by Middelboe et al. (1996).

Since dissolved DNA (d-DNA) is an important intermediate in P regeneration (Marine microbial P cycle section) it indirectly becomes an important Pi source in marine systems (see Karl & Bailiff, 1989). It has been estimated that viruses *per se* contain 0.08×10^{-16} g P per virus (for comparison there are ca. 1×10^{-16} g C and 0.4×10^{-16} g N per virus) (Mathews et al., 1983; Borsheim et al., 1990), so viral DNA itself will ultimately become a component of DOP pools. Indeed, following immunochemical quantification of DNA in size fractionated seawater samples, Maruyama et al. (1993) suggested that DNA-rich particles in the <0.2 μm fraction, attributed to viral assemblages, are a significant P reservoir in the marine environment.

d-DNA has been operationally defined as DNA which passes through a 0.2 μM Nuclepore filter (DeFluan et al., 1986), and consists of soluble and uncharacterised bound (particulate) forms and viral particles (Paul et al., 1991; Jiang & Paul 1995). Concentrations of d-DNA range from 0.6 μg l^{-1} in oligotrophic open ocean sites to 80.6 μg l^{-1} in coastal surface waters (reviewed in Karl & Bailiff, 1989). Since the discovery of high concentrations of viruses in the marine environment (Bergh et al., 1989), several studies have investigated how virus distribution correlates not only with physical and biological properties of the water column, but also concentrations of both dissolved and particulate DNA. Boehme et al. (1993) found that the distribution of viral particles closely matched that of particulate DNA. However, they concluded that DNA in viral particles was only a small proportion of the total d-DNA pool. Jiang & Paul (1995) used differential ultracentrifugation to separate viral particles from soluble DNA in seawater samples collected from a range of coastal and offshore stations. Once separated, various 16S rRNA-targeted oligonucleotides were used to probe the soluble DNA fraction in an attempt to determine the relative importance of viral particles in the d-DNA pool. Approximately 50% of the 0.2 μM-filterable DNA was found to be soluble or free and the other half was composed of bound forms (viral particles and uncharacterised colloids). However, only 17–30% of the bound forms could be accounted for by viral particles.

Weinbauer et al. (1993) correlated the concentration of d-DNA with virus and bacterial abundance along a trophic gradient in the northern Adriatic Sea, a region potentially P-limited. Their results indicated that d-DNA production caused by viral lysis of bacteria was greater in eutrophic waters compared to more mesotrophic waters. However, measured d-DNA concentrations did not vary significantly between trophic conditions. Weinbauer et al. (1993) suggested that bacterial utilisation of d-DNA, derived from these viral lysis products, may be enhanced in P-limited systems, and that the comparatively low concentrations of d-DNA in the eutrophic stations of their study were a result of an increased bacterial demand for P. In a subsequent study at the same site, estimated viral DNA contribution to the total pool of d-DNA varied strongly on a seasonal scale (Weinbauer et al., 1995). The greatest contribution was found in the autumn when high viral abundance coincided with relatively low total d-DNA concentrations. How efficiently the DNA derived from viral particles is hydrolysed for subsequent P regeneration is not known, but will probably be determined by the size and form of the released DNA (i.e. the relative molecular weight of the d-DNA and whether it is single- or double-stranded). Knowledge of the distribution of such types of DNA is limited (DeFlaun et al., 1987) and clearly requires further work. Collectively, these studies demonstrate that viral particles themselves do not contribute a significant component of the oceanic d-DNA, although they are instrumental in the production of dissolved DNA through lysis of microbial populations.

Effect of P availability on viral infection

In the above section, we have established that viruses, or the products of viral lysis, may represent a significant P reservoir for microbial or phytoplankton populations in the marine environment. In addition to viruses being a source of P, conversely, the outcome of a virus-host interaction can be determined by the physiological status of the host organism, especially with regard to limitation by nutrients, and particularly P.

In the open ocean, there are three possible outcomes for a virus infecting a bacterial or phytoplankton cell. It can enter the lytic cycle, where viral nucleic acid (typically DNA) is copied by the host replication machinery, producing multiple copies of the infecting virus and ultimately releasing viral progeny from the lysed host. The cycle is repeated when a free virus infects another susceptible host. Secondly, during lysogeny, although the initial infection process is the same, the viral nucleic acid can become incorporated into host chromosomal DNA and replicate over several generations within the lysogenic host chromosome. Induction of such a temperate virus into the lytic cycle occurs following stimulation of the lysogenic cell by specific environmental factors. The relative importance of these two mechanisms in the marine environment is currently unknown and subject to considerable speculation, as are the environmental factors which determine whether lysogeny is established or whether lysogens re-enter the lytic cycle, a topic which has been recently reviewed (Wilson & Mann, 1997).

The final outcome is for a virus to enter a state of pseudolysogeny (after Romig & Brodetsky, 1961). Here, following infection of a dormant host cell, the viral nucleic acid enters an arrested stage of the lytic cycle. The virus resumes its lytic cycle when the host recommences vegetative growth, such as occurs when availability of nutrients to the host cell increases, allowing gene expression of viral proteins and leading to the lytic cycle or alternatively, establishment of true lysogeny. So, pseudolysogeny is experienced only when the host cell is in a non-active state, such as occurs during extreme nutrient limitation (Kokjohn et al., 1991; Wilson et al., 1996; Ripp & Miller, 1997). Certainly, nutrient limitation has been shown to reduce bacteriophage production in different phage/host systems isolated from the North Sea (Moebus, 1996a). In a subsequent study, Moebus (1996b) demonstrated pseudolysogeny in non-active stationary phase marine bacteria (strain H3) and suggested that this may be common among marine bacteria. Wilson et al. (1996) suggested that pseudolysogeny was established following cyanophage S-PM2 infection of the marine cyanobacterium *Synechococcus* sp. WH7803 during growth of the host under P-deplete conditions. Here, there was an apparent 80% reduction in viral burst size during host growth in P-deplete medium compared to a nutrient-replete control. Furthermore, following cyanophage infection, only 9.3% of P-deplete infected *Synechococcus* sp. WH7803 cells lysed in contrast to 100% of infected P-replete cells. No difference in viral infection kinetics was observed during N-deplete growth, or merely by a decrease in cell-specific growth rate, achieved by decreasing the ambient light intensity but in which the growth medium was nutrient replete. These results suggest that it is the P status of the *Synechococcus* cell that determines the ultimate outcome of cyanophage-host interactions, presumably reflecting the high P demand (in nucleic acids) of cyanophage replication in the host cell. Wikner et al. (1993) demonstrated that nucleic acids from within host cells are the major source of nucleotides for marine bacteriophages following infection [cf. T-even phages of enteric bacteria which derive phage DNA P from the surrounding medium (Kozloff & Putnam, 1950)], suggesting that marine bacteriophages should not be especially sensitive to P limitation. Wikner et al. (1993) further proposed that this metabolic strategy may be common among marine viruses as an adaptation to a nutrient deficient environment.

Using a different approach, several studies have manipulated P concentrations in natural seawater samples, either collected from the open ocean or from large-scale mesocosm enclosures, and the abundance and status of viral populations assessed during changes in P availability to the host organism. Tuomi et al. (1995), using seawater taken from Raunefjorden, Norway, demonstrated that when Pi was added back to P-deplete seawater there was a significant drop in bacterial abundance. They suggested that one subpopulation of bacteria lysed when viral activity was stimulated by an increase in P availability, and that increased nutrient availability induced temperate viruses from lysogenic bacterial populations. In contrast, Pomeroy et al. (1995) demonstrated that bacterial numbers actually increased following Pi addition to P-limited seawater collected from the Gulf of Mexico. Such contrasting results may reflect differences in either the water bodies sampled or experimental procedures used in each study.

One of the original studies which implicated P as a control mechanism in determining the outcome of marine virus/host interactions, was one conducted in large-scale mesocosm enclosures (see Egge & Aksnes, 1992) in a fjord south of Bergen, Norway. Bratbak et al. (1993) enumerated different phytoplankton species and virus-like particles in four separate mesocosm studies over a period of four years, with different N:P ratios added each year. These workers demonstrated that restricted P availability in the mesocosm enclosures inhibited the development of viruses infecting *Emiliania huxleyi*, whilst N availability appeared to have no effect. They suggested that such viruses may be more sensitive to P limitation compared to N limitation, since they require a high nucleic acid to protein ratio. In a more recent study at the same site, carried out over a 26 day period in the summer of 1995, Wilson et al. (1998) induced P limitation in a single mesocosm enclosure by the addition of excess N at a N:P ratio of 75:1. A large *Synechococcus* spp. bloom developed throughout the period of the study, and using the diagnostic PstS molecular marker the *Synechococcus* population was demonstrated to be phosphate deplete (Scanlan et al., 1997b). Concurrently, *Synechococcus* spp. abundance increased to a maximum of 5.05×10^5 cells ml^{-1} within the enclosure, which was followed by the largest daily increase in virus abundance, prior to the onset of an apparent *Synechococcus* spp. bloom collapse following Pi re-addition. Wilson et al. (1998) suggested that this was indirect evidence to indicate that the P status of the *Synechococcus* spp. population could determine the outcome of a cyanophage/host interaction. However, there was insufficient data to demonstrate that there was a lysogenic or pseudolysogenic population present in the P-limited mesocosm enclosure, as there was a great deal of variation in estimated virus abundance. Nonetheless, these studies from the natural environment compare favourably with the results obtained from laboratory studies and lend well to further investigation.

However, the major problem in determining marine virus/host interactions under the varying nutritional growth conditions encountered in nature, is the lack of suitable methodology for studying virus communities. In the studies reported above, the majority of virus abundance and productivity measurements have been made using transmission electron microscopy (TEM). Unfortunately, this technique is susceptible to a lack of specificity and accuracy. It is also difficult to attribute specific roles to individual viruses if they are always studied as a diverse, mixed population, and where different host/virus systems may interact very differently within a given environment. It is only recently that molecular techniques have been employed to investigate algal-virus populations (for review see Wilson & Mann, 1997), and we consider some advantages of this methodology now.

Use of molecular techniques to analyse virus communities

As we have mentioned previously, molecular biological techniques have been instrumental in changing our understanding of the composition of marine bacterial communities. Recent studies have adopted the underlying strategy to extend this approach to virus communities. However, rather than virologists having to hand an extensive 16S rRNA database which can be used for comparison of bacterial diversity, development of suitable marker genes for assessing viral genetic diversity has been necessary (see Chen & Suttle, 1995; Chen et al., 1996; Fuller et al., 1998). So, the approach has been to identify conserved regions within these virus-specific genes, for the synthesis of PCR primers which can be used for amplification of corresponding sequences from the natural environment to investigate their genetic diversity. Potential then exists to enumerate specfic groups of virus using quantitative PCR techniques, and to investigate changes in specific viral communities using the technology of denaturing gradient gel electrophoresis (DGGE) (Muyzer et al., 1993). Using marine cyanophage as a specific example, we now examine how such techniques have been used to extend the mesocosm data described above with specific regard to the effect of P.

Cyanophage-specific primers CPS1 and CPS2 were developed using sequence information derived from a gene that exhibited significant similarity to that encoding a capsid assembly protein (gp20) from the enteric coliphage T4 (Fuller et al., 1998). These primers were used to amplify cyanophage DNA from the phytoplankton size fraction (>0.3 μm), concentrated by tangential flow filtration. A 165 bp product was amplified from each mesocosm enclosure (mesocosm details outlined in Wilson et al., 1998) indicating that cyanophages were present in each (Figure 4a). In the P-limited enclosure (N:P additions at a ratio of 75:1), cyanophage DNA was amplified from the >0.3 μm fraction both before and after P addition (Figure 2a, lanes 13 & 14). Such a result contradicts the proposal that re-addition of P following P limitation causes lysogenic or pseudolysogenic cyanophage to re-enter the lytic cycle, since no cyanophage DNA would then be amplified following the PCR (it would exist as free virus in the <0.3 μm fraction). An alternative explanation is that since it is known that considerable genetic diversity exists among marine *Synechococcus* spp. (Palenik, 1994), and that different strains may possess slightly different physiological traits (Toledo & Palenik, 1997), it may be possible that lysogeny is only established in specific *Synechococcus* strains. In order to test this hypothesis, we have recently investigated the population structure of the cyanophage community using DGGE methodology to detect changes in virus diversity following P addition to the P-limited enclosure.

Preliminary DGGE analysis of cyanophage isolates determined the specific denaturation conditions required for separating the PCR products from different cyanophage strains (Figure 4b, lanes 1–4). Subsequent DGGE analysis of the mesocosm cyanophage populations, concentrated from the phytoplankton (>0.3 μm) fraction, revealed a diverse population of cyanophages from each of the mesocosm enclosures (Figure 4b, lanes 5–14). More importantly, it was evident that the diversity of cyanophages decreased

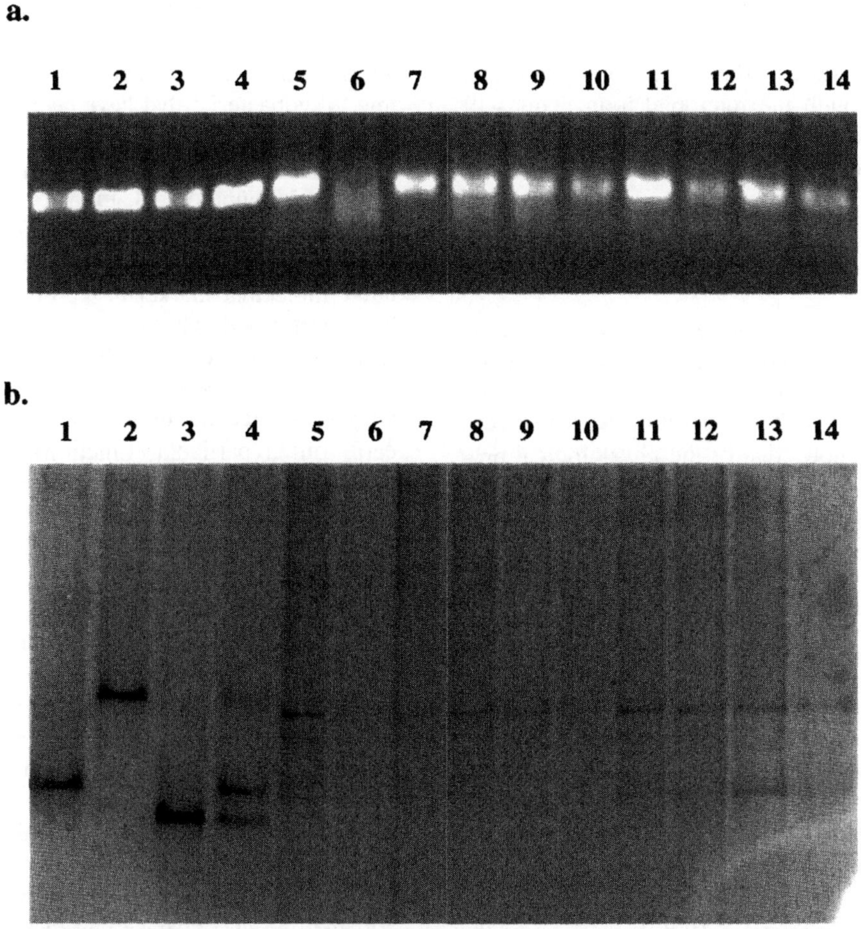

Figure 4. PCR and DGGE analysis of cyanophage strains S-BM1, S-WHM1, S-BnM1 and DNA extracted from the >0.3 μm fraction concentrated from natural seawater mesocosm enclosures located in a Norwegian fjord. (a) PCR products were subject to electrophoresis on a 1.2% agarose gel and visualised by ethidium bromide staining. The size of each product was ca. 165 bp. (b) PCR products subject to DGGE analysis using a 10–30% urea/formamide gradient at 150 v for 5 hr in 10% acrylamide and also visualised by ethidium bromide staining. Tracks:- cyanophage strains S-BM1 (lane 1); S-WHM1 (lane 2); S-BnM1 (lane 3); all 3 strains (lane 4); cellular (>0.3 μm) fractions concentrated from seawater enclosures (collected during the mesocosm study outlined in Wilson et al., 1998): enclosure 1 was a control bag with no additions (lane 5); enclosure 3 contained nutrients to induce a diatom bloom (lane 6); enclosure 5 contained nutrients to induce a flagellate (predominantly *Emiliania huxleyi*) bloom (lane 7); enclosure 7 was manipulated to induce P-limited growth conditions (lanes 8–9); Pi was added back to enclosure 7 after 3 weeks of limitation (lane 10); enclosure 8 was manipulated to induce P-limited growth conditions (N:P additions of 75:1) (lanes 11–13); Pi was added back to enclosure 8 after 3 weeks of limitation (N:P additions of 15:1) (lane 14). See Wilson et al. (1998) for details of virus and *Synechococcus* spp. counts in each of the mesocosm enclosures. Cyanophage specific primer pair CPS1 and CPS2 (Fuller et al., 1998), with the upstream primer (CPS1) altered with a 40-nucleotide GC-rich sequence (GC clamp) added to the 5′ end (see Wawer & Muyzer, 1995 for GC clamp sequence), was used for PCR amplification of all of the samples prior to DGGE.

following addition of Pi to the P-limited enclosure which contained the *Synechococcus* spp. bloom (Figure 4b, lanes 13–14). Specifically, several DNA bands disappeared providing good evidence that cyanophage DNA associated with the phytoplankton size compartment is released into the 'viral' (<0.3 μm) fraction. Such data further suggests that P limitation induces either a lysogenic or a pseudolysogenic response in marine *Synechococcus* cells infected by a temperate cyanophage. To date, there is very little direct evidence of lysogeny in marine *Synechococcus* spp., although Waterbury *et al.* (1994) found that during plaque purification it was possible to isolate, from rare colonies that grew within plaques, clones of *Synechococcus* that had apparently been lysogenised. This suggests that some cyanophage isolates can undergo a temperate cycle. If a lysogenic interaction does exist during P limiting conditions in the ocean, it will significantly reduce the amount of picoplankton-derived carbon entering the microbial loop. In order to fully understand the role of P availability in determining virus/host relationships, there needs to be more definitive studies to investigate viruses and their effect on carbon and nutrient flow through the microbial loop. This will undoubtedly involve further development of current molecular techniques.

Conclusions

The cycling of P in the marine environment is inextricably linked to the biological processes contained within. Thus, for oceanographers to fully understand the complexities of such a nutrient cycle, it is essential they have a knowledge of the physiological processes that are being carried out by marine organisms *in situ*. The previous lack of available methodology for addressing such questions as: Is a specific bacterioplankton or phytoplankton species (or a single cell) actually experiencing P-limitation *in situ*? What specific P source does this organism utilise *in situ* (dissolved/particulate inorganic/organic)? Can the organism accumulate residual P *in situ* so as to form a polyphosphate store?, may now start to be addressed using physiological markers that are diagnostic for a specific process. Such data can also then be utilised to realistically model P transport and assimilation and be combined with models that are capable of assessing interactions with other environmental variables such as light and N (Geider et al., 1996). We reiterate the phrase *in situ* since this is the important context here. To perform realistic physiological experiments relies on the culture of isolated organisms. This can then give an insight into the physiological capacity of that organism, but not necessarily what it is actually experiencing *in situ*. The complexities of the microbial community, with potential consortia of microorganisms operating within (see Paerl & Pinckney, 1996), and with the now well-evident presence of viruses, are much greater than can be simulated in a laboratory. Hence, the real need is to begin to evaluate these processes in natural phytoplankton and bacterioplankton populations. Throughout this review, we have laboured the point that the tremendous diversity of the microbial community is not mirrored by an ability to culture these organisms. Thus, we have argued that it is beneficial to first focus development of physiological molecular markers on those organisms that can dominate the euphotic zone, and for which laboratory culture is now possible. Although this includes some ubiquitous phytoplankton genera, it is more difficult for the bacterioplankton component, and certainly it is desirable to obtain more experimental data of nutrient uptake and acquisition, and cell cycle analysis in those few 'oligobacteria' that have been isolated (Button et al., 1993; Eguchi et al., 1996).

The benefits of the use of physiological, and ultimately immunological, markers, as a diagnostic of a cells nutrient status will ultimately only be gained when a thorough knowledge of the regulation of a putative molecular marker in the organism of study has been performed. The underlying regulatory mechanisms will determine whether the researcher is using such a marker as an absence/presence indication of stress (or capacity to utilise a particular substrate, carry out a particular function), or if a quantitative assessment of expression is being sought. Once achieved, the technology can then be transferred to natural populations. The particular use of antibody probes to assess expression of a given marker have the potential advantage over DNA probes of being specific for the 'end-product' of the regulatory network, though unfortunately not distinguishing active from inactive protein. In the future, it would be hoped to be able to monitor expression of a particular protein in a single cell using a fluorescently-tagged secondary antibody and flow cytometry. This approach will certainly be aided by the rapid progress that is being made in the development of sensitive fluorescent labels. Indeed, a new insoluble fluorogenic substrate for alkaline phosphatase has recently been reported, which is capable of yielding a highly fluorescent and stable product at the site of enzyme activity. This enzyme-labeled fluorescence (ELF) has been shown to be useful for assessing the P status of single eukaryotic phytoplankton cells (González-Gil et al., 1998). More extensive testing of the utility of ELF in a wide range of microorganisms in laboratory culture will ultimately allow its use in natural populations, which should provide extremely valuable information on the extent of P-limitation, as well as complement the molecular probe or antibody work described above. Alternative techniques for monitoring expression of a particular gene or groups of genes in natural microbial communities rely on the use of *in situ* hybridisation methodologies and reverse transcriptase (Hodson et al., 1995; Chen et al., 1997), or the use of reporter gene technology (Albertson et al., 1996). The former techniques rely on the use of a solid surface, be it a microscope slide or the surface of a filter. The use of reporter genes to monitor gene expression *in situ* requires the re-introduction of the modified organism into the natural environment. Thus far, the use of such biosensors is in its infancy, but potential certainly exists with specific regard to P, to monitor P-regulated genes by coupling a suitable reporter (e.g. luciferase or the green fluorescent protein) to the *pho* box regulatory region.

With regards to the impact of viruses on marine P cycling, again new techniques are helping to evaluate their role. From the evidence so far, it seems that viruses may have a direct effect in mediating P regeneration, on the one hand by acting as a potential source of P, but also by the fact that the outcome of a virus/host relationship may be directly linked to the P status of the host cell. Further use of molecular techniques in assessing the diversity of viral populations and the presence of lysogenic populations will certainly clarify their role.

Acknowledgements

This work was supported in part by NERC grants GST/02/1081 and GST/02/1053 via the PRIME (Plankton Reactivity in the Marine Environment) thematic program. This is PRIME contribution number 20. Work described in Figure 2 was performed by Dr Nyree West (Warwick University) and in collaboration with Dr Jim Cotner (Minnesota) and Dr Jim Ammerman (Texas A & M). NSF Grant OCE-9416614 provided support for this Bermuda cruise. D. J. S. is a Royal Society University Research Fellow. W. H. W. is a Marine Biological Association Research Fellow.

References

Albertson, N. H., S. Stretton, S. Pongpattanakitshote, J. Östling, K. C. Marshall, A. E. Goodman & S. Kjelleberg, 1996. Construction and use of a new vector/transposon, pLBT::miniTn*10:lac:kan*, to identify environmentally responsive genes in a marine bacterium. FEMS Microbiol. Letts. 140: 287–294.

Ammerman, J. W., 1993. Microbial cycling of inorganic and organic phosphorus in the water column. In Kemp P. F., B. F. Sherr, E. B. Sherr, & J. J. Cole (eds), Handbook of Methods in Aquatic Microbial Ecology. Lewis, Boca Raton: 649–660.

Ammerman, J. W., 1991. Role of ecto-phosphohydrolases in phosphorus regeneration in estuarine and coastal ecosystems. In Chrost R. J. (ed.), Microbial Enzymes in Aquatic Environments. Springer-Verlag, New York: 165–186.

Ammerman, J. W. & F. Azam, 1991a. Bacterial 5′-nucleotidase activity in estuarine and coastal marine waters: Characterisation of enzyme activity. Limnol. Oceanogr. 36: 1427–1436.

Ammerman, J. W. & F. Azam, 1991b. Bacterial 5′-nucleotidase activity in estuarine and coastal marine waters: Role in phosphorus regeneration. Limnol. Oceanogr. 36: 1437–1447.

Ammerman, J. W. & F. Azam, 1985. Bacterial 5′–nucleotidase in aquatic ecosystems: a novel mechanism of phosphorus regeneration. Science 227: 1338–1340.

Ammerman, J. W., E. Bentzen, E. R. Peele, J. B. Cotner & W. H. Jeffrey, 1994. Phosphorus cycling at the Bermuda time-series station: an overview. Eos, Trans. am. Geophys. Union, 75:100.

Anderson, D. & N. Lindquist, 1985. Time-course measurements of phosphorus depletion and cyst formation in the dinoflagellate *Gonyaulax tamerensis lebour*. J. exp. mar. Biol. Ecol. 86: 1–13.

Azam, F., T. Fenchel, J. G. Field, J. S. Gray, L. A. Meyer-Reil & F. Thingstad, 1983. The ecological role of water-column microbes in the sea. Mar. Ecol. Prog. Ser. 10: 257–263.

Barber, R. T. & F. P. Chavez, 1991. Regulation of primary productivity rate in the equatorial Pacific Ocean. Limnol. Oceanogr. 36: 1803–1815.

Barber, R. T., 1992. Geologic and climatic time scales of nutrient variability. In Falkowski P. G. & A. D. Woodhead (eds), Primary Productivity and Biogeochemical Cycles in the Sea. Plenum Press, New York: 89–106.

Bengis-Garber, C., 1985. Membrane-bound 5′-nucleotidase in marine luminous bacteria: Biochemical and immunological properties. Can. J. Microbiol. 31: 543–548.

Bengis-Garber, C. & D. J. Kushner, 1981. Purification and properties of 5′-nucleotidase from the membrane of *Vibrio costicola*, a moderate halophilic bacterium. J. Bact. 146: 24–32

Berdalet, E., M. Latasa & M. Estrada, 1994. Effects of nitrogen and phosphorus starvation on nucleic acid and protein content of *Heterocapsa* sp. J. Plankton Res. 16: 303–316.

Bergh, O., K. Y. Børsheim, G. Bratbak & M. Heldal, 1989. High abundance of viruses found in aquatic environments. Nature, 340: 467–468.

Berland, B. R., D. J. Bonin & S. Y. Maestrini, 1980. Azote ou phosphore? Considérations sur le 'paradoxe nutritionnel' de la mer méditerranée. Oceanol. Acta. 3: 135–142.

Berner, R. A., K. C. Ruttenberg, E. D. Ingall & J. L. Rao, 1993. The nature of phosphorus burial in modern marine sediments. In Wollast R., F. T. Mackenzie & L. Chou (eds), Interactions of C, N, P & S Biochemical Cycles and Global Change. NATO ASI Series, I4, Springer, Berlin: 365–378.

Bhaya, D., R. Schwarz & A. R. Grossman, 1999. Molecular responses to environmental stress. In Whitton B. A. & M. Potts (eds), The Ecology of Cyanobacteria: Their Diversity in Time and Space. Kluwer, Dordrecht, The Netherlands (in press).

Björkman, K. & D. M. Karl, 1994. Bioavailability of inorganic and organic phosphorus compounds to natural assemblages of microorganisms in Hawaiian coastal waters. Mar. Ecol. Prog. Ser. 111: 265–273.

Björkman, O., 1981. Responses to different quantum flux densities. In O. L. Lange, P. S. Nobel, C. B. Osmond & H. Ziegler (eds), Encyclopedia of Plant Physiology, New Series, vol. 12A. Springer-Verlag, Berlin: 57–107.

Boehme, J., M. E. Frischer, S. C. Jiang, C. A. Kellogg, S. Pichard, J. B. Rose, C. Steinway & J. H. Paul, 1993. Viruses, bacterioplankton and phytoplankton in the southeastern Gulf of Mexico: distribution and contribution to oceanic DNA pools. Mar. Ecol. Prog. Ser. 97: 1–10.

Børsheim, K. Y., G. Bratbak & M. Heldal, 1990. Enumeration and biomass estimation of planktonic bacteria and viruses by transmission electron microscopy. Appl. envir. Microbiol. 56: 352–356.

Boyer, G. L., J. J. Sullivan, R. J. Anderson, P. J. Harrison & F. J. R. Taylor, 1987. Effetcs of nutrient limitation on toxin production and composition in the marine dinoflagellate *Protogonyaulax tamarensis*. Mar. Biol. 96: 123–128.

Brahamsha, B., 1996. A genetic manipulation system for oceanic cyanobacteria of the genus *Synechococcus*. Appl. envir. Microbiol. 62: 1747–1751.

Bratbak, G., J. K. Egge & M. Heldal, 1993. Viral mortality of the marine alga *Emiliania huxleyi* (Haptophyceae) and termination of algal blooms. Mar. Ecol. Prog. Ser. 93: 39–48.

Bratbak, G., F. Thingstad & M. Heldal, 1994. Viruses and the microbial loop. Microb. Ecol. 28: 209–221.

Bratbak, G., M. Heldal, T. F. Thingstad, B. Riemann & O. H. Haslund. 1992. Incorporation of viruses into the budget of microbial C-transfer. A first approach. Mar. Ecol. Prog. Ser. 83: 273–280.

Bun-Ya, M., M. Nishimura, S. Harashima & Y. Oshima, 1991. The *pho84* gene of *Saccharomyces cerevisiae* encodes an inorganic phosphate transporter. Mol. cell. Biol. 11: 3229–3238.

Button, D. K., F. Schut, P. Quang, R. Martin & B. R. Robertson, 1993. Viability and isolation of marine bacteria by dilution culture: theory, procedures and initial results. Appl. envir. Microbiol. 59: 881–891.

Campbell, L., H. A. Nolla & D. Vaulot, 1994. The importance of *Prochlorococcus* to community structure in the central North Pacific Ocean. Limnol. Oceanogr. 39: 954–961.

Capone, D. G., A. Subramanian, E. J. Carpenter, J. Montoya & C. Humborg, 1996. A spatially and temporally extensive bloom of the diazotrophic cyanobacterium, *Trichodesmium*, in the central Arabian Sea during the spring intermonsoon. Eos, Trans. am. geophys. Union, 76: OS11A–13.

Capone, D. G., J. P. Zehr, H. W. Paerl, B. Bergman & E. J. Carpenter, 1997. *Trichodesmium*, a globally significant marine cyanobacterium. Science 276: 1221–1229.

Caron, D. A., 1990. Evolving role of protozoa in aquatic nutrient cycles. In Reid P. C., C. M. Turley & P. H. Burkhill (eds), Protozoa and their Role in Marine Processes. NATO ASI Series, Springer-Verlag, Berlin: 387–415.

Cembella, A. D., N. J. Antia & P. J. Harrison, 1984. The utilisation of inorganic and organic phosphorus compounds as nutrients by eukaryotic microalgae: a multidisciplinary perspective: Part 1. Crit. Rev. Microbiol. 10: 317–391.

Cembella, A. D., N. J. Antia & P. J. Harrison, 1984. The utilisation of inorganic and organic phosphorus compounds as nutrients by eukaryotic microalgae: a multidisciplinary perspective: Part 2. Crit. Rev. Microbiol. 11: 13–81.

Chavez, F. P., K. R. Buck, K. H. Coale, J. H. Martin, G. R. DiTullio, N. A. Welschmeyer, A. C. Jacobson & R. T. Barber, 1991. Growth rates, grazing, sinking, and iron limitation of equatorial Pacific phytoplankton. Limnol. Oceanogr. 36: 1816–1833.

Chen, F. & C. A. Suttle, 1995. Amplification of DNA polymerase gene fragments from viruses infecting microalgae. Appl. envir. Microbiol. 61: 1274–1278.

Chen, F., C. A. Suttle & S. M. Short, 1996. Genetic diversity in marine algal virus communities as revealed by sequence analysis of DNA polymerase genes. Appl. envir. Microbiol. 62: 2869–2874.

Chen, F., J. M. González, W. A. Dustman, M. A. Moran & R. E. Hodson, 1997. *In situ* reverse transcription, an approach to characterise genetic diversity and activities of prokaryotes. Appl. envir. Microbiol. 63: 4907–4913.

Chrost, R. J. (ed.), 1991. Microbial Enzymes in Aquatic Environments. Springer-Verlag, New York.

Coale, K. H., K. S. Johnson, S. E. Fitzwater, R. M. Gordon, S. Tanner, F. P. Chavez, L. Ferioli, C. Sakamoto, P. Rogers, F. Millero, P. Steinberg, P. Nightingale, D. Cooper, W. P. Cochlan, M. R. Landry, J. Constantinou, G. Rollwagen, A. Trasvina & R. Kudela, 1996. A massive phytoplankton bloom induced by an ecosystem-scale iron fertilisation experiment in the equatorial Pacific Ocean. Nature 383: 495–501.

Codispoti, L. A., 1989. Phosphorus vs. nitrogen limitation of new (export) production. In Berger W. H., V. S. Smetacek & G. Wefer (eds), Productivity of the Ocean, Present and Past. Wiley, Chichester: 377–394.

Collier, J. L. & A. R. Grossman, 1992. Chlorosis induced by nutrient deprivation in *Synechococcus* sp. strain PCC7942: not all bleaching is the same. J. Bact. 174: 4718–4726.

Cotner, J. B., J. W. Ammerman, E. R. Peele & E. Bentzen, 1997. Phosphorus-limited bacterioplankton growth in the Sargasso Sea. Aquat. Microb. Ecol. 13: 141–149.

Cotner, J. B., J. W. Ammerman, G. Van Dijken & W. Lehman, 1998. Cell specific dissolved inorganic and organic P incorporation in the Sargasso Sea. Eos, Trans. am. Geophys. Union.

Cullen, J. J., X. Yang & H. L. MacIntyre, 1992. Nutrient limitation of marine photosynthesis. In P. G. Falkowski & A. D. Woodhead (eds), Primary Productivity and Biogeochemical Cycles in the Sea. Plenum Press, New York: 69–88.

Davies, A. G., 1988. Nutrient interactions in the marine environment. In L. J. Rogers & J. R. Gallon (ed.), Biochemistry of the Algae and Cyanobacteria. Proc. Phytochem. Soc. Eur. 28. Clarendon Press, Oxford: 241–256.

DeFlaun, M. F., J. H. Paul & W. H. Jeffrey, 1987. Distribution and molecular weight of dissolved DNA in subtropical estuarine and oceanic environments. Mar. Ecol. Prog. Ser. 38: 65–73.

DeFlaun, M. F., J. H. Paul & W. H. Jeffrey, 1986. Simplified method for dissolved DNA determination in aquatic environments. Appl. envir. Microbiol. 52: 654–659.

DeLong, E. F., 1997. Marine microbial diversity: the tip of the iceberg. Trends Biotechnol. 15: 203–207.

DiTullio, G. R., D. A. Hutchins & K. W. Bruland. 1993. Interaction of iron and major nutrients controls phytoplankton growth and species composition in the tropical North Pacific Ocean. Limnol. Oceanogr. 38: 495–508.

Donald, K. M., D. J. Scanlan, N. G. Carr, N. H. Mann & I. Joint, 1997. Comparative phosphorus nutrition of the marine cyanobacterium *Synechococcus* WH7803 and the marine diatom *Thalassiosira weissflogii*. J. Plankton Res. 19: 1793–1813.

Downing, J. A., 1997. Marine nitrogen: phosphorus stoichiometry and the global N:P cycle. Biogeochemistry 37: 237–252.

Duce, R. A. & N. W. Tindale, 1991. Atmospheric transport of iron and its deposition in the ocean. Limnol. Oceanogr. 36: 1715–1726.

Dyhrman, S. T. & B. P. Palenik, 1998. The identification of cell surface proteins as markers of phosphate stress in *Emiliania huxleyi*. Eos, Trans. am. Geophys. Union.

Dyhrman, S. T. & B. P. Palenik, 1997. The identification and purification of a cell-surface alkaline phosphatase from the dinoflagellate *Prorocentrum minimum* (Dinophyceae). J. Phycol. 33: 602–612.

Eberl, L., M. Givskov, C. Sternberg, S. Møller, G. Christiansen & S. Molin, 1996. Physiological responses of *Pseudomonas putida* KT2442 to phosphate starvation. Microbiology 142: 155–163.

Eccleston-Parry, J. D. & B. S. Leadbeater, 1995. Regeneration of phosphorus and nitrogen by four species of heterotrophic nanoflagellates feeding on three nutritional states of a single bacterial strain. Appl. envir. Microbiol. 61:1033–1038.

Edvardsen, B., F. Moy & E. Paasche, 1990. Hemolytic activity in extracts of *Chrysochromulina polylepis* grown at different levels of selenite and phosphate. In E. Granéli, B. Sundström, L. Edler & D. M. Anderson (eds), Toxic Marine Phytoplankton, Elsevier, New York: 284–289.

Egge, J. K. & D. L. Aksnes, 1992. Silicate as regulating nutrient in phytoplankton competition. Mar. Ecol. Prog. Ser. 83: 281–289.

Eguchi, M., T. Nishikawa, K. MacDonald, R. Cavicchioli, J. C. Gottschal & S. Kjelleberg, 1996. Responses to stress and nutrient availability by the marine ultramicrobacterium *Sphingomonas* sp. strain RB2256. Appl. envir. Microbiol. 62: 1287–1294.

Escoubas, J. M., M. Lomas, J. LaRoche & P. G. Falkowski, 1995. Light-intensity regulation of *cab* gene transcription is signaled by the redox state of the plastoquinone pool. Proc. natn. Acad. Sci. U.S.A. 92: 10237–10241.

Eymann, C., H. Mach, C. R. Harwood & M. Hecker, 1996. Phosphate-starvation-inducible proteins in *Bacillus subtilis*: a two-dimensional gel electrophoresis study. Microbiology 142: 3163–3170.

Falkowski, P. G., 1997. Evolution of the nitrogen cycle and its influence on the biological sequestration of CO_2 in the ocean. Nature 387: 272–275.

Falkowski, P. G. & J. LaRoche, 1991. Molecular biology in studies of ocean processes. International review of cytology – a survey of cell biology 128: 261–303.

Falkowski, P. G., R. M. Greene & R. J. Geider, 1992. Physiological limitations on phytoplankton productivity in the oceans. Oceanography 5: 84–91.

Falkowski, P. G., A. Sukenik & R. Herzig, 1989. Nitrogen limitation in *Isochrysis galbana* (Haptophyceae). II. Relative abundance of chloroplast proteins. J. Phycol. 25: 471–478.

Fanning, K. A., 1992. Nutrient provinces in the sea: concentration ratios, reaction rate ratios and ideal covariation. J. Geophys. Res. 97: 5693–5712.

Fisher, T R., E. R. Peele, J. W. Ammerman & L. W. J. Harding, 1992. Nutrient limitation of phytoplankton in Chesapeake Bay. Mar. Ecol. Prog. Ser. 82: 51–63.

Flynn, K. J., H. Opik & P. J. Syrett, 1986. Localisation of the alkaline phosphatase and 5′-nucleotidase activities of the diatom *Phaeodactylum tricornatum*. J. gen. Microbiol. 132: 289–298.

Flynn, K., J. M. Franco, P. Fernandéz, B. Reguera, M. Zapata, G. Wood & K. J. Flynn, 1994. Changes in toxin content, biomass and pigments of the dinoflagellate *Alexandrium minutum* during nitrogen refeeding and growth into nitrogen or phosphorus stress. Mar. Ecol. Prog. Ser. 111: 99–109.

Föllmi, K. B., 1996. The phosphorus cycle, phosphogenesis and marine phosphate-rich deposits. Earth-Science Reviews 40: 55–124.

Fourqurean, J. W., R. D. James & J. C. Zieman, 1993. Processes influencing water column nutrient characteristics and phosphorus limitation of phytoplankton biomass in Florida Bay, FL, U.S.A.: Inferences from spatial distributions. Estuar. coast. Shelf Sci. 36: 295–314.

Fuhrman, J., 1992. Bacterioplankton roles in cycling of organic matter: the microbial food web. In Falkowski, P. G. & A. D. Woodhead (eds), Primary Productivity and Biogeochemical Cycles in the Sea. Plenum Press, New York: 361–383.

Fuller, N. J., W. H. Wilson, I. R. Joint & N. H. Mann, 1998. Occurrence of a sequence in marine cyanophages similar to T4 g20 and its application to PCR-based detection and quantification techniques. Appl. envir. Microbiol. 64: 2051–2060.

Gage, M. A. & E. Gorham, 1985. Alkaline phosphatase activity and cellular phosphorus as an index of the phosphorus status of phytoplankton in Minnesota lakes. Freshwat. Biol. 15: 227–233.

Gallagcher, J. C., 1982. Physiological variation and electrophoretic banding patterns of genetically different seasonal populations of *Skeletonema costatum* (Bacillariophyceae). J. Phycol. 18: 148–162.

Geider, R. J., 1987. Light and temperature dependence of the carbon to chlorophyll *a* ratio in microalgae and cyanobacteria: implications for physiology and growth of phytoplankton. New Phytol. 106: 1–34.

Geider, R. J., H. L. MacIntyre & T. M. Kana, 1996. A dynamic model of photoadaptation in phytoplankton. Limnol. Oceanogr. 41: 1–15.

Geider, R. J., J. LaRoche, R. M. Greene & M. Olaizola, 1993. Response of the photosynthetic apparatus of *Phaeodactylum tricornatum* (Bacillariophyceae) to nitrate, phosphate, or iron starvation. J. Phycol. 29: 755–766.

Giovannoni, S. J., T. B. Britschgi, C. L. Moyer & K. G. Field, 1990. Genetic diversity in Sargasso Sea bacterioplankton. Nature 345: 60–63.

Givskov, M., L. Eberl & S. Molin, 1994. Responses to nutrient starvation in *Pseudomonas putida* KT2442: two-dimensional electrophoretic analysis of starvation- and stress-induced proteins. J. Bact. 176: 4816–4824.

Gobler, C. J., D. A. Hutchins, N. S. Fisher, E. M. Cosper & S. A. Sanudo-Wilhelmy, 1997. Release and bioavailability of C, N, P, Se and Fe following viral lysis of a marine chrysophyte. Limnol. Oceanogr. 42: 1492–1504

Goldstein, A. H., 1994. Involvement of the quinoprotein glucose dehydrogenase in the solubilisation of exogenous mineral phosphates by Gram-negative bacteria. In A. Torriani-Gorini, E. Yagil & S. Silver (eds), Phosphate in Microorganisms – Cellular and Molecular Biology ASM Press, Washington, D.C.: 197–203.

González-Gil, S., B. A. Keafer, R. V. M. Jovine, A. Anguilera & D. M. Anderson, 1998. Detection and quantification of alkaline phosphatase in single cells of phosphorus-starved marine phytoplankton. Mar. Ecol. Prog. Ser. 164: 21–35.

Granéli, E., 1987. Nutrient limitation of phytoplankton biomass in a brackish water bay highly influenced by river discharge. Estuar. coast. Shelf Sci. 25: 555–565.

Graziano, L. M., R. J. Geider, W. K. W. Li & M. Olaizola, 1996a. Nitrogen limitation of North Atlantic phytoplankton: analysis of physiological condition in nutrient enrichment experiments. Aquat. microb. Ecol. 11: 53–64.

Graziano, L. M., J. LaRoche & R. J. Geider, 1996b. Physiological responses to phosphorus limitation in batch and steady-state cultures of *Dunaliella tertiolecta* (Chlorophyta): a unique stress protein as an indicator of phosphate deficiency. J. Phycol. 32: 825–838.

Hancock, R. E. W., K. Poole & R. Benz, 1982. Outer membrane protein P of *Pseudomonas aeruginosa*: regulation by phosphate deficiency and formation of small anion-specific channels in lipid bilayer membranes. J. Bact. 150: 730–738.

Hancock, R. E. W., C. Egli, R. Benz & R. J. Siehnel, 1992. Overexpression in *Escherichia coli* and functional analysis of a novel PP_i-selective porin, OprO, from *Pseudomonas aeruginosa*. J. Bact. 174: 471–476.

Harrison, M. J. & M. L. van Buuren, 1995. A phosphate transporter from the mycorrhizal fungus *Glomus versiforme*. Nature 378: 626–629.

Harrison, P. J., M. H. Hu, Y. P. Yang & X. Lu, 1990. Phosphate limitation in estuarine and coastal waters of China. J. exp. mar. Biol. Ecol. 140: 79–87.

Healey, F. P. & L. L. Hendzel, 1975. Effect of phosphorus deficiency on two algae growing in chemostats. J. Phycol. 11: 303–309.

Healey, F. P., 1973. Inorganic nutrient uptake and deficiency in algae. Crit. Rev. Microbiol. 3: 69–113.

Hecky, R. E. & P. Kilham, 1988. Nutrient limitation of phytoplankton in freshwater and marine environments: a review of recent evidence on the effects of enrichment. Limnol. Oceanogr. 33: 796–822.

Hernández, I., J. A. Fernández & F. X. Niell, 1995. A comparative study of alkaline phosphatase activity in two species of Gelidium (Gelidiales, Rhodophyta). Eur. J. Phycol. 30: 69–77.

Hodson, R. E., W. A. Dustman, R. P. Garg & M. A. Moran, 1995. *In situ* PCR for visualisation of microscale distribution of specific genes and gene products in prokaryotic communities. Appl. envir. Microbiol. 61: 4074–4082.

Holligan, P. M., E. Fernández, J. Aiken, W. M. Balch, P. Boyd, P. H. Burkill, M. Finch, S. B. Groom, G. Malin, K. Muller, D. A. Purdie, C. Robinson, C. C. Trees, S. M. Turner & P. van der Wal, 1993. A biogeochemical study of the coccolithophore, *Emiliania huxleyi*, in the North Atlantic. Glob. Biogeochem. Cycl. 7: 879–900.

Ikeya, T., K. Ohki, M. Takahashi & Y. Fujita, 1997. Study on phosphate uptake of the marine cyanophyte *Synechococcus* sp. NIBB 1071 in relation to oligotrophic environments in the open ocean. Mar. Biol. 129: 195–202.

Jiang, S. C. & J. H. Paul, 1995. Viral contribution to dissolved DNA in the marine environment as determined by differential centrifugation and kingdom probing. Appl. envir. Microbiol. 61: 317–325.

Joint, I. (ed.), 1995. Molecular Ecology of Aquatic Microbes. NATO ASI Series, G38, Springer, Berlin.

Karl, D. M., 1996. Alternation of N and P control of new and export production in the north Pacific gyre: a hypothesis based on the HOT program data set. Eos, Trans. am. Geophys. Union, 76: OS221–12.

Karl, D. M. & M. D. Bailiff, 1989. The measurement and distribution of dissolved nucleic acids in aquatic environments. Limnol. Oceanogr. 34: 543–558.

Karl, D. M. & G. Tien, 1992. MAGIC: a sensitive and precise method of measuring dissolved phosphorus in aquatic environments. Limnol. Oceanogr. 37: 105–116.

Karl, D. M., R. Letelier, D. V. Hebel, D. F. Bird & C. D. Winn, 1992. *Trichodesmium* blooms and new nitrogen in the north Pacific gyre. In E. J. Carpenter, D. G. Capone & J. G. Rueter (eds), Marine Pelagic Cyanobacteria: *Trichodesmium* and Other Diazotrophs. NATO ASI Series, C362, Kluwer, Dordrecht, The Netherlands: 219–237.

Karl, D. M., R. Letelier, D. Hebel, L. Tupas, J. Dore, J. Christian & C. Winn, 1995. Ecosystem changes in the North Pacific subtropical gyre attributed to the 1991–1992 El Ninö. Nature 373: 230–234.

Kilham, P. & R. E. Hecky, 1988. Comparative ecology of marine and freshwater phytoplankton. Limnol. Oceanogr. 33: 776–795.

Kjelleberg, S., K. B. G. Flärdh, T. Nyström & D. J. W. Moriarty, 1993a. Growth limitation and starvation of bacteria. In T. E. Ford (ed.), Aquatic Microbiology: an Ecological Approach, Blackwell, Oxford: 289–320.

Kjelleberg, S., N. Albertson, K. Flärdh, L. Holmquist, A. Jouper-Jaan, R. Marouga, J. Östling, B. Svenblad & D. Weichart, 1993b. How do non-differentiating bacteria adapt to starvation? Antonie van Leeuwenhoek 63: 333–341.

Kokjohn, T. A., G. S. Sayler & R. V. Miller, 1991. Attachment and replication of *Pseudomonas aeruginosa* bacteriophages under conditions simulating aquatic environments. J. gen. Microbiol. 137: 661–666.

Kolber, Z., J. Zehr & P. G. Falkowski, 1988. Effects of growth irradiance and nitrogen limitation on photosynthetic energy conversion in photosystem II. Plant Physiol. 88: 923–929.

Kozloff, L. M. & F. W. Putnam, 1950. Biochemical studies of virus reproduction: III. The origin of virus phosphorus in the *Escherichia coli* T_6 bacteriophage system. J. Biol. Chem. 182: 229–243.

Krom, M. D., N. Kress, S. Brenner & L. I. Gordon, 1991. Phosphorus limitation of primary productivity in the eastern Mediterranean Sea. Limnol. Oceanogr. 36: 424–432.

LaRoche, J., H. Murray, M. Orellana & J. Newton, 1995. Flavodoxin expression as an indicator of iron limitation in marine diatoms. J. Phycol. 31: 520–530.

LaRoche, J., R. J. Geider, L. M. Graziano, H. Murray & K. Lewis, 1993. Induction of specific proteins in eukaryotic algae grown under iron-, phosphorus-, or nitrogen-deficient conditions. J. Phycol. 29: 767–777.

Laudenbach, D. E., D. Ehrhardt, L. Green & A. Grossman, 1991. Isolation and characterisation of a sulfur-regulated gene encoding a periplasmically localised protein with sequence similarity to rhodanese. J. Bact. 173: 2751–2760.

Leopold, K., S. Jacobsen & O. Nybroe, 1997. A phosphate-starvation-inducible outer membrane protein of *Pseudomonas fluorescens* Ag1 as an immunological phosphate-starvation marker. Microbiology 143: 1019–1027.

Li, H., M. J. W. Veldhuis & A.F. Post, 1998. Alkaline phosphatase activities among planktonic communities in the northern Red Sea. Mar. Ecol. Prog. Ser. 173: 107–115.

Lin, S. & E. J. Carpenter, 1996. An empirical protocol for whole-cell immunofluorescence of marine phytoplankton. J. Phycol. 32: 1083–1094.

Lugtenberg, B., 1987. The pho regulon in *Escherichia coli.* In A. Torriani-Gorini, F. G. Rothman, S. Silver, A. Wright & E. Yagil (eds), Phosphate Metabolism and Cellular Regulation in Microorganisms. ASM Press, Washington, D.C.: 1–2.

Macrae, M. J. D., W. Glover, J. W. Ammerman & R. H. Sada-Ruvalcaba, 1994. Seasonal phosphorus deficiency in the Mississippi River plume: unusual large areal extent during the record flood of 1993. Eos, Trans. am. Geophys. Union, 75/3: 30.

Mann, N. H., 1994. Protein phosphorylation in cyanobacteria. Microbiology 140: 3207–3215.

Martin, J. H., 1990. Glacial-interglacial CO_2 change: the iron hypothesis. Paleoceanography 5: 1–13.

Martin, J. H., K. H. Coale, K. S. Johnson, S. E. Fitzwater, R. M. Gordon, S. J.Tanner, C. N. Hunter, V. A. Elrod, J. L. Nowicki, T. L. Coley, R. T. Barber, S. Lindley, A. J. Watson, K. Van Scoy, C. S. Law, M. I. Liddicoat, R. Ling, T. Staunton, J. Stockel, C. Collins, A. Anderson, R. Bidigare, M. Ondrusek, M. Latasa, F. J. Millero, K. Lee, W. Yao, J. Z. Zhang, G. Friederich, C. Sakamoto, F. Chavez, K. Buck, Z. Kolber, R. Greene, P. Falkowski, S. W. Chisholm, F. Hoge, R. Swift, J. Yungel, S. Turner, P. Nightingale, A. Hatton, P. Liss & N. W. Tindale, 1994. Testing the iron hypothesis in ecosystems of the equatorial Pacific Ocean. Nature 371: 123–129.

Martinez, J. & F. Azam, 1993a. Periplasmic aminopeptidase and alkaline phosphatase activities in a marine bacterium: implications for substrate processing in the sea. Mar. Ecol. Prog. Ser. 92: 89–97.

Martinez, J. & F. Azam, 1993b. Aminopeptidase activity in marine chroococcoid cyanobacteria. Appl. envir. Microbiol. 59: 3701–3707.

Maruyama, A., M. Oda & T. Higashihara, 1993. Abundance of virus-sized non-DNase-digestable DNA (coated DNA) in eutrophic seawater. Appl. envir. Microbiol. 59: 712–717.

Mathews, C. K., E. M. Kutter, G. Mosig & P. B. Berget, (eds), 1983. Bacteriophage T4. American Society for Microbiology, Washington, DC.

McGrady-Steed, J., P. M. Harris & P. J. Morin, 1997. Biodiversity regulates ecosystem predictability. Nature 390: 162–165.

McKay, R. M. L., R. J. Geider & J. LaRoche, 1997. Physiological and biochemical response of the photosynthetic apparatus of two marine diatoms to Fe stress. Plant Physiol. 114: 615–622.

Michaels, A. F., D. Olson, J. L. Sarmiento, J. W. Ammerman, K. Fanning, R. Jahnke, A. H. Knap, F. Lipschultz & J. M. Prospero, 1996. Inputs, losses and transformations of nitrogen and phosphorus in the pelagic North Atlantic Ocean. Biogeochemistry 35: 181–226.

Middelboe, M., N. O. G. Jorgensen & N. Kroer, 1996. Effects of viruses on nutrient turnover and growth efficiency of noninfected marine bacterioplankton. Appl. envir. Microbiol. 62: 1991–1997.

Miyata, K., A. Hattori & A. Ohtsuki. 1986. Variation of cellular phosphorus composition of *Skeletonema costatum* and *Heterosigma akashiwo* grown in chemostats. Mar. Biol. 93: 291–297.

Moebus, K., 1996a. Marine bacteriophage reproduction under nutrient-limited growth of host bacteria. I. Investigations with six phage-host systems. Mar. Ecol. Prog. Ser. 144: 1–12.

Moebus, K., 1996b. Marine bacteriophage reproduction under nutrient-limited growth of host bacteria. II. Investigations with phage-host system [H3:H3/1]. Mar. Ecol. Prog. Ser. 144: 13–22.

Muchhal, U. S., J. M. Pardo & K. G. Raghothama, 1996. Phosphate transporters from the higher plant *Arabidopsis thaliana*. Proc. natn. Acad. Sci. U.S.A. 93: 10519–10523.

Muyzer, G., E. C. De. Waal & A. G. Uitterlinden, 1993. Profiling of complex microbial populations by denaturing gradient gel electrophoresis analysis of polymerase chain reaction-amplified genes coding for 16S rRNA. Appl. envir. Microbiol. 59: 695–700.

Naeem, S. & S. B. Li, 1997. Biodiversity enhances ecosystem reliability. Nature 390: 507–509.

Nyström, T., R-M. Olsson & S. Kjelleberg, 1992. Survival, stress resistance, and alterations in protein expression in the marine *Vibrio* sp. strain S14 during starvation for different individual nutrients. Appl. envir. Microbiol. 58: 55–65.

Olson, R. J. & E. R. Zettler, 1995. Potential of flow cytometry for 'pump and probe' fluorescence measurements of phytoplankton photosynthetic characteristics. Limnol. Oceanogr. 40: 816–820.

Olson, R. J., A. M. Chekalyuk & H. M. Sosik, 1996. Phytoplankton photosynthetic characteristics from fluorescence induction assays of individual cells. Limnol. Oceanogr. 41: 1253–1263.

Orellana, M. V. & M. J. Perry, 1995. Optimisation of an immunofluorescent assay of the internal enzyme ribulose-1,5-bisphosphate carboxylase (rubisco) in single phytoplankton cells. J. Phycol. 31: 785–794.

Orellana, M. V. & M. J. Perry, 1992. An immunoprobe to measure Rubisco concentrations and maximal photosynthetic rates of individual phytoplankton cells. Limnol. Oceanogr. 37: 478–490.

Orrett, K. & D. M. Karl, 1987. Dissolved organic phosphorus production in surface seawaters. Limnol. Oceanogr. 32: 383–395.

Overbeck, J., 1991. Early studies on ecto- and extracellular enzymes in aquatic environments. In Chrost R. J. (ed.), Microbial Enzymes in Aquatic Environments. Springer Verlag, Berlin: 1–5.

Paasche, E. & S. Brubak, 1994. Enhanced calcification in the coccolithophorid *Emiliania huxleyi* (Haptophyceae) under phosphorus limitation. Phycologia 33: 324–330.

Paasche, J. S. & S. R. Erga, 1988. Phosphorus and nitrogen limitation of phytoplankton in the inner Oslofjord (Norway). Sarsia 73: 229–243.

Paerl, H. & J. L. Pinckney, 1996. A mini-review of microbial consortia: their roles in aquatic production and biogeochemical cycling. Microb. Ecol. 31: 225–247.

Palenik, B., 1994. Cyanobacterial community structure as seen from RNA polymerase gene sequence analysis. Appl. envir. Microbiol. 60: 3212–3219.

Palenik, B. & A. M. Wood, 1997. Molecular markers of phytoplankton physiological status and their application to individual cells. In Cooksey K. E. (ed.), Molecular Approaches to the Study of the Ocean. Chapman & Hall, London: 187–205.

Palenik, B. & J. A. Koke, 1995. Characterization of a nitrogen-regulated protein identified by cell surface biotinylation of a marine phytoplankton. Appl. envir. Microbiol. 61: 3311–3315.

Pan, Y., D. V. Subba Rao & K. H. Mann, 1996. Changes in domoic acid production and cellular chemical composition of the toxigenic diatom *Pseudo-nitzschia multiseries* under phosphate limitation. J. Phycol. 32: 371–381.

Parpais, J., D. Marie, F. Partensky, P. Morin & D. Vaulot, 1996. Effect of phosphorus starvation on the cell cycle of the photosynthetic prokaryote *Prochlorococcus* spp. Mar. Ecol. Prog. Ser. 132: 265–274.

Parsons, T. R., Y. Maita & C. M. Lalli, 1984. A Manual of Chemical and Biological Methods for Seawater Analysis. Pergamon Press.

Paul, J. H., S. C. Jiang & J. B. Rose, 1991. Concentration of viruses and dissolved DNA from aquatic environments by vortex flow filtration. Appl. envir. Microbiol. 57: 2197–2204.

Perry, M. J. & R. W. Eppley, 1981. Phosphate uptake by phytoplankton in the central North Pacific Ocean. Deep Sea Res. 28A: 39–49.

Pomeroy, L. R., J. E. Sheldon, W. M. Sheldon & F. Peters, 1995. Limits to growth and respiration of bacterioplankton in the Gulf of Mexico. Mar. Ecol. Prog. Ser. 117: 259–268.

Prospero, J. M., K. Barrett, T. Church, F. Dentener, R. A. Duce, J. N. Galloway, H. Levy, J. Moody & P. Quinn, 1996. Atmospheric deposition of nutrients to the North-Atlantic basin. Biogeochemistry 35: 27–73.

Rappé, M. S., M. T. Suzuki, K. L. Vergin & S. J. Giovannoni, 1998. Phylogenetic diversity of ultraplankton plastid small-subunit rRNA genes recovered in environmental nucleic acid samples from the Pacific and Atlantic coasts of the United States. Appl. envir. Microbiol. 64: 294–303.

Rath, J., C. Schiller & G. J. Herndle, 1993. Ectoenzyme activity and bacterial dynamics along a trophic gradient in the Caribbean Sea. Mar. Ecol. Prog. Ser. 102: 89–96.

Reddy, K. J., G. S. Bullerjahn, D. M. Sherman & L. A. Sherman, 1988. Cloning, nucleotide sequence, and mutagenesis of a gene (*irpA*) involved in iron-deficient growth of the cyanobacterium *Synechococcus* strain PCC 7942. J. Bact. 170: 4466–4476.

Redfield, A. C., 1958. The biological control of chemical factors in the environment. Am. Sci. 46: 205–222.

Reinfelder, J. R. & N. S. Fisher, 1991. The assimilation of elements ingested by marine copepods. Science 251: 794–796.

Riegman, R., A. A. M. Noordeloos & G. C. Cadeé, 1992. *Phaeocystis* blooms and eutrophication of the continental coastal zones of the North Sea. Marine Biology 112: 479–484.

Ripp, S. & R. V. Miller, 1997. The role of pseudolysogeny in bacteriophage-host interactions in a natural freshwater environment. Microbiology 143: 2065–2070.

Rivkin, R. B. & M. R. Anderson, 1997. Inorganic nutrient limitation of oceanic bacterioplankton. Limnol. Oceanogr. 42: 730–740.

Romig, W. & A. M. Brodetsky, 1961. Isolation and preliminary characterisation of bacteriophages for *Bacillus subtilis*. J. Bact. 82: 135–141.

Sakshaug, E., K. Andersen, S. Myklestad & Y. Olsen, 1983. Nutritional status of phytoplankton communities in Norwegian waters (marine, brackish, and fresh) as revealed by their chemical composition. J. Plankton Res. 5: 175–196.

Scanlan, D. J., J. A. Bourne & N. H. Mann, 1997a. A putative transcriptional activator of the Crp/Fnr family from the marine cyanobacterium *Synechococcus* sp. WH7803. J. appl. Phycol. 8: 565–567.

Scanlan, D. J., N. H. Mann & N. G. Carr, 1993. The response of the picoplanktonic marine cyanobacterium *Synechococcus* species WH7803 to phosphate starvation involves a protein homologous to the periplasmic phosphate-binding protein of *Escherichia coli*. Mol. Microbiol. 10: 181–191.

Scanlan, D. J., N. H. Mann & N. G. Carr, 1989. Effect of iron and other nutrient limitations on the pattern of outer membrane proteins in the cyanobacterium *Synechococcus* PCC7942. Arch. Microbiol. 152: 225–228.

Scanlan, D.J., N. J. Silman, K. M. Donald, W. H. Wilson, N. G. Carr, I. Joint & N. H. Mann, 1997b. An immunological approach to detect phosphate stress in populations and single cells of photosynthetic picoplankton. Appl. envir. Microbiol. 63: 2411–2420.

Smith, M. W. & J. W. Payne, 1992. Expression of periplasmic binding proteins for peptide transport is subject to negative regulation by phosphate limitation in *Escherichia coli*. FEMS Microbiol. Letts. 100: 183–190.

Smith, S. M. & G. L. Hitchcock, 1994. Nutrient enrichments and phytoplankton growth in the surface waters of the Louisiana Bight. Estuaries 17: 740–753.

Smith, S. V., 1984. Phosphorus versus nitrogen limitation in the marine environment. Limnol. Oceanogr. 29: 1149–1160.

Smith, S. V. & M. J. Atkinson, 1984. Phosphorus limitation of net production in a confined aquatic ecosystem. Nature 307: 626–627.

Sobecky, P. A., M. A. Schell, M. A. Moran & R. E. Hodson, 1996. Impact of a genetically engineered bacterium with enhanced alkaline phosphatase activity on marine phytoplankton communities. Appl. envir. Microbiol. 62: 6–12.

Solorzano, L. & J. H. Sharp, 1980. Determination of total dissolved phosphorus and particulate phosphorus in natural waters. Limnol. Oceanogr. 25: 754–758.

Suttle, C. A., 1994. The significance of viruses to mortality in aquatic microbial communities. Microb. Ecol. 28: 237–243.

Tamminen, T., 1989. Dissolved organic phosphorus regeneration by bacterioplankton: 5′-nucleotidase activity and subsequent phosphate uptake in a mesocosm enrichment experiment. Mar. Ecol. Prog. Ser. 58: 89–100.

Tanoue, E., M. Ishii & T. Midorikawa, 1996. Discrete dissolved and particulate proteins in oceanic waters. Limnol. Oceanogr. 41: 1334–1343.

Tanoue, E., S. Nishiyama, M. Kamo & A. Tsugita, 1995. Bacterial membranes: possible source of a major dissolved protein in seawater. Geochimica Cosmochimica Acta. 59: 2643–2648.

Thingstad, T. F. & R. Lignell, 1997. Theoretical models for the control of bacterial growth rate, abundance, diversity and carbon demand. Aquat. Microb. Ecol. 13: 19–27.

Thingstad, T. F. & F. Rassoulzadegan, 1995. Nutrient limitations, microbial food webs, and 'biological C-pumps': suggested interactions in a P-limited Mediterranean. Mar. Ecol. Prog. Ser. 117: 299–306.

Thingstad, T. F., E. F. Skjodal & R. A. Bohne, 1993. Phosphorus cycling and algal-bacterial competition in Sandsfjord, western Norway. Mar. Ecol. Prog. Ser. 99: 239–259.

Toledo, G. & B. Palenik, 1997. *Synechococcus* diversity in the California current as seen by RNA polymerase (*rpoC1*) gene sequences of isolated strains. Appl. envir. Microbiol. 63: 4298–4303.

Torriani-Gorini, A., E. Yagil & S. Silver (eds), 1994. Phosphate in microorganisms – cellular and molecular biology, ASM Press, Washington, D.C.

Tuomi, P., K. M. Fagerbakke, G. Bratbak & M. Heldal, 1995. Nutritional enrichment of a microbial community: The effects on activity, elemental composition, community structure and virus production. FEMS Microb. Ecol. 16: 123–34.

Turk, V., A. S. Rehnstam, E. Lundberg & Å. Hagström, 1992. Release of bacterial DNA by marine nanoflagellates, an intermediate step in phosphorus regeneration. Appl. envir. Microbiol. 58: 3744–3750.

Tyrell, T. & C. S. Law, 1997. Low nitrate: phosphate ratios in the global ocean. Nature 387: 793–796.

van Bleijswijk, J. D. L. & M. J. W. Veldhuis, 1995. *In situ* growth rates of *Emiliania huxleyi* in enclosures with different phosphate loadings revealed by diel changes in DNA content. Mar. Ecol. Prog. Ser. 121: 271–277.

van Bleijswijk, J. D. L., R. S. Kempers, P. van der Wal, P. Westbroek, J. K. Egge & T. Lukk, 1994. Standing stocks of PIC, POC, PON and *Emiliania huxleyi* coccospheres and liths in sea water enclosures with different phosphate loadings. Sarsia 79: 207–317.

van Boekel, W. H. M., 1991. Ability of *Phaeocystis* sp. to grow on organic phosphates: direct measurement and prediction with the use of an inhibition constant. J. Plankton Res. 13: 959–970.

VanBogelen, R. A., E. R. Olson, B. L. Wanner & F. C. Neidhardt, 1996. Global analysis of proteins synthesised during phosphorus restriction in *Escherichia coli*. J. Bact. 178: 4344–4366.

van Cappellen, P. V. & E. D. Ingall, 1996. Redox stabilisation of the atmosphere and oceans by phosphorus-limited marine productivity. Science 271: 493–496.

Vaulot, D. & F. Partensky, 1992. Cell cycle distributions of prochlorophytes in the North western Mediterranean Sea. Deep Sea Res. 39: 727–742.

Vaulot, D., N. LeBot, D. Marie & E. Fukai, 1996. Effect of phosphorus on the *Synechococcus* cell cycle in surface Mediterranean waters during summer. Appl. envir. Microbiol. 62: 2527–2533.

Veldhuis, M. J. W., L. A. H. Venekamp & T. Ietswaart, 1987. Availability of phosphorus sources for blooms of *Phaeocystis pouchetii* (Haptophyceae) in the North Sea: impact of the river Rhine. Neth. J. Sea Res. 21:219–229.

Vrieling E. G. & D. M. Anderson, 1996. Immunofluorescence in phytoplankton research: applications and potential. J. Phycol. 32: 1–16.

Ward, D. M., R. Weller & M. M. Bateson, 1990. 16S rRNA sequences reveal numerous uncultured microorganisms in a natural community. Nature 345: 63–65.

Waterbury, J. B., F. W. Valois & D. G. Franks, 1994. The role of cyanophages in the dynamics of marine *Synechococcus* populations. In book of abstracts on VIII International Symposium on Phototrophic Prokaryotes, Urbino, Italy, Sept. 10–15, 9 pp.

Waterbury, J. B., S. W. Watson, F. W. Valois & D. G. Franks, 1986. Biological and ecological characterisation of the marine unicellular cyanobacterium *Synechococcus*. In T. Platt & W. K. W. Li (eds), Photosynthetic picoplankton, Canadian Bull. Aquat. Sci. 214: 71–120.

Watson, G. M. F., D. J. Scanlan & N. H. Mann, 1996. Characterisation of the genes encoding a phosphate-regulated two component sensory system in the marine cyanobacterium *Synechococcus* sp. WH7803. FEMS Microbiol. Letts. 142: 105–109.

Wawer, C. & G. Muyzer, 1995. Genetic diversity of *Desulfovibrio* spp. in environmental samples analysed by denaturing gradient gel electrophoresis of [NiFe] hydrogenase gene fragments. Appl. envir. Microbiol. 61: 2203–2210.

Weinbauer, M. G., D. Fuks & P. Peduzzi, 1993. Distribution of viruses and dissolved DNA along a coastal trophic gradient in the northern Adriatic sea. Appl. envir. Microbiol. 59: 4074–4082.

Weinbauer, M. G., D. Fuks, S. Puskaric & P. Peduzzi, 1995. Diel, seasonal and depth related variability of viruses and dissolved DNA in the northern Adriatic Sea. Microb. Ecol. 30: 25–41.

Westbroek, P., C. W. Brown, J. van Bleijswijk, C. Brownlee, G. J. Brummer, M. Conte, J. Egge, E. Fernández, R. Jordan, M. Knappertsbusch, J. Stefels, M. Veldhuis, P. van der Wal & J. Young, 1993. A model system approach to biological climate forcing. The example of *Emiliania huxleyi*. Global and Planetary Change 8: 27–46.

Wikner, J., F. Rassoulzadegan & Å. Hagström, 1990. Periodic bacterivore activity counterbalances bacterial growth in the marine environment. Limnol. Oceanogr. 35: 313–324.

Wikner, J., J. J. Vallino, G. F. Steward, D. C. Smith & F. Azam, 1993. Nucleic acids from the host bacterium as a major source of nucleotides for three marine bacteriophages. FEMS Microb. Ecol. 12: 237–248.

Wilson, W. H. & N. H. Mann, 1997. Lysogenic and lytic production in marine microbial communities. Aquat. Microb. Ecol. 13: 95–100.

Wilson, W. H., N. G. Carr & N. H. Mann, 1996. The effect of phosphate status on the kinetics of cyanophage infection in the oceanic cyanobacterium *Synechococcus* sp. WH7803. J. Phycol. 32: 506–516.

Wilson W. H., S. Turner & N. H. Mann, 1998. Population dynamics of phytoplankton and viruses in a phosphate-limited mesocosm and their effect on DMSP and DMS production. Estuar. coast. Shelf Sci. 46 (Supplement A): 49–59.

Wood, A. M., 1988. Molecular biology, single cell analysis, and quantitative genetics: new evolutionary genetic approaches in phytoplankton ecology. In Yentsch C. M., F. C. Mague & P. K. Horan (eds), Immunochemical Approaches to Coastal, Estuarine, and Oceanographic Questions. Springer-Verlag: 41–73.

Yentsch, C. S., 1980. Phytoplankton growth in the sea – a coalescence of disciplines. In P. G. Falkowski (ed.), Primary Productivity in the Sea. Plenum Press, New York: 17–32.

Hydrobiologia **401:** 177–198, 1999.
J.P. Zehr & M.A. Voytek (eds), Molecular Ecology of Aquatic Communities.

Immunological and molecular probes to detect phytoplankton responses to environmental stress in nature

J. La Roche[1], R. M. L. McKay[2] & P. Boyd[3]
[1]*Department of Applied Science, Brookhaven National Laboratory, Upton NY 11973, U.S.A.*
Current address for J. La Roche: Institut für Meereskunde an der Universität Kiel, Düstenbrooker Weg 20, 24105 Kiel, Germany
[2]*Department of Biological Sciences, Bowling Green State University, Bowling Green, OH 43403, U.S.A.*
[3]*NIWA Centre for Chemical and Physical Oceanography, Department of Chemistry, University of Otago, Dunedin, New Zealand*

Key words: phytoplankton physiological status, probes for phytoplankton, nutrient deficiency, iron limitation in phytoplankton, nitrogen limitation in phytoplankton, flavodoxin, plastocyanin, trace metal toxicity, MAA, superoxide dismutase

Abstract

Recent advances in using immunological and nucleic acid probes to detect the effects of environmental stress on phytoplankton growth rate and yield are reviewed here. The rationale for this approach is discussed in the context of the general response observed from microorganisms grown under stress imposed by various environmental factors. Retrenchment, or the observed down-regulation of metabolic processes under nutrient deficiency, is categorized as a general response that is of limited use in designing probes to stress induced by a specific nutrient. In contrast, compensation and the increase capacity for nutrient acquisition are specific responses that appear more promising for the development of such probes. Methods and approaches used to design immunological and nucleic acid probes for stress imposed by nutrient deficiency are reviewed. Specific examples for iron and nitrogen limitation are presented to demonstrate the potential use of nutrient stress indicators in natural populations of phytoplankton. Finally, the limitations of this approach and the importance of understanding the regulation of the genes and proteins used to prepare the probes are emphasized.

Introduction

Phytoplankton growth is dependent on a balance between nutrients and energy from the environment. Under balanced growth, the major macronutrients are generally assimilated by phytoplankton in a stochiometric ratio of 106 C:16 N:1 P (Redfield, 1958), reflecting cellular composition. Micronutrient requirements, in contrast, are more variable between species and taxonomic groups and vary further as a function of other environmental factors such as light (Raven, 1990). Ironically, both deficiency and excess of an essential nutrient can lead to cellular stress. In photoautotrophs, the depletion of an essential nutrient will shift the dynamic balance between light and nutrient supply such that the sinks for products of photochemistry at a given light intensity are reduced, ultimately affecting growth rate. Alternatively, stress caused by the excess of an essential or non-essential nutrient can also reduce growth via cytotoxic effects. The response of phytoplankton to toxic levels of a substance is not the major goal of this review and will be discussed briefly in the last section in terms of trace metal toxicity and potential probes that could be developed to study this phenomenon. Rather, this review will focus primarily on approaches to study environmental stress induced by deficiency of essential nutrients.

The concept of limitation was initially introduced by Liebig (von Liebig, 1840) and was originally related to crop yield. The concept that the factor limiting crop yield is the nutrient which is least available for biosynthesis subsequently became a central tenet of ecology (Odum, 1971). Liebig's "Law of the Minimum", however, was not directly related to growth rate and it was later proposed that a nutrient can also limit the rate of growth of photoautotrophs (Blackman, 1905). Both concepts are now used in ecology with the realization that Liebig's "Law of the Minimum" which addresses yield and Blackman's concept of growth rate limitation are more relevant to biogeochemical cycles and selection of species in ecosystems, respectively (Falkowski & Raven, 1997). However, yield and growth rate limitation are not mutually exclusive and are likely to be related to the same nutrient in a given environment (Cullen, 1991). In several of the examples presented here we refer to 'nutrient deficiency' as a general term encompassing both yield and growth rate limitation, as well as cases where no clear distinction has been made between these two possibilities. Nutrient deficiency nevertheless implies either morphological or physiological changes that reduce the overall performance of the phytoplankton as primary producers.

Molecular biology and immunological techniques have vastly improved our ability to examine differences in community structure of microorganisms in nature. The development of probes for specific algae and other microorganisms has intrinsic scientific merit in answering questions related to species diversity, distribution and population succession. Moreover, recently developed sensitive tests for the direct identification of toxic phytoplankton will be of benefit in the area of public health (Adachi et al., 1996; Liu et al., 1996, Miller & Scholin, 1996; Simon et al., 1997).

Here, we review recent advances in the design and characterization of functional probes to assess the physiological status of microorganisms, which in conjunction with probes for species-specific resolution, should provide insight to elucidate factors that constrain productivity in diverse environments, and also affect community structure and species succession. Furthermore, functional probes may have the added advantage to help elucidate factors that control the dominance of a particular toxic microorganism in a given environment.

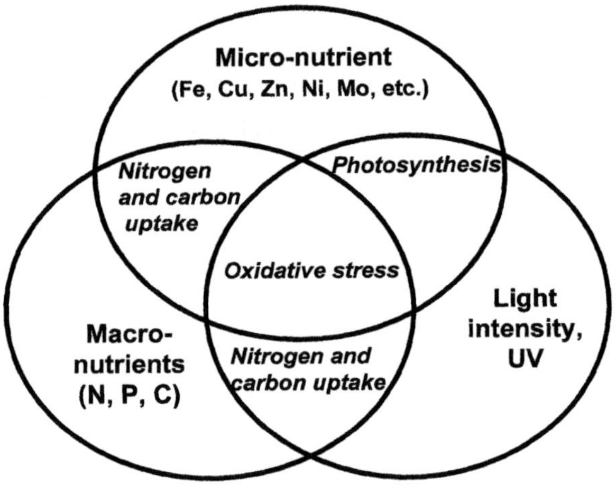

Figure 1. Venn diagram showing the interaction of macronutrients, micronutrients and light energy in a photosynthetic cell. Bulk biological elements, listed here as macronutrients, include H, O, S, Cl, Na, K, Mg, Ca, in addition to N, P and C. Various trace elements believed to be essential for life, listed here as micronutrients, include F, Si, V, Cr, Mn, Co, Ni, Se, I in addition to those listed in the figure (da Silva & Williams, 1991). General biological processes that are affected by imbalances between macro- or micro-nutrient and light availability are in italics at the intersection of any two rings. Oxidative stress (at the intersection of all three rings may play a central role during nutrient limitation and/or excess light energy as a stress response induced by several environmental factors.

Synergistic effects between stress factors

One of the major problems faced by phytoplankton ecologists is to identify the factor that is most likely to limit growth in a specific environment at any given time. The dynamic environment in which phytoplankton live complicates this task. Although there are no reports that two macro-nutrients (e.g. P and N) or two micronutrients (e.g. Zn and Fe) can be co-limiting, several reports suggest co-limitation of a micro- and a macro-nutrient via their synergistic interactions. For example, it is postulated that cells can be simultaneously limited by Zn and inorganic C since Zn serves as a prosthetic group of carbonic anhydrase, an enzyme central to the C acquisition efforts of many phytoplankton (Price & Morel, 1990; Morel et al. 1994). Similarly, nitrogen limitation may accompany deficiencies by both Ni and Fe, the former as a result of its role in the enzyme urease (Price & Morel, 1991) and the latter as a result of its association with key enzymes of nitrogen fixation and both nitrate and nitrite reduction (Raven, 1988; Timmermans et al., 1994; Maldonado & Price, 1996; Muggli & Harrison, 1996; Muggli et al., 1996; Falkowski, 1997). A similar scenario can be found for a synergistic effect between essential nutrients and light (e.g. iron or manganese and light via proteins of PS I and PS II, (Raven, 1990; Geider & La Roche, 1994)). Theoretical considerations suggest that low light intensity can exacerbate the effects of Fe limitation (Raven, 1990) and there exists limited field data that support this claim (reviewed by Geider & La Roche, 1994).

In addition, low light will result in reduced rates of photoreduction for Fe and Mn which further exacerbates the deficiency (Sunda, 1994). Examples of possible interactions between light, macro-nutrients and micro-nutrients are shown in Figure 1. Excess light energy relative to macro- or micro-nutrient, as well as excess of certain trace metals can lead to oxidative stress and this is likely to result also in general stress responses that are not necessarily specific for the nutrient limiting phytoplankton productivity in a given environment. The design and use of molecular probes will therefore require an understanding of universal stress responses as well as the specific response to each nutrient.

Defining the responses to nutrient deficiency

Phytoplankton exhibit three major responses to stress imposed by nutrient limitation:

1. retrenchment or down regulation of physiological rates,
2. compensation, which generally results in the synthesis of new proteins, and
3. acquisition, which includes the development of more efficient uptake systems for the limiting nutrient.

These have been described briefly in the context of iron limitation in cyanobacteria (Straus, 1994) but can be extended here to general responses of nutrient limitation and reviewed in the context of each response's usefulness in designing probes to detect environmental stress.

Retrenchment or down-regulation of metabolic processes is a progressive response that is expected to increase in severity as nutrient deficiency increase. As a result, down-regulation of physiological rates can be observed over a range of metabolic functions and can occur over both short (picoseconds) to longer term (days) time scales. A change in the balance between nutrient and light will be rapidly reflected by changes in the biophysical properties of the cell on a very short time scale (nanoseconds to hours). Persistence of the nutrient deficiency will result in further down regulation of physiological processes that will be reflected in changes in growth rates, chemical composition, gene expression and cell morphology. Ultimately, severe nutrient deficiency will lead to a complete shut-down of physiological processes that will result either in cell death (Falkowski & Berges, 1998) or the formation of a stress-resistant stage (e.g. spore, cyst or dormant stage; Hershkovits et al., 1997). However, within the tolerance range for an given organism, retrenchment is a reversible process and a return to normal physiological conditions is expected once the limiting nutrient is replenished.

The general response of retrenchment in microalgae is critical for maintaining the balance between light energy absorption, the photochemical production of NADPH and ATP, and other cellular processes. Thus, the response of phytoplankton to nutrient limitation will range from decreases in the ratio of variable to maximum fluorescence (Fv/Fm) to changes in biochemical composition and in morphology. Low Fv/Fm are characteristic of nutrient limitation in general and this parameter has been used to assess the severity of both nitrogen and iron limitation (Kolber et al., 1988; Greene et al., 1992). Although measurements of Fv/Fm alone cannot be used to identify the limiting nutrient, this parameter has one of the fastest response time to *in vitro* and *in situ* nutrient addition experiments (Greene et al., 1992; Geider et al., 1993; La Roche et al., 1993; Vassiliev et al., 1995; Behrenfeld et al., 1996). Intermediate retrenchment responses often involve the degradation of photosynthetic proteins (Geider et al., 1993; Collier & Grossman, 1994). At the other end of the spectrum of responses, structural changes induced by nutrient limitation are shown for a diatom and a chlorophyte species grown under iron, phosphorus and nitrogen stress (Figure 2). Particularly noticeable, are the atrophied thylakoid membranes of the chloroplasts in Fe and nitrogen-limited cells, and the accumulation of storage compounds such as starch and lipids in nitrogen and phosphorus-limited cells. In cyanobacteria, a response regulator that controls the general acclimation response to many environmental factors is responsible for the degradation of the light harvesting complex under nutrient limitation (Schwarz & Grossman, 1998).

Compensation includes all cellular responses that alleviate the effects of nutrient stress and thereby retard or reduce the extent of down-regulation of physiological processes. Although compensation generally results from the synthesis of new proteins, it can sometimes be accomplished by alternate means, such as pigment synthesis and modification (Olaizola et al., 1994; Demmig-Adams, 1990). For example, the carotenes antheraxanthin, zeaxanthin and lutein have been identified as the specific xanthophylls necessary for the protection of the photosynthetic apparatus from photo-oxidative damage in *Chlamydomonas* cells exposed to excessive light (Niyogi et al., 1997).

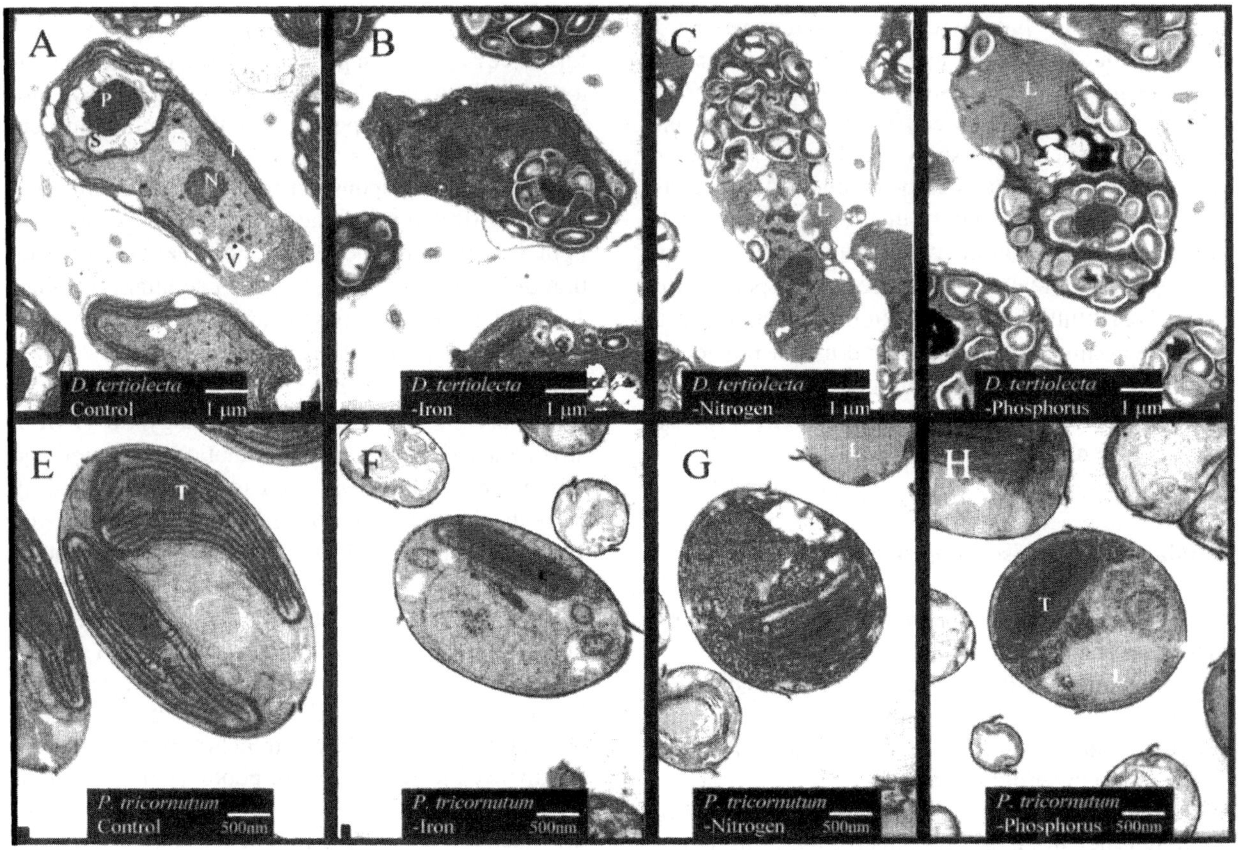

Figure 2. Transmission electron micrographs of *Dunaliella tertiolecta* (A–D) and *Phaeodactylum tricornutum* (E–H) showing structural differences between nutrient replete cells (A, E), iron-limited (B, F), nitrogen-limited (C, G), and phosphorus-limited (D, H) algae. P, perynoid; N, nucleus; V, vacuole; S, starch; T, thylakoids; L, lipids. Of interests are the predominance of starch granules and, to a lesser extent, lipids in the N- and P-limited *D. tertiolecta*, and the degeneration of thylakoid membranes in N-limited *D. tertiolecta*. In contrast, *P. tricornutum* accumulated mainly lipids when N- or P-limited, and showed atrophied chloroplasts in Fe-, N- and P-limited cells. The predominance of various cellular components such as starch and lipids were confirmed with independent biochemical measurements (Richard Geider, personal communication). Data on impairment of photosynthetic function and chemical compositions have been presented elsewhere (Geider et al., 1993; La Roche et al., 1993).

Chlorophyll-binding proteins induced by high light in cyanobacteria (Dolganov et al., 1995) and higher plants (Adamska et al., 1993) are believed to play a role in the acclimation to phototoxic conditions. Superoxide dismutases are metallo-enzymes also involved in the prevention against oxidative damage to living cells. Although these enzymes are generally expressed at a basal level for the detoxification of superoxide radicals produced through normal cell metabolism, they are generally up-regulated in photoautotrophs in response to excess light or nutrient limitation (Campbell & Laudenbach, 1995), both of which can lead to increased incidence of oxidative stress. Protein substitution is another successful mechanism to achieve compensation for stress imposed by nutrient deficiency. Several examples of protein substitution have been identified for stress induced by deficiency of trace elements but are less common for macronutrients. Under iron deficiency, flavodoxin which contains flavin mononucleotide (FMN) can successfully replace the iron sulfur protein ferredoxin in several metabolic processes, the most notable being photosynthetic electron transport (Straus, 1994; Geider & La Roche, 1994; Razquin et al., 1995). In addition, flavodoxin may substitute for ferredoxin in some steps of N assimilation (Vigara et al., 1995; Vigara et al., in press), but notably not nitrogen fixation (Razquin et al., 1995; Masepohl et al., 1997). Similarly, under copper deficiency, replacement of the copper containing plastocyanin with cytochrome c6 is reported for various cyanobacteria (Sandmann, 1986; Navarro et al., 1997), a prochlorophyte (Arudchandran

& Bullerjahn, 1996), and some green algae (Merchant et al., 1991; Hill & Merchant, 1992).

Substitutions of one metal by another is an additional mean by which an organism can compensate for stress induced by trace metal deficiency. For example, Cd and Co can substitute for Zn in many proteins (da Silva & Williams, 1991).The substitution of Zn by Cd and Co in certain proteins may explain the surface depletion of these trace metals in the ocean (Price & Morel, 1990). It has been recently established that Co and Cd can substitute Zn in different isoform of carbonic anhydrase (CA), an enzyme involved in C acquisition (Lee et al., 1995; Yee & Morel, 1996).

Mechanisms that enhance the acquisition of a limiting nutrient will also help reduce the extent to which down regulation of cellular physiological processes occurs. There are several ways in which increased uptake capacity can be achieved. Under limitation by a given element, an organism may de-repress a high affinity uptake system for the preferred form of the element (Scanlan et al., 1997). Alternatively, the organisms may express several enzymes and uptake systems designed to take advantage of other chemical forms of the limiting nutrient that are more energetically costly to assimilate. The latter strategy can be achieved through use of membrane-bound cell-surface enzymes or extracellular enzymes, which are often located in the periplasm (Palenik & Morel, 1990b; Martinez & Azam, 1993; Quisel et al., 1996). Extracellular enzymes are generally used to convert an alternate nutrient source into its preferred chemical form which is then assimilated by the cell (Martinez & Azam 1993, Quisel et al. 1996). Nutrients such as phosphorus and nitrogen are found in various forms, including inorganic and organic compounds. The inorganic forms of these elements are generally the preferred sources of nutrients for most photosynthetic organisms. In prokaryotic organisms, a common response to phosphorus limitation includes the de-repression of high affinity uptake systems for inorganic phosphorus (P_i) (Scanlan et al., 1993; Mann, 1995; Scanlan et al., 1997). The de-repression of this uptake system occurs when the external Pi falls below 50 nM. This type of induction pattern is the most useful in designing general markers for nutrient limitation. However, proteins that are expressed preferentially during growth with specific alternate forms of a limiting nutrient are likely to be useful in resolving issues related to community structure. For example, different forms of nitrogen are apparently preferred by different taxonomic groups, and utilization of alternate sources of nitrogen may shape the outcome of species competition, succession and community structure (Berg et al., 1997; La Roche et al., 1997; Maestrini et al., 1997).

One common way in which microorganisms address Fe deficiency is via the production of siderophores. Siderophores are low molecular mass ligands exhibiting a high affinity for ferric iron (Guerinot, 1994). Once excreted from the cell, they complex ferric iron and transport it to specific receptors located on the outer membrane. The utilization of siderophores by cyanobacteria has been studied extensively (Wilhelm & Trick, 1994; Wilhelm, 1995) and will be discussed below only in the context of uptake mechanisms by cell surface receptors that may be useful as indicators of micronutrient limitation.

Each stress response described above can be considered an acclimation response that is reversible upon alleviation of the stress. However, chronic stress, such as extreme low light condition, high temperature, radiation levels or chronic nutrient deficiency may eventually lead to a genetic adaptation to a stressful environment. Genetic adaptation can be manifested as the constitutive expression of a group of genes and deletion of others, or yet major modification of proteins (Mazel & Marliere, 1989), thereby conferring a competitive advantage in a given environment. Examples for genetic adaptation are more commonly found from organisms living in extreme environment (e.g. hot springs, Brock, 1967) but may also be present in organisms living in chronically nutrient-limited areas of the ocean. The characterization of such adaptation may lead to the identification of indicator species for specific ecosystems.

Targeting and designing probes

Several approaches have been used to design probes to assess environmental stress. Frequently, it is possible to target specific genes in phytoplankton when it is known that these are expressed under a specific environmental stress. In such cases, there may be protein or nucleic acid sequences available, enzyme assays or protein isolation methods that have been used with other organisms that will facilitate the isolation of a similar protein or DNA sequence from the desired organism. Alternatively, genes of specific function can be isolated in species for which there are well developed genetic systems available. Genetic transformation systems are well developed for use with a variety of freshwater cyanobacteria (e.g. Buikema &

Haselkorn, 1993) and several marine strains (Sode et al., 1994; Brahamsha, 1996). Among eukaryotes, genetic transformation is broadly applicable for use with the green alga *Chamydomonas* sp. in addition to several other species of chlorophytes (Davis et al., 1994; Stevens & Purton, 1997). By comparison, genetic transformation systems are poorly resolved for marine eukaryotes, although some success has been recently achieved using diatoms (Apt et al., 1993; Dunahay et al., 1995).

An alternate approach to targeting specific proteins is a 'shot gun' approach which allows the detection of uncharacterized genes that may be induced by a given environmental stress. This approach can yield probes targeted against protein or RNA depending on the method used for screening. Although the shot gun approach may require a greater investment of time on the part of the researcher, it is also likely to lead to discoveries of new genes or to designation of function to previously uncharacterized open reading frames (ORFs). The recent success at obtaining the sequences for complete genomes demonstrates that a major challenge remaining will be the assignment of function to putative ORFs (Kaneko & Tabata, 1997; Nakamura et al., 1998).

Immunological probes

Specific proteins

Immunological probes to specific proteins can be prepared from synthetic peptides, recombinant proteins or naturally-occurring proteins isolated to homogeneity from a given organism. The first two approaches require at least partial direct or derived amino acid sequence information. The third approach requires large initial quantities of algal material in order to obtain sufficient quantities (ranging from several hundred μg to a few mg) of the purified protein necessary for antibody preparation. Depending on the cellular abundance of the protein of interest, the preparation of antisera therefore often requires 10–50 g wet weight of unicellular alga. Probes for proteins that are difficult to isolate or those that are of low abundance may be more successfully prepared from synthetic peptides or recombinant proteins. Using this approach, an immunological probe directed against nitrogenase from the marine cyanophyte *Trichodesmium* has been prepared using an antigen synthesized from a pMAL expression system (Zehr & McReynolds, 1989). Because of the highly conserved nature of nitrogenase, it was possible to use degenerate nucleotides to isolate a segment of this gene from *Trichodesmium*. Similarly, this approach has been used to obtain antisera directed against prochlorophyte plastocyanin (Arudchandran & Bullerjahn, 1996). The recombinant protein approach also has the added advantage of providing reference material that can be used to standardize ELISA or western blot assays.

Shot-gun approach

Proteins induced by a specific environmental stress have been detected by comparing protein profiles separated by one- or two- dimensional SDS-PAGE. However, it is difficult to detect less abundant proteins in this manner. Radioactive labeling or partial biochemical fractionation of cellular components to yield a subset of total protein are useful means to overcome this problem (La Roche et al., 1993). As an alternative, biotinylating reagents can be used to label cell surface proteins in order to detect the extracellular domains of membrane proteins (Meier et al., 1992). This technique has been used to label cell surface proteins of the cosmopolitan eukaryotic alga *Emiliania huxleyi* grown under different conditions and revealed the presence of a nitrogen-regulated 82-kDa protein (Palenik & Koke, 1995). Because of the low abundance of cell surface proteins, these may not be detectable unless they are preferentially labeled. An outline of this technique and an example of cell surface labeling of the brown tide-causing chrysophyte *Aureococcus* are presented in Figure 3.

Once a potential marker protein has been identified, it is necessary to purify it for the preparation of antibodies. Since the identity and function of the proteins selected by shot-gun methods are rarely known, trial and error with standard biochemical techniques is generally required to further purify the protein of interest (La Roche et al., 1993; Graziano et al. 1997). Occasionally, definitive identification of the protein may be provided by N-terminal microsequencing techniques (Matsudaira, 1990) or, alternatively, if the protein is modified (blocked) at its N-terminus, by sequencing tryptically digested protein fragments (Van der Staay et al., 1995). In instances where the protein function is not elucidated, the protein can still be used as a marker for a specific stress but a more detailed study of its regulation will be required in order to identify its function (e.g. Reddy et al., 1988).

Nucleic acid probes

Specific genes

DNA probes targeted at specific genes can be prepared

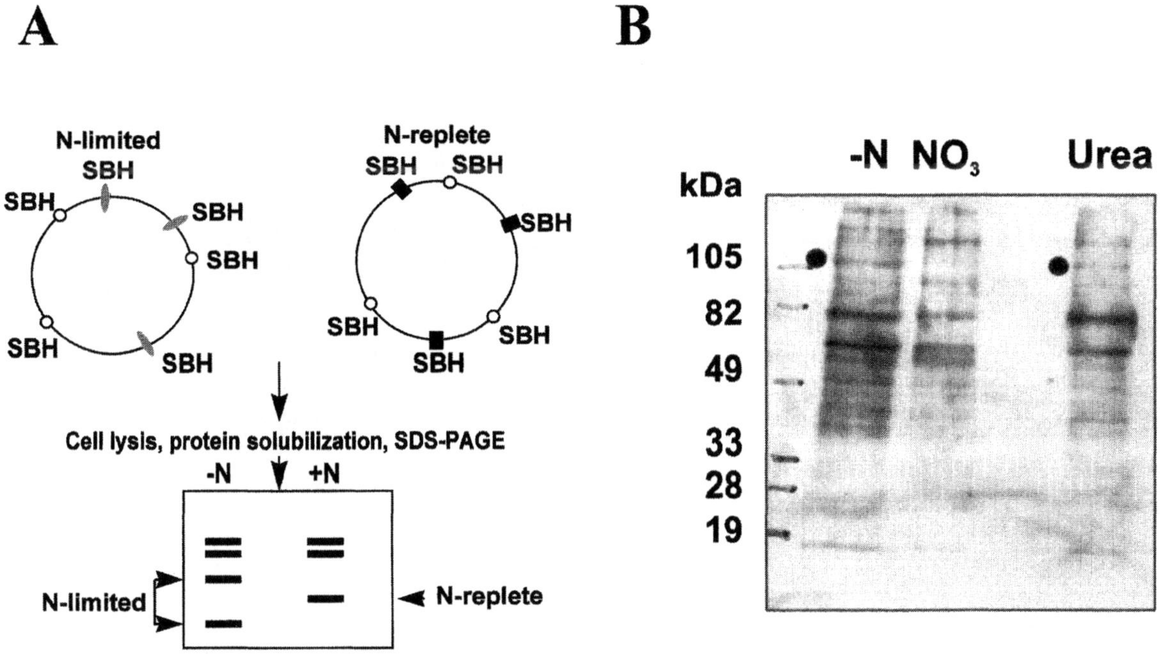

Figure 3. Detection of cell surface proteins using the biotinylating reagent succinimidyl 6-(biotinamido) hexanoate (SBH). (A) Diagram illustrating the steps followed for the detection of cell surface proteins specific to nitrogen-limited (left) or nitrogen replete (right) cells. Common proteins (○) and unique proteins (■, ⧫) are labeled on the cell surface. (B) Example of cell surface labeling of the brown tide alga, *Aureococcus anophagefferens*, to detect cell surface protein expressed under nitrogen-limitation (-N), but not nitrogen-replete conditions (NO3). Note that the protein induced under N-limitation (●) is also detected in urea grown cell (right most lane). A similar pattern of expression has been observed for nrp1, a nitrogen-regulated cell surface protein from the prymnesiophyte *Emiliania huxleyi* (Palenik & Koke, 1995).

by isolating genes of interest via PCR using degenerate primers corresponding to regions of conserved amino acid sequences within a protein (Gueneau et al., 1998). Alternatively, it may be possible also to use cloned genes from other species as heterologous probes on southern blots of genomic DNA from the species of interest. It is important to keep in mind that because of the degeneracy in the genetic code, the use of heterologous probes between distant taxa is not always successful.

Shot-gun approach

A promising technique for studying the expression of specific genes under a given environmental stress is the use of RNA differential display (Liang & Pardee, 1992; Liang & Pardee, 1997). This technique has been used successfully to identify infectious agents in plants and animals (Benito et al., 1996; Truesdell & Dickman, 1997) and in developmental biology (Li & Gray, 1997). These techniques have also been applied to the identification of salt stress induced proteins in tomato (Wei & MA, 1996) and compression stress induced in loblolly pine (Li et al., 1997). The approach is based on the same principle as DNA fingerprinting using random primers (RAPD; Williams et al., 1990; Barker et al., 1994) however, it uses RNA as starting material instead of DNA. The RNA is reverse transcribed and then amplified using a single random primer, usually 10 mers, or a pair of primers. Although originally designed for the detection of eukaryotic genes, the method has been successfully adapted to prokaryotic organisms (Fislage et al., 1997). The number of primers necessary to sample the population of transcripts expressed at any one time for a given genome size has been calculated on theoretical grounds and will be dependent on the design and complexity of the primers used for the amplification (Bauer et al., 1993). Figure 4 outlines the technique and provides an example using *Prochlorococcus*, a chl *a*/*b* containing prokaryote, grown under bifactorial conditions of high and low light and iron. Drawbacks of this technique, like the shotgun approach for protein, include the possibility that the gene has no similarity with other sequences in the available sequence data bases. In addition, this approach requires the use of clonal axenic cultures to prevent the amplification of DNA belonging to contaminating organisms. False positives can be a problem with this approach and it is important to verify the correct expression pattern of the transcript on northern blots. One way to reduce

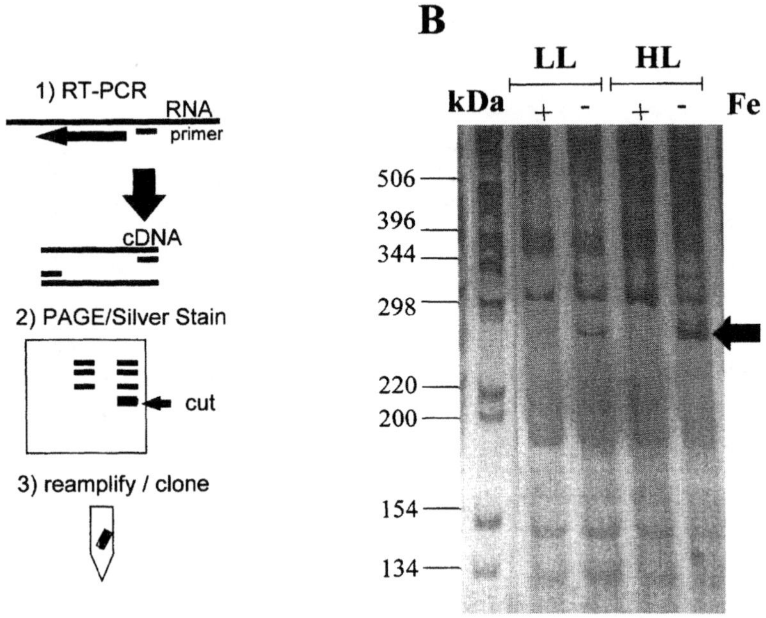

Figure 4. Detection of genes expressed under a specific environmental stress. (A) Diagram illustrating the steps followed to synthesize, isolate and clone stress-induced specific cDNAs, using a non-radioactive method. Step 1 consist of reverse transcription using an arbitrary primer (10 mer) to synthesize first strand cDNA, followed by amplification with the same primer. Double stranded cDNAs are then separated by SDS-PAGE and gels are silver stained (Step 2). Stained gels are dried on filter paper and can be stored in this way. Bands of interest can be cut out from the gels, re-hydrated in a microcentrifuge tube and amplified for cloning and sequencing. (B) Example of RNA differential display (DDRT-PCR) using RNA from the prokaryote *Prochlorococcus marinus* grown under a factorial combination of high (HL) and low light (LL), and high (+) and low (−) iron. cDNAs were separated by SDS-PAGE and the gel silver stained. The arrow points to a cDNA which is specifically expressed under iron-limited conditions and is present under high and low light.

the possibility of false positives would be to establish the expression of a specific transcript over an increasing gradient of the environmental stress. For example, observing an increasing abundance of a Fe-induced transcript with decreasing iron concentration is more convincing then simply looking at the transcript level under iron-replete and iron-limited conditions.

The advent of whole genome sequencing also opens up many possibilities with regards to differential gene expression. In fact, once a whole genome is sequenced, much of the remaining work will be related to understanding the regulation and function of the identified genes. New techniques are currently being developed to tackle these questions and several may be directly applicable to the study of stress induced by nutrient-limitation. For example, high-density DNA microarrays of yeast ORFs can now be used for the study of differential gene expression (Lashkari et al., 1997). Arrays containing up to 2479 yeast ORFs have been printed on a single slide. Fluorescently labeled samples from cells grown under various conditions can then be hybridized to the array and gene expression of specific ORFs detected and quantitated using laser confocal scanning microscopy. There are currently no complete genomes that have been sequenced from a marine phototautotroph although chloroplastic genomes have been resolved from a marine diatom (Kowallik et al., 1995) and a red algae (Reith & Munholland, 1995). Additionally, the complete genome sequence of *Synechococcus* PCC6803 may be of some limited use for studies of marine cyanobacteria (Kaneko & Tabata, 1997). However, given the speed at which whole genomes are now being sequenced it is not inconceivable that these new techniques will be available to marine scientists as well in the near future.

Specificity to environmental conditions and taxa

Once a protein has been identified as a potential marker of a specific environmental stress, it is important to confirm that other conditions do not also induce its expression. As indicated in Figure 1, stress induced by nutrient limitation may increase the cells' susceptibility to oxidative stress. Immunological analyses of photosynthetic proteins as well as biophysical assays (Geider et al., 1993) suggest that the additional oxidative stress encountered by nutrient-deficient algae is in part due to a reduction in the potential for energy dissipation relative to excitation energy delivered from the light-harvesting apparatus. Oxidative stress can in turn lead to the expression of a suite of proteins which are used to protect the cell from the deleterious effects of free oxygen radicals. Similarly, many heat shock proteins have complex expression patterns such that these proteins are also expressed during wounding in plants or at various stages of development and play a role in protein assembly or degradation (Sherman & Goldberg, 1996). It is especially important to verify the expression of a protein when it has been obtained by a shotgun approach and has an unknown function. Certain environmental stress such as heat shock, salt and pH stress are unlikely to be relevant under normal conditions in open oceanic basins where salinity and pH have narrow ranges and temperature remains within physiological limits for protein and enzyme activity. However, these may become important in shallow coastal embayments with restricted flow (temperature and salinity), during phytoplankton bloom conditions where O_2, CO_2 and pH may change significantly, or in coastal bodies of water influenced by heavy rain or ice melt that can result in substantial changes in salinity. Commonly, probes for a given nutrient deficiency, for example iron, nitrogen or phosphorus, are not expressed as a function of stress resulting from other nutrient deprivation (Palenik & Koke, 1995; La Roche et al., 1995; Dyhrman & Palenik, 1997; Graziano et al., 1997).

Probes to detect environmental stress, whether they are immunological or nucleic acid probes, may vary in their specificity within various taxonomic groups, depending on the level to which they are conserved. The taxonomic specificity of each probe must be verified against a large group of species belonging to several taxonomic groups before its use can be extended to field studies. A probe raised against flavodoxin from the diatom *Phaeodactylum* shows strong cross-reactivity with all diatom species tested to date, whereas it does not cross-react with other major algal taxonomic groups (La Roche et al., 1995). In contrast, mixed results have been obtained for antibody probes raised against ferredoxin isolated from *Thalassiosira weisflogii* with one laboratory reporting a probe of very narrow specificity (Erdner 1997) and another probe generated with the same species having much wider applicability to diatoms (McKay et al., 1999). Specificity of an immunological probe can be at the level of the species, genus or taxa, as well as being nearly universal (Orellana & Perry, 1992; Lin et al., 1994). The problem of probe specificity is important for field work and will have to be continuously revised as more species are tested. One way to get a handle on specificity in field samples is to use a single-cell immunofluorescence assay, which may allow partial species identification as well as a positive test for specificity. This was employed in field studies to verify the diatom specificity of an anti- flavodoxin immunological probe (La Roche et al., 1996a).

Case studies

Iron deficiency

Although iron is the second most abundant metal in the earth crust, the equilibrium concentration of free ferric ion (FeIII) in an oxygenic, aqueous environment, is approximately 10^{-17} M. The specific chemistry of Fe in the current oxygenic atmosphere, therefore, makes this element among the least biologically available (da Silva & Williams, 1991). Yet, the chemical properties of iron together with the broad redox potential observed in Fe-containing enzymes make Fe an invaluable component of diverse, but fundamental metabolic processes such as nitrogen fixation and assimilation, in addition to the important role it plays in the electron transport reactions of photosynthesis and respiration. It is clear from the abundance of iron in virtually all organisms that this element has played a crucial role in the evolution of life on earth. The distribution of iron and other trace metals in extant organisms indicate that iron must have been selected for its specific chemical properties early in the evolution of life, presumably more than 10^9 years ago, before the solubility of Fe was affected by significant quantities of oxygen in the atmosphere.

Iron limitation of cell growth is observed throughout biological systems from pathogenic bacteria where virulence has been tied to the ability to scavenge iron, to iron deficiency in humans, a common nutritional disorder (Weinberg, 1974; Bullen & Griffith, 1987; Klausner et al., 1993). By comparison, the pioneering work of Martin and colleagues (Martin & Fitzwater, 1988; Martin et al., 1990; Martin et al., 1991) indicated that iron could be a limiting factor constraining phytoplankton productivity and standing crop in vast regions of the oceans, and in particular, those regions characterized as 'high nutrient and low chlorophyll' (HNLC). Notably, results from a recent mesoscale iron

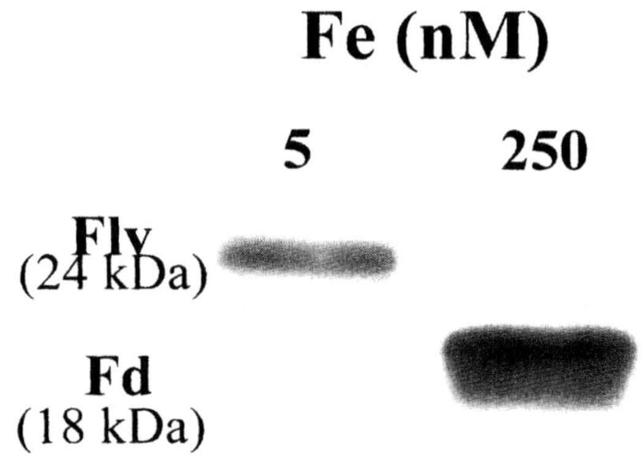

Figure 5. Western blot of flavodoxin (Flv, left lane) and ferredoxin (Fd, right lane) from total proteins extracted from *Phaeodactylum tricornutum* grown with 5 nM and 250 nM iron. Total proteins were separated by SDS-PAGE and the gels transferred to nitrocellulose. Western blots were sequentially challenged with anti-flavodoxin and anti-ferredoxin antisera raised against the same species. The immunological reactions were detected by chemiluminescence using horseradish peroxidase conjugate with the ECL reagent (Amersham).

fertilization effort in the HNLC equatorial Pacific have been compelling in their support for the 'iron hypothesis' (Behrenfeld et al., 1996; Coale et al., 1996a). Iron may also be a limiting factor in some coastal regions, particularly in those regions lacking an extensive continental shelf (Hutchins & Bruland, 1998). In addition, iron may play a role in the onset of certain red tide blooms. Laboratory and field measurements indicate that many red tide dinoflagellate species possess unusually high iron requirements and as a result, will bloom only when sufficient Fe exist with which to satisfy this demand (Doucette & Harrison, 1990; Wells et al., 1991).

As for the general response to nutrient-limitation, the response of most microorganisms, including photoautotrophs, to iron limitation can be divided into three categories: retrenchment, compensation and acquisition (Straus, 1994). Of these three categories, the last two are potentially useful in designing probes to detect iron deficiency in oceanic phytoplankton. A list of proteins induced under iron limitation is presented in Table 1 and as noted, most genes help alleviate iron stress either via improving iron acquisition or by compensation. Most of the work on iron limitation has been done in heterotrophic bacteria (including several pathogens), fresh-water cyanobacteria and yeast, but it is likely that systems with similar structure and function exist also in marine cyanobacteria and in eukaryotic unicellular algae.

Fe compensation responses

Flavodoxin Microorganisms confronted with Fe deficiency commonly respond by replacing ferredoxin, a non-heme iron-sulfur protein with flavodoxin, a protein that contains flavin-mononucleotide as a prosthetic group (Geider & La Roche, 1994; Table 5). Flavodoxin can replace ferredoxin in many enzymatic reactions including the downstream side of Photosystem I (Smillie, 1965), and as an electron donor to both nitrite reductase (Zumft, 1972) and glutamate synthase (Vigara et al., 1995). Although the major attributes of flavodoxin and ferredoxin as they pertain to phytoplankton have been discussed elsewhere (Geider & La Roche, 1994; Straus, 1994), it is important to note that there are added biosynthetic cost associated with the production of flavodoxin, which is twice as large as ferredoxin (15–24 kDa *versus* 6–15 kDa). This in itself, and the slightly higher efficiency of ferredoxin as an electron donor in enzymatic reactions (Vigara et al., 1995), can explain why ferredoxin has been retained despite selective pressure created as a result of low iron availability in most aqueous environments. Because of its mode of regulation by Fe, flavodoxin appears to be a good marker for iron limitation in a number of organisms. In the marine environment, the earliest attempt to assess iron limitation using flavodoxin was reported by Entsch and colleagues (Entsch et al., 1983). Their work suggested that the presence of flavodoxin in the cyanobacterium *Phormidium* sp. indicated iron limitation in an Australian coral reef ecosystem. More recent analysis has demonstrated that flavodoxin in marine diatoms, in particular, is strongly induced by iron deficiency, whereas it is usually absent from iron-replete cultures (La Roche et al., 1995; Doucette et al., 1996; McKay et al., 1997). Although there are reports of constitutive flavodoxin expression in cyanobacteria (Sandman et al., 1990; Crespi et al., 1972) and heterotrophic bacteria (Mayhew & Ludwig, 1975), it is not clear at present whether there exists any eukaryotic algal species that constitutively express flavodoxin. All species of diatoms examined to date by our laboratory have showed a similar pattern of repression of flavodoxin by high iron and de-repression in low iron culture medium (La Roche et al., 1993). An early report of a constitutive flavodoxin in *Thalassiosira* sp. (Doucette et al., 1996) was later demonstrated to be completely repressed by very high iron concentrations (60 μM; Erdner, 1997). We have recently prepared polyclonal antisera directed against diatom ferredoxin. Consistent with prior observations of flavodoxin accumulation

Table 1. List of selected Fe-repressible genes in microorganisms also present in cyanobacteria

Gene	Protein	Size	Organism	Function	Location	Homology
mapA[1]	MapA	34 kDa	*Synechococcus* PCC7942	Fe acquisition	cell envelope, thylakoids	*fbp, sfuA*
irpA[2]	IrpA	36 kDa	*Synechococcus* PCC7942	Fe acquisition or storage		
fbp[3]	Fbp	40 kDa	*N. menengitidis, Synechocystis* PCC6803	iron acquisition		*mapA*
frpC[4]	FrpC	198 kDa	*N. menengitidis, Synechocystis* PCC6803	possibly a cytotoxin	outer membrane	rtx family
feoB[5]	FeoB		*E. coli, Synechocystis* PCC6803	Fe(II) transport system	transmembrane protein	
isiA[6]	CP43'	36 kDa	freshwater cyanobacteria	Chla binding protein	thylakoid	*psbC, pcb*[10]
IsiB[6]	Flv[9]	24 kDa	freshwater cyanobacteria most eukaryotic algae	redox protein	chloroplast	rubridoxin
idiA[7]	IdiA	34 kDa	*Synechococcus* PCC6301 and PCC 7942	PS II O_2 evolution	thylakoids	*sfuA, fbp, mapA*
sfuA[8]	SfuA		*Serratia marcescens, Synechococcus* PCC6301	iron transport	periplasm	

Sources of information for this table are as follows: [1]Webb et al., 1994, [2]Reddy et al., 1988, [3]Zhou & Spratt, 1992, [4]Thompson et al., 1993, [5]Kammler et al., 1993, [6]Ferreira & Straus, 1994, [7]Michel et al., 1996, [8]Angerer et al., 1990, [9]Flv is flavodoxin, [10]La Roche et al., 1996b.

(McKay et al., 1997), there is an inverse relationship between flavodoxin and ferredoxin abundance in batch cultures of the diatom *Phaeodactylum tricornutum*, with only ferredoxin detected at 250 nM Fe, and only flavodoxin detectable at 5 nM Fe (Figure 5). Field data demonstrated that diatom assemblages from estuarine and coastal regions did not contain flavodoxin *in situ*, yet the populations were easily induced to accumulate flavodoxin by increasing the ratio of macronutrients relative to dissolved iron in nutrient enrichment bioassays (La Roche et al., 1995, 1996a; Boyd et al., 1996). However, there may be some exception to this tight regulation of flavodoxin and ferredoxin by iron, especially among some large species of diatoms for which iron acquisition may be ultimately limited by diffusion (Sunda & Huntsman, 1997). In support of this hypothesis, preliminary data for *Rhizosolenia formosa*, an isolate from the Sargasso Sea, suggests that flavodoxin may be constitutively expressed in this large diatom (McKay & Villareal, unpublished). Further studies are required to confirm this hypothesis but a similar situation may exist for large species of dinoflagellates that have unusually high iron requirements (Doucette & Harrison, 1990).

The immunological probe raised against flavodoxin purified from *P. tricornutum* (La Roche et al., 1995) has recently been used to assess the iron status of diatoms in the Northeastern subarctic Pacific (La Roche et al., 1996a). Along an east to west transect, flavodoxin was mainly undetectable in coastal areas and increased westward, where it reached maximal levels at the former weather station Papa, the archetypal HNLC locale (Martin & Fitzwater, 1988). The patterns of flavodoxin distribution across the oceanic basin corroborate previous work supporting that iron is the major factor limiting phytoplankton growth in the Northeastern subarctic Pacific (Martin et al., 1989; Coale, 1991; Boyd et al. 1996). The amount of flavodoxin markedly increased at stations where the Fe concentration was below 1 nM, the concentration at which growth becomes saturated by Fe in open ocean regions (Coale et al., 1996a). Additional data along the same transect has demonstrated a consistent pattern of increasing flavodoxin abundance as dissolved iron is depleted from the surface waters, as well as an increase in flavodoxin abundance with depth sampled in the euphotic zone at station Papa (McKay, Boyd & La Roche, unpublished data).

With the exception of two early reports on cyanobacteria (Entsch et al., 1983; Jones, 1988), the use of flavodoxin as a marker for iron limitation in the field has been limited to diatoms in recent years (La Roche et al., 1995, 1996a; Doucette et al., 1996; Erdner, 1997). However, laboratory work suggest that other algal taxonomic groups such as chlorophytes, dinoflagellates and prymnesiophytes may also show similar substitution of flavodoxin for ferredoxin under iron limitation (La Roche et al., 1993, 1995; Erdner 1997). Although immunological probes against flavodoxin and/or ferredoxin may also work for these taxa, other analytical techniques such as detection of flavodoxin and ferredoxin by HPLC may provide valuable information with field samples (Doucette et al., 1996; Erdner, 1997). However, the HPLC technique is approximately 3 orders of magnitude less sensitive then the immunological assay, thereby requiring a larger sample volume. Moreover, the HPLC technique does not discriminate between algal taxa, a feature that can be advantageous when looking at ecosystem-level response to available Fe. In contrast, immunological probes frequently offer the possibility to look at the response from discrete taxa and even from single cells, where species can be identified by morphological characteristics (La Roche et al., 1996a).

IsiA/B operon The high abundance of cyanobacteria in many HNLC regions (Coale et al., 1996b) is somewhat paradoxical because of the high iron requirement characteristic of this taxonomic groups (Brand et al., 1983; Brand, 1991; Wilhelm et al., 1996). It is of considerable interest, therefore to determine the mechanism(s) by which this group remains a major component of the biomass in many of the iron-poor waters. Unfortunately, the study of iron limitation in natural populations of cyanobacteria using molecular markers is more difficult than for eukaryotic algae. In freshwater cyanobacteria (*Synechococcus* sp. PCC7002, *Synechococcus* sp. PCC7942 and *Synechocystis* sp. PCC6803), *isiB*, the gene for flavodoxin, is co-transcribed with the *isiA* gene which encodes a chlorophyll *a* binding protein (Laudenbach et al., 1988a; Laudenbach & Straus, 1988b; Straus, 1994). This dicistronic operon is tightly regulated by iron with the *isiA* transcript being approximately 10 times more abundant than *isiB*. Furthermore, the *isiAB* operon is under the control of Fur regulatory elements (Ghassemian & Straus, 1996). At first glance, this dicistronic operon, and *isiA* in particular, appear to be ideal candidates for molecular markers of iron limitation in cyanobacteria in general. This may not be the case for several reasons. First, the function of *isiA* in relieving iron stress is poorly understood. *isiA* does not appear to be directly involved in light harvesting although it has been suggested that it may function as a chlorophyll-storage protein used to speed up recovery from chlorosis when iron becomes available (Falk et al., 1995). However, limitation by other nutrients, such as nitrogen, phosphorus and sulfur can induce chlorosis in cyanobacteria (Collier et al., 1994) without concomitant induction of the *isiA* gene.

Second, and perhaps of more serious concern than the unknown function of *isiA*, is the fact that it shares considerable amino acid identity with the *psbC* gene product (CP43), a chlorophyll-binding protein that together with CP47 constitutes the proximal antenna of the PS II reaction center (Laudenbach & Straus, 1988b). In addition the *pcb* gene family present in prochlorophytes also has extensive amino acid identity to *isiA* genes from phycobilisome-containing cyanobacteria (La Roche et al., 1996b). In prochlorophytes, the *pcb* genes encode for chl *a*/*b* binding proteins (Pcb) which appear to be constitutively expressed and which serve as the major light-harvesting antenna. Therefore, probes targeting the *isiA* gene or its protein product would have to be carefully designed and characterized to avoid detection of other members of this possibly large and diverse gene family.

It is not clear yet how common the occurrence of *isiA* is spread among freshwater cyanobacteria, let alone marine strains. Likewise, flavodoxin, the *isiB* gene product has yet to be detected or purified from marine *Synechococcus* (Chadd et al., 1996a; Erdner, 1997; McKay & La Roche, unpublished results), a genus that is prevalent in the oligotrophic open ocean (Waterbury & Watson, 1979; Carr & Mann, 1994).

Plastocyanin Plastocyanin is a Cu-containing protein that is functionally interchangeable with cytochrome c6 in photosynthetic electron transfer. Vascular plants contain only plastocyanin and this protein alone can account for up to 80% of leaf Cu. Cu-deficiency results in lower levels of plastocyanin and to subsequent impairment of photosynthetic electron transport. In contrast, numerous green algae and cyanobacteria remain photosynthetically competent under Cu-deficiency because of the ability to replace plastocyanin with cytochrome c6, an heme Fe containing protein. The regulation of plastocyanin and cytochrome c6 by Cu is well understood (Hill & Merchant, 1992) and is likely to involve a Cu-dependent regulator. In chlorophytes, the Cu-dependent accumulation of plastocyanin is probably controlled by the differential stability of the holoprotein relative to the apoprotein. This dictates that plastocyanin will accumulate as long as Cu is available for its synthesis. In contrast, cytochrome c6 is transcriptionally repressed in the presence of Cu, possibly mediated via a Cu-binding regulatory protein (Hill & Merchant, 1992). Although replacement of plastocyanin and cytochrome c6 is common among chlorophytes, prochlorophytes and cyanobacteria, not all photoautrophs contain both proteins. Plastocyanin has not been reported in chromophytic algae, whereas cytochrome c6, is reportedly absent from higher plants. The unusual distribution of plastocyanin and cytochrome c6 among photosynthetic organisms has been hypothesized to be a result of the selection of transition metals other than iron, after the accumulation of a significant dioxygen concentration in the atmosphere and the ocean (da Silva & Williams, 1991; Falkowski & Raven, 1997). Therefore in cyanobacteria and chlorophytes, it is tempting to predict that expression of plastocyanin will contribute to the alleviation of iron deficiency, despite evidence that cytochrome c6 is directly regulated by Cu rather then Fe. The presence of plastocyanin may confer an advantage to chlorophytes and cyanophytes who are often the dominant taxa in HNLC waters (Binder et al., 1996; Coale et al., 1996b). However, this may be regarded as a species adaptation rather than an acclimation response, since it is unlikely to find Cu limitation in the ocean simply based on it biological availability and cellular requirements (da Silva & Williams, 1991). In addition, it has been argued that under conditions of iron limitation strong enough to induce the production of siderophores by cyanobacteria, most of the dissolved Cu ions will be bound to siderophores, since these do not discriminate between iron and Cu (McKnight & Morel, 1980). As the Cu-siderophore complex is not taken up by cyanobacteria, this may create an artificial condition of Cu deficiency and induce synthesis of cytochrome c6 (Straus, 1994). Both fresh-water and marine cyanobacterial species are known to produce siderophores (Wilhelm & Trick, 1994; Wilhelm, 1995) and it has been argued that under iron limitation, the advantage gained from a high affinity transport system and its siderophores, are more economical then the replacement of cytochrome c6 by plastocyanin (Ferreira & Straus, 1994).

Increase Fe acquisition

Until recently, the evidence for high affinity iron transport in phytoplankton was limited to physiological measurements with radioactive Fe and to the presence of siderophores in certain species of cyanobacteria. Genes with homology to Fe uptake systems from heterotrophic bacteria, including some pathogens, have now been sequenced from several species of cyanobacteria and this has shed some light into the structure of these uptake systems. Notably, the sequence of the entire genome of *Synechocystis* PCC6803, has revealed that at least freshwater cyanobacteria contain several genes homologous to known bacterial Fe transport system (Table 1). It is not clear yet whether the Fe transport system from eukaryotic algae will be similar to those present in prokaryotic systems, but based on the lack of similarity between bacterial and yeast Fe uptake systems, we can expect that it may be different. In *Saccharomyces cerevisiae*, there appears to be both a low affinity system (Fet4; Dix et al., 1994) and a two-component high affinity system comprised of a copper-containing ferroxidase (FET3; (De Silva et al., 1995) and a permease (FTR1; Stearman et al., 1996). Yeast also contains the FRE1 protein, a ferric reductase that is repressed by iron (Dancis et al., 1992). To our knowledge, there are so far no reported homologies of the yeast systems with higher plants or other eukaryotic algae. Notably, some algae similarly possess cell surface redox enzymes capable of reducing Cu and Fe complexes (Jones et al., 1987; Lynnes et al., 1998).

Recently, a cell surface, transferrin-like protein has been isolated from the green unicellular alga *Dunaliella salina*. This 150 kDa, highly hydrophilic protein (p150) was found to accumulate under high salt concentrations but further investigation revealed that it was also induced by iron deficiency as expected from its similarity with other transferrins (Fisher et al., 1997). Until the discovery of p150, tranferrins had been found only in animals. However, several periplasmic proteins involved in iron transport in bacteria show similarity to transferrin in their three-dimensional structure but not in their amino acid sequences. It is interesting to note here that another report has noted the expression of a iron induced protein in the cyanobacterium *Synechocystis* grown under high salt concentration (Fulda & Hagemann, 1995). It is not clear at this point whether the up-regulation of these Fe-repressible genes under high salt is due to indirect effect on the biological availability of Fe or to independent control by high salt as well as Fe.

With the exception of flavodoxin/ferredoxin, further research is required to determine whether any of the proteins listed in Table 1 will also be useful as markers of iron deficiency in field studies.

Macronutrient (N and P) limitation

Traditionally, nitrogen has been widely considered to be the nutrient most likely to limit primary productivity in the ocean (Ganeshram et al., 1995), whereas lacustrine environments are presumed more susceptible to phosphorus limitation (Wetzel, 1983). However, certain marine environments may be subjected to geographical and temporal gradients of nitrogen and phosphorus limitation. For example, the limiting nutrient in some estuaries has been shown on the basis of nutrient distribution and bioassays to alternate between phosphorus and nitrogen limitation both with distance from the freshwater river discharge to the mouth and seasonally (Maestrini et al., 1997). Phosphorus has also been shown to be the major limiting nutrient in the eastern Mediterranean Sea (Krom et al., 1991). However, the fact that both inorganic and organic forms of these two nutrients can be taken up by both bacterioplankton and phytoplankton (Palenik & Morel, 1990a; Cotner & Wetzel, 1992; Dyhrman & Palenik, 1997) complicates the traditional approaches of determining the limiting nutrient by analyzing inorganic nutrient ratios and bioassays using primarily inorganic forms of phosphorus and nitrogen. Diagnostic markers would therefore be of great value for identifying *in situ* the possibility of nitrogen or phosphorus limitation in phytoplankton populations.

The search for diagnostic markers for nitrogen and phosphorus limitation is more difficult then for iron, primarily for two reasons. First, nitrogen and phosphorus are major constituents of biological macromolecules and polymers such as proteins, nucleic acid and lipids, unlike iron which is present only in selected proteins. As a result, there are to our knowledge no analogous examples of protein substitution as are found for the Fe-containing proteins ferredoxin and cytochrome c6. Second, both of these macronutrients are accessible in a multitude of organic and inorganic forms having various oxidation states. Many organisms have the ability to use a variety of nitrogen sources for growth and in a given taxonomic group, there often exists a hierarchy in the preference for which various compounds are utilized (Lomas et al., 1996). Nitrogen can be found either as dissolved inorganic (DIN), or dissolved organic (DON) nitrogen and it is generally believed that inorganic sources such as NO3 and NH4 are preferred amongst many taxonomic groups. Until recently, it was assumed that DON was mainly utilized by heterotrophic bacteria. However, the realization that many algal species possess cell-surface oxidases that use a variety of amino acids and other simple organic nitrogen compounds as substrate challenges this view (Palenik & Morel, 1990a, 1991). Likewise, urea is a source of organic nitrogen that is readily utilized by many phytoplankton species (Lomas et al., 1996). A similar situation may be encountered with the uptake of phosphorus with the possibility that alternate organic sources of phosphorus may be taken up when inorganic phosphate is depleted (Dyhrman & Palenik, 1997).

The types of response to nitrogen and phosphorus limitation that may be useful in designing markers for stresses imposed by these nutrients will most likely fall within the category of nutrient acquisition. Within this category, the response can be subdivided between the induction of high affinity uptake systems for the preferred form of the nutrient and a switch to the utilization of an alternate form of the nutrient that may be more energetically costly to acquire. The response to phosphorus limitation is discussed elsewhere in this volume (Scanlan, this volume) and the remaining of this section will focus on nitrogen limitation.

Switch to alternate nitrogen sources

Microorganisms that are deprived of their preferred source of nitrogen often induce or derepress systems that utilize alternate sources of this essential nutrient. Such systems are well understood in yeast and heterotrophic bacteria in which a number of genes are under the control of a regulator protein (Stock et al., 1989).

Nrp1 protein (Palenik & Koke, 1995) have recently reported that in the prymnesiophyte *Emiliania huxleyi* a nitrogen-regulated protein, Nrp1, is induced under nitrate- and NH^{4+}-limitation and also in nitrogen-replete, urea-grown cells. Nrp1 is a cell membrane or cell wall-associated protein of unknown function. It has been hypothesized that Nrp1 may play a function in protecting against nitrogen limitation or in transport of alternate nitrogen sources such as amino acids or urea. The authors believe that the protein is not a cell surface amino acid oxidase or an amine oxidase because of the lack of activity for this enzyme detected in their nitrogen-limited cultures (Palenik & Koke, 1995). As pointed out by others (Scanlan et al., 1997), an ideal probe should be induced strictly under limiting conditions of a nutrient and not during growth with alternate sources, however, given the role that utilization of alternate sources of nitrogen may play in the outcome of species competition, probes such as Nrp1 may provide valuable information on nutrient utilization.

NtcA *gene* *ntcA* is a regulator that is involved in the global regulation of nitrogen in cyanobacteria and prochlorophytes (Vega-Palas et al., 1992; Frias et al., 1993; Bradley & Reddy, 1997; Lindell et al., 1998). Many species of microorganisms, including cyanobacteria, prefer NH_4^+ as a nitrogen source and genes involved in the uptake and assimilation of NH_4^+ are constitutively expressed within the cell, regardless of the predominant source of nitrogen available in the natural environment. *NtcA* encodes a DNA binding protein that acts as a transcriptional activator of genes subject to nitrogen control. Genes under the control of *ntcA* are presumed to include those involved in the acquisition and assimilation of nitrate, nitrite and dinitrogen gas. The *ntcA* gene however, appears to be repressed in the presence of NH_4^+ and as such, it has been suggested that expression of this gene may provide an indictor of 'new' (i.e. using nitrate) vs. 'regenerated' (i.e. using NH_4^+) primary production (Lindell et al., 1998). However, the presence of *ntcA* alone is not sufficient to induce alternate uptake and assimilation systems such as those required for nitrate and nitrite uptake (Lindell et al., 1998). The expression patterns of *ntcA* may indicate whether or not there is sufficient NH_4^+ for growth but it is not clear at this stage if it can provide information about which alternate source of inorganic nitrogen is utilized instead.

Light and oxidative stress

High-light compensation response

High-light induced photosynthetic proteins Early-light induced proteins (ELIPs) belong to the larger family of light-harvesting chl *a*/*b* binding proteins (LHC) and have extensive homology with the other members of this nuclear-encoded gene family (Adamska & Kloppstech, 1991). The three membrane-spanning helices (MSH) characteristic of this family are also conserved in the ELIPs. In contrast to constitutively expressed members of this gene family, the accumulation of ELIPs appears to be confined to a transient increase during greening of chloroplasts. In addition, ELIPs are accumulated in mature chloroplasts exposed to high light conditions and are hypothesized to play a role in the acclimation response to high light (Adamska et al., 1992; Adamska et al., 1993). Unlike other members of the LHC family, the ELIPs are stable in the absence of photosynthetic pigments (Funk et al., 1995) and are degraded in low light (Adamska et al., 1996). Cyanobacteria also possess small (72 amino acids), high-light induced proteins (HLIPs) that have similarity to the MSH of the LHC proteins (Dolganov et al., 1995). Similarly to the ELIPs, the HLIPs are accumulated under high light and UV-A radiation. Although HLIPs have not yet been isolated in *Prochlorococcus*, a close relative to marine cyanobacteria, a high-light induced thylakoid membrane protein has been noted in two strains of *Prochlorococcus* (Partensky et al., 1997). It is not clear yet whether marine cyanobacterial strains also possess HLIPs, but if present, these proteins may be useful indicators of high-light stress.

UV-absorbing compounds The accumulation of UV-absorbing compounds such as mycosporine-like amino acids (MAA) is a common response observed in cells exposed to high-levels of UV radiation (Carreto et al., 1990; Wachi et al., 1995). This response has been well characterized in terrestrial cyanobacterial species predominant in cryptobiotic crusts where the detrimental effects of UV radiation are reinforced by desiccation (Garcia-Pichel et al., 1992; Hill et al., 1994; Ehling-Schulz et al., 1997). MAAs are also synthesized and excreted by marine phytoplankton. It has been suggested that MAAs as a component of the DOM pool may contribute to attenuate UV radiation in the surface waters of the ocean (Vernet & Whitehead, 1996). The detection of MAAs and other UV-induced chemicals may be useful to identify UV-induced stress in phytoplankton.

Superoxide dismutase Superoxide dismutases (SODs) are metalloenzymes that are involved in the protection of living cells from oxidative damage. In prokaryotes, superoxide dismutases most often contain Fe or Mn, whereas in higher organisms, Cu/Zn superoxide dismutases are most common, although there are some exceptions to this general phylogenetic distribution (Kroll et al., 1995; Chadd et al., 1996b). The regulation of SODs in some cyanobacteria provide a good example of how one type of stress can influence another. *Phormidium boryanum*, a filamentous cyanobacteria, contains at least four superoxide dismutases, including one Fe- and three Mn- containing enzymes (Campbell & Laudenbach, 1995). The Fe- and one of the Mn-SODs are constitutively expressed and are believed to be required for the detoxification of superoxide radicals produced as a result of normal cell metabolism. Under conditions of Fe limitation, however, the Fe-SOD transcript (*sodB*) decreases dramatically. This Fe-regulatory response of an otherwise constitutively expressed SOD would be expected to increase the level of oxidative stress encountered by the cell. However, repression of *sodB* transcription is accompanied by the induction of an additional Mn-containing SodA2 in what appears to be a compensatory response to the potential for increased oxidative stress. However, SodA2 is probably not a good indicator of iron stress as it is also induced under conditions of nitrogen limitation, and is repressed under anaerobic conditions where oxidative stress will be at a minimum (Campbell & Laudenbach, 1995). The increase in oxidative stress in the case of iron-limitation is two fold: via the decreased availability of Fe for photosynthetic proteins which will lead to lesion in the electron transport chain and imbalance between light harvesting and photochemical reactions, and via the decrease in the expression of the Fe-SOD. On the other hand, the increase in oxidative stress during nitrogen limitation can possibly be explained by the imbalance between light harvesting and the reduced biosynthetic capacity due to the lack of nitrogen. The regulation of the *sod* genes is therefore complex and can be overridden at several levels.

Trace metal toxicity

There exists three major strategies to counteract trace metal toxicity:

1. intracellular sequestration of the toxic metals with metal binding proteins,
2. exclusion by extracellular binding of metals,
3. export of intracellular trace metals by active pumping out of the cell.

Very few of these mechanisms have been identified in phytoplankton, but there exists counterparts in both prokaryotes and eukaryotes in which these strategies have been well studied (reviewed by Palenik & Wood, 1997). The role of metallothionins in conferring trace metal tolerance in cyanobacteria has been well studied (Robinson, 1989; Daniels et al., 1998; Thelwell et al., 1998). Recent work on phytochelatin synthesis in eukaryotic microalgae indicate that the detection of these molecules in the field may provide indices of trace metal pollution in marine and aquatic environments (Ahner et al., 1994; Jervis et al., 1997; Morelli & Pratesi, 1997). In addition, potential markers of Cu toxicity have been recently identified by cell surface labeling of marine diatoms (Palenik, 1998).

Conclusions

We have reviewed here the current status of diagnostic probes to detect environmental stress in phytoplankton. This field of research is still in its infancy and considerable efforts is required to identify and test potential markers of environmental stress in a variety of phytoplankton species. By combining single cell techniques such as fluorescent *in situ* hybridization (FISH) or single cell immunofluorescence with microscopy and flow cytometry, it may be possible to greatly improve our understanding of species interaction within natural populations.

Similarly to the species-specific probes, functional probes will have to be extensively tested to determine their taxonomic specificity. However, Figure 1 demonstrated an added level of complexity relative to species-specific probes via interactions that occur between various environmental stress. Additional work to understand the expression of the targeted genes and proteins will be essential for the correct interpretation of the results obtained with these probes in the natural environment. After the initial probe development, parallel laboratory and field manipulation (e.g. bioassay) designed to test the applicability of these probes may provide a good basis for the interpretation of routine field results. Nevertheless, the applicability of a specific probes will need to be reevaluated as more information becomes available regarding the regulation of targeted genes by environmental factors.

The issues of calibration, quantitation and normalization remain to be addressed in order to fully exploit the potential of functional probes. Up to now, only semi-quantitative methods have been used in the field to evaluate relative changes in proteins or nucleic acid transcripts. In a mixed population, the additional problem of species diversity complicates the interpretation of relative abundance, especially when the intensity of detection with a given probe differs between species. While it is not possible to give answers to all of these questions at this stage, we believe that these challenges can be met, on the one hand, through the development of single cell techniques in conjunction with flow cytometry. Alternatively, the development of monoclonal antibodies and quantitative ELISA may also improve the quantitative analysis of functional markers.

Through the study of phytoplankton response to environmental stress, we can expect to learn basic information concerning the mechanisms of adaptation to various environmental stress, as well as developing methods to interrogate the natural environment without the need for manipulation.

Acknowledgements

We thank Richard Geider, Tracy Villareal and Deana Erdner for sharing unpublished results. We are grateful to Helen Murray for technical assistance. We are indebted to Jose Manuel Garcia Fernandez and Frederic Partensky for providing the RNA samples from *Prochlorococcus* used in the mRNA differential display. JLR acknowledges the support of the Department of Energy (contract # DE-ACO2-98CH10886).

References

Adachi, M., Y. Sako & Y. Ishida, 1996. Identification of the toxic dinoflagellates Alexandrium catenella and A. tamarense (Dinophyceae) using DNA probes and whole-cell hybridization. J. Phycol. 32: 1049–1052.

Adamska, I. & K. Kloppstech, 1991. Evidence for an association of the early light-inducible protein (ELIP) of pea with Photosystem II. Plant mol. Biol. 16: 209–23.

Adamska, I., K. Kloppstech & I. Ohad, 1992. UV light stress induces the synthesis of the early light-inducible protein and prevents its degradation. J. Biol. Chem. 267: 24732–7.

Adamska, I., K. Kloppstech & I. Ohad, 1993. Early light-inducible protein in pea is stable during light stress but is degraded during recovery at low light intensity. J. Biol. Chem. 268: 5438–44.

Adamska, I., M. Lindahl, M. Roobol-Boza & B. Andersson, 1996. Degradation of the light-stress protein is mediated by an ATP- independent, serine-type protease under low-light conditions. Eur. J. Biochem. 236: 591–9.

Ahner, B. A., N. M. Price & F. M. Morel, 1994. Phytochelatin production by marine phytoplankton at low free metal ion concentrations: laboratory studies and field data from Massachusetts Bay. Proceedings of the National Academy of Science U.S.A. 91: 8433–6.

Angerer, A., S. Gaisser & V. Braun, 1990. Nucleotide sequences of the sfuA, sfuB, and sfuC genes of Serratia marcescens suggest a periplasmic-binding-protein-dependent iron transport mechanism. J Bact. 172: 572–8.

Apt, K. E., P. G. Kroth-Pancic & A. R. Grossman, 1993. Stable nuclear transformation of the diatom *Phaeodactylum* tricornutum. Molecular and General Genetic 252: 572–579.

Arudchandran, A. & G. S. Bullerjahn, 1996. Expression of the petE gene encoding plastocyanin in the photosynthetic prokaryote, Prochlorothrix hollandica. Biochem. Biophys. Res. Comm. 226: 626–630.

Barker, G. L. A., J. C. Green, P. K. Hayes & L. K. Medlin, 1994. Preliminary results using the RAPD analysis to screen bloom populations of Emiliania huxleyi (Haptophyta). Sarsia 79: 301–306.

Bauer, D., H. Muller, J. Reich, H. Riedel, V. Ahrenkiel, P. Warthoe & M. Strauss, 1993. Identification of differentially expressed mRNA species by an improved display technique (DDRT-PCR). Nucleic Acids Res. 21: 4272–80.

Behrenfeld, M., A. J. Bale, Z. S. Kolber, J. Aiken & P. G. Falkowski, 1996. Confirmation of iron limitation of phytoplankton photosynthesis in the equatorial Pacific Ocean. Nature 383: 508–511.

Benito, E. P., T. Prins & J. A. van Kan, 1996. Application of differential display RT-PCR to the analysis of gene expression in a plant-fungus interaction. Plant mol. Biol. 32: 947–57.

Berg, G. M., P. M. Glibert, M. W. Lomas & M. A. Burford, 1997. Organic nitrogen uptake and growth by the chrysophyte *Aureococcus anophagefferens* during a brown tide event. Mar. Biol. 129: 377–387.

Binder, B. J., S. W. Chisholm, R. J. Olson, S. L. Frankel & A. Z. Worden, 1996. Dynamics of picoplankton, ultraplankton and bacteria in the central equatorial Pacific. Deep Sea Res. 43: 907–931.

Blackman, F. F., 1905. Optima and limiting factors. Ann. Bot. 19: 281–298.

Boyd, P. W., D. L. Muggli, D. E. Varela, R. H. Goldblatt, R. Chretien, K. J. Orians & P. J. Harrison, 1996. *In vitro* iron enrichment experiments in the NE subarctic Pacific. Mar. Ecol. Prog. Ser. 136: 179–193.

Bradley, R. L. & K. J. Reddy, 1997. Cloning, Sequencing and Regulation of the Global NItrogen Regulator Gene *ntcA* in the Unicellular Diazotrophic Cyanobacterium Cyanothece sp. Strain BH68K. J. Bact. 179: 4407–4410.

Brahamsha, B., 1996. A genetic manipulation system for oceanic cyanobacteria of the genus *Synechococcus*. Appl. envir. Microbiol. 62: 1747–51.

Brand, L. E., 1991. Minimum iron requirements of the marine phytoplankton and the implications for the biogeochemical control of new production. Limnol. Oceanogr. 36: 1756–1771.

Brand, L. E., W. G. Sunda & R. R. L. Guilklard, 1983. Limitation of marine photoplankton reproductive rates by zinc, manganese and iron. Limnol. Oceanogr. 28: 1182–1198.

Brock, T. D., 1967. Life at high temperatures. Evolutionary, ecological and biochemical significance of organisms living in hot springs is discussed. Science 158: 1012–9.

Buikema, W. J. & R. Haselkorn, 1993. Molecular genetics of cyanobacteria development. Ann. Rev. Plant Physiol. mol. Biol. 44: 33–52.

Bullen, J. J. & E. Griffith, 1987. Iron and infection. Wiley-Interscience, New York.

Campbell, W. S. & D. Laudenbach, 1995. Characterization of four superoxide dismutase genes from a filamentous cyanobacterium. J. Bact. 177: 964–972.

Carr, N. G. & N. H. Mann, 1994. The Oceanic Cyanobacterial Picoplankton. In Bryant D. A. (ed.). The Molecular Biology of Cyanobacteria. Kluwer: 27–48.

Carreto, J. I., M. O. Carignan, G. Daleo & 1990. Occurence of mycosporine -like amino acids in the red dinoflagellate *Alexandrium excavatum*: UV-protective compounds. J. Plankton Res. 12: 909–921.

Chadd, H. E., I. R. Joint, N. H. Mann & N. G. Carr, 1996a. The marine phytoplankter Synechococcous Sp. WH 7803 exhibits an adaptive response to restricted iron availability. FEMS Microbiol. Ecol. 21: 69–76.

Chadd, H. E., J. Newman, N. H. Mann & N. G. Carr, 1996b. Identification of iron superoxide dismutase and a copper/zinc superoxide dismutase enzyme activity within the marine cyanobacterium Synnechococcus sp. WH7803. FEMS Microbiol. Let. 138: 161–165.

Coale, K. H., 1991. Effects of iron, manganese, copper and zinc enrichments on productivity and biomass in the subarctic Pacific. Limnol. Oceanogr. 36: 1851–1864.

Coale, K. H., S. E. Fitzwater, R. M. Gordon, K. S. Johnson & R. Barber, 1996a. Control o f community growth and export production by upwelled iron in the equatorial Pacific Ocean. Nature 379: 621–624.

Coale, K. H., K. S. Johnson, S. E. Fitzwater, R. M. Gordon, S. Taner, F. P. Chavez, L. Ferioli, C. Sakamoto, P. Rogers, F. Millero, P. Steinberg, P. Nightingale, D. Cooper, W. P. Cochlan, M. R. Landry, J. Constantinou, G. Rollwagen, A. trasvina & R. Kudela, 1996b. A massive phytoplankton bloom induced by an ecosystem-scale iron fertilization experiment in the Equatorial Pacific Ocean. Nature 383: 495–501.

Collier, J. L. & A. Grossman, 1994. A small polypeptide triggers complete degradation of light-harvesting phycobiliproteins in nutrient-deprived cyanobacteria. The EMBO Journal 13: 1039–1047.

Collier, J. L., S. K. Herbert, D. C. Fork & A. R. Grossman, 1994. Changes in the cyanobacterial photosynthetic apparatus during acclimation to macronutrient deprivation. Photo. Res. 42: 173–183.

Cotner, J. B. J. & R. G. Wetzel, 1992. Uptake of dissolved inorganic and organic phosphorus compounds by phytoplankton and bacterioplankton. Limnol. Oceanogr. 37: 232–243.

Crespi, H. L., U. Smith, L. Gajda, T. Tisue & R. M. Ammeraal, 1972. Extraction adn purification of 1H, 2H, and isotope hybrid algal cytochrome, ferredoxin and flavodoxin. Biochem. Biophys. Acta 256: 611–618.

Cullen, J. J., 1991. Hypotheses to expalin high-nutrient conditions int he open sea. Limnol. Oceanogr. 36: 1578–1599.

da Silva, F. J. J. R. & R. J. P. Williams, 1991. The biological chemistry of the elements. Clarendon Press, Oxford.

Dancis, A., D. D. Roman, G. J. ANderson, A. G. Hinnebusch & R. D. Klausner, 1992. Ferric reductase of *Saccharomyces cerevisiae*: Molecular characterization, role in iron uptake, and transcriptional control by iron. Proc. natn. Acad. Sci. U.S.A. 89: 3869–3873.

Daniels, M. J., J. S. Turner-Cavet, R. Selkirk, H. Sun, J. A. Parkinson, P. J. Sadler & N. J. Robinson, 1998. Coordination of Zn2+ (and Cd2+) by prokaryotic metallothionein. Involvement Of his-imidazole [In Process Citation]. J. Biol. Chem. 273: 22957–61.

Davis, J. P., F. Yidlz & A. Grossman, 1994. Mutants of Chlamydomonas with aberrant responses to sulfur deprivation. Plant Cell 6: 53–63.

De Silva, D., C. C. Askwith, D. Eide & J. Kaplan, 1995. The FET3 gene product required for high affinity Iron transport in yeast is a cell surface ferroxidase. J. Biol. Chem. 270: 1098–1101.

Demmig-Adams, B., 1990. Carotenoids and photoprotectioin in plants: a role for the xanthophyll zeaxanthin. Biochem. Biophys. Acta 1020: 1–24.

Dix, D. R., J. T. Bridgham, M. A. Broderius, C. A. Byersdorfer & D. J. Eide, 1994. The FET4 Gene encodes the low affinity Fe(II) transport protein of *Saccharomyces cerevisiae*. J. Biol. Chem. 269: 26092–26099.

Dolganov, N. A. M., D. Bhaya & A. R. Grossman, 1995. Cyanobacterial protein with similarity to the chlorophyll *a/b* binding proteins of higher plants: Evolution and regulation. Proc. natn. Acad. Sci. U.S.A. 92: 636–640.

Doucette, G. J., D. L. Erdner, M. L. Peleato, J. J. Hartman & D. M. Anderson, 1996. Quantitative analysis of iron-stress related proteins in *Thalassiosira* weissflogii: measurement of flavodoxin and ferredoxin using HPLC. Mar. Ecol. Prog. Ser. 130: 269–276.

Doucette, G. J. & P. J. Harrison, 1990. Some effects of iron and nitrogen stress on the red tide dinoflagellate Gymnodinium sanguineum. I. Effects of iron depletion and nitrogen storage on biochemical composition. Mar. Biol. 110: 293–306.

Dunahay, T. G., E. E. Jarvis & P. G. Roessler, 1995. Genetic transformation of the diatoms *Cyclotella cryptica* and *Navicula saprophila*. J. Phycol. 33: 602–612.

Dyhrman, S. T. & B. P. Palenik, 1997. The identification and purification of a cell-surface alkaline phosphatase from the dinoflagellate *Prorocentrum minimum* (Dinophyceae). J. Phycol. 33: 602–612.

Ehling-Schulz, M., W. Bilger & S. Scherer, 1997. UV-B-induced synthesis of photoprotective pigments and extracellular polysaccharides in the terrestrial cyanobacterium Nostoc commune. J. Bact. 179: 1940–5.

Entsch, B., R. G. Sim & B. G. Hatcher, 1983. Indications from photosynthetic components that iron is a limiting nutrient in primary producers on coral reefs. Mar. Biol. 73: 17–30.

Erdner, D. L., 1997. MIT-WHOI.

Falk, S., G. Samson, D. Bruce, N. P. A. Huner & D. E. Laudenbach, 1995. Functional analysis of the iron-stress induced CP43' polypeptide of PS II in the cyanobacterium *Synechococcus* sp. PCC7942. Photo. Res. 45: 51–60.

Falkowski, P. G., 1997. Evolution of the nitrogen cycle and its influence on the biological sequestration of CO_2 in the ocean. Nature 387: 272–275.

Falkowski, P. G. & J. A. Berges, 1998. Phytoplankton proteases as models of human immune disorders. EOS, Transactions, American Geophisical Union 79: OS103.

Falkowski, P. G. & J. A. Raven, 1997. Aquatic Photosynthesis, Blackwell Science, Malden, MA, U.S.A.

Ferreira, F. & N. A. Straus, 1994. Iron deprivation in cyanobacteria. J. appl. Phycol. 6: 199–210.

Fisher, M., I. Gokhman, U. Pick & A. Zamir, 1997. A structurally novel transferrin-like protein accumulates in the plasma membrane of the unicellular green alga *Dunaliella salina* grown in high salinities. J. Biol. Chem. 272: 1565–1570.

Fislage, R., M. Berceanu, Y. Humboldt, M. Wendt & H. Oberender, 1997. Primer design for a prokaryotic differential display RT-PCR. Nucleic Acids Res. 25: 1830–5.

Frias, J. E., A. Merida, A. Herrero, J. Martin-Nieto & E. Flores, 1993. General Distribution of the Nitrogen Control Gene *ntcA* in Cyanobacteria. J. Bact. 175: 5710–5713.

Fulda, S. & M. Hagemann, 1995. Salt treatment induces accumulation of flavodoxin in the cyanobacterium *Synechocystis* PCC 6803. J. Plant Physiol. 146: 520–526.

Funk, C., I. Adamska, B. R. Green, B. Andersson & G. Renger, 1995. The nuclear-encoded chlorophyll-binding Photosystem II-S protein is stable in the absence of pigments. J. Biol. Chem. 270: 30141–7.

Ganeshram, R. S., T. F. Pedersen, S. E. Calvert & J. W. Murray, 1995. Large changes in oceanic nutrient inventories from glacial to interglacial periods. Nature 376: 755–758.

Garcia-Pichel, F., N. D. Sherry & R. W. Castenholz, 1992. Evidence for an ultraviolet sunscreen role of the extracellular pigment scytonemin in the terrestrial cyanobacterium Chlorogloeopsis sp. Photochem. Photobiol. 56: 17–23.

Geider, R. J. & J. La Roche, 1994. The role of iron in phytoplankton photosynthesis and the potential for iron-limitation of primrary productivity in the sea. Photo. Res. 39: 275–301.

Geider, R. J., J. La Roche, R. M. Greene & M. Olaizola, 1993. Response of the photosynthetic apparatus of Pheaodactylum tricornutum (Bacilliariophyceae). J. Phycol. 29: 755–766.

Ghassemian, M. & N. A. Straus, 1996. Fur regulates the expression of iron-stress genes in the cyanobacterium *Synechococcus* sp. strain PCC 7942. Microbiol. 142: 1469–1476.

Graziano, L. M., J. La Roche & R. J. Geider, 1997. Physiological responses to phhosphorus limitation in batch and steady-state cultures of Dunaliella tertiolecta (Chlorophyta): A unique stress protein as an indicator of phosphate deficiency. J. Phycol. 32: 825–838.

Greene, R. M., R. J. Geider, Z. Kolber & P. G. Falkowski, 1992. Iron-induced changes in light harvesting and photochemical energy conversion processes in eukaryotic algae. Plant Physiol. 100: 565–575.

Gueneau, P., F. M. M. Morel, J. La Roche & D. Erdner, 1998. The *petF* region of thte chloroplastic genome from the diatom *Thalassiosira weissflogii* (Grun.): sequence, organization, and phylogeny. Eur. J. Phycol.: in Press.

Guerinot, M. L., 1994. Microbial iron transport. Ann. Rev. Microbiol. 48: 743–772.

Hershkovits, G., Z. Dubinsky & D. J. Katcoff, 1997. A novel homologue of the prokaryotic htrA gene is differentially expressed in the alga Haematococcus pluvialis following stress. Mol. Gen. Genet. 254: 254–345.

Hill, D. R., S. L. Hladun, S. Scherer & M. Potts, 1994. Water stress proteins of Nostoc commune (Cyanobacteria) are secreted with UV-A/B-absorbing pigments and associate with 1,4-beta-D-xylanxylanohydrolase activity. J. Biol. Chem. 269: 7726–34.

Hill, K. L. & S. Merchant, 1992. In vivo competition between plastocyanin and a copper-dependent regulator of the *Chlamydomonas reinhardtii* cytochrome c6 gene. Plant Physiol. 100: 319–326.

Hutchins, D. A. & K. W. Bruland, 1998. Iron limitation and the silicate pump in the California upwelling region. Eos, Transactions, American Geophysical Union 79: OS21.

Jervis, L., R. Rees-Naesborg & M. Brown, 1997. Biochemical responses of the marine macroalgae Ulva lactuca and Fucus vesiculosus to cadmium and copper-from sequestration to oxidative stress. Biochem. Soc. Trans. 25: 63S.

Jones, G. J., B. P. Palenik & F. M. M. Morel, 1987. Trace metal reduction by phytoplankton: the role of plasmalemma redox enzymes. J. Phycol. 23: 237–244.

Jones, K. L., 1988. "Analysis of ferredoxin and flavodoxin in Anabaena and *Trichodesmium* using fast protein liquid chromatography" M.S. Thesis, Portland State University.

Kammler, M., C. Schon & K. Hantke, 1993. Characterization of the ferrous iron uptake system of Escherichia coli. J. Bact. 175: 6212–9.

Kaneko, T. & S. Tabata, 1997. Complete genome structure of the unicellular cyanobacterium *Synechocystis* sp. PCC6803 [In Process Citation]. Plant Cell Physiol. 38: 1171–6.

Klausner, R. D., T. A. Rouault & J. B. Harford, 1993. Regulating the fate of mRNA: The control of cellular iron metabolism. Cell 72: 19–28.

Kolber, Z., J. Zehr & P. G. Falkowski, 1988. Effects of growth irradiance and nitrogen limitation on photosynthetic energy conversion in Photosystem II. Plant Physiol. 88: 72–79.

Kowallik, K. V., B. Stoebe, J. Schaffran, P. Kroth-Pancic & U. Freier, 1995. The chloroplast genome of a chlorophyll a+c-containing alga, *Odontella sinensis*. Plant mol. Biol. rep. 13: 336–342.

Kroll, J. S., P. R. Langford, K. E. Wilks & A. D. Keil, 1995. Bacterial [Cu, Zn]-superoxide dismutase:phylogenetically distinct from the eukaryotic enzyme and not so rare after all! Microbiol. 141: 2271–2279.

Krom, M. D., N. Kress & S. Brenner, 1991. Phosphorus limitation of primray productivity in the eastern Mediterranean Sea. Limnol. Oceanogr. 36: 424–432.

La Roche, J., P. W. Boyd, R. M. L. McKay & R. J. Geider, 1996a. Flavodoxin as an *in situ* marker for iron stress inphytoplankton. Nature 382: 802–805.

La Roche, J., R. J. Gieder, L. M. Graziano, H. Murray & K. Lewis, 1993. Induction of specific proteins in eukaryotic algae grown under iron-, phosphorus or nitrogen-deficient conditions. J. Phycol. 29: 767–777.

La Roche, J., H. Murray, M. Orellana & J. Newton, 1995. Flavodoxin expression as an indicator of iron limitation in marine diatoms. J. Phycol. 31: 520–530.

La Roche, J., R. Nuzzi, R. Waters, K. Wyman, P. G. Falkowski & D. W. R. Wallce, 1997. Brown tide blooms in Long Island's Coastal waters linked to interannual variability in groundwater flow. Global Change Biol. 3: 397–410.

La Roche, J., G. W. M. van der Staay, F. partensky, A. Ducret, R. Aebersold, R. Li, S. S. Golden, R. G. Hiller, P. M. Wrench, A. W. D. Larkum & B. Green, 1996b. Independent evolution of the prochlorophyte and the green plant chlorophyll *a*/*b* light-harvesting proteins. Proc. natn. Acad. Sci. U.S.A. 93: 15244–15248.

Lashkari, D. A., J. L. DeRisi, J. H. McCusker, A. F. Namath, C. Gentile, S. Y. Hwang, P. O. Brown & R. W. Davis, 1997. Yeast microarrays for genome wide parallel genetic and gene expression analysis. Proc. natn. Acad. Sci. U.S.A. 94: 13057–62.

Laudenbach, D. E., M. E. Reith & N. A. Straus, 1988a. Isolation , sequence analysis, and transcriptional studies of the flavodoxin gene from *Anacystis nidulans* R2. J. Bact. 170: 258–265.

Laudenbach, D. E. & N. A. Straus, 1988b. Characterization of a cyanobacterial iron stress-induced gene similar to *psbC*. J. Bact. 170: 5018–5026.

Lee, J. G., S. B. Roberts & F. M. M. Morel, 1995. Cadmium: a nutrient for the marine diatom *Thalassiosira weissflogii*. Limnol. Oceanogr. 40: 1056–1063.

Li, H. Y. & J. E. Gray, 1997. Pollination-enhanced expression of a receptor-like protein kinase related gene in tobacco styles. Plant mol. Biol. 33: 653–65.

Li, L. G., X. H. Zhang, C. P. joshi & V. L. Chiang, 1997. Differential display mediated identification of compression stress-induced cDNA from Loblolly pine (*Pinus taeda* L.). Plant Physiol. 114: 840.

Liang, P. & A. B. Pardee, 1992. Differential display of eukaryotic messenger RNA by means of the polymerase chain reaction [see comments]. Science 257: 967–71.

Liang, P. & A. B. Pardee, 1997. Differential display. A general protocol. Methods mol. Biol. 85: 3–11.

Lin, S., J. Chang & E. J. Carpenter, 1994. Detection of proliferating cell nuclear antigen analog in four species of marine phytoplankton. J. Phycol. 30: 449–456.

Lindell, D., E. Padan & A. F. Post, 1998. Regulation of *ntcA* expression and nitrite uptake in the marine *Synechococcus* sp. strain WH 7803. J Bact. 180: 1878–86.

Liu, B.-H., F.-Y. Yu & F. S. Chu, 1996. Anti-idiotype and anti-anti-idiotype antibodies generated from polyclonal antibodies against microcystin-LR. J. Agric. Food Chem., 44: 4037–4042.

Lomas, M. W., P. M. Glibert & G. M. Berg, 1996. Characterization of nitrogen uptake by natural populations of *Aureococcus* anophagefferens (Chrysophyceae) as a function of incubation duration, substrate concentration, light and temperature. J. Phycol. 32: 907–916.

Lynnes, J. A., T. L. M. Derzaph & H. G. Weger, 1998. Iron limitation results in induction of ferricyanide reductase and ferric chelate reductase activities in Chlamydomonas reinhardtii. Planta 204: 360–365.

Maestrini, S. Y., M. Balode, C. Bechemin, I. Purina & C. Verite, 1997. Nutrients limiting the algal growth potential (AGP) in the Gulf of Riga, eastern Baltic Sea, in spring and early summer 1996. La mer 35: 49–68.

Maldonado, M. T. & N. M. Price, 1996. Influence of N substrate on Fe requirements of marine centric diatoms. Mar. Ecol. Prog. Ser. 141: 161–172.

Mann, N. H., 1995. How do cells express nutrient limitation at the molecular level? In Joint I. (ed.), Molecular Ecology of Aquatic Microbes. Springer-Verlag, Berlin.

Martin, J. H. & S. E. Fitzwater, 1988. Iron deficiency limits phytoplankton growth in the north-east Pacific subarctic. Nature 331: 341–343.

Martin, J. H., S. E. Fitzwater & e. al., 1989. WERTEX: phytoplankton/iron studies in the Gulf of Alaska. Deep Sea Res. 36: 649–680.

Martin, J. H., R. M. Gordon & S. E. Fitzwater, 1990. Iron in Antarctic waters. Nature 345: 156–158.

Martin, J. H., R. M. Gordon & S. E. Fitzwater, 1991. The case for iron. Limnol. Oceanogr. 36: 1793–1802.

Martinez, J. & F. Azam, 1993. Aminopeptidase Activity in Marine Chroococcoid Cyanobacteria. Appl. envir. Microbiol. 59: 3701–3707.

Masepohl, B., K. Scholisch, K. Gorlitz, C. Kutzki & H. Bohme, 1997. The heterocyst-specific fdxH gene product of the cyanobacterium *Anabaena* sp. PCC 7120 is importnat but not essential for nitrogen fixation. Mol. Gen. Genet. 253: 770–776.

Matsudaira, P., 1990. limited N-terminal sequence analysis. Methods Enzym. 182: 602–613.

Mayhew, S. G. & M. L. Ludwig, 1975. Flavodoxin and Electron-Transferring Proteins. In Boyer P. D. (ed.), The Enzymes. Academic Press Inc., New York.

Mazel, D. & P. Marliere, 1989. Adaptive eradication of methionine amd cysteine from cyanobacterial light-harvesting proteins. Nature 341: 245–248.

McKay, R. M. L., R. J. Geider & J. La Roche, 1997. Physiological and Biochemical Repsonse of the Photosynthetic Apparatus of Tow Marine Diatoms to Fe Stress. Plant Physiol. 114: 615–622.

McKay, R. M. L., J. LaRoche, A. F. Yakumin, D. G. Dunford & R. J. Geider, 1999. Accumulation of ferredoxin and flavodorin in a marine diatom in response to Fe. J. Phycol. 35: 510–519.

McKnight, D. M. & F. M. M. Morel, 1980. Copper complexation by siderophores from filamentous blue-green algae. Limnol. Oceanogr. 25: 62–71.

Meier, T., S. Arni, S. Malarkannan, M. Poincelet & D. Hoessli, 1992. Immunodetection of biotinylated lymphocyte-surface proteins by enhanced chemiluminescence: A non-radioactive method for cell-surface protein analysis. Anal. Biochem. 204: 220–226.

Merchant, S., K. Hill & G. Howe, 1991. Dynamic interplay between two copper-titrating components in the transcriptional regulation of cyt c_6. The EMBO Journal 10: 1383–1389.

Michel, K.-P., H. H. Thole & E. K. Pistorius, 1996. IdiA, a 34 kDa protein in the cyanobacteria SYnechococcus sp. strain PCC6301 and PCC7942 is required for growth under iron and manganese limitations. Microbiol. 142: 2635–2645.

Miller, P. E. & C. A. Scholin, 1996. Identification of cultured Pseudo-nitzschia (Bacilliariophyceae) using species-specific LSU rRNA-targetted fluorescent probes. J. Phycol. 32: 646–655.

Morel, F. M. M., J. R. Reinfelder, S. B. Roberts, C. P. Chamberlain, J. G. Lee & D. Yee, 1994. Zinc and carbon co-limitation of marine phytoplankton. Nature 369: 740–742.

Morelli, E. & E. Pratesi, 1997. Production of phytochelatins in the marine diatom *Phaeodactylum* tricornutum in response to copper and cadmium exposure. Bull envir Contam Toxicol 59: 657–64.

Muggli, D. L. & P. J. Harrison, 1996. Effects of nitrogen source on the physiology and metal nutrition of Emiliana huxleyi grown under different iron and light conditions. Mar. Ecol. Prog. Ser. 130: 255–267.

Muggli, D. L., M. Lecourt & P. J. Harrison, 1996. Effects of iron and nitrogen source on the sinking rate, physiology and metal composition of an oceanic diatom from the subarctic Pacific. Mar. Ecol. Prog. Ser. 132: 215–227.

Nakamura, Y., T. Kaneko, M. Hirosawa, N. Miyajima & S. Tabata, 1998. CyanoBase, a www database containing the complete nucleotide sequence of the genome of *Synechocystis* sp. strain PCC6803. Nucleic Acids Res. 26: 63–7.

Navarro, J. A., M. Hervas & M. A. De la Rosa, 1997. Co-evolution of cytochrome c6 and plastocyanin, mobile proteins transferring electrons from cytochrome b_6f to photosystem I. J. Biol. Inorg. Chem. 2: 11–22.

Niyogi, K. K., O. Bjorkman & A. R. Grossman, 1997. The roles of specific xanthophylls in photoprotection [In Process Citation]. Proc. natn. Acad. Sci. U.S.A. 94: 14162–7.

Odum, E. P., 1971. Fundamentals of ecology. W. B. Saunders, Philadelphia.

Olaizola, M., J. La Roche, Z. Kolber & P. G. Falkowski, 1994. Non-photochemical fluorescence quenching and the diadinoxanthin cycle in a marine diatom. Photo. Res. 41: 357–370.

Orellana, M. V. & M. J. Perry, 1992. An immunoprobe to measure Rubisco concentration and maximal photosynthetic rates of individual phytoplankton cells. Limnol. Oceanogr. 21: 479–490.

Palenik, B., 1998. Copper-induced cell surface proteins in a marine diatom. EOS Trans. 79.

Palenik, B. & J. A. Koke, 1995. Characterization of a nitrogen-regulated protein identified by cell surface biotinylation of a marine phytoplankton. Appl. envir. Microbiol. 61: 3311–3315.

Palenik, B. & F. M. M. Morel, 1990a. Amino acid utilization by marine phytoplankton: a novel mechanism. Limnol. Oceanogr. 35: 260–269.

Palenik, B. & F. M. M. Morel, 1990b. Comparison of cell-surface L-amino acid oxidases from several marine phytoplankton. Mar. Ecol. Prog. Ser. 59: 195–201.

Palenik, B. & F. M. M. Morel, 1991. Amine oxidases of marine phytoplankton. Appl. envir. Microbiol. 57: 2440–2443.

Palenik, B. & A. M. Wood, 1997. Molecular Markers of Phytoplankton Physiological Status and their Application at the Level of Individual Cells. In Cooksey K. E. (ed.), Molecular Approaches to the Study of the Ocean. Chapman and Hall, London: 187–205.

Partensky, F., J. La Roche, K. Wyman & P. G. Falkowski, 1997. The divinyl-chlorophyll *a*/*b*-protein complexes of two strains of the oxyphototrophic marine prokaryote *Prochlorococcus*-characterization and response to changes in growth irradiance. Photo. Res. 51: 209–222.

Price, N. M. & F. M. M. Morel, 1990. Cadmium and cobalt substitution for zinc in a marine diatom. Nature 344: 658–660.

Price, N. M. & F. M. M. Morel, 1991. Co-limitation of phytoplankton growth by nickel and nitrogen. Limnol. Oceanogr. 36: 1071–1077.

Quisel, J. D., D. D. Wykoff & A. R. Grossman, 1996. Biochemical characterization of the extracellular phosphatases produced by phosphorus-deprived Chlamydomonas reinhardtii. Plant Physiol. 111: 839–48.

Raven, J. A., 1988. The iron and molybdenum use efficicencies of plant growth with different energy, carbon and nitrogen sources. New Phytol. 109: 279–287.

Raven, J. A., 1990. Predictions of Mn and Fe use efficiencies of phototrophic growth as a function of light availability for growth and C assimilation pathway. New Phytol. 116: 1–18.

Razquin, P., S. Schmitz, M. L. Peleato, M. F. Fillat, C. Gomez-Moreno & H. Bohme, 1995. Differential activities of heterocyst ferredooxin, vegetative cell ferredoxin, and flavodoxin as electron carriers in nitrogen fixation and photosynthesis in *Anabaena* sp. Photo. Res. 43: 35–40.

Reddy, K. J., G. S. Bullerjahn, D. M. Sherman & L. A. Sherman, 1988. Cloning, nucleotide sequence, and mutagenesis of a Gene (*irpA*) involved in iron-deficient growth of the cyanobacterium *Synechococcus* Sp. strain PCC7942. J. Bact. 170: 4466–4476.

Redfield, A. C., 1958. The biological control of chemical factors in the environment. Am. Sci. 46: 205–221.

Reith, M. E. & J. Munholland, 1995. Complete nucleotide sequence of the *Porphyra purpurea* chloroplast genome. Plant mol. Biol. rep. 13: 333–335.

Robinson, N. J., 1989. Algal metallothioneins: secondary metabolites and proteins. J. Appl. Phycol. 1: 5–18.

Sandman, G., M. L. Peleato, M. F. Fillat, M. C. Lazaro & C. Gomez-Moreno, 1990. Consequences of the iron-deoendent formation of ferredoxin and flavodoxin on photosynthesis and nitrogen fixation on *Anabaena* strains. Photo. Res. 26: 119–125.

Sandmann, G., 1986. Arch. Microbiol., 145: 76–79. Scanlan, D. J., N. H. Mann & , 1993. The response of the picoplanktonc marine cyanobacterium *Synechococcus* species WH7803 to phosphate starvation involves a protein homologous to the phosphate-binding protein of *Escherichia coli*. Mol. Microbiol. 10: 181–191.

Scanlan, D. J., N. J. Silman, K. M. Donald, W. H. Wilson, N. G. Carr, I. Joint & N. H. Mann, 1997. An immunological approach to detect phosphate stress in populations and singel cells of photosynthetic phytoplankton. Appl. environm. Microbiol. 63: 2411–2420.

Schwarz, R. & A. A. Grossman, 1998. A response regulator of cyanobacteria integrates diverse environmental signals and is critical for survival under extreme conditions. Proc. natn. Acad. Sci. U.S.A. 95: 11008–11013.

Sherman, M. Y. & A. L. Goldberg, 1996. Involvement of molecular chaperones in intracellular protein breakdown. Exs 77: 57–78.

Simon, N., J. Brenner, B. Edvardsen & L. K. Medlin, 1997. The identification of Chrysochromulina and Prymnesium species (Haptophyta, Prymnesiophyceae) using fluorescent or chemiluminescent oligonucleotide. Eur. J. Phycol. 32: 393–401.

Smillie, R. M., 1965. Isolation of phytoflavin, a flavoprotein with chloroplast ferredoxin activity. Plant Physiol. 40: 1124–1128.

Sode, K., M. Tatara, N. Hatano & T. Matsunaga, 1994. Foreign gene expression in marine cyanobacteria under pseudo-continuous culture. J. Biotech. 33: 243–248.

Stearman, R., D. S. Yuan, Y. Yamaguchi-Iwai, R. D. Klausner & A. Dancis, 1996. A permease-oxidase complex involved in high-affinity iron uptake in yeast. Science 271: 1552–1556.

Stevens, D. R. & S. Purton, 1997. Genetic engineering of eukaryotic algae:Progress and prospects. J. Phycol. 33: 713–722.

Stock, J. B., A. F. Ninfa & A. M. Stock, 1989. Protein phosphorylation and regulation of adaptive response in bacteria. Microbiol. reviews 53: 450–490.

Straus, N. A., 1994. Iron Deprivation: Physiology and Gene Regulation. In Bryant D. A. (ed.), The Molecular Biology of Cyanobacteria. Kluwer, Dordrecht: 731–735.

Sunda, W. G., 1994. Trace Metal/Phytoplankton Interactions in the Sea. In G. Bidoglio & W. Stumm (eds.), Chemistry of Aquatic Systems: Local and Global Perspectives. Brussels and Luxembourg: 213–247.

Sunda, W. G. & S. Huntsman, 1997. Interrelated influence of iron, light and cell size on marine phytoplankton growth. Nature 390: 389–392.

Thelwell, C., N. J. Robinson & J. S. Turner-Cavet, 1998. An SmtB-like repressor from *Synechocystis* PCC 6803 regulates a zinc exporter [In Process Citation]. Proc. natl Acad. Sci. U.S.A. 95: 10728–33.

Thompson, S. A., L. L. Wang & P. F. Sparling, 1993. Cloning and nucleotide sequence of frpC, a second gene from Neisseria menengitidis encoding a protein similar to RTX cytoxins. Mol. Microbiol. 9: 85–96.

Timmermans, K. R., W. Stolte & H. J. W. de Baar, 1994. Iron-mediated effects on nitrate reductase in marine phytoplankton. Mar. Biol. 121: 389–396.

Truesdell, G. M. & M. B. Dickman, 1997. Isolation of pathogen/stress-inducible cDNAs from alfalfa by mRNA differential display. Plant Mol. Biol. 33: 737–43.

Van der Staay, G. W. M., A. Ducret, R. Aebersold, R. Li, S. S. Goldenn, R. G. Hiller, P. M. Wrench & B. R. Green, 1995. The Chlorophyll *a/b* Antenna from Prochlorophytes is Related to the Iron Stress-Induced Chl *a* Antenna (*isiA*) from Cyanobacteria. In P. Mathis (ed.), From Light to Biosphere. Kluwer Academic Publishers, Dordrecht.

Vassiliev, K. R., Z. Kolber & , 1995. Effects of iron limitation on Photosystem II composition and light utilization in *Dunaliella tertiolecta*. Plant Physiol. 109: 963–972.

Vega-Palas, M. A., E. Flores & A. Herrero, 1992. NtcA, a global nitrogen regulator from the cyanobacterium *Synechococcus* that belongs to the Crp family of bacterial regulators. Mol. Microbiol. 6: 1853–1859.

Vernet, M. & K. Whitehead, 1996. Release of ultraviolet-absorbing compounds by the red-tide dinoflagellate *Lingulodinium polyhedra*. Mar. Biol. 127: 35–44.

Vigara, A. J., C. Gomez-Moreno & J. M. Vega, 1995. The role of flavodoxin in the reaction catalysed by the glutamate synthetase from *Monoraphidium braunii*. Biolectrochem. Bioenerg. 38: 21–24.

Von Liebig, J., 1840. Chemistry and its application to agriculture and physiology. Taylor and Walton, London.

Wachi, Y., J. G. Burgess, K. Iwamoto, N. Yamada, N. Nakamura & T. Matsunaga, 1995. Effect of ultraviolet-A (UV-A) light on growth, photosynthetic activity and production of biopterin glucoside by the marine UV-A resistant cyanobacterium Oscillatoria sp. Biochim Biophys Acta 1244: 165–8.

Waterbury, J. B. & S. W. Watson, 1979. Widespread occurence of a unicellular, marine, planktonic, cyanobacterium. Science 277: 293–294.

Webb, R., T. Troyan, D. Sherman & L. Sherman, 1994. MapA, an iron-regulated, cytoplasmic membrane protein in the cyanobacterium *Synechococcus* sp. Strain PCC7942. J. Bact. 176: 4906–4913.

Wei, T. & O. C. MA, 1996. Structure and characterization of a putative drought-inducible H1 histone gene. Plant Mol. Biol. 30: 255–68.

Weinberg, E. D., 1974. Iron susceptibility to infectious disease. Science 184: 952–956.

Wells, M. L., L. M. Mayer & R. R. L. Guillard, 1991. Evaluation of iron as a triggering factor for red tide blooms. Mar. Ecol. Prog. Ser. 69: 93–102.

Wetzel, R. G., 1983. Limnology. Saunders, Philadelphia.

Wilhelm, S. W., 1995. Ecology of iron-limited cyanobacteria – a review of physiological responses and implications for aquatic systems. Aquat. Microbial Ecol. 9: 295–303.

Wilhelm, S. W., D. P. Maxwell & C. G. trick, 1996. Growth, iron requirements and siderophore production in iron-limited *Synechococcus* PCC 7002. Limnol. Oceanogr. 41: 89–97.

Wilhelm, S. W. & C. G. Trick, 1994. Iron-limited growth of cyanobacteria: Multiple siderophore production is a common response. Limnol. Oceanogr. 39: 1979–1984.

Williams, J. G. K., A. R. Kubelic, K. J. Livak, J. A. Rafalski & S. V. Tingey, 1990. DNA polymorphisms amplified by arbitrary primers are useful as genetic markers. Nucleic Acids Res. 18: 6531–6535.

Yee, D. & F. M. M. Morel, 1996. *In vivo* substitution of zinc by cobalt in carbonic anhydrase of a marine diatom. Limnol. Oceanogr. 41: 573–577.

Zehr, J. P. & L. A. McReynolds, 1989. Use of degenerate oligonucleotides for amplification of the nifH gene from the marine cyanobacterium *Trichodesmium* thiebautii. Appl. envir. Microbiol. 55: 2522–6.

Zhou, J. & B. G. Spratt, 1992. Sequence diversity within the argF, fbp and recA genes of natural isolates of Neisseria meningitidis: interspecific recombinatioin within the argF gene. Mol. Microbiol. 6: 2135–2146.

Zumft, W. G., 1972. Ferredoxin:nitrite oxidoreductase from Chlorella. Purification and properties. Biochem. Biophys. Acta 276: 363–375.

Hydrobiologia **401:** 199–206, 1999.
J.P. Zehr & M.A. Voytek (eds), Molecular Ecology of Aquatic Communities.

Spatial scale and the diversity of benthic cyanobacteria and diatoms in a salina

U. Nübel[1], F. Garcia-Pichel[2,*], M. Kühl[3] & G. Muyzer[4]
[1]*Max Planck Institute for Marine Microbiology, Bremen, Germany.*
[2]*Max Planck Institute for Marine Microbiology, Celsiusstr. 1, D-28359 Bremen, Germany*
Tel: [+49]421-2028-838; Fax: [+49]421-2028-580; E-mail: fgarcia@mpi-bremen.de
[3]*Present Address: Marine Biological Laboratory, University of Copenhagen, Strandpromenaden 5, 3000 Helsingør, Denmark*
[4]*Present Address: Netherlands Institute for Sea Research (NIOZ), P. O. Box 59. 1790 AB Den Burg (Texel), The Netherlands*

Key words: hypersaline, microbial mat, cyanobacteria, diatom, microbial richness, biodiversity, 16S rRNA

Abstract

We characterized the richness of benthic cyanobacteria and diatoms in a salina system using traditional and molecular biological methods. After determining the different morphotypes and 16S rRNA genes present in various localities within this hypersaline system, an analysis of the increase of organismal richness as a function of numbers of samples considered was carried out. We found that the spatial scales of sampling yielding significant increases in cumulative richness were those at which significant variations in environmental parameters (salinity, vertical microgradients) are known to exist, indicating that the presence of environmental gradients contributes to increased biodiversity. Additionally, we could use this type of cumulative analysis for the estimation, through asymptotic extrapolation, of the total richness of oxygenic phototrophs present in the entire salina system, and for the estimation of the average degree of dissemination of community members within the system. We found interesting differences between analyses based on morphotypes or 16S rRNA genes. The cumulative number of rRNA gene sequences exceeded that of morphotypes by more than two-fold. This indicates that many organisms possessing distinct 16S rRNA gene sequences could not be distinguished on the basis of morphology. Thus, some of the apparently widely distributed morphotypes may in fact conceal several ecologically independent genotypes.

Introduction

Reflecting on the notion that most microorganisms may have a cosmopolitan distribution, early microbiologists coined the expression that with respect to bacteria, 'everything is everywhere' (Baas-Becking, 1934). This hypothesis has pervaded microbiology until today. It is based on the assumption that microscopic organisms are efficiently dispersed over geographic distances, resulting in high local, but comparatively low global diversity (Fenchel, 1993). Indeed, small spatial areas may provide many ecological niches for microorganisms, and high local microbial richness has been described for many habitats (Tiedje, 1995; Pace, 1997). However, very limited data exist on the distribution of microorganisms in nature beyond the microscale. In fact, the impression of widespread occurrence of many microorganisms may to a considerable extent originate from poor taxonomic resolution.

We have investigated the richness and distribution of cyanobacteria and diatoms in a variety of microbial mats in a salina system in Baja California, Mexico. Detailed descriptions of this salina (Javor, 1989; Garcia-Pichel et al., 1994; Des Marais, 1995) and of the morphology of the mats studied (Garcia-Pichel et al., 1999; Nübel et al., 1999) have been published elsewhere. Cyanobacteria and diatoms dis-

* Author for correspondence.

play considerable morphological variability, which is the basis for their current classification (Anagnostidis & Komárek, 1985; Castenholz & Waterbury, 1989; Round et al., 1990). Accordingly, 'floristic' studies have described the spatial distribution of morphology-defined taxa. For example, patterns of occurrence of morphotypes along environmental gradients in various hypersaline environments have been reported (Jørgensen et al., 1983; Ehrlich & Dor, 1985; Montoya T. & Golubic, 1991; Clavero et al., 1994). Moreover, for thermophilic cyanobacteria, the absence of particular morphotypes in some geothermal habitats suggests restrictions to dispersal at the global scale (Castenholz, 1996). However, limitations of morphology-based classification have also become apparent, particularly for morphologically more simple cyanobacteria (Castenholz, 1992; Komárek, 1996). Morphological groupings may (Garcia-Pichel et al., 1996) or may not (Ferris et al., 1996) represent phylogenetically coherent entities, and pleiomorphism may cause additional difficulties for microscopic identification (Garcia-Pichel et al., 1998). Thus, depending on the organisms and on the habitat studied, the analysis of cell and colony morphology may be misleading to an unknown extent, if the study of organismic diversity and distribution is intended. We investigated the richness of oxygenic phototrophs as reflected by their morphology and 16S rRNA genes. We compared richness differences among and within localities, and, using cumulative analyses, evaluated the use of such techniques to determine the spatial scales at which diversity increases. Additionally, we used comparisons of data based on either morphotypes or rRNA gene sequences to evaluate the power of morphological classification to delineate ecologically coherent units in this habitat across environmental gradients.

Materials and methods

Sampling

Microbial mats were sampled during the second to fourth weeks of April 1996 (mats P2, P4, P6, NC2, NC3) and 1997 (mats P3/4, P5, NC52). Sampling sites were located in evaporation ponds of the saltern in Guerrero Negro, Baja California Sur, Mexico, and in the salt marsh of the Ojo de Liebre Lagoon, as illustrated in Figure 1. Within the salina system, the samples analyzed covered the entire spectrum of salinity at which microbial mats occured, ranging from 6 to 16% (Figure 1). In addition, mats from upper tidal channels that had been exposed to natural diel variations of salinity were included (see Figure 1 for locations and ranges of salinity variation). Each of these sampling sites is referred to hereafter as a locality. Within each locality, triplicate cores 10–20 cm apart were subsampled. For light-microscopy, mat samples (core diameter, 4 mm) were fixed in 5% formaldehyde (wt/vol) and stored at 4 °C. For extractions of nucleic acids, mat samples (core diameter, 25 mm) were frozen on site, transported to the laboratory in liquid nitrogen, and stored at −80 °C until processed. Measurements of photosynthesis were performed in a field laboratory with microsensor techniques (Revsbech & Jørgensen, 1986). The maximum depth where gross photosynthesis was detectable varied from 1.5 to 4.8 mm (mean, 2.4 mm). These photic zones were sampled for the investigation of richness. Further results of these supplementary studies are reported elsewhere (Garcia-Pichel et al., 1999).

Analysis of morphotypes

The layers corresponding to photic zones were cut from formaldehyde-fixed mat samples with scalpel blades, and sectioned vertically into subcores of approximately 0.5×0.5 mm mat surface area. These pieces were placed on glass slides in one drop of water and chopped and stirred to achieve even distributions. Of each subcore, 25–40 randomly chosen phase-contrast microscopic fields were photographed at 400-fold magnification and cell counts were performed on projections of the resulting slides. Depending on the respective mat, 2000–3000 cells needed to be investigated to detect all rare and localized morphotypes in a single subcore, and thus to achieve stable richness estimates (Nübel et al., 1999).

Analysis of 16S rRNA genes

The layers corresponding to photic zones were aseptically cut from mat cores (100–400 mg, representing approximately 60 mm^2 of mat surface) and homogenized in Dounce tissue homogenizers (Novodirect, Kehl, Germany). Cell lysis and DNA extraction were performed as described previously (Nübel et al., 1997). Briefly, the suspensions were repeatedly frozen and thawed, and subsequently incubated in the presence of sodium dodecyl sulfate and proteinase K. Cell lysis was controlled microscopically. DNA was extracted by applying hexadecylmethylammonium bromide, phenol, chloroform, and

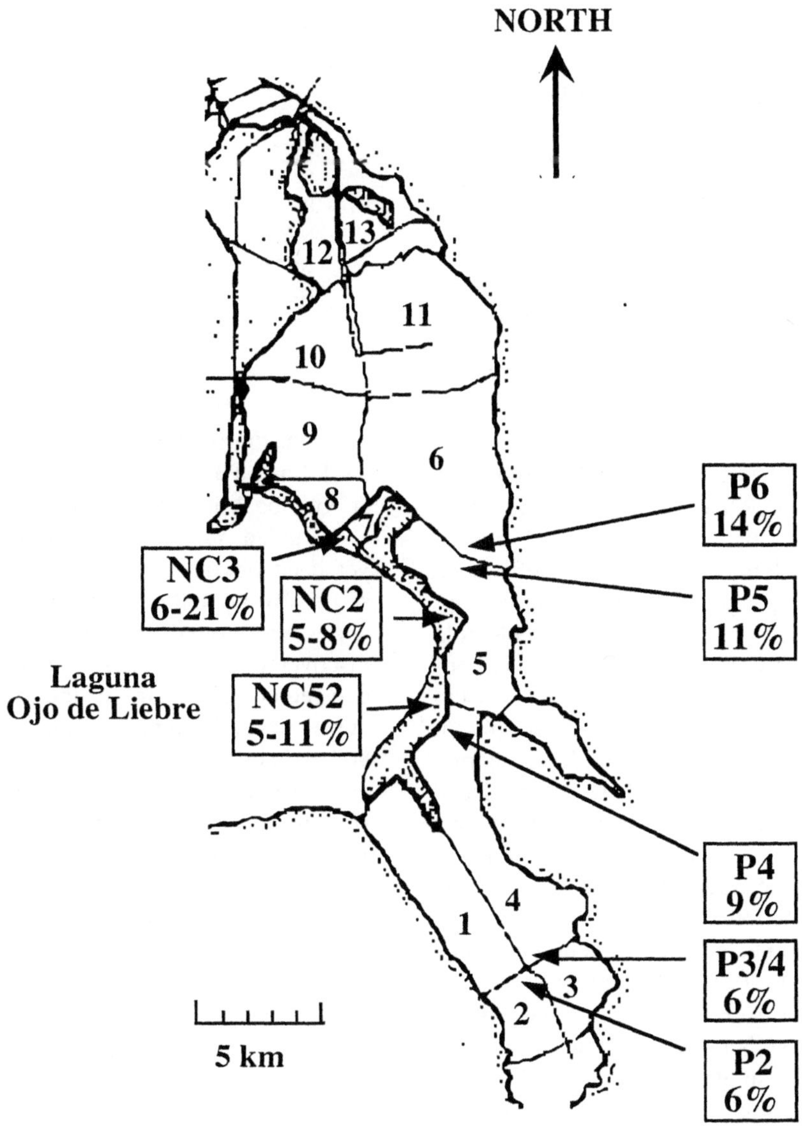

Figure 1. Map showing concentrator ponds of the saltern Exportadora de Sal, S. A. de C. V., Baja California Sur, Mexico. Sampling sites (localities) and salinities of brines, measured at time of collection, are indicated. In most localities salinitiy was virtually constant. In localities situated within tidal channels salinity varied with tidal cycles within the ranges indicated.

isoamylalcohol, and precipitated by addition of isopropyl alcohol. 10 ng of DNAs extracted from mat samples were added as templates to each 100 μl polymerase chain reaction (PCR) mixture. Oligonucleotide primers CYA359F and CYA781R were applied to selectively amplify 16S rRNA gene segments from cyanobacteria and plastids (Nübel, et al., 1997). Numbers in primer designations refer to 5′-ends of target signature sites in 16S rRNA genes (*Escherichia coli* nucleotide numbering (Brosius et al., 1981)). A 40-nucleotide GC-rich sequence was attached to the 5′-end of the primer CYA359F to improve the detection of sequence variation in amplified DNA fragments by subsequent denaturing gradient gel electrophoresis (DGGE (Nübel et al., 1997)). Amplification products generated by duplicate polymerase chain reactions with the same template DNAs were pooled and subsequently purified and concentrated by using the QIAquick PCR purification kit (Diagen, Düsseldorf, Germany). DNA concentrations in resulting solutions were determined by comparisons to the Gibco low DNA mass standard (Gibco, Eggenstein, Germany) after agarose gel electrophoresis. 500 ng of DNA was applied onto denaturing gradient gels. DGGE separates DNA molecules with identical length but different sequences on the basis of their differential melting behaviour in a gradient of denaturants. It was performed as described previously (Nübel et al., 1997). Briefly, polyacrylamide gels with a denaturant gradient from 20% to 60% were used, electrophoreses were run for 3.5 h at 60 °C and 200 V, and subsequently DNA was stained by using ethidium bromide.

Estimation of richness

The estimation of the richness and diversity of morphotypes and 16S rRNA genes of oxygenic phototrophic microorganisms is described and discussed in detail elsewhere (Nübel et al., 1999). Briefly, cells and molecules were grouped into classes defined by unique cell and colony morphologies or nucleotide sequences, respectively. Ecological richness is considered to be the number of different classes. Cells and molecules from organisms other than oxygenic phototrophs were excluded from the analyses. Microscopically, diatoms and cyanobacteria could be distinguished from most other microorganisms due to their size, characteristic morphologies, and natural autofluorescence. 16S rRNA gene fragments were amplified by polymerase chain reaction (PCR) from cyanobacterial and plastid DNA after nucleic acid extraction from mat samples (Nübel et al., 1997). Phylum-specific amplification enabled the exclusion from the analyses of DNA from organisms other than oxygenic phototrophs (Nübel et al., 1997). Numbers of unique rRNA gene segments amplified were estimated after DGGE analysis of PCR products. With increasing numbers of bands detected in electrophoretic analyses the probability increases that classes cannot be discerned because they run at identical positions in the gel. However, based on the total distance of separation, average peak width, and the low numbers of peaks detected, this crowding effect was estimated to have only negligible effect on our DGGE results (Nübel et al., 1999).

Cumulative richness and its extrapolation

With increasing number of samples investigated, novel classes (morphotypes or 16S rRNA gene sequences) were detected that resulted in an increase of the cumulative richness. For example, the total number of different morphotypes detected in mats from localities P2 and P3/4 was 16 (five morphotypes occurred at both localities). Inclusion of locality P4 in the analysis added another five morphotypes that had not yet been observed in mats from P2 and P3/4, and so on. We pooled data obtained from individual localities and plotted the cumulative number of classes discovered as a function of the number of localities considered. The shape of the resulting accumulation curves depends on the sequence in which samples are added. In order to avoid this arbitrariness we computed and considered all possible accumulation sequences (Figure 2). The morphotype and gene sequence accumulation curves approach asymptotic maxima (S_{max}) which we estimated by extrapolation. For extrapolation two-parameter hyperbolas were fitted through means of cumulative richness calculated for an increasing number of localities as:

$$S(n) = \frac{S_{max} n}{K + n}$$

In this equation, S is the mean cumulative richness and n is the number of localities pooled. S_{max} and K are fitted constants, which were calculated by double inverse linear regression ($1/S(n)$ on $1/n$, Figure 2). $S(n) = S_{max}$ is the asymptote and S_{max} can be interpreted as the total richness to be expected in the system under study (Colwell & Coddington, 1994). We did not attempt to estimate confidence intervals for S_{max} and K because of questionable statistical validity of such a computation (Colwell and Coddington, 1994). The above equation is well known as the Michaelis-Menten equation of enzyme kinetics.

Table 1. Numbers of morphotypes

Locality	Triplicate subsamples			Cumulative total
P2	11	10	11	12
P3/4	7	7	8	9
P4	11	10	10	12
P5	7	6	8	8
P6	10	10	9	10
NC2	13	11	12	13
NC52	11	13	12	13
NC3	13	12	13	13

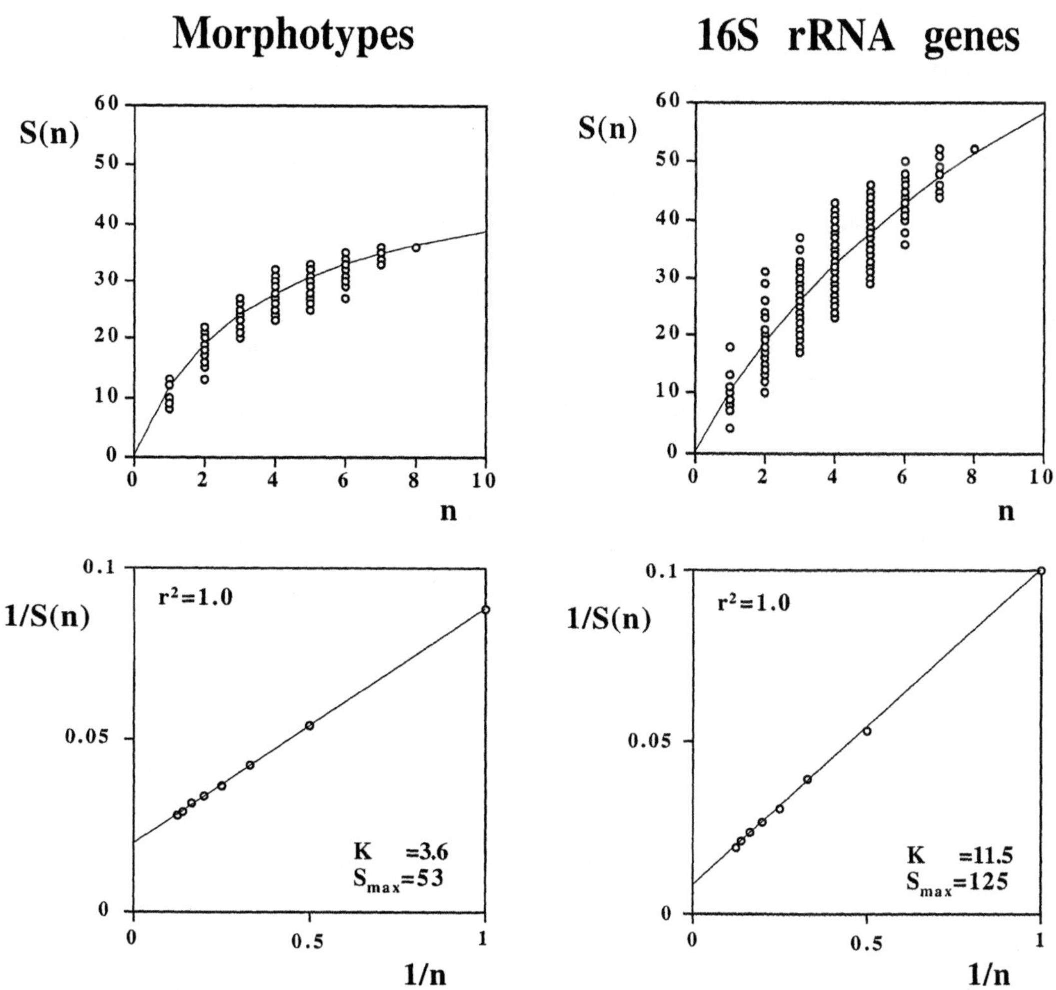

Figure 2. Cumulative number of classes, $S(n)$, plotted as functions of the number of localities, data from which were pooled, n. The left hand graphs show data from morphotype analyses, the right hand graphs show data from 16S rRNA gene analyses. In the upper graphs results from all possible combinations of localities were plotted to eliminate variation in curve shape due to accumulation order. Two-parameter hyperbolas were fitted through means of $S(n)$ based on double-reciprocal plots as shown in the lower graphs. Pearson correlation coefficients, r, estimates of asymptotic total regional richness, S_{max}, and half saturation constants, K, are given.

Table 2. Numbers of 16S rRNA genes

Locality	Triplicate subsamples			Cumulative total
P2	10	12	12	12
P3/4	4	4	4	4
P4	10	10	10	10
P5	7	7	6	7
P6	10	10	10	10
NC2	12	12	12	13
NC52	13	12	12	13
NC3	14	17	17	18

Results and discussion

Spatial heterogeneity

The analyses performed permitted investigation of the heterogeneity of the distribution and diversity of cyanobacteria and diatoms at various spatial scales. Considerable patchiness in the organismal distribution was observed at the scale of cyanobacterial colonies and filaments (10–100 μm). Detailed microscopic analysis indicated that at least 2000–3000 cells per subsample needed to be investigated to achieve stable and repeatable estimates of morphotype richness (Nübel et al., 1999). In few cases, and to a lesser extent, this type of heterogeneity was still apparent at the scale of millimeters. For example, tufts of diatoms were macroscopically visible on the surfaces of some of the mats. However, for all eight localities investigated, richness estimates based on either morphotypes or 16S rRNA genes were consistent among triplicate subsamples that had been taken several centimeters apart. The coefficients of variation for triplicates within localities were typically 5% and maximally 8%, indicating that heterogeneity was low within localities. As a result of this homogeneity, cumulative richness within localities was not much higher than that of the single subsamples (Tables 1 and 2). In contrast, we found marked differences among localities several kilometers apart. This is consistent with the assumption that the concentration of total salts in the brines overlying the microbial mats directly or indirectly determined the horizontal distribution of cyanobacteria and diatoms (Golubic, 1980; Ehrlich & Dor, 1985).

Estimation of regional richness

In the eight mat communities investigated, a total of 36 different morphotypes and 52 different 16S rRNA gene segments from oxygenic phototrophs were detected. Pooling of the data from an increasing number of localities resulted in increasing, but saturating, cumulative richness (Figure 2). Extrapolation of accumulation curves estimated the total numbers of different morphotypes and 16S rRNA genes from benthic oxygenic phototrophs probably present in the whole salina, S_{max}, to be 53 and 125, respectively. The mean richness detected at single localities, *meanS*, was 10.2 and 10.3 in the microscopic and molecular biological analyses, and the ratios $meanS/S_{max}$ accordingly were 0.19 and 0.08, respectively. These quotients are relative measures of the average dissemination of classes, theoretically ranging from zero (all classes are locally endemic) to one (all classes are ubiquitous). A dissemination quotient close to one may be reached in habitats such as soil, in which microbial community composition can be very homogeneous at a geographical scale of kilometers (Felske & Akkermans, 1998). The dissemination quotients determined in our study must be considered rather low, and are probably related to the marked physicochemical differences among sampling sites. However, it was observed that some morphotypes and rRNA genes were widespread among the various mat communities analysed, whereas others were found at few or single localities only (our unpublished results). While widespread occurrance through a wide range of salinity clearly indicates that an organism is euryhaline, restriction to certain communities may be caused by various other physicochemical factors or biological interactions. *K*, analogous to the half saturation constant K_m in enzyme kinetics, can be interpreted as the average number of samples needed to detect half of the richness in the system under study. These constants were found to be 3.6 and 11.5 for the analyses of morphotypes and 16S rRNA genes, respectively. In general, such estimates should have the potential for optimizing efforts in diversity surveys.

Morphotypes and 16S rRNA genes

We compared the richness of oxygenic phototrophs in microbial mats as reflected by their morphology and 16S rRNA gene sequences. Both approaches must be considered biased to an unknown extent, especially with respect to abundance determinations (discussed in detail in (Nübel et al., 1999)). However, when investigating samples within localities, both approaches repeatedly yielded similar results. Richness estimates based on 16S rRNA genes within localities correlated significantly with those obtained by morphotype analyses (Pearson correlation coefficient, r=0.940, p=0.001). Interestingly, the ratio of both richness estimates (means of triplicate analyses) equaled 1 in six of eight localities (Nübel et al., 1999). Thus, when analysing the diversity of cyanobacteria and diatoms at small spatial scale, morphotypes and 16S rRNA genes may seem equivalent. However, we found that morphology analysis detected significantly less richness than gene analysis when organism distribution was considered over larger spatial scales. The number of sequence-different 16S rRNA gene segments found in the eight microbial mats clearly exceeded the number of morphotypes identified. The lower degree of saturation (*K*, Figure 2) found for the latter analysis caused the estimates of S_{max} to differ more than two-fold.

The number of different 16S rRNA gene sequences found should be considered a minimum estimate of the underlying richness of 16S rRNA genes because oxygenic phototrophs may exist in the samples that do not contain the signature sequences for efficient PCR amplification (Nübel et al., 1997), and because amplified DNA molecules potentially may occur at the same position in the denaturing gradient gel despite differential sequences (Nübel et al., 1999). In turn, sequence analysis for the majority of bands indicated that most likely, none of the bands was derived from any undesired amplification product (data not shown). Slightly different 16S rRNA gene sequences may exist in some organisms (Nübel et al., 1996), but this phenomenon seems to be rather rare among cultivated cyanobacteria and diatoms from hypersaline environments (our unpublished results).

With continued sampling sequence-different 16S rRNA genes accumulated at a faster rate and with a higher saturation value than morphotypes (Figure 2). Since the presence of strong limits to dispersal within this system of contiguous flow seem unlikely, the differences must indicate that many of the morphotypes concealed genotypes that are ecologically distinct and therefore occur at different sampling sites. Two reasons are plausible to explain the observation that these organisms have indistinguishable morphology, and probably both apply to some extent. First, morphology analysis may detect diversity at a phylogenetic level too high to correlate with ecophysiological diversity. In fact, once they are established, morphological features are rather conserved in phylogenetic lineages of bacteria in general (Siefert & Fox, 1998), and physiological differentiation may have occurred relatively later in the evolutionary history of the respective organisms. Alternatively, convergent evolution of morphology may have followed, or may have been part of, ecophysiological specializations. In any case, our results reflect the limitations of morphology-based classification, and indicate that the analysis of nucleic acid sequences may be more sensitive for the detection of the ecophysiological diversity of oxygenic phototrophic microorganisms. Clearly, there is a need for new cyanobacterial taxa, integrating information on the ecology and physiology of the respective organisms (Golubic, 1980; Anagnostidis & Komárek, 1985; Castenholz, 1992). For in-depth characterisations of ecophysiological capabilities of microorganisms, however, information retrievable from field observations will be insufficient in most cases. At best, differences in occurrence patterns may indicate some, but largely cryptic, divergence among distinguishable populations. In contrast, the bacteriological system of cyanobacterial classification (Castenholz & Waterbury, 1989) allows much more detailed experimental investigations on individual strains, but severely suffers from the difficulties inherent in the need for isolation and cultivation and currently is far from being complete (Castenholz, 1992). Molecular biology, however, offers powerful tools for the comparison of cultivated strains and their counterparts in the wild, as well as for guiding future isolation attempts towards the discovery of as yet unknown microorganisms.

Conclusions and outlook

It seems unlikely that limits to dispersal may isolate microbial populations only a few kilometers apart. The differential distribution of organisms must have been determined by their specific ecological demands. Accordingly, large increases in cumulative richness, detected by continued sampling, occurred at a spatial scale which included measurable variations in environmental parameters. We cannot assert at this point if this apparent relationship between microbial richness and environmental heterogeneity will hold in general for microbial communities, but it seems plausible in the light of niche specialization theory. If it does, then it should indicate that the spatial scales important for the accumulation of diversity may depend on the spatial structure of environmental gradients or boundaries. In habitats which show considerable physical heterogeneity at smaller scales (millimeter, centimeter), but not necessarily at larger scales (meter, kilometer), microbial richness may increase at different scales than those that were important in our system. It should prove interesting to extend this type of study to the global scale, as this may offer important information on extant global microbial diversity, for which data are simply non-existent. If cumulative analysis of similar habitats in widely separated geographical localities (for example, salina systems worldwide) should reveal small cumulative richness increases, this would speak for efficient dispersal at planetary scales. The dissemination quotients should give some information on the potential existence of limits to dispersal for a portion of the community members. Additionally, this type of analysis may facilitate the distinction of the presence of climatic borders for microbial communities or the determination of the space/time scales at which microbial dispersal occurs. It must be noted, however, that the primary structure of ribosomal RNA genes is highly conserved, and to identify ecologically coherent populations, it may even be necessary to analyse more variable nucleic acid sequences, such as those encoding proteins (Palys et al., 1997). Generally, ecologically and evolutionarily more meaningful species concepts for microorganisms may be developed in the future (Ward, 1998). Only then, when based on more equivalent units, the diversity and distribution in space of plants, metazoa and unicellular organisms can be meaningfully compared.

Acknowledgements

We gratefully acknowledge Exportadora de Sal, S. A. de C. V., Baja California Sur, Mexico, for continued logistic support and we thank E. Clavero for her introduction to the classification of the diatoms. The research described in this paper was financially supported by the Max Planck Society and the Deutsche Forschungsgemeinschaft.

References

Anagnostidis, K. & Komárek, J., 1985. Modern approach to the classification system of cyanophytes. 1. Introduction. Arch. Hydrobiol. Suppl. 71: 291–302.

Baas-Becking, L. G. M., 1934. Geobiologie of inleiding tot de milieukunde. Den Haag, The Netherlands.

Brosius, M., T. Dull, D. D. Sleeter, & H. F. Noller, 1981. Gene organization and primary structure of a ribosomal RNA operon from *Escherichia coli*. J. Mol. Biol. 148: 107–127.

Castenholz, R. W., 1992. Species usage, concept, and evolution in the Cyanobacteria (blue-green algae). J. Phycol. 28: 737–745.

Castenholz, R. W., 1996. Endemism and biodiversity of thermophilic cyanobacteria. Nova Hedwigia 112: 33–47.

Castenholz, R. W. & J. B. Waterbury, 1989. Oxygenic phototrophic bacteria, group I. Cyanobacteria. In Bryant, M. P., N. Pfennig & J. G. Holt. Baltimore, Md. (eds), Bergey's Manual of Systematic Bacteriology. The Williams and Wilkins Co.: 1710–1728.

Clavero, E., J. O. Grimalt, V. Merino, & M. Hernández-Mariné, 1994. Structure of the sediment at depositional saline environments. In Stal L. J. & P. Caumette (eds), Microbial Mats – Structure, Development and Environmental Significance. Springer Verlag: 91–96.

Colwell, R. K. & J. A. Coddington, 1994. Estimating terrestrial biodiversity through extrapolation. Phil. Trans. R. Soc. Lond. B 345: 101–118.

Des Marais, D. J., 1995. The biogeochemistry of hypersaline microbial mats. Adv. Microb. Ecol. 14: 251–274.

Ehrlich, A. & I. Dor, 1985. Photosynthetic microorganisms of the Gavish Sabkha. In Friedman G. M. & W. E. Krumbein (eds), Hypersaline Ecosystems. Berlin Heidelberg, Germany: Springer Verlag.

Felske, A. & A. D. L. Akkermans, 1998. Spatial homogeneity of abundant bacterial 16S rRNA molecules in grassland soils. Microb. Ecol. 36: 31–36.

Fenchel, T., 1993. There are more small than large species? OIKOS 68: 375–378.

Ferris, M. J., A. L. Ruff Roberts, E. D. Kopczynski, M. M. Bateson & D. M. Ward, 1996. Enrichment culture and microscopy conceal diverse thermophilic *Synechococcus* populations in a single hot spring microbial mat habitat. Appl. environ. Microbiol. 62: 1045–1050.

Garcia-Pichel, F., M. Kühl, U. Nübel & G. Muyzer, 1999. Salinity-dependent limitation of photosynthesis and oxygen exchange in microbial mats. J. Phycol. 35: 227–238.

Garcia-Pichel, F., M. Mechling & R. W. Castenholz 1994. Diel migrations of microorganisms within a benthic, hypersaline mat community. Appl. environ. Microbiol. 60: 1500–1511.

Garcia-Pichel, F., U. Nübel & G. Muyzer 1998. The phylogeny of unicellular, extremely halotolerant cyanobacteria. Arch. Microbiol. 169: 469–482.

Garcia-Pichel, F., L. Prufert-Bebout & G. Muyzer 1996. Phenotypic and phylogenetic analyses show *Microcoleus chthonoplastes* to be a cosmopolitan cyanobacterium. Appl. environ. Microbiol. 62: 3284–3291.

Golubic, S., 1980. Halophily and halotolerance in cyanophytes. Origins of life 10: 169–183.

Javor, B., 1989. *Hypersaline environments*. Berlin, Germany: Springer Verlag.

Jørgensen, B. B., N. P. Revsbech & Y. Cohen, 1983. Photosynthesis and structure of benthic microbial mats: microelectrode and SEM studies of four cyanobacterial communities. Limnol. Oceanogr. 28: 1075–1093.

Komárek, J., 1996. Towards a combined approach for the taxonomy and species delimitation of picoplanktic cyanoprokaryotes. Arch. Hydrobiol. Suppl. 117: 377–401.

Montoya T. H. & S. Golubic, 1991. Morphological variability in natural populations of mat-forming cyanobacteria in the salinas of Huacho, Lima, Peru. Algol. Studies 64: 423–441.

Nübel, U., B. Engelen, A. Felske, J. Snaidr, A. Wieshuber, R. I. Amann, W. Ludwig & H. Backhaus, 1996. Sequence heterogeneities of genes encoding 16S rRNAs in *Paenibacillus polymyxa* detected by temperature gradient gel electrophoresis. J. Bact. 178: 5636–5643.

Nübel, U., F. Garcia-Pichel & G. Muyzer, 1997. PCR primers to amplify 16S rRNA genes from cyanobacteria. Appl. environ. Microbiol. 63: 3327–3332.

Nübel, U., F. Garcia-Pichel, M. Kühl & G. Muyzer, 1999. Quantifying microbial diversity: morphotypes, 16S rRNA genes, and carotenoids from oxygenic phototrophs in microbial mats. Appl. environ. Microbiol. in press.

Pace, N. R., 1997. A molecular view of microbial diversity and the biosphere. Science 276: 734–740.

Palys, T., L. K. Nakamura & F. M. Cohan, 1997. Discovery and classification of ecological diversity in the bacterial world: the role of DNA sequence data. Int. J. syst. Bacteriol. 47: 1145–1156.

Revsbech, N. P. & B. B. Jørgensen, 1986. Microelectrodes: their use in microbial ecology. Adv. Microb. Ecol. 9: 293–352.

Round, F. E., R. M. Crawford & D. G. Mann, 1990. The diatoms: morphology and biology of the genera. Cambridge, UK: Cambridge University Press.

Siefert, J. L. & G. E. Fox, 1998. Phylogenetic mapping of bacterial morphology. Microbiology 144: 2803–2808.

Tiedje, J. M., 1995. Approaches to the comprehensive evaluation of prokaryote diversity of a habitat. In Allsopp D., R. R. Colwell & D. L. Hawksworth (eds), Microbial Diversity and Ecosystem Function. CAB International. Oxon, UK: 73–87.

Ward, D. M., 1998. A natural species concept for prokaryotes. Curr. Opin. Microbiol. 1: 271–277.

Hydrobiologia **401:** 207–214, 1999.
J.P. Zehr & M.A. Voytek (eds), Molecular Ecology of Aquatic Communities.

A rapid method to score plastid haplotypes in red seaweeds and its use in determining parental inheritance of plastids in the red alga *Bostrychia* (Ceramiales)

G.C. Zuccarello[1], J.A. West[2], M. Kamiya[3] & R.J. King[1]
School of Biological Science, University of New South Wales, Sydney, NSW 2052, Australia
Tel: [+61]-2-9385-2075; Fax: [+61]-2-9385-1558; E-mail: g.zuccarello@unsw.edu.au
[2]*School of Botany, University of Melbourne, Parkville, VIC 3052, Australia*
[3]*Kobe University Research Center for Inland Seas, Iwaya, Awaji-cho, Tsunagun, Hyogo-ken 656-2401, Japan*

Key words: plastid haplotypes, SSCP, population variation, uniparental inheritance, *Bostrychia*, *Caloglossa*, Rhodophyta, Ceramiales

Abstract

A method for the rapid identification of plastid haplotypes in red algae is presented. This method is based on single-stranded conformation polymorphism (SSCP) to detect variation in DNA sequence of PCR-amplified plastid-encoded ribulose bisphosphate carboxylase/oxygenase large and small subunit intergenic spacer. The method detects variation within a population of the mangrove red alga *Caloglossa leprieurii* from New South Wales, Australia. Three haplotypes (plastid variants) were easily scored in our samples, and a spatial variation of haplotypes within the intertidal is suggested. This method also reveals maternal inheritance of plastids in crosses between isolates of *Bostrychia radicans* and in multiple crosses between isolates of *B. moritziana*. Fluorescence microscopy shows that plastids are contained within the spermatia of these species and that bi- or paternal inheritance of plastids is possible. SSCP is a simple and rapid method for the detection of plastid variation (haplotypes) within populations of red algae that should also be applicable to brown seaweeds.

Introduction

Molecular methods for the study of evolution and population genetics are becoming increasingly prevalent in biology (Avise, 1994; Hillis et al., 1996), though their use in macroalgae is as yet limited. Population studies are important in community ecology research as aspects such as population variation, population subdivision, geneflow and recruitment can be addressed. The application of DNA-based molecular studies to the investigation of seaweeds has focused mostly on the use of DNA sequence to elucidate systematic and taxonomic problems. There are relatively few studies that have addressed issues below the species level and most of these have had a biogeographic focus (for example see Van Oppen et al., 1995; Van Oppen et al., 1996; Stache-Crain et al., 1997; Zuccarello & West, 1997; Kamiya et al., 1998). DNA sequence methods have proved useful for phylogenetic reconstruction but they have the drawback that the large samples required for studies of diversity within populations cannot be screened quickly or cheaply.

Genetic studies on seaweeds that have determined variation at the population level, have relied mostly on protein electrophoresis (see Pearson & Murray, 1997; Benzie et al., 1997), although Van Oppen et al. (1995) were able to demonstrate the use of RAPD (randomly amplified polymorphic DNA) data in a study on the red alga *Phycodrys rubens*. Single-locus microsatellite markers have also been used successfully in a study of the population genetics of the red alga, *Gracilaria gracilis* (Wattier et al., 1997). Microsatellite methods are ideal for population genetic studies due to the high variability of microsatellite loci (i.e. multiple alleles). The development, however, of single-locus microsatellite markers can be expensive and time consuming. These studies all point to the need for a generally applicable PCR-based method that can assess variation within populations.

DNA derived from organelles, and especially mitochondrial DNA (mtDNA), has proved to be extremely useful in population studies and phylogenetic reconstructions in animals (see Avise, 1994). The non-recombinant mode and rapid evolution of this genome often provides multiple haplotypes that can be ordered phylogenetically within a species and provide useful population genetic and demographic data. PCR primers for the highly variable control region of the mitochondrial genome are readily available (Hillis et al., 1996). Similar variable DNA regions are not known in the organelles of plants or algae with the usual method for scoring plastid haplotypes in plants being restriction fragment length polymorphism (RFLP) analyses (Avise, 1994; Hillis et al., 1996). This procedure, however, usually involves elaborate extraction procedures and protocols.

The red algae (Rhodophyta) and the brown algae (Phaeophyceae), have both the large subunit of the ribulose bisphosphate carboxylase/oxygenase gene (*rbc*L) and the small subunit of the ribulose bisphosphate carboxylase/oxygenase gene (*rbc*S) cotranscribed in the plastid (Kostrzewa et al., 1990; Fujiwara et al., 1993; Bhattacharya & Medlin, 1995), whereas the *rbc*S in green algae and land plants is encoded in the nucleus. The two plastid genes are separated by a short intergenic spacer region (Rubisco spacer). DNA sequence variations in this region have provided useful data for the reconstruction of algal inter- and intra-species phylogenies (Destombe & Douglas, 1991; Maggs et al., 1992; Bird et al., 1994; Goff et al., 1994; Stache-Crain et al., 1997; Zuccarello & West, 1997; Kamiya et al., 1998). In these studies, the intergenic region has been used to analyze isolates from different biogeographic locations, but in only one case did sequence data reveal variation at the population level (Zuccarello & West, 1997).

Hybridization is an important component in increasing the evolutionary potential of organisms (Rieseberg, 1997) and has been demonstrated in marine seaweeds (Scott & Hardy, 1994; Lewis & Neushul, 1995). Organelle DNA, mitochondrial and plastid, is thought to be inherited uniparentally in most organisms (Birky, 1995). Though plastid DNA is reported to be maternally inherited in most flowering plants (Corriveau & Coleman, 1988), its mode of transmission can vary and needs to be verified experimentally (Milligan, 1992). The pattern of organelle inheritance in algae is less well understood. In the model green alga *Chlamydomonas*, plastid inheritance can vary (Sager, 1974) with certain species exhibiting predominantly uniparental plastid inheritance (mating-type plus strain in *C. reinhardtii*, Sears & Vanwinkle-Swift, 1994) while others have a predominantly biparental inheritance of plastids, (in *C. moewusii*, Lee & Lemieux, 1986).

The mode of plastid inheritance in the red algae has not been studied. Ultrastructural studies of male gametes (spermatia) are limited, and while some studies of spermatia reveal plastids, albeit with reduced or absent thylakoids (Kugrens & West, 1972; Cole & Sheath, 1980; Kugrens, 1980; Broadwater & Scott, 1983), others detail spermatia in which there are no plastids (Fetter & Neushul, 1981; Broadwater et al., 1991). The evidence, however, is ambiguous at best, as the presence of plastids in spermatia does no more than indicate the possibility of bi- or male plastid inheritance. As a result of this, only a DNA-based method, targeting plastid DNA, can be conclusive.

Methods for screening allelic variation in polymerase chain reaction amplified DNA, from the large sample sizes needed for studies of population biology, demography and phylogeny, are available (see Lessa & Applebaum, 1993). One of the simpler and less costly methods is single-stranded conformation polymorphism or SSCP. This method was first described by Orita et al. (1989). The method involves no special equipment beyond standard sequencing apparatus and has been improved and applied non-radioactively (Glavac & Dean, 1993; Hongyo et al., 1993), Although some of the latter techniques do involve special equipment (i.e. circulating temperature baths). The ability of the SSCP method to distinguish between mutant DNA fragments relies on the principles that the molecular conformation of single stranded DNA is nucleotide-sequence specific, that this conformation is altered by point substitutions, insertions and deletions, and that the altered conformation affects the mobility of the molecule to migrate through a polyacrylamide gel. This method has been reported to reveal up to 100% of the mutations in fragments of the 150–400 base pairs size.

In this study, we apply a rapid method of DNA extraction, Rubisco spacer PCR and SSCP techniques, to the study of genetic variation in the plastids of field collected material of *Caloglossa leprieurii* (Montagne) J. Agardh. Furthermore, this method was used to assess the inheritance of plastids in crosses within two different species of *Bostrychia* Montagne.

Materials and methods

Sample collection, preparation and DNA extraction

Caloglossa leprieurii was collected from Woolooware Bay, New South Wales (N.S.W.) on 10 June 1998. Pneumatophores of *Avicennia marina*, with attached algae, were collected from three 30 cm^2 quadrats 5 m apart perpendicular to the shoreline on an early morning low tide. Samples were placed in plastic bags and returned to the laboratory. Pneumatophores were washed in seawater and one plant of *C. leprieurii* was removed from each pneumatophore. The *C. leprieurii* used had approximately 4–5 blades each of 5–10 mm length. Due to the stoloniferous growth habit of this alga, individuals could not be recognized and only by sampling different pneumatophores could we be sure of collecting different individuals.

Bostrychia isolates [*Bostrychia moritziana* (Sonder *ex* Kützing) J. Agardh and *B. radicans* (Montagne) Montagne] were collected, isolated in unialgal culture and crossed as outlined previously in Zuccarello & West (1995).

DNA extractions from samples were performed using the Chelex-100 (Biorad, Hercules, California) protocol of Goff & Moon (1993). Thalli, approximately 5–15 mm of an epiphyte-free axis were placed in 100–200 μl of a 10 mM Tris (pH 8.0) solution containing 5% Chelex^{TM}-100 resin (BIORAD, Hercules, Ca, U.S.A.) in a 1.5 ml microfuge tube. The Chelex solution was shaken vigorously before each aliquot was removed to ensure proper proportions of the resin. Tissue was ground in the microcentrifuge tube using Kontes^{TM} microcentrifuge tube grinders until the solution became colored (cells disrupted), and only small fragments of tissue were visible. To remove exogenous DNA grinders were sterilized by placing them in 10% bleach for an hour or more, and washing repeatedly in autoclaved double-distilled water until all bleach odor was eliminated. The samples were kept on ice before and after grinding. After grinding, samples were placed in a floating rack in boiling water for 10 min and then immediately placed on ice. Samples were centrifuged for 10 min and then the supernatant was carefully removed and placed in new marked microcentrifuge tubes. Care was taken not to remove the Chelex resin or the unground tissue.

PCR reactions

Two sets of Rubisco spacer primers were used. For the *Caloglossa leprieurii* amplification the primers were: forward primer 5′ -TATACTTCTACAGACACAGCTG A - 3′ (rbcF1), reverse primer 5′ -ATTTCACACAGGA AACAGCTATGACATGTCAAATAATGGTAGTCCC CA - 3′ (rbcR2-M2). The sequence complementary to the M13 reverse sequencing primer was added to the latter primer to aid in future DNA sequencing (Kamiya et al., 1998). For *Bostrychia* amplification the primers were: forward primer 5′ -TGTGGACCTCTACAAACAGC - 3′, and reverse primer 5′ -CCCCATAGTTCCCAAT - 3′ (Maggs et al., 1992).

PCR was performed using radioactively end-labelled forward primer. Final end-labelling conditions were: 100 pmol of primer, 1–2 μl of gamma-[^{33}P]ATP (0.27 10^9 Bq ml^{-1}), 1X reaction buffer and 3 units of T4 polynucleotide kinase in a 10 μl volume, incubated at 37 °C for 20 min and heat-inactivated at 95 °C for 5 min. Amplification reactions were performed in 10 μl volumes using the following final concentrations: 1X PCR buffer (Promega^{TM} buffer A), 120 nmol dNTP's, 2 mM $MgCl_2$, 2.5 pmol of each primer (labelled forward, unlabelled reverse), 0.25% BSA (Sigma, Cat no. A-2153) and 0.5 U *Taq* polymerase (Promega^{TM}) and 1 μl Chelex-extracted template DNA. Amplifications were performed under a variety of conditions, the final optimal conditions were: an initial denaturation was carried out at 94 °C for 4 min, followed by 5 cycles of 93°/45°/72° for 1 min each, 20 cycles of 93°/55°/72° for 1 min each and a final step 72 °C for 5 min.

SSCP (single-strand conformation polymorphism) analysis.

SSCP gels were poured in a standard sequencing gel box (IBI-Kodak STS-45 sequencing apparatus, 350×420×0.4 mm dimensions), with 64-well sharktooth combs.

For SSCP, 2 μl of PCR product was mixed with 10 μl denaturing solution (98% formamide, 10 mM NaOH, 0.025% bromophenol blue, 0.025% xylene cyanol), heated to 95 °C for 5 min and placed directly on ice. Two μl of this denatured DNA was loaded onto a non-denaturing acrylamide gel consisting of 0.5 X MDE acrylamide (FMC Bioproducts) in 0.6 X TBE and electrophoresis carried out at 30 W for 12 hours at 4 °C cold room. In addition, the surface of the gel was cooled continuously by a blowing fan. Other SSCP conditions used included gels consisting of 4.5% acrylamide [37.5:1 acrylamide: bisacrylamide], 1 X TBE) and electrophoresis, in 1 X TBE, at 50W for approximately 5 h (4 °C). Although these latter conditions also produced resolvable haplotypes, the degree of separation (i.e. resolution) was reduced. After electrophoresis, gels were dried on Whatmann paper and exposed to X-ray film (Hyperfilm MP, Amersham) for 12–24 h.

Results and discussion

The advantages of the Chelex/PCR/SSCP methodology over traditional methods of DNA extraction/PCR/Sequencing are multiple. Samples of both *Caloglossa* and *Bostrychia* from the field were easily selected and extracted, even from relatively small thalli (less than 1 cm in man axis length). This methodology is also very rapid, as large numbers of specimens (50) could be routinely collected, sorted and extracted in one half day. The Chelex^{TM} extraction procedure is rapid and has been very successful in producing PCR amplifiable template DNA with various mangrove and non-mangrove red algae, predominantly within the order Ceramiales. PCR reactions can be performed the same day and SSCP gels run overnight. For PCR, it is also important to avoid too much tissue in the initial extraction as this can inhibit amplification. The addition of 0.25% BSA to the PCR mixture greatly increased the reliability of successful PCR amplifications from the field collected specimens which is probably due to the known effects of BSA in circumventing the inhibition effects of contaminants from environmental samples (Wintzingeorde et al., 1997).

Gels made with MDE acrylamide provided the best resolution but are more expensive than normal acrylamide gels. Acrylamide gels (4.5% [37.5:1 acrylamide: bisacrylamide]) also gave satisfactory results, and occasionally, for other species, were able to resolve variation not apparent in the MDE gels. Both sets of gels were routinely used on our PCR amplifications. Room temperature gels, with and without 10% glycerol, did not resolve our particular products well, although not all combinations of concentrations (%) of acrylamide or cross-linker (bis-acrylamide) were tested (Glavac & Dean, 1993). It is also important to avoid heating of the gel by cooling with a fan and performing electrophoresis at low wattage.

Amplification products were between 250–350 base pairs in size, depending on the primer set used, which is an ideal size for SSCP analysis. Figure 1 shows a SSCP gel of three specimens of *Caloglossa leprieurii* collected from a single transect in Woolooware Bay. Three haplotypes were easily resolved within this population. Not only was plastid variation uncovered by this method but a spatial variation in haplotype distribution is also indicated. Algal samples with three different haplotypes, in roughly equal proportions, were found in quadrat A, corresponding to the quadrat furthest from the low water mark. Two haplotypes are found in samples from quadrat B, though one haplotype is clearly dominant. In quadrat C, corresponding to the quadrat closest to the low water mark and only 10 m from quadrat A, only a single haplotype was observed from sampled algae. Though preliminary, this population and spatial variation of haplotypes shows that SSCP is able to detect haplotypes quickly and easily in the populations. These different haplotypes could correlate to genotypic/phenotypic adaptations (e.g. tol-

Table 1. Haplotypes of parental strains and hybrid sporophytes of *B. radicans*. Culture numbers correspond to isolates from the J.A. West culture collection (for further details on these isolates see Zuccarello & West, 1997). Haplotype = arbitrary haplotype designation. Lane = gel lanes in Figure 3. n.a. = not applicable, i.e. parental strain. X = indicates cross between cultures, multiple crosses indicate different sporophytes derived from different released carpospores

Culture numbers			
Female	Male	Haplotype	Lane
3136VZ	n.a	A	a
n.a.	3206MX	B	b
n.a.	3207MX	B	c
3136VZ X	3206MX	A	d
3136VZ X	3207MX	A	e
3207MX X	3136VZ	B	f
3017BZ	n.a.	C	g
n.a.	3043PE	D	h
3017BZ X	3043PE	C	i
3017BZ X	3043PE	C	j
3017BZ X	3043PE	C	k
3017BZ X	3043PE	C	l
3017BZ X	3043PE	C	m
3043PE X	3017BZ	D	n
3043PE X	3017BZ	D	o
3043PE X	3017BZ	D	p
3043PE X	3017BZ	D	q
3043PE X	3017BZ	D	r
3043PE X	3017BZ	D	s

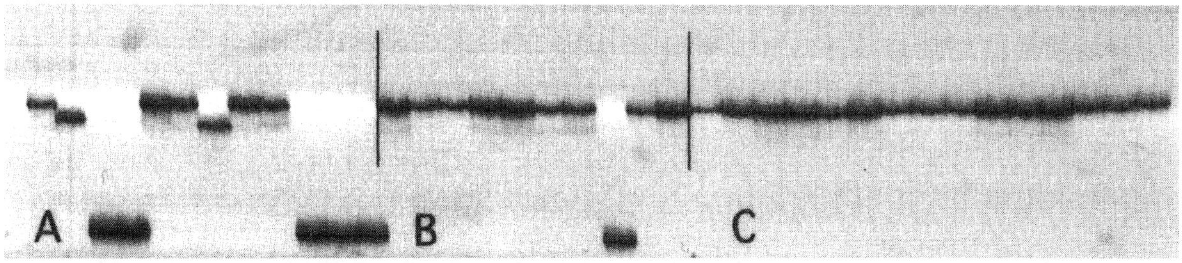

Figure 1. SSCP analysis of amplified Rubisco spacer region from *Caloglossa leprieurii* collected from Woolooware Bay, NSW. Lines indicate separation between samples from the 3 different quadrats (A, B, C).

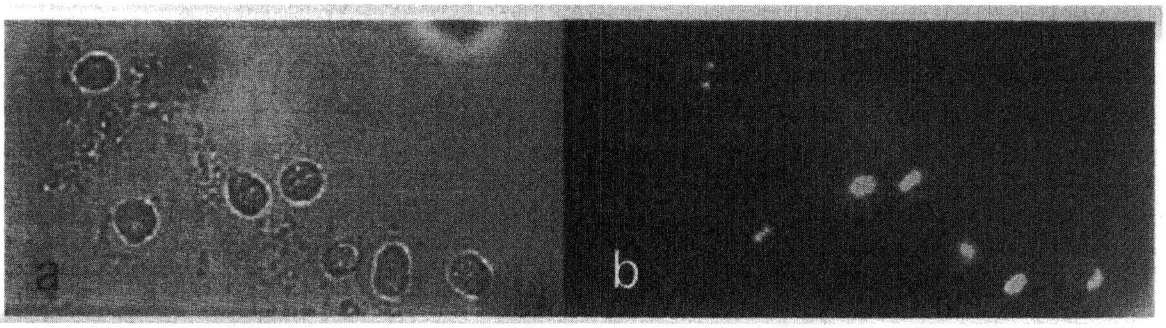

Figure 2. Micrograph of released spermatia from *Bostrychia moritziana* (live specimen). a= Bright field image. b= same image seem under fluorescence. Autofluorescent plastids are visible.

erance to desiccation) or may suggest very limited dispersal/recruitment of thalli in the field. Work in currently in progress to answer these questions.

Previous studies have shown that isolates of *Bostrychia radicans* from different locations around the world are able to hybridize (Zuccarello & West, 1995, 1997). DNA sequence data from these isolates has also revealed sequence variation in the Rubisco spacer. For example, isolate 3017 from Brazil differed from isolate 3043 from Peru by a single base pair substitution. Fluorescence microscopy of the released spermatia of *Bostrychia moritziana* clearly shows that plastids are found within the spermatia (Figure 2). Spermatia autofluorescence was identical in color to vegetative cell autofluorescence. Interestingly, only ca. 70% of the observed spermatia contained autofluorescent plastids which varied considerably in size. Fixation and DAPI staining of the spermatia, to visualize plastid genome DNA was unsuccessful, possibly due to plastid autofluorescence and small size of the plastid genome. The presence of autofluorescent bodies does not definitely prove the presence of plastids, but as they are a similar color to vegetative plastids this interpretation appears to be reasonable. Nor can we be sure that they contain plastid DNA, but if they do they allow for the possibility that the spermatia once fused to the egg (trichogyne) could transfer a plastid to the future zygote cytoplasm. Plastid presence, based on endogenous autofluorescence in spermatia, has been demonstrated likewise in *B. radicans* and *Caloglossa leprieurii*. These data are consistent with ultrastructural data that indicate plastids within many red algal spermatia (Kugrens & West, 1972; Cole & Sheath, 1980; Kugrens, 1980; Broadwater & Scott, 1983). Only molecular data using a plastid marker will confirm the mode of inheritance of this organellar genome.

To test for the parental mode of plastid inheritance in the red alga *Bostrychia radicans*, crosses were performed between isolates of *B. radicans* from which different plastid haplotypes had already been demonstrated (Table 1, Figure 3) (Zuccarello & West, 1997). SSCP analysis showed maternal inheritance of the PCR-amplified plastid gene in reciprocal crosses. Furthermore the analysis of 5 different hybrid sporophytes derived from different carpospores was consistent, in all cases, with maternal inheritance. In the limited samples analyzed here, it does not appear that the nuclear genotype (different isolate) or plastid genotype (haplotype) in any way affects the plastid inheritance mode, as is seen in other organelle inheritance patterns in plants and animals (Birky, 1995). 'Leakage',

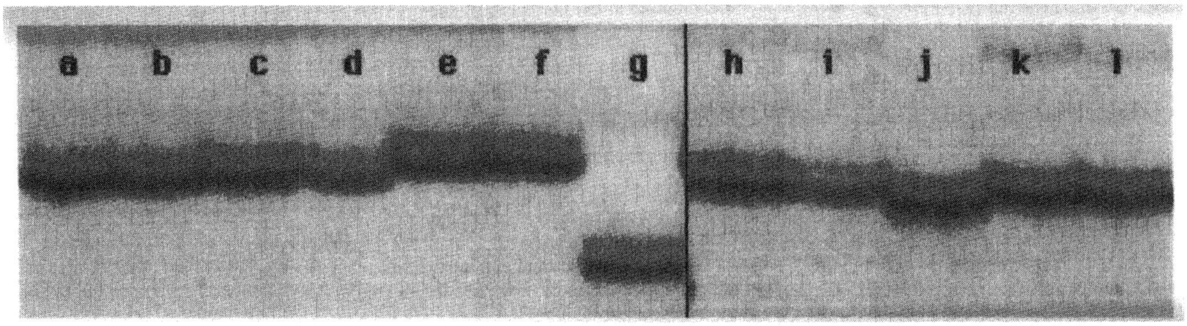

Figure 3. SSCP of amplified Rubisco spacer region of parental isolates and hybrid F_1 tetrasporophytes of *Bostrychia radicans*. For lane designation see Table 1.

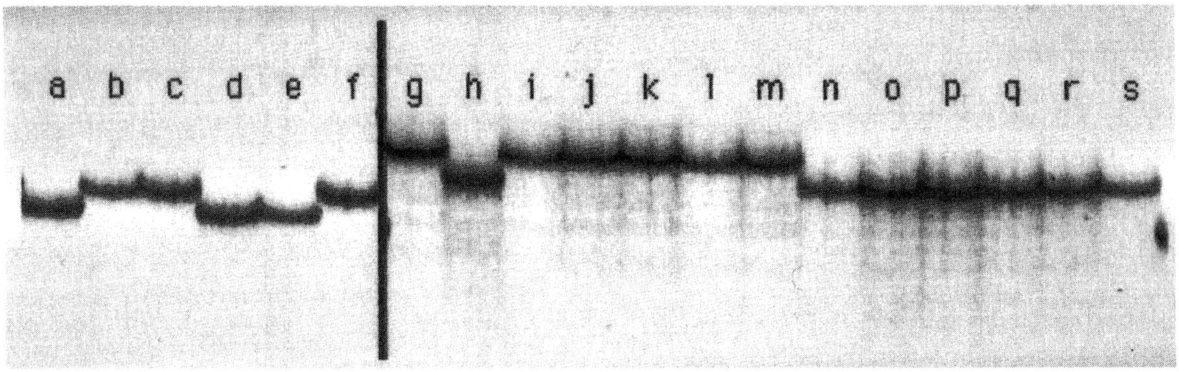

Figure 4. SSCP of amplified Rubisco spacer region of parental isolates and hybrid F_1 tetrasporophytes of *Bostrychia moritziana*. For lane designation see Table 2.

```
A.3738FJ     ATTGGATCTATGGAAAGATATAACGTTTAACTATACTTCTACAGATACAGCAGACTTTGT
B.3209SA     ............................................................
C.3275SA     ..............................T.............................

A.3738FJ     AGAAACTCCAACAGCTAATGTTTAAATAAAAAATTTTCTTCACTAGTATTTTTTAATTAT
B.3209SA     ............................................................
C.3275SA     ............................................................

A.3738FJ     TTGTATGTATACAAATAATTTAAATCTTTTATATACAGATCAATTCTCATAAGGAGTGAA
B.3209SA     ............................................................
C.3275SA     ............................................................

A.3738FJ     TAGTGAGACTAACACAAGGAACTTTTTCCTTTTTACCAGACTTAACTGATGAACAAATCA
B.3209SA     .....................................C......................
C.3275SA     .....................................C......................

A.3738FJ     AAAAACAAATTGACTACGCAATTTCTCAAAATTGGGCAGTAAATATTGAATATACAGAAG
B.3209SA     ............................................................
C.3275SA     ............................................................

A.3738FJ     ATCCGCATCCTAGAAATAGCT
B.3209SA     .....................
C.3275SA     .....................
```

Figure 5. DNA sequence of the Rubisco spacer from *Bostrychia moritziana*isolates. See Table 2. Haplotype A= A.3738FJ. Haplotype B= B.3209SA. Haplotype C= C.3275SA.

Table 2. Haplotypes of parental strains and hybrid sporophytes of *B. moritziana*. Culture numbers correspond to isolates from the J.A. West culture collection. Haplotype = arbitrary haplotype designation. Lane = gel lanes in Figure 4. n.a. = not applicable, i.e. parental strain. X = indicates cross between two isolates

Culture numbers			
Female	Male	Haplotype	Lane
3738FJ X	3197SA	A	a
3738FJ X	3209SA	A	b
3738FJ X	3275SA	A	c
3738FJ	n.a.	A	d
n.a.	3197SA	B	e
n.a.	3209SA	B	f
n.a.	3275SA	C	g
3197SA X	3738FJ	B	h
3235SA X	3738FJ	B	i
n.a.	3738FJ	A	j
3197SA	n.a.	B	k
3235SA	n.a.	B	l

the usual term for the alternate type of organelle inheritance (usually paternal) pattern, is rare in most organisms (Avise, 1994)), and was not detected in the limited number of samples analyzed in this study.

Our studies on the biogeography and reproductive patterns of the mangrove alga *Bostrychia moritziana* have led to the screening using SSCP and sequencing of the Rubisco spacer in many isolates, and to crosses between different isolates. In this species no haplotype variation could be found within the western Pacific (Australia, New Zealand, Fiji-unpublished data). Crosses between isolates that did show Rubisco spacer variation are shown in Table 2 and Figure 4. These crosses again show solely maternal inheritance of plastids. Sequencing of these haplotypes also shows that the SSCP method was able to distinguish variation of just a single base pair difference (Figure 5). Haplotype A differs from haplotype B by a single base change, from an adenosine to a cytidine in position 218, while haplotype A differs from haplotype C by 2 base changes, a cytidine to a thymidine in position 31 and an adenosine to a cytidine in position 218. SSCP screening appears to be extremely sensitive in detecting extremely low levels of sequence variation. This is confirmed by complete sequencing of identical haplotypes which, as yet, have not revealed a single case of DNA sequence variation not observed by SSCP (unpublished data).

The application of SSCP is currently the best approach available for the rapid screening of plastid variation in the seaweeds. We believe this method should also be applicable to the Rubisco spacer region of brown algae through the use of different primer combinations (Stache-Cairns et al., 1997). In the initial use of this methodology it is recommended that concurrent sequence data be gathered to confirm the SSCP results. A drawback of this method for population genetic studies may be the limited degree of plastid DNA sequence variation. The plastid genome is believed to mutate more infrequently at the DNA sequence level than the mitochondrial genome (Avise, 1994). Data from this study support this contention, within a single population of *Caloglossa leprieurii* and *Bostrychia radicans* only a few haplotypes have been revealed (Zuccarello & West, 1997, this study and unpublished data).

This is the first study to conclusively show maternal inheritance of plastids in marine red algae. We are still ignorant to the fate of mitochondria in crosses between red algae, although studies in this area are currently in progress. Future work in our laboratory utilizing SSCP is aimed at demonstrating the utility of this region and these methods in describing genetic variation, demography and systematics of algal populations.

Acknowledgements

The research was supported by a grant from the Australian Research Council and the UNSW Vice-Chancellor's Research Fellowship to GCZ.

References

Avise, J. C., 1994. Molecular Markers, Natural History and Evolution. Chapman & Hall, New York, 511 pp.

Benzie, J. A. H., I. R. Price & E. Ballament, 1997. Population genetics and taxonomy of *Caulerpa* (Chlorophyta) from the Great Barrier Reef, Australia. J. Phycol. 33: 491–504.

Bhattacharya, D. & L. Medlin, 1995. The phylogeny of plastids: A review based on comparisons of small-subunit ribosomal RNA coding regions. J. Phycol. 31: 489–498.

Bird, C. J., P. A. Sosa & R. M. Mackay, 1994. Molecular evidence confirms the relationship of *Petrocelis* in the western Atlantic to *Mastocarpus stellatus* (Rhodophyta, Petrocelidaceae). Phycologia 33: 134–137.

Birky, C. W., 1995. Uniparental inheritance of mitochondrial and chloroplast genes- mechanism and evolution. Proc. natn. Acad. Sci. U.S.A. 92: 11331–11338.

Broadwater, S. & J. Scott, 1983. Fibrous vacuole associated organelles (FVAO's) in the Florideophyceae: a new interpretation of the 'appareil cinétique'. Phycologia 1983: 225–233.

Broadwater, S., J. L. Scott & J. A. West, 1991. Spermatangial appendages of *Spyridia filamentosa* (Ceramiaceae, Rhodophyta). Phycologia 30: 189–195.

Cole, K. & R. G. Sheath, 1980. Ultrastructural changes in major organelles during spermatial differentiation in *Bangia* (Rhodophyta). Protoplasma 102: 253–279.

Corriveau, J. L. & A. W. Coleman, 1988. Rapid screening method to detect potential biparental inheritance of plastid DNA and results for over 200 angiosperm species. Am. J. Bot. 75: 1443–1458.

Destombe, C. & S. E. Douglas, 1991. Rubisco spacer sequence divergence in the rhodophyte alga *Gracilaria verrucosa* and closely related species. Curr. Genet. 19: 395–398.

Fetter, R. & M. Neushul, 1981. Studies on developing and released spermatia in the red alga, *Tiffaniella snyderae* (Rhodophyta). J. Phycol. 17: 141–159.

Fujiwara, J., H. Iwahashi, J. Someya & S. Nishikawa, 1993. Structure and cotranscription of the plastid-encoded *rbcL* and *rbcS* genes of *Pleurochrysis carterae* (Prymnesiophyta). J. Phycol. 29: 347–355.

Glavac, D. & M. Dean, 1993. Optimization of the single-stranded conformation polymorphism (SSCP) technique for the detection ofpoint mutations. Human Mutation 2: 404–414.

Goff, L. J. & D. A. Moon, 1993. PCR amplification of nuclear and plastid genes from algal herbarium specimens and algal spores. J. Phycol. 29: 381–384.

Goff, L. J., D. A. Moon & A. W. Coleman, 1994. Molecular delineation of species and species relationships in the red algal agarophytes *Gracilariopsis* and *Gracilaria* (Gracilariales). J. Phycol. 30: 521–537.

Hillis, D. M., C. Moritz & B. K. Mable, 1996. Molecular Systematics Sinauer Associates Inc., Sunderland, Massachusetts, 655 pp.

Hongyo, T., G. S. Buzard, R. J. Calvert & C. M. Weghorst, 1993. 'Cold SCCP': a simple, rapid and non-radioactive method for optimized single-stranded conformation polymorphism analyses. Nucleic Acids Research 21: 3637–3642.

Kamiya, M., J. A. West, R. J. King, G. C. Zuccarello, J. Tanaka & Y. Hara, 1998. Evolutionary divergence in the red algae *Caloglossa leprieurii* and *C. apomeiotica*. J. Phycol. 34: 361–370.

Kooistra, W. H. C. F., W. T.Stam, J. L. Olsen & C. van den Hoek, 1992. Biogeography of the green alga *Cladophoropsis membranacea* (Chlorophyta) based on nuclear rDNA ITS sequences. J. Phycol.28: 660–668.

Kostrzewa, M., K. Valentin, U. Maid, R. Radetzky & K. Zetsche, 1990. Structure of the Rubisco operon from the multicellular red alga *Antithamnion* spec. Curr. Genet. 18: 465–469.

Kugrens, P., 1980. Electron microscopic observations on the differentiation and release of spermatia in the marine red alga *Polysiphonia hendryi* (Ceramiales, Rhodophyta). Am J. Bot. 67: 519–528.

Kugrens, P. & J. A. West, 1972. Ultrastructure of spermatial development in the parasitic red algae *Levringiella gardneri* and *Erythrocystis saccata*. J. Phycol. 8: 331–343.

Lee, R. W. & C. Lemieux, 1986. Biparental inheritance of non-Mendeliangene markers in *Chlamydomonas moewusii*. Genetics. 113: 589–600.

Lessa, A. P. & G. Applebaum, 1993. Screening techniques for detecting allelic variation in DNA sequences. Mol. Ecol. 2: 119–129.

Lewis, R. J. & M. Neushul, 1995. Intergeneric hybridization among five genera of the family Lessoniaceae (Phaeophyceae) and evidence for polyploidy in a fertile *Pelagophycus* X *Macrocystis* hybrid. J. Phycol. 31: 1012–1017.

Maggs, C. A., S. E. Douglas, J. Fenety & C. J. Bird, 1992. A molecular and morphological analysis of the *Gymnogongrus devoniensis* (Rhodophyta) complex in the north Atlantic. J. Phycol. 28: 214–232.

Milligan, B. G., 1992. Is organele DNA strictly maternally inherited? Power analysis of a binomial distribution. Am. J. Bot. 79: 1325–1328.

Orita, M., M. Suzuki, T. Sekiya & K. Hayashi, 1989. Rapid and sensitive detection of point mutations and DNA polymorphisms using the polymerase chain reaction. Genomics. 5: 874–879.

Pearson, E. A. & S. N. Murray, 1997. Patterns of reproduction, genetic diversity, and genetic differentiation in California populations of the geniculate coralline alga *Lithothrix aspergillum* (Rhodophyta). J. Phycol. 33: 753–763.

Rieseberg, L. H., 1997. Hybrid origins of plant species. Ann. Rev. Ecol. Syst. 28: 359–389.

Sager, R., 1974. Nuclear and cytoplasmic inheritance in green algae. In Stewart, W. D. P. (ed.), Algal Physiology and Biochemistry. University of California Press, Berkeley: 314–345.

Sears, B. B. & K. Vanwinkle-Swift, 1994. The Salvage-turnover-repair (STOR) model for uniparental inheritance in *Chlamydomonas*: DNA as a source of sustenance. J. Heredity. 85: 366–376.

Scott, G. W. & F. G. Hardy, 1994. Observations of the occurrence of hybrids between 2 sympatric species of fucoid algae. Crypt. Algol. 15: 297–305.

Stache-Crain, B., D. G. Muller & L. J. Goff, 1997. Molecular systematics of *Ectocarpus* and *Kuckuckia* (Ectocarpales, Phaeophyceae) inferred from phylogenetic analysis of nuclear- and plastid-encoded DNA sequences. J. Phycol. 33: 152–168.

van Oppen, M. J. H., J. L. Olsen & W. T. Stam, 1995. Genetic variation within and among North Atlantic and Baltic populations of the benthic alga *Phycodrys rubens* (Rhodophyta). Eur. J. Phycol. 30: 251–260.

van Oppen, M. J. N., H. Klerk, M. de Graaf, W. T. Stam & J. L. Olsen, 1996. Hidden diversity in the marine algae: some examples of genetic variation below the species level. J. mar. biol. Ass. U.K. 76: 239–242.

Wattier, R., J. F. Dallas, C. Destombe, P. Saumitou-Laprade & M. Valero, 1997. Single locus microsatellites in Gracilariales (Rhodophyta): High level of genetic variability within *Gracilaria gracilis* and conservation in related species. J. Phycol. 33: 868–880.

Wintzingeorde, F., U. B. Göbel & E. Stackebrandt, 1997. Determination of microbial diversity in environmental samples: pitfalls of PCR-based rRNA analysis. FEMS Microb. Rev. 21: 213–229.

Zuccarello, G. C. & J. A. West, 1995. Hybridization studies in *Bostrychia*. 1: *B. radicans* (Rhodomelaceae, Rhodophyta) from Pacific and Atlantic North America. Phycol. Res. 43: 233–240.

Zuccarello, G. C. & J. A. West, 1997. Hybridization studies in *Bostrychia*. 2: Correlation of crossing data and plastid DNA sequence data within *B. radicans* and *B. moritziana* (Rhodophyta, Ceramiales). Phycologia. 36: 293–304.

Hydrobiologia **401:** 215–227, 1999.
J.P. Zehr & M.A. Voytek (eds), Molecular Ecology of Aquatic Communities.

Protistan community structure: molecular approaches for answering ecological questions

D.A. Caron, R.J. Gast, E.L. Lim & M.R. Dennett
Biology Department, Woods Hole Oceanographic Institution, Woods Hole, MA 02543, U.S.A.

Key words: 18S rRNA, protistan biogeography, protists, RFLP, oligonucleotide probe, protistan community structure, *in-situ* hybridization, microbial ecology, molecular ecology

Introduction

Protists are essential to the function of most aquatic food webs. Photosynthetic protists ranging in size from a few micrometers to more than 1 mm produce much of the organic material available in these ecosystems, particularly in pelagic ecosystems. In addition, protistan predators ranging up to more than 1 cm in size are major consumers of bacterial, cyanobacterial and protistan populations (Sherr & Sherr, 1994).

Our knowledge of the structure and function of protistan assemblages has advanced considerably in recent years, particularly with regard to planktonic communities. Estimates of abundance and biomass for a variety of protistan groups have been reported from diverse environments (Fenchel, 1967; Davis et al., 1985; Porter et al., 1985; Garrison et al., 1986; Caron & Swanberg, 1990; Stoecker et al., 1994). Major trophic modes for many of these species have been examined, and a number of generalized models of organic carbon production and utilization have been proposed (Figure 1). These depictions typically involve grouping species by trophic mode (i.e. phototrophy, heterotrophy, mixed nutrition) and size-dependent grazing by consumers.

It can be argued that protistan community structure among the larger protists is considerably better characterized than for the smaller protists. For example, extensive species lists of phototrophic protists such as diatoms and dinoflagellates exist, as well as distributional information. Likewise, the distributions and abundances of large heterotrophic protists in freshwater and marine ecosystems have been fairly well established. Probably the most thoroughly documented distributions of marine protists exist for the large, shelled species of Foraminifera and some Actinopoda which are used in paleoclimatological studies (Bé, 1959; Bé, 1960; Anderson, 1980; Dworetzky & Morley, 1987; Hemleben et al., 1988; Caron & Swanberg, 1990). Likewise, the use of the Utermöhl settling method, and the recent development and application of methods such as the quantitative protargol stain for the identification and enumeration of ciliates, have enabled studies of the distributions of these latter species (Utermöhl, 1958; Stoecker et al., 1989; Montagnes & Lynn, 1993).

Our knowledge concerning the diversity, abundance and distribution of large protists is more complete than for small species in part because of the presence of morphological features that are taxonomically useful, visible by light and/or electron microscopy, and which remain intact throughout sampling, preservation and examination procedures. These features may include cell size and shape, skeletal structures (loricae, frustules, tests) patterns of ciliature, thecal plates and body scales. Distinguishing attributes typically are more numerous and/or more apparent at low magnification for large protists, thus facilitating the acquisition of accurate identifications and quantitative information on their abundances.

In contrast, species diversity within assemblages of small protists (<20 μm in size) is not well understood, and the distributions of most of these species are poorly described. It is clear from studies of small protists in laboratory cultures, however, that these assemblages are composed of species that differ markedly in their rates of growth and trophic activity. A logical next step in improving our understanding of microbial trophodynamics, and the biogeochemical significance of small protists in aquatic ecosystems, is to begin to understand the biodiversity of these assemblages and the factors controlling their distributions and activities. Characterization of the biodiversity of small protistan assemblages, and the func-

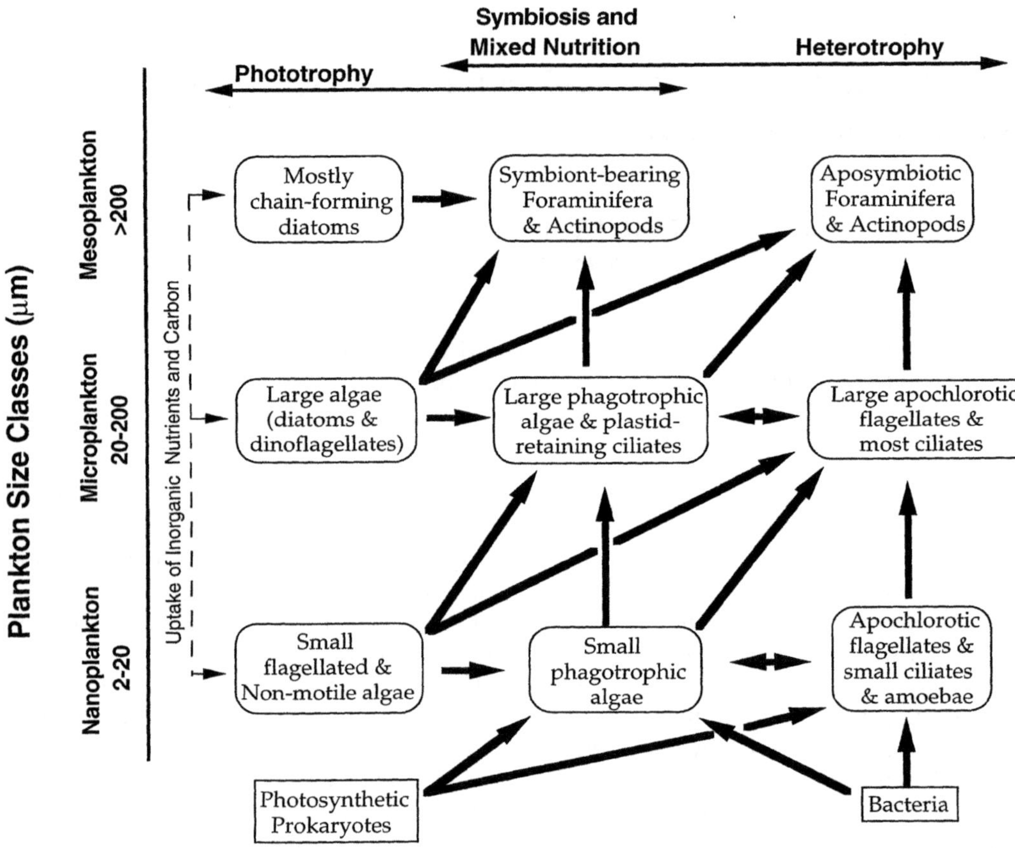

Figure 1. Diagrammatic representation of a microbial marine food web, with generalized roles for various protistan taxa. Taxa are separated by trophic mode (phototrophy, mixotrophy or photosymbiosis, heterotrophy) and size. Solid lines and arrows between the compartments indicate probable predator-prey relationships. This model assumes size-dependent trophic relationships (i.e. grazers eat prey smaller than themselves).

tional diversity that they represent, will provide new insights into how microbial food webs are structured and how they function in nature.

In this manuscript, we briefly describe some of the issues involved in the identification of small protists in natural mixed assemblages, explain the limitations of extant methods, and discuss several molecular approaches that have been applied, or are being developed, to address these issues. Much of this work is ground-breaking. Therefore, most of the impact of these approaches on our understanding of the biodiversity and biogeography of small protists is only beginning to be realized.

HNAN species identification: inherent problems

Techniques for examining assemblages of small protists in natural waters have included a variety of light microscopical techniques, electron microscopy and culture-dependent methods (Kopylov & Sazhin, 1989; Wehr, 1990; Rogerson & Laybourn-Parry, 1992; Vørs, 1993; Stoecker et al., 1994; Vørs et al., 1995). Generally, these methods strike a compromise between the ability to accurately enumerate individuals in a sample, and to identify species within the sample. For example, culture-dependent methods and electron microscopy provide information which aid in the identification of small protists, but these methods generally underestimate their abundances in natural samples (Caron et al., 1989). The use of light microscopy (most notably epifluorescence microscopical techniques) with preserved samples can provide more accurate abundance estimates, but these methods provide limited taxonomic information.

The most commonly employed methods in use today for counting natural assemblages of small planktonic protists entail the use of fluorochrome staining and epifluorescence microscopy to visualize these microorganisms, and the autofluorescence of chlorophyll to differentiate phototrophic individuals. Typically, 2–20 μm algae are counted as a group (the phototrophic nanoplankton, usually abbreviated PNAN), while the heterotrophic nanoplankton also are grouped into a single entity (the HNAN, also sometimes called HNF = heterotrophic nanoflagellates). Epifluorescence microscopical techniques have been used widely to demonstrate abundances of small protists ranging from several hundred ml^{-1} to tens of thousands ml^{-1} in freshwater and marine planktonic ecosystems, on suspended detrital particles and in benthic communities (Davis et al., 1985; Caron et al., 1986; Pick & Caron, 1987; Sanders et al., 1989; Alongi, 1991; Bak et al., 1991; Sherr et al., 1993).

These latter methods for enumerating protists $<$20 μm in size provide relatively quick and accurate estimates of abundance, but they also act to aggregate species into a few "black boxes" that are then each represented by a single behavior. Species and even generic distinctions often cannot be made because of the lack of distinguishing morphological features apparent at this level of examination. Within these broad HNAN/PNAN characterizations, however, there is considerable variability among species with respect to light preferences and nutrient uptake capabilities (for PNAN), food preferences and ingestion rates (for HNAN), or even basic nutritional modes (e.g. numerous PNAN species possess some degree of phagotrophic ability). These differing abilities result in considerable variation in the activities and growth rates of nanoplanktonic protists, and thus their relative contribution to community metabolism and other aspects of ecosystem function. A considerable amount of research continues to be directed at documenting and understanding this diversity among small protists (Fenchel, 1986; Sanders, 1991b). Incorporation of these varied behaviors into descriptions of microbial food webs would considerably increase our accuracy of predicting and explaining energy production and flow among microbial species in nature.

There are several impediments in reaching this goal. Two key problems are the accurate determination of the species diversity (i.e. number of species) within natural communities of small protists, and the related problem of identifying and quantifying the abundances of individual species within natural, mixed assemblages. Small protists usually have few morphological features that are distinguishable using light microscopy. It is therefore difficult, and often impossible, to identify species or even genera of PNAN and HNAN using epifluorescence microscopical techniques that have become common for quantifying these assemblages. Electron microscopy and direct microscopical examination of living specimens are often sufficient for identification (Vørs, 1993; Vørs et al., 1995), but this approach is cost- and labor-intensive and does not easily provide quantitative information about these species.

The accurate identification of minute protists can be problematic even in pure culture because of the scarcity (or variability) of distinctive morphological features. Therefore, uncertainties persist regarding the taxonomy and systematics of a number of these species (Patterson & Larsen, 1991; Knauber et al., 1996). For example, the identification of species within the chrysomonad genera *Ochromonas* and *Spumella* is very difficult. These genera contain scaleless phototrophic and heterotrophic species, respectively, that are identified based largely on features of size, shape and details of the flagella and chloroplasts (when present).

Some of these characters can vary with physiological condition of the species. For example, cell size varies considerably with growth state in some phagotrophic chrysomonads (Fenchel, 1982). Average cell volume of *Paraphysomonas imperforata* varied by a factor of approximately six between different growth phases and also as a consequence of prey type (Caron et al., 1985). Chloroplast size and activity can change in some mixotrophic chrysophytes. The chloroplast size in *Poterioochromonas malhamensis* was greatly reduced and chlorophyll autofluorescence was not detectable when high abundances of bacteria were present as food (Caron et al., 1990). Interestingly, this ‘phototrophic’ chrysomonad was able to grow phagotrophically in continuous darkness.

Even those species that possess ‘reliable’ morphological features can produce aberrant forms of diagnostic characters under physiological stress. For example, species identifications in the scale-bearing colorless chrysomonad genus *Paraphysomonas* are based on the morphology of the siliceous scales produced by these microorganisms (Preisig & Hibberd, 1982). Nevertheless, scaleless forms of these individuals have been observed under severe silica limitation (Leadbeater & Barker, 1995). This behavior could hamper the identification of individual cells of *Paraphysomonas* in natural water samples where physiological condition is unknown.

HNAN species identification: a case study with a pedinellid flagellate

We have been enriching, isolating and culturing species of heterotrophic flagellates from marine and freshwater environments for several years. Our goal in this work has been to establish a collection of HNAN species for ecological and molecular biological investigations that represents the diversity of free-living, heterotrophic flagellates. Our culture collection now includes approximately 80 clonal cultures of phagotrophic flagellates spanning several major taxa (chrysomonads, dinoflagellates, choanoflagellates, heliozoa, pedinellids, prymnesiophytes, bodonids and bicosoecids).

Enrichment cultures in this study have been initiated using standard methods (Caron, 1993), and clonal cultures of dominant species have been established by micropipetting individual cells from these enrichments. Live and preserved specimens have been examined initially by light microscopy and identified where possible or grouped into major taxonomic categories (e.g. chrysomonads). Negatively stained (1% uranyl acetate) specimens have been examined by transmission electron microscopy to aid identifications where applicable. For example, electron microscopy allows visualization of the body scales of species of *Paraphysomonas*. These scales allow differentiation of the genus *Paraphysomonas* from *Spumella* (whose species do not produce scales) and form the basis for species identifications within the genus *Paraphysomonas* as described by Preisig & Hibberd (1982).

Species examinations conducted in this manner have provided sufficient detail for the identification of many of the clones that we have established. We have been unable, however, to easily resolve the identification of a number of isolates through the use of light and electron microscopy. For example, we obtained several clones of a flagellate that possessed a single emergent flagellum and lacked chloroplasts. Cursory light microscopy of free-swimming and preserved specimens, and transmission electron microscopy, revealed no additional useful details except for the presence of mastigonemes on the flagellum.

Nucleic acids of one of these isolates were extracted and small subunit rDNA (ssu rDNA) was amplified and sequenced (Caron et al., In press). Phylogenetic analysis of full length sequence of this gene revealed that this species was, in fact, most closely related to the pedinellid species *Pteridomonas danica* (Figure 2). Subsequent microscopical examination of our cultures in different physiological states revealed the presence of short tentacles near the flagellum and some cells attached by a stalk to particles in the cultures (Figure 3A,B). These features are characteristic of *Pteridomonas* species (Larsen & Patterson, 1990). Tentacles were not present on the specimens originally examined because they are often retracted in free-swimming or preserved cells (Figure 3C). To date, we have been unable to obtain electron micrographs of specimens with intact tentacles. Some pedinellid species are characterized by the long flagellum extending into a pronounced 'wing'. This feature also would have aided identification by electron microscopy, but it is absent in species of *Pteridomonas* (Moestrup, 1995).

The situation described above is not meant to indicate that the identification of cultures of *Pteridomonas* is difficult by morphological criteria. Nor do we suggest at this time that molecular identification of HNAN is easier or more straightforward than morphological identification. However, our experience with this isolate exemplifies the problematical nature of identifying HNAN species, even in pure culture. Clearly, the accurate identification of species in natural assemblages of small protists using light microscopical examination of preserved samples presents an extremely difficult task for some species. We obtained complete sequence information for the ssu rDNA of our *Pteridomonas* isolate in order to identify this clonal culture. This degree of thoroughness may not be necessary in the future for making distinctions at the generic (or perhaps even species) level. Partial sequence identity (established, for example, via partial sequencing or hybridization with species- or group-specific oligonucleotides) may be sufficient for 'matching' new isolates with species in sequence data bases. These latter approaches are already possible for some heterotrophic protists that parasitize humans (Uliana et al., 1991; Benavides et al., 1993; Gast & Byers, 1995), and they are widely employed for prokaryotes.

Alternatively, recent studies using restriction fragment length polymorphisms (RFLP) have proven useful for differentiating cultured isolates of some free-living protists. We have employed this approach with our collection of small (<10 μm), free-living heterotrophic flagellates as a means of grouping clonal cultures of these species (Lim 1997). Groupings based on RFLP patterns of the ssu rRNA gene were able to differentiate between genera and even many species of these protists. While this type of approach may not

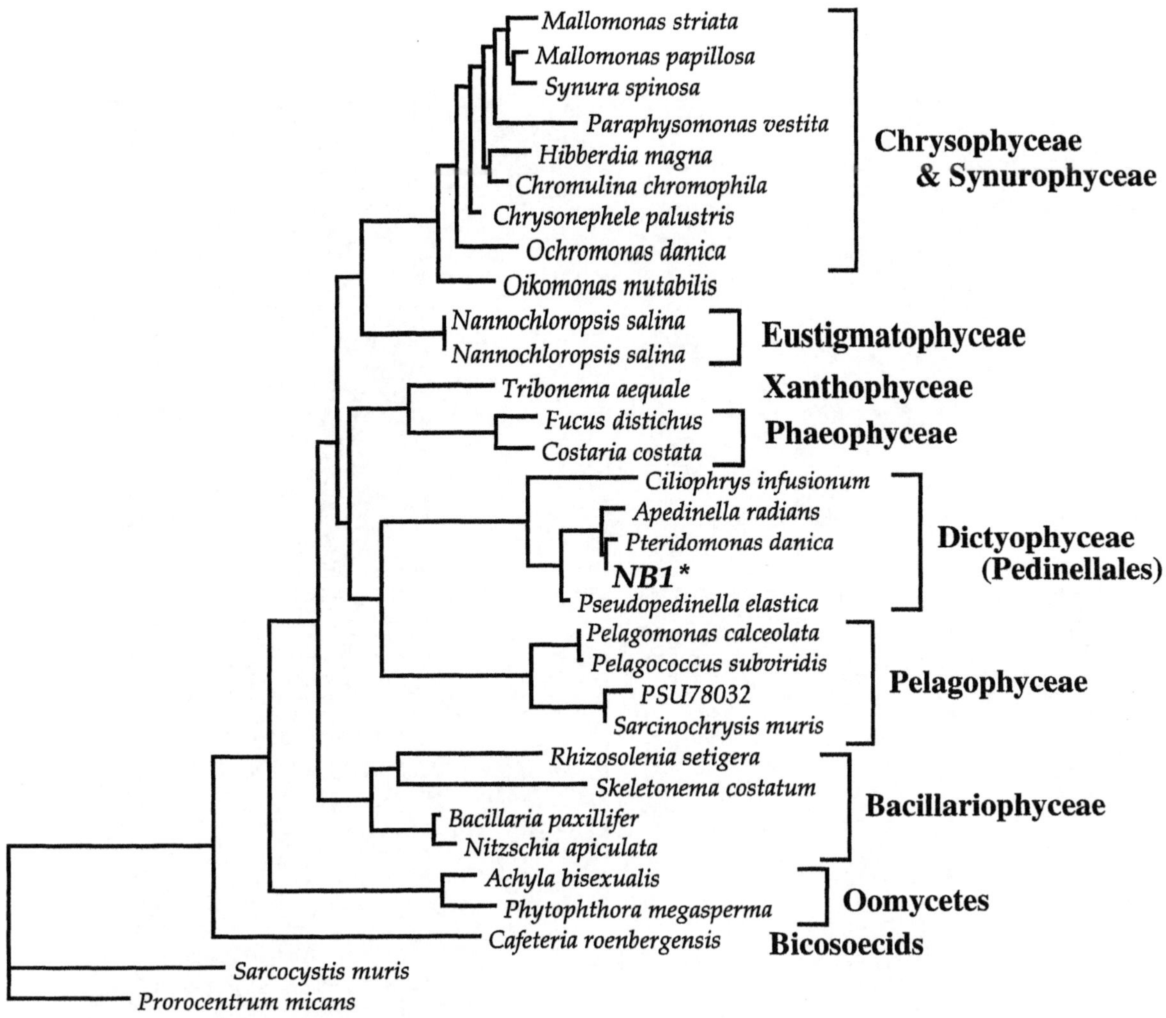

Figure 2. Phylogenetic analysis of the heterotrophic nanoflagellate clone NB1. This tree is the result of a maximum likelihood phylogenetic reconstruction using 1394 characters of which 288 were informative. This analysis places clone NB1 well within the pedinellids. The sequence for NB1 was only 0.66% different than *Pteridomonas danica* (12 out of 1814 sites) and 3.53% different than *Apedinella radians* (64 out of 1814 sites) in an alignment of these three species.

provide unequivocal species identifications, it is a very useful 'first cut' for separating isolates into related groups.

Molecular characterization of HNAN species undoubtedly will proceed hand-in-hand with morphological descriptions and identifications in the future. Ultimately, this collective approach will help to derive new identification schemes that will be useful to ecologists studying natural microbial assemblages.

HNAN species identification: detection in natural assemblages

The correct identification of cultured species of HNAN is essential for accurately cataloging autecological data for these species. In addition, the extrapolation of this information to the ecological and biogeochemical significance of small protists in nature relies on the accurate identification and enumeration of these organisms in natural assemblages. A mechanistic understanding of the occurrence and abundance of these species (i.e. biogeographical information) can be obtained only through an ability to quantitatively determine the abundance of species in natural

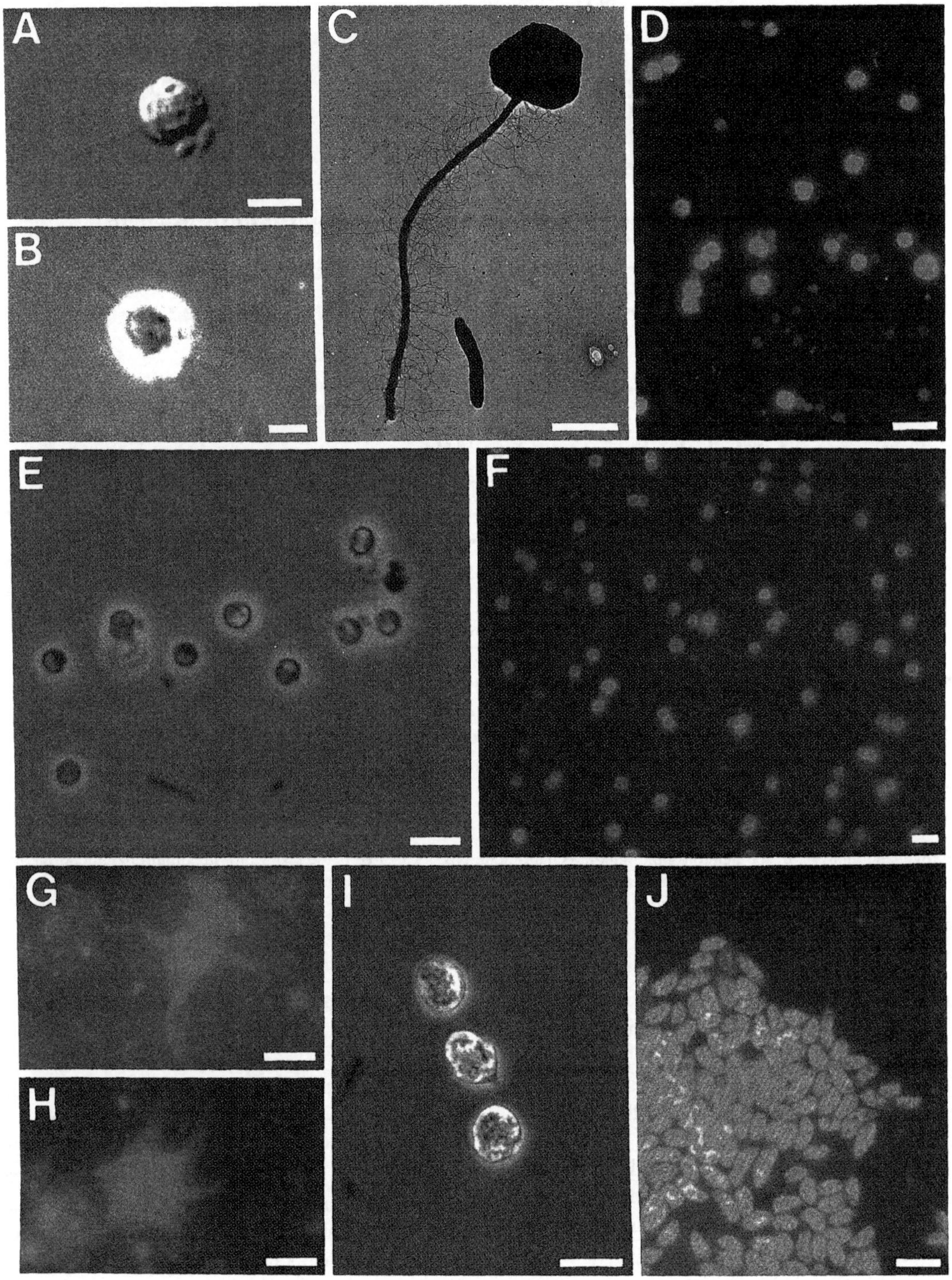

Figure 3.

assemblages. As detailed above, these latter determinations are exceedingly difficult using light microscopical methods that are used routinely for enumerating these populations. The inability of extant methods of routine sample analysis to provide *quantitative* data on the occurrence of ecologically-relevant species of small protists remains a fundamental hindrance to the study of these microorganisms.

We have adapted the use of fluorescently labeled oligonucleotide probes targeted against ssu rRNA for enumerating species of small (5–10 μm) heterotrophic flagellates in natural samples. The overall scheme involves the establishment and identification of clonal cultures of the target species (and closely related species), probe design and testing, and then application in a quantitative *in-situ* hybridization technique (Figure 4). We have taken an approach that we hope can serve as a template for probe design and application for a wide variety of small protists.

Our initial work was directed at testing the feasibility of using oligonucleotide probes in *in-situ* hybridizations using cultured species of heterotrophic flagellates. Primary considerations included signal strength and permeability of the protistan cells (Lim et al., 1993). Signal amplification using a biotin-avidin system proved necessary in order to visualize 3–10 μm cells in late stationary growth phase when ribosomal number was low.

Another major step in the development of our technique for applying oligonucleotide probes was the development of a *quantitative* method that could be employed with natural samples (i.e. preventing loss of sample during dehydration, hybridization and rinsing). Centrifugation and immobilization in gelatin have been employed, but these methods can result in significant losses of preserved cells from samples during the probing process. Our quantitative *in-situ* hybridization method employs Transwell® cell culture inserts for Corning Costar tissue culture dishes that allow solution exchange through a polycarbonate filter while retaining the target population in the Transwell chamber (Lim et al., 1996). This method is essentially an adaptation of the fluorochrome staining methods that are used routinely for counting nanoplankton assemblages (Sherr et al., 1993).

The development of species-specific oligonucleotide probes for identifying and counting small protists in natural water samples focused on the heterotrophic chrysomonad species *Paraphysomonas imperforata* (Caron, et al., In press). This phagotrophic protist is nanoplanktonic in size, its distribution is cosmopolitan, it is easily enriched and cultured in the laboratory (there is considerable physiological information for this species), and it can be readily identified by electron microscopy because of the species-specific, siliceous body scales that it produces. Oligonucleotide probes targeting the ssu rRNA of this species were designed based on sequence comparisons for this species, several congeners and a variety of other related species of protists. Testing performed in slot blots and *in-situ* hybridizations indicated specificity for *P. imperforata* even in comparison to closely related congeners (Figure 3D).

Field applications of oligonucleotide probes targeting *P. imperforata* have already altered our perception of heterotrophic flagellate assemblages of natural plankton communities (Lim et al., 1999). Although this cosmopolitan species was often the numerical dominant of our enrichment cultures initiated from a wide variety of aquatic ecosystems, thus far we have not observed situations in which *P. imperforata* constituted more than $\approx$1% of the total nanoplankton of unenriched, natural samples. Based on our study using this molecular approach, we conclude that this species dominates enrichment cultures because it is strongly favored by the cultures conditions provided in

←

Figure 3. (A–C) Differential interference contrast (A), phase (B) and transmission electron (C) micrographs of the heterotrophic nanoflagellate *Pteridomonas* sp. The stalk and some of the tentacles were visible on the live specimens in (A) and (B), and the beating flagellum was visible between the tentacles in (B). Only the flagellum was visible in preserved specimens, the stalk and tentacles were absent (C). (D) Epifluorescence micrograph of a mixture of the two heterotrophic nanoflagellates *Paraphysomonas imperforata* and *P. bandaiensis*. *P. imperforata* cells have been probed with a species-specific oligonucleotide probe that causes these cells to be highly fluorescent. Unprobed *P. bandaiensis* cells (about 1/2 the size of the *P. imperforata* cells) are faintly visible due to background fluorescence. (E,F) Light micrograph (E) and epifluorescence micrograph of immunofluorescently-labelled cells of the 'brown tide' alga *Aureococcus anophagefferens*. (G,H) Epifluorescence micrographs of two naked amoebae (star-like shapes in center of micrographs) probed with eukaroyote-specific oligonucleotide probes. (I) Light micrograph of the free-living form of the photosynthetic dinoflagellate *Scrippsiella nitricula* that occurs as an intracellular symbiont in the polycystine radiolarian *Thalassicolla nucleata*. (J) Epifluorescence micrograph of the symbiotic dinoflagellate *Gymnodinium beii* in a squashed foraminiferal specimen. The intracellular symbionts lack typical dinoflagellate morphology which is present during the free-living stage (compare to specimens in (I)). Fluorescent signal in these specimens is a result of probing with a species-specific oligonucleotide probe designed for *G. beii*. Marker bars = 5 μm (A–F), 10 μm (I,J) or 15 μm (G,H).

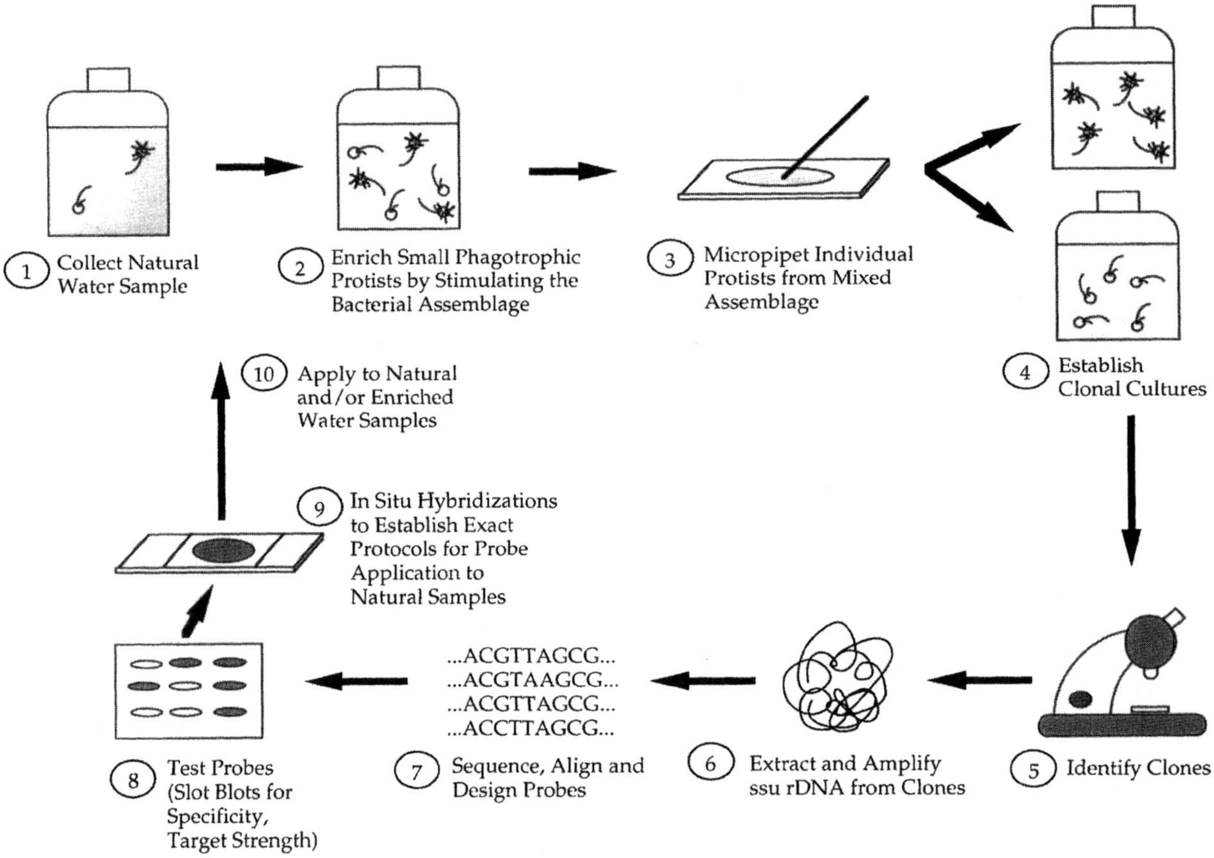

Figure 4. Generalized protocol for the design, testing and application of ssu rRNA-targeted oligonucleotide probes for heterotrophic nanoflagellates (or other small protists); (1–5) establishment of clonal cultures and identification by morphological criteria; (6 – 7) extraction, sequencing and comparison of ssu rDNA from closely- and distantly-related species for the design of oligonucleotide probes; (8 – 9) testing specificity of probes against extracted nucleid acids and with whole cell preparations of target and non-target species; (10) application of probes to natural samples.

our enrichment technique. Moreover, we postulate that natural assemblages of heterotrophic flagellates may be far more diverse than we have suspected previously. Our experiences are indicative of the poor understanding we presently have of the ecology of small protists, and the new insights that can be gained from the application of molecular biological methods to the study of these microorganisms.

PNAN and Yeasts

Species identifications among microscopic algae and fungi present much the same difficulties that are associated with the identification of HNAN. Many of the minute algae possess few morphological characters that are useful as taxonomic criteria. Pigment composition has provided a means of characterizing the major phototrophic taxa present within planktonic as well as benthic environments (Riaux-Gobin & Klein, 1993; Bidigare & Ondrusek, 1996), but further taxonomic characterization is not possible by this approach. Identification of marine fungi has been based on morphological and physiological characteristics determined in the laboratory using pure cultures. These features are of limited value for identifying species in natural microbial assemblages.

Recent applications of DNA sequence analysis to the characterization of species of marine yeasts is an excellent example of the exploitation of molecular biological approaches for studies of the biodiversity of small eukaryotes (Fell et al., 1992; Fell, 1993). The use of ssu rDNA primers has revealed a diverse assemblage of microscopic fungi, and a community complexity that far exceeds what had been demonstrated using culture-dependent methods up to that time. Extension of this approach to analyses of natural microbial communities should provide new insights into the abundances and distributions of these poorly documented microorganisms.

Molecular identification based on DNA sequence data has been reported recently as an ecological tool for species of PNAN (Knauber et al., 1996; Simon et al., 1997), and also for some larger algae of interest to human health (Miller & Scholin, 1996; Scholin & Anderson, 1996; Scholin et al., 1996). The application of this approach for small algae, however, has not yet become common. Instead, the use of immunological methods has received more attention for these species (Shapiro et al., 1989; Campbell et al., 1994). Such methods have allowed, for example, the enumeration in natural samples of *Aureococcus anophagefferens*, a 2–3 μm pelagophyte which does not possess flagella and has few distinctive morphological features (Sieburth et al., 1988; Anderson et al., 1989) (Figure 3E,F). Immunoprobes also have been used effectively to determine the occurrence of different strains of chroococcoid cyanobacteria in mixed natural cyanobacterial assemblages (Campbell & Iturriaga, 1988).

The reason for the more widespread use of antibody techniques for small algae rather than DNA-based methods is not clear, but may relate to the ease of obtaining antibodies commercially, and thus the smaller investment necessary to conduct this work (relative to equipping a lab for molecular biology). A potential advantage of the DNA-based methods, however, is that sequence information is unlikely to change over ecologically-relevant time scales, while antigenic characters of some protists can change rapidly due to physiological condition and/or life stage.

Amoeboid protists

Perhaps the most recalcitrant problems of species identification of small protists exist for the non-testate ('naked') amoebae. At present, it is not commonplace to identify or enumerate these protists from preserved specimens. This situation is due to the difficulty of distinguishing these amorphous-shaped cells by light microscopy (especially in preserved samples), and this problem is further complicated by the attachment of amoebae to particles which may obscure vision or absorb fluorochromes that are used to stain protists.

Abundances of amoebae in natural ecosystems generally are determined using dilution-enrichment approaches which are dependent upon the growth of these species in the enrichments (Davis et al., 1978; Rogerson & Laybourn-Parry, 1992; Anderson & Rogerson, 1995). Identifications (which are primarily based on features of the living cells) can be performed on amoebae growing from these enrichments but, as with some flagellates, 'species' identifications often are hampered by the morphological variability of these specimens. Some species are even capable of transformation into flagellated life stages. The prevalence of these latter forms in natural samples is unknown.

The use of culture-dependent methods for enumerating amoebae in natural samples also raises questions as to the accuracy of these abundance estimates. Some species (or individuals) of amoebae may not grow in enrichment cultures. Lack of growth would result in an underestimation of their abundances in natural samples. Conversely, excystment of encysted amoebae could overestimate the abundance of individuals that are trophically active. This latter problem is also a concern for estimating the abundances of other protistan taxa that produce cysts.

Biogeographical data for amoebae are meager relative to our knowledge of other free-living protists because of our present inability to obtain quantitative information on the abundances of these species from preserved samples. As a result, their ecological/biogeochemical roles in most aquatic ecosystems are poorly understood and probably underestimated (Arndt, 1993). Detection of amoebae using DNA-based protocols could enable new approaches for their identification and enumeration in natural samples. These methods would obviate the need to base enumeration and identification on the ability of individuals to grow in culture. The design and application of species- or group-specific oligonucleotide probes could provide a routine means of obtaining quantitative information for amoebae in preserved samples, and potentially also recognize other life stages of these species (e.g. cysts, flagellates). In addition, the use of oligonucleotide probes does not result in staining of particulate material (Lim et al., 1996). This feature has considerable advantage for visualizing protists such as amoebae that are typically associated with particles in the plankton (Figure 3G,H).

Difficulties associated with the identification of amoeboid protists are not constrained only to small species. Although the identification of larger, free-living 'amoeboid' or sarcodine protists (Foraminifera and Actinopoda) has not been considered particularly problematic, specific aspects of their biology still present unique difficulties. Some of these issues relate to the complex life cycles of these species. For example, many of the Foraminifera, Radiolaria and Acantharea form tests or central capsules as adult specimens that can attain diameters greater than 1 mm (pseudopodial networks can increase the effective size of some of these specimens to > 1 cm). The identification of the shell-bearing stages of these specimens is based largely on the skeletal structures. In contrast, these species reproduce by the release of swarmers that are flagellated cells several micrometers in size (Anderson, 1983; Bé et al., 1983). The identification of sarcodine swarmer cells presents difficulties that are similar to the identification of HNAN species; few morphological details are apparent by light microscopy that enable their accurate discrimination in natural samples. As a consequence, there is no quantitative information on the occurrence or depth distributions of these swarmer cells in oceanic plankton communities.

Other problems concerning the identification of sarcodine protists relate to variable morphology during the adult life stages of some species. Taxonomically-relevant skeletal structures vary considerably for some species of Radiolaria and benthic Foraminifera (Bjørklund & Swanberg, 1987; Bowser et al., 1995; Gooday et al., 1996). At one extreme, some species of Foraminifera are capable of complete loss of the test (Cedhagen & Tendal, 1989). The ecological significance of this variability is intriguing and may be a consequence of environmental influence, but it can result in erroneous identifications or even the description of synonyms. On the other hand, such morphological variability may mask the presence of cryptic species, as has been postulated for the solitary Radiolarian *Thalassicolla nucleata* (Amaral Zettler et al., 1998). DNA sequence information provides additional insight to classical morphological approaches for resolving these issues.

Dinoflagellates

Dinoflagellates are a highly diverse taxon of protists. These species constitute a significant fraction of both the phototrophic and heterotrophic protistan assemblages of plankton communities, and thus have tremendous ecological significance as primary producers and consumers. Heterotrophic dinoflagellates exhibit a variety of interesting and unique feeding behaviors that significantly impact phytoplankton assemblages (Jacobson & Anderson, 1986; Gaines & Elbrächter, 1987; Hansen, 1991; Lessard, 1991; Jeong, 1994), and mixotrophic (phagotrophic) activity among phototrophic dinoflagellates is common (Sanders, 1991a). In addition, some species cause harmful algal blooms (Anderson & Garrison, 1997), or occur as symbionts or parasites in a variety of protistan and metazoan hosts (Cachon & Cachon, 1987; Trench, 1987).

Dinoflagellates have received ample attention from protistan taxonomists, and morphology-based schemes for their identification are well established. An important consideration for ecological studies of dinoflagellates, however, is that many of these species possess relatively complex life cycles that pose significant problems for identification *in situ*. An extreme example of this behavior exists for the ambush-predator dinoflagellate, *Pfiesteria piscicida*, for which more than two dozen life stages have been described (Burkholder et al., 1995a,b; Burkholder & Glasgow, 1997). At present, the existence of multiple life stages makes the identification of this species in natural assemblages very difficult (if not impossible) because different methodologies must be used to establish the presence of the various life stages. In general, the distributions in nature of dinoflagellates with multiple life stages will remain enigmatic until tools are developed that will allow detection and enumeration of all of the various life stages in a single process.

Morphological variations (or life stage transformations) of dinoflagellates can be induced by environmental conditions, but they can also be the result of physical association of dinoflagellates with other organisms. The morphologies of dinoflagellate parasites and symbionts in their free-living states are sometimes radically different than their morphologies in a host. Dinoflagellate symbionts of the planktonic Foraminifera and Actinopoda, for example, do not produce thecal plates or flagella when they occur as intracellular symbionts (Anderson & Bé, 1976; Anderson, 1980). The development of DNA-based protocols for species detection of these symbionts from DNA sequence information (Rowan & Powers, 1992; Gast & Caron, 1996) will enable ecological studies of these dinoflagellates in both the free-living thecate stages and as symbionts within various sarcodine species (Figure 3I,J).

Similarly, heterotrophic dinoflagellates that cause parasitic infections typically cannot be identified when they occur in association with their hosts (Cachon & Cachon, 1987; Coats et al., 1996). The agents of these infections usually are identified only after lysis of the host cell or tissue and release of a free-living stage of the dinoflagellates. At present, there is little information concerning the impact of these parasitic infections on populations of their hosts in nature, in part because of the difficulties of identifying and counting these parasites. DNA-based protocols for the detection of dinoflagellate parasites in the cytoplasm or tissues of infected hosts, and in their free-living forms, will greatly enhance ecological studies of these relationships.

Perspectives on the Future

Traditional methodological approaches cannot easily address many unresolved issues and unanswered questions concerning the biodiversity and biogeography of small protists in aquatic food webs. As a consequence, many aquatic biologists have looked to molecular biology within the last decade as a panacea for rapidly solving many of these problems. This highly optimistic attitude has been somewhat tempered in recent years, as limitations and/or problems have arisen when trying to adapt these methods for use on natural assemblages of microorganisms. These setbacks have inspired more care in incorporating molecular methods into ecological research programs, but they have not diminished the potential for these approaches to answer long-standing ecological questions. As evidenced by this volume, molecular techniques and conceptual approaches clearly are providing new insights into the ecology of marine bacteria, protists and fungi. Expectations concerning these approaches are just beginning to be realized. The application of these new powerful molecular approaches, in concert with traditional methods of study, will continue to enlarge our base of knowledge concerning the ecology of free-living microorganisms.

Acknowledgements

The authors are grateful to Dawn M. Moran and Katie Rose Cellineri for some of the micrographs used in Figure 3. This study was supported by NSF grant OCE-9216270. Woods Hole Oceanographic Institution Contribution Number 9776.

References

Alongi, D. M., 1991. Flagellates of benthic communities: characteristics and methods of study. In Patterson D. J. & J. Larsen, (ed.), The Biology of Free-living Heterotrophic Flagellates. Clarendon Press, Oxford: 57–75.

Amaral Zettler, L., O. R. Anderson & D. A. Caron, 1998. Insights on the diversity within a 'species' of *Thalassicolla* (Spumellarida) based on 16S-like ribosomal RNA gene sequencing. J. Euk. Microbiol. 45: 488–496.

Anderson, D. M. & D. J. Garrison, 1997. The ecology and oceanography of harmful algal blooms. Limnol. Oceanogr. 42: 1009–1305.

Anderson, D. M., D. M. Kulis & E. M. Cosper, 1989. Immunofluorescent detection of the brown tide organism Aureococcus anophagefferens. In Cosper, E. M. V. M. Bricelj & E. J. Carpenter, (eds) Novel phytoplankton blooms: causes and impacts of recurrent brown tides and other unusual blooms. vol. 35. Springer-Verlag, Berlin: 265–294.

Anderson, O. R., 1980. Radiolaria. In: Levandowsky M. & S. H. Hutner, (eds), Biochemistry and Physiology of Protozoa. vol. 3. Academic Press, New York, 1–42.

Anderson, O. R. 1983. Radiolaria. Springer-Verlag, New York.

Anderson, O. R. & A. W. H. Bé, 1976. The ultrastructure of a planktonic foraminifer, *Globigerinoides sacculifer* (Brady), and its symbiotic dinoflagellates. J. Foram. Res. 6: 1–21.

Anderson, O. R. & A. Rogerson, 1995. Annual abundances and growth potential of Gymnamoebae in the Hudson Estuary with comparative data from the Firth of Clyde. Eur. J. Protistol. 31: 223–233.

Arndt, H., 1993. A critical review of the importance of rhizopods (naked and testate amoebae) and actinopods (heliozoa) in lake plankton. Mar. Microb. Food Webs. 7: 3–29.

Bak, R. P. M. F. C. van Duyl, G. Nieuwland, A. J. Kop, 1991. Benthic heterotrophic nanoflagellates in North Sea field/mesocosm bottoms and their response to algal sedimentation. Ophelia 33: 187–196.

Bé, A. W. H., 1959. Ecology of recent planktonic foraminifera. Part I. Areal distribution in the western North Atlantic. Micropaleontology 5: 77–100.

Bé, A. W. H. 1960. Ecology of recent planktonic foraminifera: Part 2 – Bathymetric and seasonal distributions in the Sargasso Sea off Bermuda. Micropaleontology 6: 373–392.

Bé, A. W. H., O. R. Anderson, W. W., Jr., Faber & D. A. Caron, 1983. Sequence of morphological and cytoplasmic changes during gametogenesis in the planktonic foraminifer *Globigerinoides sacculifer* (Brady). Micropaleontology. 29: 310–325.

Benavides, G. R., J. J. Sullivan, F. Steurer, R. A. McGraw & R. L. Tarleton, 1993. Differentiation of trypanosomatid species by hybridization to selected rRNA probes. Mol. Cell. Probes. 7: 89–96.

Bidigare, R. R. & M. E. Ondrusek, (1996). Spatial and temporal variability of phytoplankton pigment distributions in the central equatorial Pacific Ocean. Deep-Sea Res. 43: 809–834.

Bjørklund, K. R. & N. R. Swanberg, 1987. The distribution of two morphotypes of the radiolarian *Amphimelissa setosa* Cleve (Nassellarida): a result of environmental variability? Sarsia 72: 245–254

Bowser, S. S., A. J. Gooday, S. P. Alexander, J. M. Bernhard, 1995. Larger agglutinated foraminifera of McMurdo Sound, Antarctica: are *Astrammina rara* and *Notodendrodes antarctikos* allogromiids incognito? Mar. Micropaleontol. 26: 75–88.

Burkholder, J. M. & H. B. Glasgow, Jr., (1997). Trophic controls on stage transformations of a toxic ambush-predator dinoflagellate. J. Euk. Microbiol. 44: 200–205.

Burkholder, J. M., H. B. Glasgow, Jr. & C. W. Hobbs, 1995a. Fish kills linked to a toxic ambush-predator dinoflagellate: distribution and environmental conditions. Mar. Ecol. Prog. Ser. 124: 43–61.

Burkholder, J. M., H. B. Glasgow, Jr., K. A. Steidinger, 1995b. Stage transformations in the complex life cycle of an ichthyotoxic 'ambush-predator' dinoflagellate. In: Lassus P., G. Arzul, E. Erard-Le Denn, P. Gentien & C. Marcaillou-Le Baut, (eds), Harmful Marine Algal Blooms. Lavoisier Publishing, Paris: 567–572.

Cachon, J. & M. Cachon, 1987. Parasitic dinoflagellates. In: F. J. R. Taylor, (ed.), The Biology of Dinoflagellates. Blackwell Scientific Publications, Oxford: 571–610.

Campbell, L. & R. Iturriaga, 1988. Identification of *Synechococcus* spp. in the Sargasso Sea by immunofluorescence and fluorescence excitation spectroscopy performed on individual cells. Limnol. Oceanogr. 33: 1196–1201.

Campbell, L., L. P. Shapiro & E. Haugen, 1994. Immunochemical characterization of eukaryotic ultraplankton from the Atlantic and Pacific oceans. J. Plankton Res. 16: 35–51.

Caron, D. A., 1993. Enrichment, isolation, and culture of free-living heterotrophic flagellates. In: Kemp P. F., B. F. Sherr, E. B. Sherr & J. J. Cole, (eds), Handbook of Methods in Aquatic Microbial Ecology. Lewis Publishers, Boca Raton: 77–89.

Caron, D. A., P. G. Davis, L. P. Madin, J. M. Sieburth, 1986. Enrichment of microbial populations in macroaggregates (marine snow) from the surface waters of the North Atlantic. J. Mar. Res. 44: 543–565.

Caron, D. A., P. G. Davis & J. M. Sieburth, 1989. Factors responsible for the differences in cultural estimates and direct microscopical counts of populations of bacterivorous nanoflagellates. Microb. Ecol. 18: 89-104.

Caron, D. A., J. C. Goldman, O. K. Andersen & M. R. Dennett, 1985. Nutrient cycling in a microflagellate food chain. II. Population dynamics and carbon cycling. Mar. Ecol. Prog. Ser. 24: 243-254.

Caron, D. A., K. G. Porter & R. W. Sanders, 1990. Carbon, nitrogen and phosphorus budgets for the mixotrophic phytoflagellate *Poterioochromonas malhamensis* (Chrysophyseae) during bacterial ingestion. Limnol. Oceanogr. 35: 433–443.

Caron, D. A. & N. R. Swanberg, 1990. The ecology of planktonic sarcodines. Rev. aquat. Sci. 3: 147–180.

Cedhagen, T. & O. S. Tendal, 1989. Evidence of test detachment in *Astrorhiza limicola*, and two consequential synonyms: *Amoeba gigantea* and *Megamoebomyxa argillobia* (Foraminiferida). Sarsia 74: 195–200.

Coats, D. W., E. J. Adam, C. L. Gallegos, S. Hedrick, 1996. Parasitism of photosynthetic dinoflagellates in a shallow subestuary of Chesapeake Bay, U.S.A. Aquat. Microb. Ecol. 11: 1–9.

Davis, P. G., D. A. Caron, P. W. Johnson & J. M. Sieburth, 1985. Phototrophic and apochlorotic components of picoplankton and nanoplankton in the North Atlantic: geographic, vertical seasonal and diel distributions. Mar. Ecol. Prog. Ser. 21: 15–26.

Davis, P. G., D. A. Caron & J. M. Sieburth, 1978. Oceanic amoebae from the North Atlantic: culture, distribution, and taxonomy. Trans. Am. Microsc. Soc. 96: 73–88.

Dworetzky, B. A. & J. J. Morley, 1987. Vertical distribution of radiolaria in the Eastern Equatorial Atlantic: analysis of a multiple series of closely-spaced plankton tows. Mar. Micropaleontol. 12: 1–19.

Fell, J. W., 1993. Rapid identification of yeast species using three primers in a polymerase chain reaction. Mol. Mar. Biol. Biotechnol. 2: 174–180.

Fell, J. W., A. Statzell-Tallman, M. J. Lutz & C. P. Kurtzman, 1992. Partial rRNA sequences in marine yeasts: a model for identification of marine eukaryotes. Mol. Mar. Biol. Biotechnol. 1: 175–186

Fenchel, T., 1967. The ecology of marine microbenthos. I. The quantitative importance of ciliates as compared with metazoans in various types of sediments. Ophelia 4: 121–137.

Fenchel, T., 1982. Ecology of heterotrophic microflagellates. III. Adaptations to heterogeneous environments. Mar. Ecol. Prog. Ser. 9: 25–33.

Fenchel, T., 1986. The ecology of heterotrophic microflagellates. Adv. Microb. Ecol. 9: 57–97.

Gaines, G. & M. Elbrächter, 1987. Heterotrophic nutrition. In: Taylor, F. J. R. (ed.), The Biology of Dinoflagellates. vol. 21. Blackwell Scientific Publications, Oxford: 224–268.

Garrison, D. L., C. W. Sullivan & S. F. Ackley, 1986. Sea ice microbial communities in Antarctica. Bioscience 36: 243–250.

Gast, R. J. and T. J. Byers, 1995. Genus- and subgenus-specific oligonucleotide probes for Acanthamoeba. Mol. Biochem. Parasitol. 71: 255–260.

Gast, R. J. & D. A. Caron, 1996. Molecular phylogeny of symbiotic dinoflagellates from Foraminifera and Radiolaria. Mol. Biol. Evol. 13: 1192–1197.

Gooday, A. J., S. S. Bowser & J. M. Bernhard, 1996. Benthic foraminiferal assemblages in Explorers Cove, Antarctica: a shallow-water site with deep-sea characteristics. Prog. Oceanogr. 37: 117–166.

Hansen, P. J., 1991. Quantitative importance and trophic role of heterotrophic dinoflagellates in a coastal pelagial food web. Mar. Ecol. Prog. Ser. 73: 253–261.

Hemleben, C., M. Spindler & O. R. Anderson, 1988. Modern planktonic foraminifera. Springer-Verlag, New York.

Jacobson, D. M. & D. M. Anderson, 1986. Thecate heterotrophic dinoflagellates: feeding behavior and mechanisms. J. Phycol. 22: 249–258.

Jeong, H. J., 1994. Predation by the heterotrophic dinoflagellate *Protoperidinium* cf. *divergens* on copepod eggs and early naupliar stages. Mar. Ecol. Prog. Ser. 114: 203–208.

Knauber, D. C., E. S. Berry & M. W. Fawley, 1996. Ribosomal RNA-based oligonucleotide probes to identify marine green ultraphytoplankton. J. Euk. Microbiol. 43: 89–94.

Kopylov, A. I. & A. F. Sazhin, 1989. Heterotrophic nanoplankton of the aerobic zone of the Black Sea. In Nauka (ed.), Structure and Productive Characteristics of Plankton Communities of the Black Sea. Moscow. 139–142 (in Russian).

Larsen, J. & D. J. Patterson, 1990. Some flagellates (Protista) from tropical marine sediments. J. nat. His. 24: 801–937.

Leadbeater, B. S. C. & D. A. N. Barker, 1995. Biomineralization and scale production in the Chrysophyta. In C. D. Sandgren, J. P. Smol & J. Kristiansen, (ed.), Chrysophyte Algae: Ecology, Phylogeny and Development. Cambridge University Press, Cambridge 141–164.

Lessard, E. J., 1991. The trophic role of heterotrophic dinoflagellates in diverse marine environments. Mar. Microb. Food Webs. 5: 49–58.

Lim, E. E., L. A. Amaral, D. A. Caron & E. F. DeLong, 1993. Application of rRNA-based probes for observing marine nanoplanktonic protists. Appl. envir. Microbiol. 59: 1647–1655.

Lim, E. L. 1997. New insights on the ecology of free-living, heterotrophic nanoflagellates based on the use of molecular biological approaches. Ph.D. Thesis, Massachusetts Institute of Technology and Woods Hole Oceanographic Institution

Lim, E. L., D. A. Caron, E. F. DeLong, 1996. Development and field application of a quantitative method for examining natural assemblages of protists using oligonucleotide probes. Appl. environ. Microbiol. 62: 1416–1423.

Lim, E. L., D. A. Caron, M. R. Dennett, 1999. The ecology of *Paraphysomonas imperforata* based on studies employing oligonucleotide probe identification in coastal water samples and enrichment culture. Limnol. Oceanogr. 44: 37–51.

Miller, P. E. & C. A. Scholin, 1996. Identification of cultured *Pseudo-nitzschia* (Bacillariophyceae) using species-specific LSU rRNA-targeted fluorescent probes. J. Phycol. 32: 646–655.

Moestrup, Ø., 1995. Current status of chrysophyte 'splinter groups': synurophytes, pedinellids, silicoflagellates. In C. D. Sandgren, J. P. Smol & J. Kristiansen, (eds) Chrysophyte Algae: Ecology, Phylogeny and Development. Cambridge University Press, Cambridge, 75–91.

Montagnes, D. J. S. & D. H. Lynn, 1993. A quantitative protargol stain (QPS) for ciliates and other protists. In P. F. Kemp, B. F. Sherr, E. B. Sherr & J. J. Cole, (eds), Handbook of Methods in Aquatic Microbial Ecology. Lewis Publishers, Boca Raton: 229–240.

Patterson, D. J. & J. Larsen, (eds) 1991. The Biology of Free-Living Heterotrophic Flagellates. Clarendon Press, Oxford.

Pick, F. R. & D. A. Caron, 1987. Picoplankton and nanoplankton biomass in Lake Ontario: relative contribution of phototrophic and heterotrophic communities. Can. J. Fish. aquat. Sci. 44: 2164–2172.

Porter, K. G., E. B. Sherr, B. F. Sherr, M. L., Pace & R. W. Sanders, 1985. Protozoa in planktonic food webs. J. Protozool. 32: 409–415

Preisig, H. R. & D. L. Hibberd, 1982. Ultrastructure and taxonomy of *Paraphysomonas* (Chrysophyceae) and related genera 2. Nord. J. Bot. 2: 601–638.

Riaux-Gobin, C. & B. Klein, 1993. Microphytobenthic biomass measurement using HPLC and conventional pigment analysis. In: P. F. Kemp, B. F. Sherr, E. B. Sherr & J. J. Cole, (eds), Handbook of Methods in Aquatic Microbial Ecology. Lewis Publishers, Boca Raton: 369–376.

Rogerson, A. & J. Laybourn-Parry, 1992. The abundance of marine naked amoebae in the water column of the Clyde estuary. Estuar. coast. shelf Sci. 34: 187–196.

Rowan, R. & D. A. Powers, 1992. Ribosomal RNA sequences and the diversity of symbiotic dinoflagellates (zooxanthellae). Proc. nat. Acad. Sci. 89: 3639–3643.

Sanders, R. W., 1991a. Mixotrophic protists in marine and freshwater ecosystems. J. Protozool. 38: 76–81.

Sanders, R. W., 1991b. Trophic strategies among heterotrophic flagellates. In: D. J. Patterson, J. Larsen, (eds), The Biology of Free-Living Heterotrophic Flagellates. Clarendon Press, Oxford: 21–38.

Sanders, R. W., K. G. Porter, S. J. Bennett & A. E. DeBiase, 1989. Seasonal patterns of bacterivory by flagellates, ciliates, rotifers, and cladocerans in a freshwater planktonic community. Limnol. Oceanogr. 34: 673–687.

Scholin, C. A. & D. M. Anderson, 1996. LSU rDNA-based RFLP assays for discriminating species and strains of *Alexandrium* (Dinophyceae). J. Phycol. 32: 1022–1035.

Scholin, C. A., K. R. Buck, T. Britschgi, G. Cangelosi, F. P. Chavez, 1996. Identification of *Pseudo-nitzschia australis* (Bacillariophyceae) using rRNA-targeted probes in whole cell and sandwich hybridization formats. Phycol. 35: 190–197.

Shapiro, L. P., L. Campbell & E. M. Haugen, 1989. Immunochemical recognition of phytoplankton species. Mar. Ecol. Prog. Ser. 57: 219–224.

Sherr, E. B., D. A. Caron & B. F. Sherr, 1993. Staining of heterotrophic protists for visualization via epifluorescence microscopy. In Kemp P., J. Cole, B. Sherr & E. Sherr, (eds), Handbook of Methods in Aquatic Microbial Ecology. Lewis Publishers, Boca Raton: 213–227

Sherr, E. B. & B. F. Sherr, 1994. Bacterivory and herbivory: key roles of phagotrophic protists in pelagic food webs. Microb. Ecol. 28: 223–235.

Sieburth, J. M., P. W. Johnson & P. E. Hargraves, 1988. Ultrastructure and ecology of *Aureococcus anophagefferans* gen. et sp. nov. (Chrysophyceae): the dominant picoplankter during a bloom in Narragansett Bay, Rhode Island, summer 1985. J. Phycol. 24: 416–425.

Simon, N., J. Brenner, B. Edvardsen & L. K. Medlin, (1997). The identification of *Chrysochromulina* and *Prymnesium* species (Haptophyta, Prymnesiophyceae) using fluorescent or chemiluminescent oligonucleotides probes: a means for improving studies on toxic algae. Eur. J. Phycol. 32: 393–401.

Stoecker, D., A. Taniguchi, A. E. Michaels, 1989. Abundance of autotrophic, mixotrophic and heterotrophic planktonic ciliates in shelf and slope waters. Mar. Ecol. Prog. Ser. 50: 241–254.

Stoecker, D. K. 1994. Nanoplankton and protozoan microzooplankton during the JGOFS N. Atlantic Bloom Experiment. J. mar. biol. Assoc. U.K. 74: 427–443.

Trench, R. K. 1987. Dinoflagellates in non-parasitic symbioses. In Taylor, F. J. R. (ed.), The Biology of Dinoflagellates. Blackwell Scientific Publications, Oxford: 530–570

Uliana, S. R. B., M. H. T. Affonso, E. P. Camargo & L. M. Floeter-Winter 1991. *Leishmania*: genus identification based on a specific sequence of the 18S ribosomal RNA sequence. Exp. Parasitol. 72: 157–163

Utermöhl, H. 1958. Zur Vervollkommung der quantitativen phytoplankton-methodik. Mitt Int Ver Limnol. 9: 38.

Vørs, N. 1993. Heterotrophic amoebae, flagellates and heliozoa from Arctic marine waters (North West Territories, Canada and West Greenland). Polar Biol. 13: 113–126.

Vørs, N., 1995. Nanoplankton of the equatorial Pacific with emphasis on the heterotrophic protists. Deep-Sea Res. 42: 585–602

Wehr, J. D., 1990. Predominance of picoplankton and nanoplankton in eutrophic Calder Lake. Hydrobiologia. 203: 35–44.

Hydrobiologia **401**: 229–238, 1999.
J.P. Zehr & M.A. Voytek (eds), Molecular Ecology of Aquatic Communities.

Molecular and demographic measures of arsenic stress in *Daphnia pulex*

C. Y. Chen[1], K. B. Sillett[1], C. L. Folt[1], S. L. Whittemore[2] & A. Barchowsky[3]
[1]*C.Y. Chen, Department of Biological Sciences, Dartmouth College, Hanover, NH, 03755, U.S.A.*
[2]*Department of Biology, MS 2001, Keene State College, Keene, NH 03435-2001, U.S.A.*
[3]*Department of Pharmacology and Toxicology, Dartmouth Medical School, Hanover, NH 03755, U.S.A.*

Key words: biomarker, heat shock protein, arsenic, metal toxicity

Abstract

Environmental exposures to toxic metals result in ecological effects that are difficult to detect and quantify. There is a pressing need for biomarkers that can be used in the field and that are sensitive to the sublethal metal levels that are prevalent in the environment. Using a combination of molecular and demographic techniques, we developed a biomarker to detect arsenic stress in freshwater zooplankton. We measured changes in HSP 83 mRNA levels by RT-PCR analysis in *Daphnia pulex* adults and juveniles exposed to arsenite and arsenate, and we compared the HSP response to demographic effects on *Daphnia* survival, reproduction and development. HSP 83 appears promising as a biomarker for arsenic stress given that arsenic exposure induced an increased and sustained expression of mRNA. Both molecular and demographic measures of arsenic stress revealed differences between adult and juvenile responses and toxicity of arsenate and arsenite. The molecular responses are more sensitive than the demographic response of individuals exposed as adults but corroborate the reproductive responses for individuals exposed as neonates.

Introduction

Toxic metals released into the environment from anthropogenic activities, contaminate ground water and surface waters throughout the world. Arsenic, which causes human vascular disease and cancers, comes from natural geological sources as well as industrial and agricultural activities (Eisler, 1994). Elevated arsenic concentrations have been found in drinking water in the U.S., India, Argentina and Taiwan (Engel & Smith, 1994; Hopenhayn-Rich et al., 1995; Cantor, 1996; Tseng et al., 1997), and in many surface waters, arsenic reaches levels likely to impact flora and fauna (Eisler, 1994). Concentrations of arsenic in lakes generally range from 0 to 100 $\mu g\ l^{-1}$ but can reach as high as 198000–243000 $\mu g\ l^{-1}$ (USEPA, 1980; Eisler, 1994). Arsenic is found *in situ* in two organic forms, arsenate and arsenite, that differ in toxicity to aquatic organisms with arsenite generally thought to be more toxic (Ferguson & Gavis, 1972; USEPA, 1985; Anderson & Bruland, 1991; Aurillo et al., 1994; Huang & Lee, 1996; Sohrin et al., 1997). The greater toxicity of arsenite may be due to its lower oxidation state and reactivity with sulfhydryl groups on proteins causing enzyme inactivation. In contrast, arsenate toxicity appears to be diminished by phosphate which shares a common transport system (Huang & Lee, 1996). Different limnological conditions result in the predominance of different forms of arsenic necessitating that both forms be evaluated in aquatic toxicity studies.

Approaches used to evaluate arsenic toxicity are often not relevant to exposures of aquatic organisms to arsenite and arsenate in the field. The acute toxicities of arsenic have been quantified in laboratory studies (Borgmann et al., 1980; Lima et al., 1984; Passino & Novak, 1984), but most experiments concentrate on a single life stage, and use higher concentrations than usually found *in situ*. Lethal dose responses, although indicative of acute effects, are not applicable for most environmental exposures which occur at sublethal concentrations. Studies of cellular and molecular responses to arsenic exposure also use high concentrations of arsenite over short time courses and have not examined the effects of arsenate (Kothary & Candido, 1982; Kapoor & Lewis, 1987; Amaral et al., 1988; Dyer et al., 1993).

Impacts of arsenic exposure are difficult to detect in natural populations because sublethal responses are less easily observed. Also, population and community level parameters that may be altered take more time to measure than is typical of toxicity studies. Therefore, there is a need for rapid field biomarkers of environmental stress that are easy to assess, measurable after long periods of exposure, sensitive to low toxin concentrations and related to long term ecological impacts. At the molecular level, stress proteins, such as heat shock proteins (HSPs), are known to be useful biomarkers of exposure to certain environmental toxins (Bradley, 1993; Hightower, 1993; Sanders, 1993). HSPs have a structure and message that is highly conserved across taxa and their expression can be sensitive, rapid and consistent in response to stressors.

In cases where a reliable molecular response to metal exposure is desired, measuring changes in mRNA is a reasonable alternative to monitoring changes in protein levels. While protein levels better reflect a functional role of the protein, changes in the mRNA indicate that the metal either increased gene activity or altered cell signaling. In past studies, reverse transcription of stress protein mRNA followed by polymerase chain reaction (RT-PCR) has demonstrated that mRNA accumulates in cells in proportion to the concentration of a stressor (Sanders, 1990; Cochrane et al., 1994; Chan, 1995). RT-PCR analysis provides sensitive and specific measures of stress, with the advantage of requiring low amounts of sample tissue. In addition, development of species specific reagents is much more rapid and cost effective for detecting mRNA compared to raising antibodies to detect the proteins. The combination of conserved sequences and expression which is sensitive to a wide variety of stressors make stress proteins and their mRNA ideal candidates as biomarkers of exposure to environmental toxins (Sanders, 1990).

The challenge of using stress protein levels or their expression as biomarkers is to link molecular responses predictably to ecologically relevant effects at both population and community levels (Sanders, 1990; McCarthy & Shugart, 1990; Landis & Yu, 1995). Moreover, expression of HSPs must be evaluated in organisms exposed to sublethal levels of particular stressors that are characteristic of most natural systems. In this study, we establish the relationship between particular molecular and demographic responses to arsenic, with the long term goal of developing methods for detecting arsenic stress in field populations of aquatic organisms. In both molecular and demographic experiments, we compared the toxicities of arsenate and arsenite because both chemical species are present in aquatic systems and the relative sublethal toxicities of these two forms is not clearly understood particularly for molecular responses.

Our study focused on four objectives examined over a range of arsenite and arsenate exposures:

1. compare the responses of juvenile and adult *D. pulex*;
2. measure changes of *D. pulex* stress protein expression by RT-PCR analysis;
3. measure the population level response of *D. pulex* (survival, reproduction and development); and
4. compare the molecular and demographic responses of *D. pulex*.

We examined arsenic toxicity in zooplankton, specifically *Daphnia*, because cladocerans are important components of food webs in lakes and *Daphnia* are a commonly used test genus in toxicity studies. The responses of juvenile and adult *Daphnia* were compared to determine whether differences measured at the population level were paralleled in the molecular level response. Our molecular experiments examined the expression of what we have identified as HSP 83, which is one of the proteins in the HSP 90 family. A number of HSPs including HSP 70 and HSP 90 have been shown to be induced in *Daphnia* by heat exposure and chemical stressors (Bradley et al., 1994; Bond & Bradley, 1995; Bond & Bradley, 1997; S. Whittemore, unpublished data). This study is unique in its use of RT-PCR to characterize the expression of stress proteins induced by arsenic toxicity in *Daphnia*. Lastly, demographic experiments measured population-level parameters such as survival, reproduction and development, because they are often the most sensitive early indicators of stress on ecosystems (Schindler, 1987).

Materials and methods

We conducted molecular and demographic experiments at sublethal and lethal doses of arsenite (as arsenic trioxide) and arsenate (as arsenic pentoxide). The exposures used in all experiments unless otherwise mentioned were 0, 10, 100, 1000 and 3000 $\mu g\ l^{-1}$ of arsenite or arsenate. *Daphnia* cultures were maintained in 1.6 μm filtered water from Storrs Pond,

NH, U.S.A. (containing trace amounts of EDTA) at 20 °C on a 14:10 light dark cycle and fed the exponentially growing alga, *Cryptomonas erosa*. To ensure that effects of experimental treatments were not confounded with the effects of age of individuals and maternal effects, we harvested experimental animals from same-aged 'brood mothers'. Prior to each experiment, neonates were separated into 500 ml beakers and grown to maturity at which time 20 gravid individuals were kept in 1 l beakers and used as 'brood mothers'. Same-aged neonates or adults were harvested from brood mothers after their first clutch and used in both molecular and demographic experiments. Animals used in adult demographic and molecular experiments were grown to maturity prior to the start of the experiment, whereas individuals used in juvenile experiments were less than 24 h old. Animals were fed 1000 cells ml^{-1} *C. erosa* daily, unless otherwise discussed.

Molecular Measures

We used RT-PCR to amplify and compare expression of HSP 83 mRNA in control and arsenic exposed *D. pulex*. Fifteen adult *D. pulex* were placed in 1 l beakers with filtered pond water containing either 1000 μg l^{-1} arsenate (arsenic pentoxide), 1000 $\mu g\ l^{-1}$ arsenite (arsenic trioxide), or no arsenic for a period of 6 days. Experiments were terminated by rinsing the animals with EDTA-treated filtered lake water, and transferring adults and juveniles separately from each treatment to 1.5 ml sterile tubes. Juveniles were produced during the 6 days of the experiment. Water was removed and the animals were immediately ground with a sterile pestle. Total cellular RNA was then isolated with Trizol reagent according to the manufacturer's instructions. The resulting RNA pellet was suspended in 50 μl of sterile water by heating for 10 min at 60 °C, and quantified by measuring absorbance at 260 nm.

The oligonucleotide primer set used to amplify HSP 83 cDNA (forward primer 5′ TTTGGTGTGGGTTTCTACTC 3′, reverse primer 5′ CCTTTGGTTTCTCCTCATTC 3′) was synthesized and HPLC purified in the Molecular Biology Core facility at Dartmouth. These primers were derived from a partial HSP 90 family cDNA sequence cloned from *D. pulex* using a degenerate PCR screening protocol. Specificity of primers and products was determined by comparing their sequences against known sequences in Genbank by using NCBI network Blast software version 1.8 (Altschul et al., 1990). The PCR product was sequenced and demonstrated approximately 80% identity with cDNA sequences reported for chicken (Binart et al., 1989) and *D. magna* HSP 90, or *Drosophila melanogaster* HSP 83 (Hasson & Eanes, 1996). Amplification of constitutively expressed message for the intermediate filament cyclophilin was used as a control for global arsenic effects on mRNA expression, and for the efficiency of the assay. The primer set (forward 5′ CCATCGTGTCATCAAGGACTTCAT 3′, reverse 5′ TTGCCATCCAGCCAGGAGGTCT 3′) for amplifying cyclophilin was obtained from Ambion (Austin, TX), and is complementary to highly conserved regions of the cyclophilin cDNA from a variety of species including *D. pulex*. This is in contrast to actin or other potential 'housekeeping' genes, whose sequences are not as well conserved between vertebrates and invertebrates.

Total RNA (2.0 μg) was reverse-transcribed in a 30 μl reaction mixture containing Ambion 10x first strand buffer, 10 U RNAse inhibitor, 0.33 μm each dNTP, 1.7 μM oligo dT, and 100 U M-MLV reverse transcriptase. The mixtures were incubated at 44 °C for 60 min in a MJ Research PTC-100 thermocycler. After inactivating the enzymes at 95 °C, separate 5 μl aliquots of the reaction mixtures were then mixed with either 0.31 μM of HSP 83 or 0.25 μM of cyclophilin forward and reverse primers, 2.5 U of Taq polymerase and 0.125 μM each dNTP in Ambion complete PCR buffer. The total reaction volume was 50 μl. cDNA was amplified by either 26 or 35 cycles of 20 sec at 95 °C, 30 sec at 55 °C, and 40 sec at 72 °C for HSP 83 and cyclophilin, respectively. Amplified products were separated on 2% agarose gels in 1.0x TAE and detected by ethidium bromide staining, followed by UV illumination, photography of the gel and scanning densitometry of the negative image.

Demographic measures

We measured survival, reproduction and development of adults and juveniles at different levels of arsenate and arsenite. In juvenile experiments, the time to reproductive maturity and subsequent reproduction were measured for individuals first exposed to arsenic as 24 h old neonates. In adult experiments, survival and reproduction were measured for individuals first exposed to arsenic after they produced a first clutch. Experiments were terminated after most individuals produced eight clutches.

For juvenile experiments, individuals were placed in 40 ml glass vials containing 30 ml of pond water with specific arsenic concentrations. Juveniles were fed every other day and transferred to new food and water every fourth day. Once juveniles attained reproductive age (6–12 days), they were moved to beakers containing 200 ml of filtered pond water. Every other day, study individuals were transferred and new neonates and ephippial eggs were removed and counted. For each individual, age of first reproduction and neonate and ephippial production were measured over 26 days. For adult experiments, single individuals were placed in 200 ml of filtered pond water containing *C. erosa* and an appropriate arsenic treatment. Adults were transferred to new water every other day and female survivorship and cumulative reproduction were measured for 22 days.

Survival was compared using a censored nonparametric Kaplan-Maier survival analysis. Cumulative reproduction and age of first reproduction were analyzed using one-way ANOVA's with pair-wise comparisons. Cumulative reproduction was calculated as the total number of neonates produced per female over the entire experiment (Chen & Folt, 1993). This estimate of reproduction reflects both reproduction and survival in that it measures reproduction over the entire survival period of each individual. All analyses were performed using the statistical software, JMP (SAS, Inc., 1995).

Results

Molecular responses

Expression of HSP 83 mRNA in *D. pulex* exposed to arsenite and arsenate (1000 μg l^{-1}) differed from that in controls (Figure 1). The band intensity for these products was increased by all the arsenic exposures relative to control. In addition, there were differences between expression products following stressor treatments in the response of adults vs. juveniles. In control and arsenate exposures, RT-PCR amplification of HSP mRNA from adults yielded a single 400 base pair product while an additional 500 base pair product was produced in juveniles. Moreover, in arsenate exposures, the HSP 83 response of adults appeared greater than that of juveniles, whereas the adult and juvenile responses to arsenite were of similar intensity and each consisted of a single band.

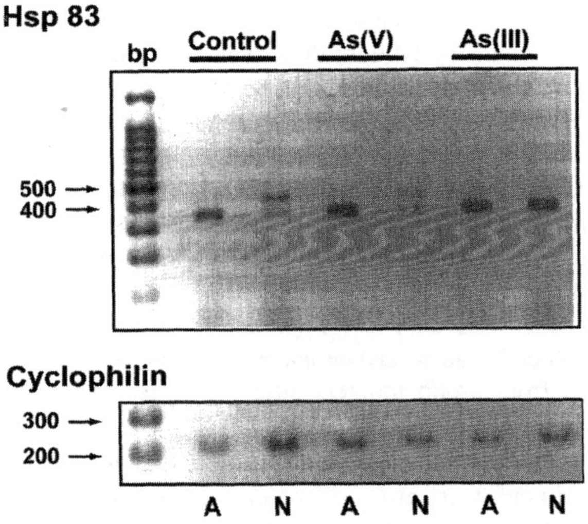

Figure 1. Effects of arsenic on HSP 83 mRNA levels in *D. pulex*. Adults (A) and their neonates (N) were exposed to either arsenic free conditions (control), 1000 μg l^{-1} arsenate (AsV), or 1000 μg l^{-1} arsenite (AsIII) for six days. Following exposure, animals were rinsed, ground, and total cellular RNA was isolated. The RNA was reverse transcribed and separate aliquots were amplified by PCR with primers specific for either Hsp 83 or cyclophilin. The products were separated on a 2% agarose gel stained with ethidium bromide.

Demographic responses

Survival was similar for juveniles and adults in that there were no differences between controls and treatments up to 1000 μg l^{-1} arsenite or arsenate. Mortality occurred within 3 days for both neonates and adults in 3000 μg l^{-1} arsenite and arsenate. However, adult survival in the 3000 μg l^{-1} arsenite treatment was significantly lower than the equivalent arsenate treatment indicating that at these acute levels, arsenite is more toxic than arsenate (Figures 2 & 3). In fact, the arsenate-exposed adults were able to reproduce minimally before dying. No differences were observed for juveniles exposed to 3000 μg l^{-1}arsenite or arsenate since neither exposure resulted in survival beyond one day.

Cumulative reproduction was a more sensitive indicator of toxicity than survival in juvenile experiments. Significant declines in reproduction were observed in exposures with 1000 μg l^{-1} arsenite and 10, 100 and 1000 μg l^{-1}arsenate relative to controls (Figure 4). However, arsenate was more toxic than arsenite in treatments with equal concentrations. In addition, reproduction in 100 and 1000 μg l^{-1} arsenate was significantly lower than controls and 10 μg l^{-1} arsenate. Production of ephippial eggs increased

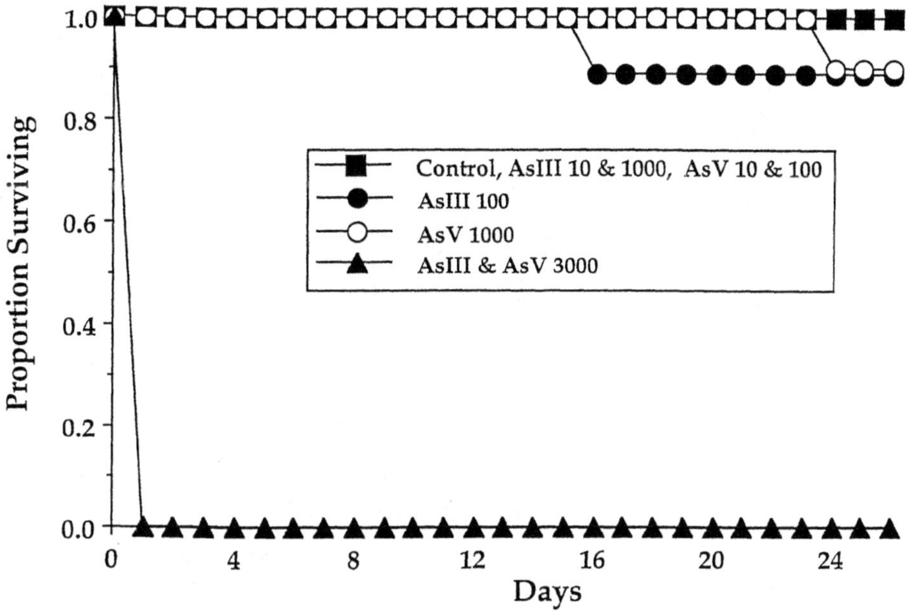

Figure 2. Survivorship of *D. pulex* individuals from juvenile experiment (see text) in control, arsenite and arsenate exposures. Experiment started on September 4, 1997 and ran for 26 days. Arsenite (AsIII) and arsenate (AsV) exposures are in $\mu g\ l^{-1}$.

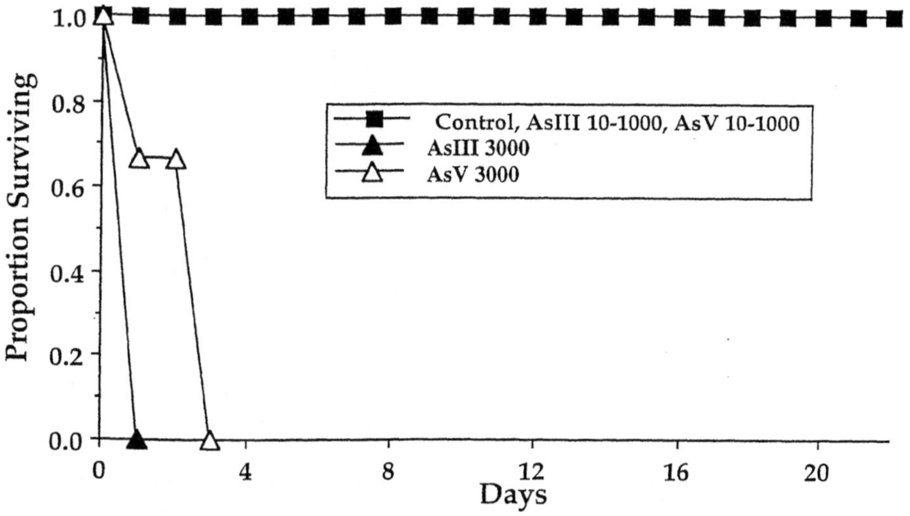

Figure 3. Survivorship of *D. pulex* individuals from adult experiment (see text) in control, arsenite and arsenate exposures. Experiment started on February 23, 1998 and ran for 22 days. Arsenite (AsIII) and arsenate (AsV) exposures are in $\mu g\ l^{-1}$. No mortality exhibited in any exposures except 3000 $\mu g\ l^{-1}$ arsenite and arsenate.

with increased arsenic stress as well. Ephippial egg production in arsenate exposures (100 and 1000 $\mu g\ l^{-1}$) was significantly greater than that of controls, whereas arsenite exposures exhibited no differences from controls (Figure 5).

Cumulative reproduction in adult experiments did not differ significantly between controls and arsenite or arsenate exposures although control animals produced the most offspring per individual (Figure 6). Only reproduction in the 3000 $\mu g\ l^{-1}$ arsenate concentration was significantly lower than in all other treatments where reproduction occurred because individuals only survived for 3 days. Similar to cumulative reproduction, ephippial egg production in adult experiments did not differ across arsenic exposures.

In juvenile experiments, the effect of arsenic ex-

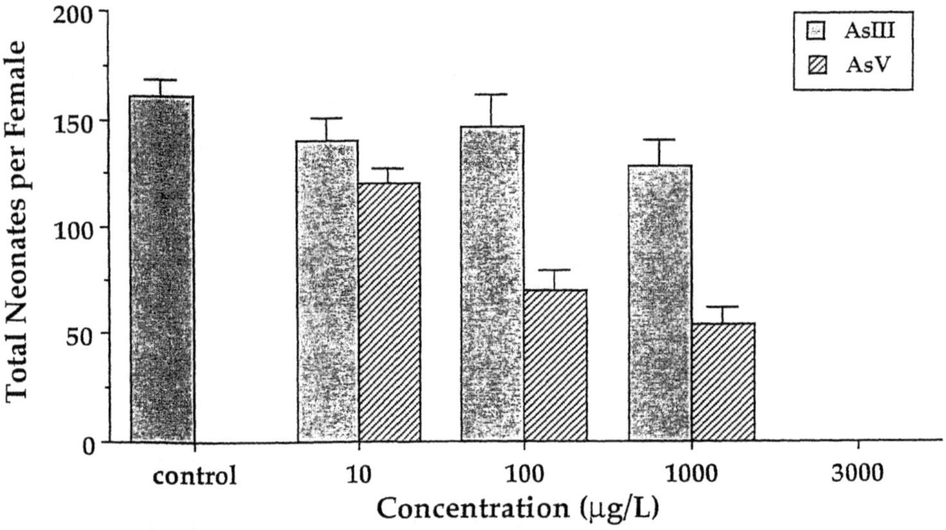

Figure 4. Figure 4. Cumulative egg production of *D. pulex* individuals from juvenile experiment (see text) in control, arsenite and arsenate exposures. Experiment started on September 4, 1997 and ran for 26 days. Plotted values are means±SE of the mean.

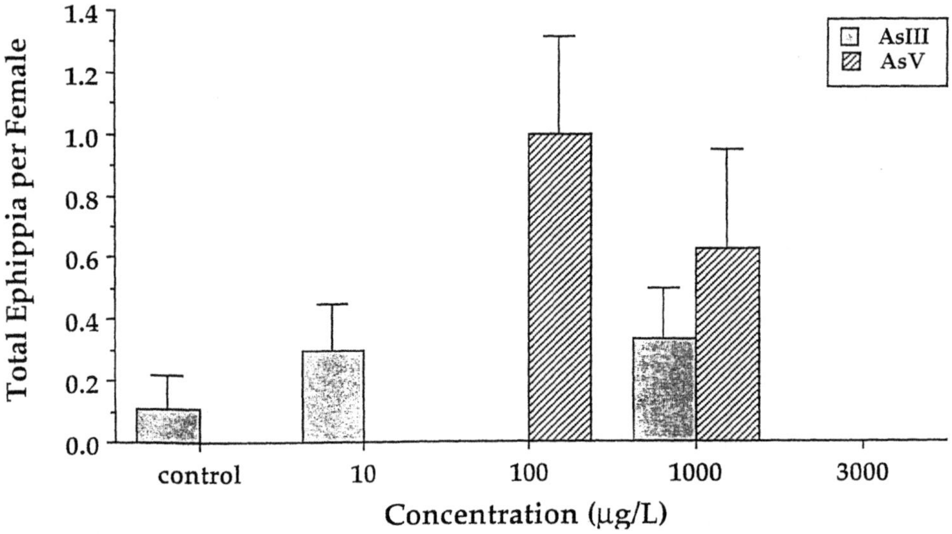

Figure 5. Ephippial production of *D. pulex* individuals from juvenile experiment (see text) in control, arsenite and arsenate exposures. Experiment started on September 4, 1997 and ran for 26 days. Plotted values are means±SE of the mean.

posure on development time was less consistent than was cumulative reproduction as an indicator of arsenic stress. The 100 $\mu g\, l^{-1}$ arsenate exposure was the only treatment that exhibited a significantly greater age of first reproduction (AFR) than the controls (Figure 7).

Discussion

The results of this study demonstrate the utility of molecular measures, such as HSPs, as biomarkers for arsenic stress in aquatic communities. Not only was there a sustained increase in heat shock protein mRNA expression in the cladoceran zooplankton species, *D. pulex*, following sublethal arsenic exposures, but this response paralleled key demographic responses to arsenic. Taken together the molecular and demographic measures also identified differences in the toxicities of arsenate and arsenite that have not been previously shown. Combining molecular and demographic studies to develop biomarkers for environmental stress appears feasible and is the necessary first step in the

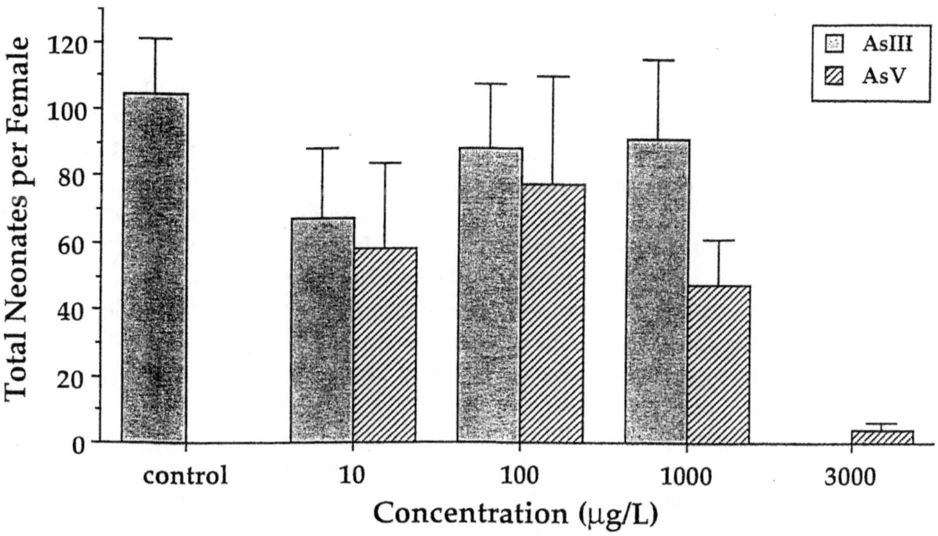

Figure 6. Cumulative egg production of *D. pulex* individuals from adult experiment (see text) in control, arsenite and arsenate exposures. Experiment started on February 23, 1998 and ran for 22 days. Plotted values are means±SE of the mean.

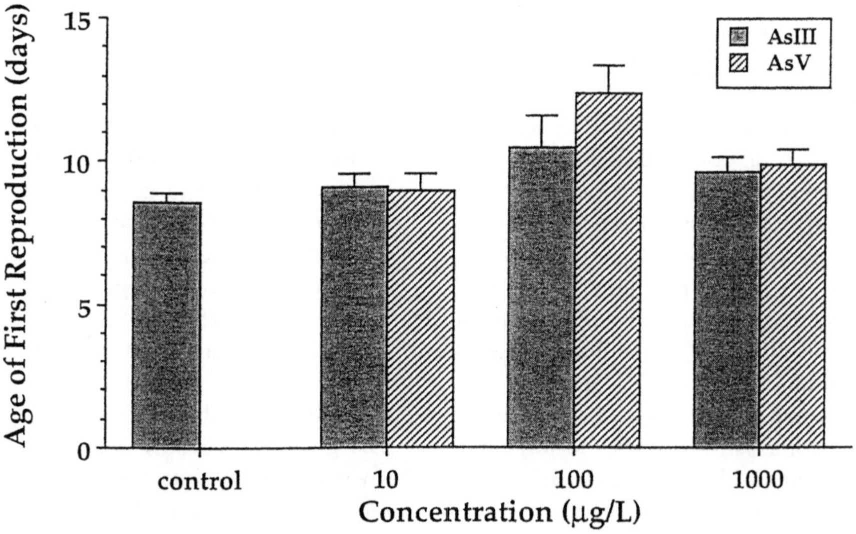

Figure 7. Age of first reproduction of *D. pulex* individuals from juvenile experiment (see text) in control, arsenite and arsenate exposures. Experiment started on September 4, 1997 and ran for 26 days. Plotted values are means±SE of the mean.

application of such biomarkers to identifying stress in field populations.

Molecular response

HSP 83 mRNA exhibited several characteristics that make it a potentially useful biomarker for arsenic stress. Sublethal arsenic exposures induced expression of HSP 83 mRNA to a greater degree than controls and its expression was detected after arsenate and arsenite exposures of 6 days. In addition, the molecular responses of juveniles to arsenate and arsenite differed from one another which paralleled the demographic response.

Stress proteins of many types including heat shock proteins are induced by chemical stressors in a variety of aquatic organisms (Lindquist, 1986; Bradley & Ward, 1989; McLennan & Miller, 1990; Hakimzadeh & Bradley, 1990; Brown et al., 1995). Specifically, arsenic has induced heat shock protein response in protozoans and fish (Kothary & Candido, 1982; Kapoor & Lewis, 1987; Amaral et al., 1988; Dyer et al.,

1993). However, all of these studies have measured protein responses to extremely high concentrations of arsenite over short periods of exposure. For example, Amaral et al. (1988) exposed *Tetrahymena pyriformis* to concentrations of 26 000 $\mu g\ l^{-1}$ arsenite for 30 min and Dyer et al. (1993) exposed fish gill tissue cells to 35 000 $\mu g\ l^{-1}$ arsenite for 8 h. Although these studies are useful in demonstrating stress protein response, they do not help us identify potential biomarkers which need to exhibit responses after long term exposures to sublethal concentrations. In contrast, our results show that stress protein mRNA in *D. pulex* can be detected from sublethal arsenate or arsenite exposures after much longer time periods.

The HSP 83 response of *D. pulex* to arsenic differed between adults and juveniles in both intensity and form, suggesting that the response varies with the age of individuals exposed. First, the adult HSP 83 response was greater in terms of band intensity than in juveniles. In past studies, protein patterns of individuals exposed to heat shock have been shown to vary with age of individuals (Bedard & Brandhorst, 1986; Bond et al., 1993). Bond et al. (1993) found that some of the proteins expressed in 1 – 3 day old *D. magna*, but not 4 – 6 day old individuals, may be growth proteins or heat shock proteins between 65 and 75 kDa (HSP70 family). In sea urchin embryogenesis, Bedard & Brandhorst (1986) found that HSP 90 is also developmentally regulated. HSP 90 appears to be synthesized throughout embryogenesis but undergoes a striking increase during the 32 – 64 cell stage due to the selective activation or translation of a stored maternal mRNA. Although no previous work has been conducted on the synthesis of HSP 90 during early embryogenesis in *Daphnia*, HSP 83 in *D. pulex* may undergo an increase during a later stage prior to reproductive maturity which could increase tolerance to certain stressors in later life.

The HSP 83 response of adults also differed from juveniles in the number of bands exhibited depending on the species of arsenic exposure; juveniles exposed to arsenite exhibited single bands similar to adults but juveniles exposed to arsenate exhibited higher intensity double bands. The double banding may indicate that there are additional transcriptional start sites or alternate splicing in neonates. Kapoor & Lewis (1987) also observed double bands in HSP 80 and HSP 90 in *Neurospora crassa* under conditions of heat shock and arsenite exposure, however, the authors did not discuss the implications of these 'doublets'.

Demographic response

At concentrations known to occur in aquatic environments (10–100 $\mu g\ l^{-1}$), arsenic had significant effects on *D. pulex* reproduction and development. In fact, arsenate was more toxic than arsenite at these levels, while arsenite appeared to be more toxic at lethal doses. In most comparisons of the relative toxicities of these inorganic species to aquatic organisms, arsenite is found to be more toxic (Eisler, 1994). For example, Spehar et al. (1980) found arsenite to be more toxic than arsenate at 1000 $\mu g\ l^{-1}$ for amphipod survival, but no differences to survival and reproduction of *D. magna* were found in this study at 100 or 1000 $\mu g\ l^{-1}$. Results similar to ours were observed in an earlier study of copepod assemblage growth rates and arsenic exposure (Borgmann et al., 1980). The study found that arsenate had a slightly greater effect in decreasing growth rates than arsenite at low concentrations ($<4000\ \mu g\ l^{-1}$), whereas arsenite had a greater effect at higher concentrations. Thus, the relative toxicities may be dependent on the range of concentrations. For exposures representative of field concentrations, it appears that arsenate is more toxic.

Individuals exposed as juveniles exhibited a greater sensitivity to arsenic than adults in terms of survival and reproduction. This pattern was not consistent with the expression of HSP 83 where adults appeared to exhibit a higher intensity heat shock response to arsenate than juveniles. However, the greater induction of heat shock mRNA in adults could be providing greater protection against arsenic stress and result in reduced effects of arsenic on reproduction. The greater effect of arsenic on reproduction by individuals exposed as juveniles may be due to lower reproduction resulting from a reduction in body size. Organisms exposed to chemical stress during development have been shown to have smaller body sizes and produce smaller offspring and clutch sizes (Moore & Folt, 1993; Havens & Hanazato, 1993). Other arsenic studies also document that juveniles are more vulnerable than adults. Sanders (1986) found juvenile *Eurytemora affinis* to be more vulnerable than adults to arsenate toxicity at 100 $\mu g\ l^{-1}$. However, *E. affinis* showed no reductions in maturity, survival or reproductive success at 25 $\mu g\ l^{-1}$ arsenate, whereas in our study, *D. pulex* exhibited reduced reproduction at 10 $\mu g\ l^{-1}$ arsenate.

Conclusion

We conclude that HSP 83 mRNA in *D. pulex* is a promising biomarker for environmental stresses such as arsenic toxicity. Unlike previous studies of arsenic induced HSPs, we have shown that the induction of HSP 83 mRNA in *D. pulex* is sustained over longer term arsenic exposures, which are more representative of conditions in the field. Moreover, our results link molecular and population level responses to both arsenate and arsenite exposures, and show that arsenate is more toxic than arsenite for concentrations likely to be found in the field. The ultimate utility of HSP 83 expression as a field biomarker, however, will require additional development of a quantitative approach to relate levels of HSP 83 expression to levels of arsenic stress.

Acknowledgements

We would like to thank K. Baumgartner, M. Calvi, C. Ware, J. Mitchell, G. Long and J. Connolly for their help culturing phytoplankton and zooplankton and conducting laboratory experiments. We also thank L. Klei and C. Coons for sharing their expertise in RT-PCR and gene cloning. This research was supported by NIEHS Superfund ES07373 to C. L. Folt, C. Y. Chen and R. S. Stemberger.

References

Altschul, S. F., W. Gish, W. Miller, E. W. Meyers & D. J. Lipman, 1990. Basic local alignment search tool. J. mol. Biol. 215: 403–410.

Anderson, L. C. D. & K. W. Bruland, 1991. Biogeochemistry of arsenic in natural waters: the importance of methylated species. Envir. Sci. Technol. 25: 420–427.

Amaral, M. D., S. Galego & C. Rodrigues-Pousada, 1988. Stress response of Tetrahymena pyriformis to arsenite and heat shock: differences and similarities. Eur. J. Biochem. 171: 463–470.

Aurillo, A. C., R. P. Mason & H. F. Hemond, 1994. Speciation and fate of arsenic in three lakes of the Aberjona Watershed. Envir. Sci. Technol. 28: 577–585.

Bedard, P. A. & B. P. Brandhorst, 1986. Translational activation of maternal mRNA encoding the heat-shock protein hsp90 during sea urchin embryogenesis. Dev. Biol. 117: 286–293.

Binart, N., B. Chambraud, B. Dumas, D. A. Rowlands, C. Bigogne, J. M. Levin, J. Garnier, E. E. Baulieu & M. G. Catelli, 1989. The cDNA-derived amino acid sequence of chick heat shock protein Mr 90,000 (HSP 90) reveals a "DNA like" structure: potential site of interaction with steroid receptors. Biochem. Biophys. Res. Comm. 159: 140–147.

Bond, J. A., C. R. M. Gonzalez & P. B. Bradley, 1993. Age-dependent expression of proteins in the cladoceran *Daphnia magna* under normal and heat-stress conditions. Comp. Biochem. Physiol. 106B: 913–917.

Bond, J. A. & B. P. Bradley, 1995. Heat-shock reduces the toxicity of malathion in *Daphnia magna*. Mar. envir. Res. 39: 209–212.

Bond, J. A. & P. Bradley, 1997. Resistance to malathion in heat-shocked *Daphnia magna*. Envir. Toxicol. Chem. 16: 705–712.

Borgmann, U., R. Cove & C. Loveridge, 1980. Effects of metals on the biomass production kinetics of freshwater copepods. Can. J. Fish. aquat. Sci. 37: 567–575.

Bradley, B. P., 1993. Are the stress proteins indicators of exposure or effect? Mar. envir. Res. 35: 85–88.

Bradley, B. P., C. M. Gonzalez, J. A. Bond & B. E. Tepper, 1994. Complex mixture analysis using protein expression as a qualitative and quantitative tool. Envir. Toxicol. Chem. 13: 1043–1050.

Bradley, B. P. & J. B. Ward, 1989. Detection of a major stress protein using a peptide antibody. Mar. envir. Res. 28: 471–475.

Brown, D. C., B. P. Bradley & M. Tedengren, 1995. Genetic and environmental regulation of HSP70 expression. Mar. envir. Res. 39: 181–184.

Cantor, K. P., 1996. Arsenic in drinking water: how much is too much? Epidemiology 7: 113–114.

Chan, K. M., 1995. Metallothionein: potential biomarker for monitoring heavy metal pollution in fish around Hong Kong. Mar. Poll. Bull. 11(4–12): 411–415.

Chen, C. Y. & C. L. Folt, 1993. Measures of food quality as demographic predictors in freshwater copepods. J. Plankton Res. 15: 1247–1261.

Cochrane, B. J., Y. D. Mattley & T. W. Snell, 1994. Polymerase chain reaction as a tool for developing stress protein probes. Envir. Toxicol. Chem. 13: 1221–1229.

Dyer, S. D., G. L. Brooks, K. L. Dickson, B. M. Sanders & E. G. Zimmerman, 1993. Synthesis and accumulation of stress proteins in tissues of arsenite-exposed fathead minnows (*Pimephales Promelas*). Envir. Toxicol. Chem. 12: 913–924.

Eisler, R., 1994. A review of arsenic hazards to plants and animals with emphasis on fishery and wildlife resources. In J. O. Nriagu (ed.), Arsenic in the Environment, Part II: Human Health and Ecosystem Effects. John Wiley & Sons, New York (NY): 185–258.

Engel, R. R. & A. H. Smith, 1994. Arsenic in drinking water and mortality from vascular disease: an ecologic analysis in 30 counties in the United States. Arch. envir. Health 49: 418–427.

Ferguson, J. F. & J. Gavis, 1972. A review of the arsenic cycle in natural waters. Water Res. 6: 1259–1274.

Gonzalez, C. R. M. & B. P. Bradley, 1994. Salinity stress proteins in *Eurytemora affinis*. Hydrobiologia 292/293: 461–468.

Hakimzadeh, R. & B. P. Bradley,1990. The heat shock response in a copepod, *Eurytemora affinis*. J. ther. Biol. 15: 67–77.

Hasson, E. & W. F. Eanes, 1996. Contrasting histories of three gene regions associated within (3L)Payne of *Drosophila melanogaster*. Genetics 144: 1565–1575.

Havens, K. E. & T. Hanazato, 1993. Zooplankton community responses to chemical stressors: a comparison of results from acidification and pesticide contamination research. Envir. Pollut. 82: 277–288.

Hightower, L. E., 1993. A brief perspective on the heat-shock response and stress proteins. Mar. envir. Res. 35: 79–83.

Hopenhayn-Rich, C., M. L. Biggs, A. Fuchs, R. Bergoglio, E. E. Tello, H. Nicolli & A. H. Smith, 1996. Bladder cancer mortality associated with arsenic in drinking water in Argentina. Epidemiology 7: 117–124.

Huang, R. N. & T. C. Lee, 1996. Cellular uptake of trivalent arsenite and pentavalent arsenate in KB cells cultured in phosphate-free medium. Toxicol. appl. Pharmacol. 136: 243–249.

Kapoor, M. & J. Lewis, 1987. Alteration of the protein synthesis pattern in *Neurospora crassa* cells by hyperthermal and oxidative stress. Can. J. Microbiol. 33: 162–168.

Kothary, R. K. & E. P. M. Candido, 1982. Induction of a novel set of polypeptides by heat shock or sodium arsenite in cultured cells of rainbow trout, *Salmo gairdnerii*. Can. J. Biochem. 60: 347–355.

Landis, W. G. & M. H. Yu, 1995. In Introduction to Environmental Toxicology, Impacts of Chemicals Upon Ecological Systems. Lewis Publishers, Boca Raton (FL.): 328 pp.

Lima, A. R., C. Curtis, D. E. Hammermeister, T. P. Markee, C. E. Northcott & L. T. Brooke, 1984. Acute and chronic toxicities of arsenic(III) to Fathead Minnows, Flagfish, Daphnids and an Amphipod. Arch. envir. Contam. Toxicol. 13: 595–601.

Lindquist, S., 1986. The heat-shock response. Ann. Rev. Biochem. 55: 1151–91.

McCarthy, J. F. & L. R. Shugart, 1990. Biological markers of environmental contamination. In McCarthy J. F. & L.R. Shugart (eds), Biomarkers of Environmental Contamination. Lewis Publishers, Boca Raton (FL.): 3–14.

McLennan, A. G. & D. Miller, 1990. A biological role for the heat shock response in crustaceans. J. ther. Biol. 15: 61–66.

Moore, M. V. & C. L. Folt, 1993. Zooplankton body size and community structure: effects of thermal and toxicant stress. Trends Ecol. Evolut. 8: 178–183.

Passino, D. R. M. & A. J. Novak, 1984. Toxicity of arsenate and DDT to the cladoceran *Bosmina longirostris*. Bull. envir. Contam. Toxicol. 33: 325–329.

SAS Institute Inc, 1995. JMP Statistics and Graphics Guide. Version 3.1 of JMP, Cary (NC), 593 pp.

Sanders, B. M., 1990. Stress proteins: potential as multitiered biomarkers. In McCarthy J. F. & L. R. Shugart (eds), Biomarkers of Environmental Contamination. Lewis Publishers, Boca Raton (FL.): 165–191.

Sanders, B. M., 1993. Stress proteins in aquatic organisms: an environmental perspective. Crit. Rev. Toxicol. 23: 49–75.

Sanders, J. G., 1986. Direct and indirect effects of arsenic on the survival and fecundity of estuarine zooplankton. Can. J. Fish. aquat. Sci. 43: 694–699.

Schindler, D. W., 1987. Detecting ecosystem responses to anthropogenic stress. Can. J. Fish. aquat. Sci. 44: 6–25.

Sohrin, Y., M. Matsui, M. Kawashima, M. Hojo & H. Hasegawa, 1997. Arsenic biogeochemistry affected by eutrophication in Lake Biwa, Japan. Envir. Sci. Technol. 31: 2712–2720.

Spehar, R. L., J. T. Fiandt, R. L. Anderson & D. L. DeFoe, 1980. Comparative toxicity of arsenic compounds and their accumulation in invertebrates and fish. Arch. envir. Contam. Toxicol. 9: 53–63.

Tseng, C. H., C.-K. Chong, C.-J. Chen & T.-Y. Tai, 1997. Lipid profile and peripheral vascular disease in arseniasis-hyperendemic villages in Taiwan. The Journal of Vascular Diseases 48: 321–335.

U.S. Environmental Protection Agency, 1985. Ambient water quality criteria for arsenic - 1984. EPA 440/5-84-033, 66 pp.

U.S. Environmental Protection Agency, 1980. Ambient water quality criteria for arsenic. EPA 440/5-80-021, 165 pp.

Hydrobiologia **401:** 239–254, 1999.
J.P. Zehr & M.A. Voytek (eds), Molecular Ecology of Aquatic Communities.

Taxonomic and systematic assessment of planktonic copepods using mitochondrial COI sequence variation and competitive, species-specific PCR

A. Bucklin[1,2,3], M. Guarnieri[1], R.S. Hill[1,3], A.M. Bentley[1] & S. Kaartvedt[4]
[1]*Ocean Process Analysis Laboratory,*
[2]*Department of Zoology, and*
[3]*Graduate Program in Genetics, University of New Hampshire, Durham, NH 03824, U.S.A.*
[4]*Department of Marine Zoology and Marine Chemistry, University of Oslo, P.O. Box 1064, N-0316 Blindern, Oslo 3, Norway*

Key words: mitochondrial DNA, cytochrome oxidase I, competitive PCR, copepoda, taxonomy, marine zooplankton

Abstract

Accurate taxonomic identification of species at all life stages is critical to understand and predict the processes that together determine marine community dynamics. However, zooplankton assemblages may include numerous sibling and congeneric species distinguished by subtle morphological characteristics. Molecular systematic databases, including DNA sequences of homologous gene regions for selected taxonomic groups, allow the design of rapid protocols to determine species' diversity and identify individuals. In this study, the DNA sequence of a 300 base-pair region of the mitochondrial cytochrome oxidase I (COI) gene was determined for eight species of three genera of calanoid copepods: *Calanus finmarchicus*, *C. glacialis* and *C. helgolandicus*; *Neocalanus cristatus*, *N. flemingeri* and *N. plumchrus*; and *Pseudocalanus moultoni* and *P. newmani*. The DNA sequences differed between congeneric species by 13–22% of the nucleotides; the protein sequences differed by zero to five amino acid substitutions. Both the DNA and amino acid sequences resolved the evolutionary relationships among congeneric species; relationships among the genera were not well-resolved by this region of mtCOI. Using the same conserved primers, the only amplification product for *C. finmarchicus* was an aberrant sequence (and putative pseudogene) which differed from the *C. finmarchicus* COI sequence by 36% of the nucleotides and 32 amino acid substitutions. Species-specific oligonucleotide primers were designed for *Calanus* spp. (which cannot be distinguished at larval stages) and *Pseudocalanus* spp. (which are difficult to distinguish even as adults). Individual copepods were identified using competitive, multiplexed species-specific polymerase chain reactions (PCR) in two studies of co-occurring sibling species. The first study confirmed the presence of three *Calanus* spp. in Oslofjord, Norway and found a predominance of *C. helgolandicus*. The second study determined patterns of distribution and abundance of *Pseudocalanus* spp. on Georges Bank in the NW Atlantic and showed that *P. moultoni* predominated in shallow and coastal waters, while *P. newmani* was more abundant in offshore regions flanking the Bank. Competitive, species-specific PCR is a useful tool for biological oceanographers. This simple, rapid, and inexpensive assay may be used to identify morphologically-similar individuals of any size and life stage, and to determine a species' presence or absence in pooled samples.

Introduction

The systematics of the copepoda is plagued by numerous sibling species groups that exhibit very subtle morphological differentiation. The North Atlantic zooplankton assemblage includes several ecologically important copepod species that co-occur with sibling species that are difficult to distinguish, especially at larval and juvenile life stages. Accurate species identification is the cornerstone of any ecological study –

yet even this fundamental step is not always possible for marine zooplankton. For planktonic crustaceans, in particular, it may be impossible to discriminate larval and juvenile stages of closely related species. Consequently, the species' life histories, ecological roles and relationships and the dynamics of the community may be incompletely known.

Larval and juvenile stages of the abundant and ubiquitous species, *Calanus finmarchicus*, are not readily discriminated from those of a sibling species, *C. glacialis* and a non-sibling species, *C. helgolandicus*, which have overlapping geographic ranges (Frost, 1974). Discrimination of all life stages of these species is essential to understanding recruitment events and other population dynamic processes of the North Atlantic zooplankton assemblage. In coastal regions of the NW Atlantic *Pseudocalanus moultoni* and *P. newmani* are sympatric for large portions of their ranges (Frost, 1989). The species differ in characteristic distribution, seasonal pattern of abundance and spatial/temporal pattern of reproduction (Davis, 1987), yet *Pseudocalanus* spp. are difficult to distinguish even as adults (Frost, 1989).

One means of identifying species' boundaries and revealing cryptic species is by phylogenetic analysis. In addition, evolutionary patterns within speciose groups may be reconstructed using molecular phylogenetic analysis. The speciose genera (such as *Calanus*, *Pseudocalanus* and *Neocalanus*) present opportunities to examine evolutionary patterns among morphologically similar species (Bradford & Jillett, 1974; Miller, 1988). These analyses also yield information on typical levels of genetic divergence among sibling species. This information may be used to design rapid protocols for species' identification.

The value of molecular systematic assessment

A molecular approach to taxonomy is particularly useful for species that cannot be readily discriminated using morphological characters, including cryptic or sibling species and non-sibling congeneric species. Through concerted efforts to build regional molecular databases, it is possible to recognize the existence of genetically distinctive groups within species and to evaluate their degree of taxonomic divergence by comparison with typical levels of interspecific divergence. Understanding marine speciation processes – and the causative factors for the sibling species assemblages noted in the marine fauna (Knowlton, 1993) – remains a challenge for marine biogeographers. Are cryptic species common in the planktonic realm? Why have some species, notably many crustacean species (Palumbi & Benzie, 1991; Bucklin et al., 1992, 1995, 1998b), diverged in genetic character without becoming morphologically differentiated? These questions will require comparison of molecular evolutionary characteristics of plankton with those of organisms with adequate paleontological records.

The significance of molecularly-based systematic approaches for calanoid copepods is that knowledge of the population dynamics, including reproductive processes, of important marine species will be needed to develop predictive capabilities for marine ecosystems. Unless we can recognize all developmental stages of closely related or cryptic species, we cannot hope to understand the forces that control the distribution and abundance of populations, the dynamics of communities and their response to anthropogenic insult and climatic variation.

Species identification using allele-specific primers and probes

DNA sequence variation may be used to design rapid, molecularly-based protocols to discriminate individuals of different species, based on species-specific oligonucleotide primers and probes. This approach has been used to identify a wide variety of marine species, including yeast (Fell, 1995), phytoplankton (DeLong et al., 1989) and the larvae (Olson et al., 1991; Dixon et al., 1995; Medeiros-Bergen et al., 1995) and adults (Banks et al., 1993) of marine invertebrates. For recently-diverged or older species that have not differentiated morphologically, mitochondrial DNA (mtDNA) is an appropriate molecular systematic character by virtue of its patterns of inter- and intra-specific variability (see Avise 1994).

A possible molecular systematic method and the one we have selected for discrimination of planktonic copepods, is allele-specific amplification by the polymerase chain reaction (PCR; see Charlieu, 1994), which has advantages in cost and efficiency. For marine zooplankton species, which are frequently numerous and geographically widespread, hundreds to thousands of individuals must be assayed to document a species' distribution and abundance pattern at high spatial and temporal resolution. Thus, it is essential that molecular protocols for species' identification be rapid, simple, inexpensive and reliable, to be useful for ecological or oceanographic studies.

Individuals of any size and life stage may be identified by a suite of PCR reactions, each using a common primer and a species-specific primer. Allele-specific amplification has a further advantage: multiple reactions can be 'multiplexed', i.e. carried out simultaneously and competitively in a single tube (Gibbs et al., 1989). In some cases, the competitive reaction may increase the accuracy and reliability of molecular systematic identifications (Bucklin, unpublished study). Competitive species-specific PCR has been used for a variety of marine organisms (Banks et al., 1993; Dixon et al., 1995; Fell, 1995; Bucklin et al., 1998a); in some cases, the same PCR primers yield different-sized products in different species (Dixon et al., 1995).

In this study, competitive, species-specific PCR was used to discriminate morphologically-similar co-occurring species of calanoid copepods. [Note: We will refer to the method as 'species-specific PCR'; the original term in the medical literature is "allele-specific PCR"; Charlieu, 1994.] In the first study, zooplankton samples collected from the Oslofjord, Norway, were examined for three species of *Calanus* that are easily confused at larval and juvenile stages: *C. finmarchicus*, *C. glacialis* and *C. helgolandicus*. In the second study, geographic patterns of distribution and abundance of two sibling species of *Pseudocalanus* were determined over Georges Bank in the NW Atlantic Ocean.

Calanus spp. in the N. Atlantic Ocean

Calanus currently consists of 14 species distinguished primarily by secondary sexual characteristics (Bradford, 1988; DeDecker et al., 1991; Hulsemann 1991). They have been divided into two groups of sibling species, a *finmarchicus* group and a *helgolandicus* group, in addition to three distinct species. The predominant species of the N. Atlantic include *C. finmarchicus* and *C. glacialis* (in the *finmarchicus* group) and *C. helgolandicus*. The species overlap in range despite their distinct hydrographic affinities: *C. finmarchicus* is a characteristic northern boreal species that borders *C. glacialis*, which is primarily an Arctic species (Frost, 1974). The apparent sympatry of *C. finmarchicus* and *C. glacialis* may result from an interleaving of water masses; whether both species reproduce in both water masses has not been definitively determined. The species are known to co-occur in both the western N. Atlantic (including the Gulf of St. Lawrence, Hudson and Davis Straits and into the Gulf of Maine; Grainger, 1961; Fleminger & Hulsemann, 1977) and the eastern N. Atlantic (including the region of the Polar Front north of Norway; Skjoldal & Rey, 1989). Farther south, the range of *C. helgolandicus*, a transition zone species, overlaps the southern edge of the distribution of *C. finmarchicus* in the western N. Atlantic (Jaschnov, 1955). In the eastern N. Atlantic, *C. finmarchicus* and *C. helgolandicus* are sympatric along the coast of Norway to about 40° N (Fleminger & Hulsemann, 1977).

The three N. Atlantic *Calanus* species are difficult to distinguish using morphological characters, especially at immature stages: *C. finmarchicus*, *C. glacialis*, and *C. helgolandicus* are very similar in the structure of the fifth thoracic leg, the morphological trait of taxonomic import for calanoid copepods (Frost, 1971, 1974). Adult females of *C. finmarchicus* and *C. glacialis* cannot be readily distinguished, although Frost (1971) identified additional, non-overlapping characters using morphometric analysis of traits used by Jaschnov (1955). Fleminger and Hulsemann (1977) discriminated N. Atlantic *Calanus* species on the basis of positions of integumental pores on the urosome of females. One solution to the difficulty of species identification has been to rely, at least partially, on the geographic location of collection as an indicator of species identity. In other cases, individual size is the primary character, which may not be diagnostic (Grainger, 1961). These approaches are not sufficient in regions of co-occurrence.

Despite the morphological similarity of the *Calanus* species, there is considerable genetic divergence between even the sibling species. Previous examination of molecular diversity and phylogenetic relationships within *Calanus* has suggested the usefulness of mitochondrial sequences for molecular systematic assessments (Bucklin et al., 1992, 1995). These studies examined intra- and inter-specific variation of a 440 bp region of the mitochondrial 16S rRNA, which resolved the evolutionary history of all but the *helgolandicus* sibling species group. The sequences differed by 12 – 15% of the nucleotides (Bucklin et al., 1992), consistent with speciation events millions of years ago (Cunningham et al., 1992).

Systematic relationships of Neocalanus *spp.*

The genus *Neocalanus* was redefined by Bradford & Jillett (1974), who also discussed the systematic relationships among the species. Miller (1988) described a new species, *N. flemingeri*, and redescribed *N. plum-*

chrus. The genus now contains six species, which are morphologically similar but ecologically distinct (Miller, 1988; Mackas et al., 1993). The genus is somewhat typical among calanoid copepods in that continuing effort is needed to understand systematic relationships among sibling species – and even to define specific and generic boundaries.

Mitochondrial DNA sequence variation has been shown to be useful to resolve evolutionary relationships among closely-related species groups for a wide range of taxa (Avise, 1994). Copepod sibling species swarms are cases in point (Bucklin et al., 1992, 1995, 1997a, 1998b). The species of *Neocalanus* exhibit subtle morphological variation, from which it is challenging to infer evolutionary pattern. We will examine the usefulness of COI sequence variation for this purpose.

Pseudocalanus *spp. on Georges Bank*

Pseudocalanus comprises a suite of sibling species that exhibit exceptional morphological similarity; they lack even the differences in secondary sexual characteristics that discriminate many sibling species of copepods (Frost, 1989). Two sibling species of *Pseudocalanus* occur sympatrically over Georges Bank, in the western North Atlantic Ocean: *P. moultoni* and *P. newmani* differ in size distribution (*P. moultoni* is larger), but individual size alone is not a diagnostic character for the two species (Frost, 1989). Due to the taxonomic confusion of the two species, fundamental aspects of their geographic distribution, population ecology and life history are incompletely known.

Our current understanding is that *P. newmani* is circumboreal and is abundant in both the N. Atlantic and N. Pacific coastal waters of the U.S. and Asia (Frost, 1989). *Pseudocalanus moultoni* is more coastally distributed and is absent from Asian waters (Frost, 1989). Although *P. moultoni* occurs on both U.S. coasts, the distribution is apparently disjunct (Frost, 1989) and these populations may be geographically isolated (Sevigny et al., 1989). The latitudinal range of the species is also more restricted; on the Atlantic coast, *P. moultoni* is restricted to waters between the New York Bight and Nova Scotia (Frost, 1989).

In this study, species of three genera, *Calanus*, *Neocalanus* and *Pseudocalanus*, were analyzed for DNA sequence variation of a 300 base-pair sequence of the mitochondrial gene encoding cytochrome oxidase I (COI). The results of this analysis were used to demonstrate the usefulness of mtDNA sequence variation for several types of studies: to reconstruct evolutionary relationships among closely-related species, to infer specific boundaries and to identify morphologically-similar species based on simple molecular protocols.

Materials and methods

Collection and preservation of zooplankton samples

Samples of *Calanus* spp. were collected by vertically stratified sampling of the Bedford Institute of Oceanography Net and Environmental Sensing System (BIONESS; Sameoto et al., 1980) in the Oslofjorden, Norway (collection coordinates: 59° 48′ N; 10° 34′ W) in a time-series study from August, 1996 to January, 1997 (Table 1). The *Pseudocalanus* spp. samples were collected by vertically-stratified tows of the Multiple Opening and Closing Net and Environmental Sensing System (MOCNESS: Wiebe et al., 1985) during the Broadscale Survey cruise of the *R/V*

Table 1. Location, date and depth of zooplankton collections from the Oslo-fjord, Norway. Samples were sorted for *Calanus* spp.; individuals were assayed by species-specific PCR to determine the relative abundances of *C. finmarchicus*, *C. glacialis* and *C. helgolandicus*. The frequency of unsuccessful PCRs (NP = no product) is also shown

Collection	Date	Depth	*C. fin.*	*C. hel.*	*C. gla.*	NP
			Numbers of Individuals			
1	8/14/96	50–75 m	1	8	-	1
1	8/14/96	75–100 m	-	14	1	1
1	8/14/96	100–114 m	-	8	-	2
2	9/3/96	0–25 m	-	4	-	-
2	9/3/96	50–75 m	-	4	-	-
2	9/3/96	75–100 m	-	4	-	-
3	10/16/96	0–114 m	-	5	-	-
3	10/16/96	75–100 m	2	4	-	1
4	11/28/96	0–25 m	-	4	-	-
4	11/28/96	50–75 m	-	2	-	2
4	11/28/96	75–100 m	-	4	-	-
5	1/23/97	25–50 m	-	4	-	-
5	1/23/97	50–75 m	-	4	-	-
5	1/23/97	75–100 m	-	4	-	-

Endeavor (EN-282) over Georges Bank between 9 – 19 April, 1996 (Table 2). The samples of *Neocalanus* spp. examined in this study were collected by integrated tows of a ring net from the Oyashio region off the east coast of Japan (collection site location: 42° 30′N; 145° E) on 21 October, 1997.

All samples were preserved in 95% ethyl alcohol; the alcohol was changed after 24 h. The copepods were prepared for molecular analysis (both PCR and sequencing) by rehydrating individual copepods in 0.5 ml distilled water in a microcentrifuge tube for up to 24 h (see Bucklin et al., 1995, 1996a,b). In some cases, individual copepods were boiled in distilled water for 10 – 15 min to evaporate the alcohol before molecular analysis.

DNA sequence and phylogenetic reconstruction

The DNA sequencing was done by direct sequencing of PCR amplification products, according to published protocols by Bucklin et al. (1995, 1996a,b). A ~700 bp region of the mitochondrial COI gene was amplified using consensus primers made according to published sequences (Folmer et al., 1994):

LCO-1490 5′ - GGT CAA CAA ATC ATA AAG ATA TTG G - 3′

HCO-2198 5′ - TAA ACT TCA GGG TGA CCA AAA AAT CA - 3′.

The COI primers are named, based on position numbers of the *Drosophila yakuba* sequence (Clary & Wolstenholme, 1985). The amplification protocol was: 94 ° C (1 min), 37 ° C (2 min), 72 ° C (3 min) carried out for 40 cycles. The sequencing primer was LCO-1490. Sequencing was carried out on an American Biotechnology, Inc., Model 373, automated DNA sequencer (see Bucklin, 1995; 1996a,b).

The DNA sequence of a 300 base-pair region of the mitochondrial cytochrome oxidase I (COI) gene was analyzed for individuals of *Calanus finmarchicus* (N = 15), *C. glacialis* (N = 5), and *C. helgolandicus* (N = 15); *Neocalanus cristatus* (N = 3), *N. flemingeri* (N = 8) and *N. plumchrus* (N = 3); and *Pseudocalanus moultoni* (N = 89) and *P. newmani* (N = 21). A consensus sequence was determined for each species using sequence editing protocols in SeqEd Ver. 1.03. The *Pseudocalanus* spp. sequences resulted from a larger population genetic study (Bucklin et al., 1998a); the most frequent haplotype was used for these species.

Analyses of the phylogenetic relationships among sibling species of *Calanus* and *Neocalanus* were done based on both the nucleotide and amino acid COI sequences. Numerous methods were used to examine the systematic relationships among the species using the Molecular Evolutionary Genetics Analysis (MEGA), Ver. 1.0, software package (Kumar et al., 1993). The trees shown were reconstructed by Neighbor Joining (Saitou & Nei, 1987) based on the nucleotide sequences for each species, using Tamura-Nei distances (Tamura & Nei, 1993) and an alpha parameter of 2.0. The trees were bootstrapped 1000 times.

Species-specific oligonucleotide primer design

Species-specific oligonucleotide primers were designed for *Calanus* spp. and *Pseudocalanus* spp. from regions of the COI gene that are conserved within each species but variable between them. Primer sequences were evaluated for suitable base composition, temperature of dissociation and self-compatibility using the software programs OLIGO (Rychlik, 1992) and *Amplify* (Engels, 1992). In order to allow multiplexing of the PCR reactions (i.e. simultaneous and competitive PCR using multiple species-specific primers and one common primer; Gibbs et al., 1989), the primers for the species were selected at different sites along the 300 bp sequence. The resultant amplification products were thus of different lengths and were distinguishable by agarose gel electrophoresis.

Calanus spp.: A 300 bp region of mitochondrial COI was sequenced for individuals of *Calanus finmarchicus*, *C. glacialis* and *C. helgolandicus*, using individuals collected from as much of the geographic range of each species as possible, with particular emphasis on the Norwegian Sea. Primers were designed from the consensus sequences by identifying diagnostic regions of ~20 bp that differed in position between the species, to allow multiplexing of the species-specific PCR reactions.

The species-specific primers were used with the common primer LCO-1490 (Folmer et al., 1994). The sequences and positions of the species-specific primers were:

C. finmarchicus (COI-1701)
5′-GGT GCT CCT ATG TAT AAT GGT-3′
C. glacialis (COI-1782)
5′-GAC TAA TGA CCT TGA GAG C-3′
C. helgolandicus (COI-1578)
5′-CAT CTC CAA TTA AAG ACC CA-3′

The amplification protocol for the multiplexed PCR reaction was: 94 °C (1min); 52 °C (2min); 72 °C (3 min); for 40 cycles.

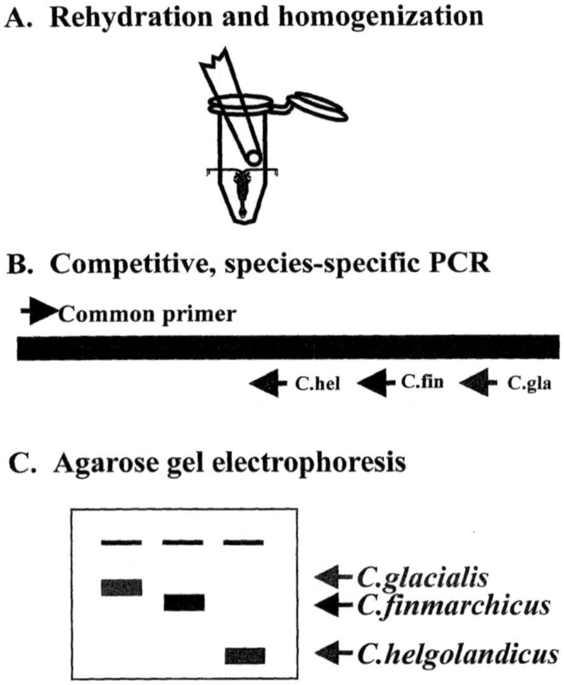

Figure 1. Diagram of the molecular protocol for competitive, species-specific PCR of *Calanus* spp. Copepods are rehydrated in distilled water and homogenized in the PCR buffer. The PCR reaction is done without purification of the DNA. The result of the competitive species-specific PCR reaction, and the species' identity of the copepod, are determined by agarose gel electrophoresis.

All PCRs included positive controls for each species using DNA extracted from identified individuals (sorted from samples collected from regions where only a single species was known to occur). The specificity and reliability of the individual and multiplexed PCR reactions were confirmed using individuals collected from diverse regions across each species' range, using archived samples.

The results we report here were obtained using individual reactions for each primer, dividing each copepod into three sections. Subsequently, the reactions were multiplexed to allow assay of individual copepods by a single, competitive reaction of one common primer and three competitive, species-specific primers. Since the lengths of the amplified products for each species differ (211 bp for *C. finmarchicus*; 292 bp for *C. glacialis*; and 88 bp for *C. helgolandicus*), the amplified products of each species were discriminated by electrophoresing the products of the multiplexed, competitive reaction on a 1.5% agarose gel (Figures 1 and 2).

Pseudocalanus spp.: For the same 300 bp region of COI, diagnostic regions of ~20 bp were identified that differed between *P. moultoni* and *P. newmani*, but showed little or no intraspecific variation. These regions were used to design species-specific oligonucleotide primers for amplification by PCR. The identification protocol was multiplexed, using a reaction cocktail with the common primer LCO-1490 (Folmer et al., 1994) and the two competing species-specific primers. The *Pseudocalanus* species-specific primer names, numbers and sequences are:
P. moultoni (COI-1615)
5′-CCC GAT TAG TG A TCC AGC-3′
P. newmani (COI-1736)
5′-CCG CAC CCA G TA TTA GAG GT-3′
The amplification protocol was: 94 °C (30sec); 40 °C (30sec); 72 °C (1min); for 30 cycles. PCR results were determined by electrophoresing the amplification products on a 2% agarose gel. Size differences in the amplified products (246 bp for *P. newmani*; 125 bp for *P. moultoni*) allowed accurate and reliable identification of the two species (Figure 3).

Calanus *spp. in the Oslofjord, Norway*

Zooplankton samples were collected from a single location in the Oslofjord (coordinates: 59° 48′ N; 10° 34′ E) between August, 1996 and January, 1997, as part of a time-series study of the planktonic community of Norwegian fjords (Table 1). Vertically-stratified samples were collected at all sampling times except one; portions of each sample were preserved in alcohol for molecular discrimination of the three co-occurring species of *Calanus*. The goal of the molecular analysis was to spot-check relative abundances of the three species. Since adult females of *Calanus* spp. can be discriminated morphologically (Frost, 1971, 1974), the definitive determination of species' distribution and abundance is now being done by microscopic viewing of *Calanus* spp. in all of the samples collected during the study.

Pseudocalanus *spp. on Georges Bank*

Samples collected during the April, 1997 Broadscale Survey cruise of the U.S. GLOBEC Georges Bank Study were examined for *Pseudocalanus* spp. using competitive, multiplexed PCR. The samples were collected at 16 stations; at all but two stations, vertically-stratified samples were taken from two depth intervals: 0–15 m and 15–40 m (Table 2). Female *Pseudocalanus* spp. were removed from the samples

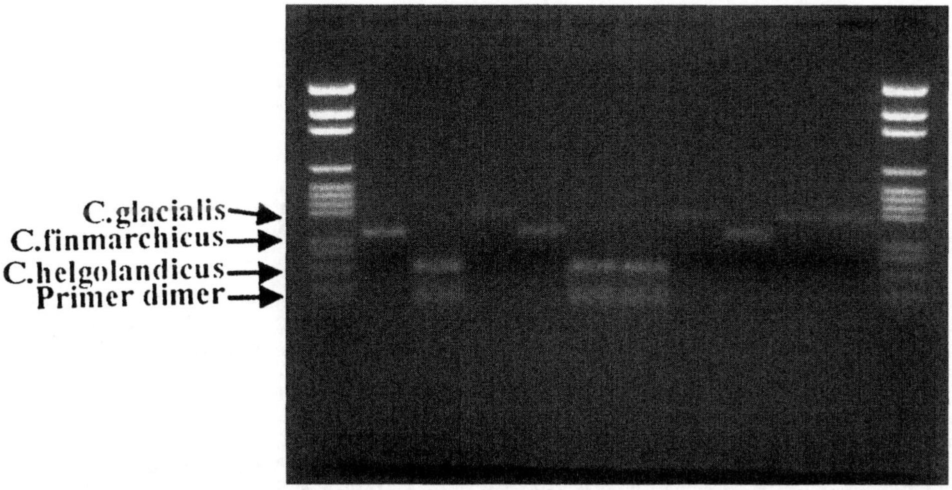

Figure 2. Photograph of the amplification products from the multiplexed species-specific PCR reactions for *Calanus* spp. The reaction cocktail included the common primer, LCO-1490 (Folmer et al., 1994), and the three species-specific primers. The amplification products differ in length (211 bp for*C. finmarchicus*; 292 bp for *C. glacialis*; and 88 bp for *C. helgolandicus*) and can be discriminated by agarose gel electrophoresis. First (#1) and last (#12, left to right) lanes are molecular size markers; lanes #2, #3 and #4 are positive contols; lanes #5 – #11 include PCR products for all three species from individual copepods.

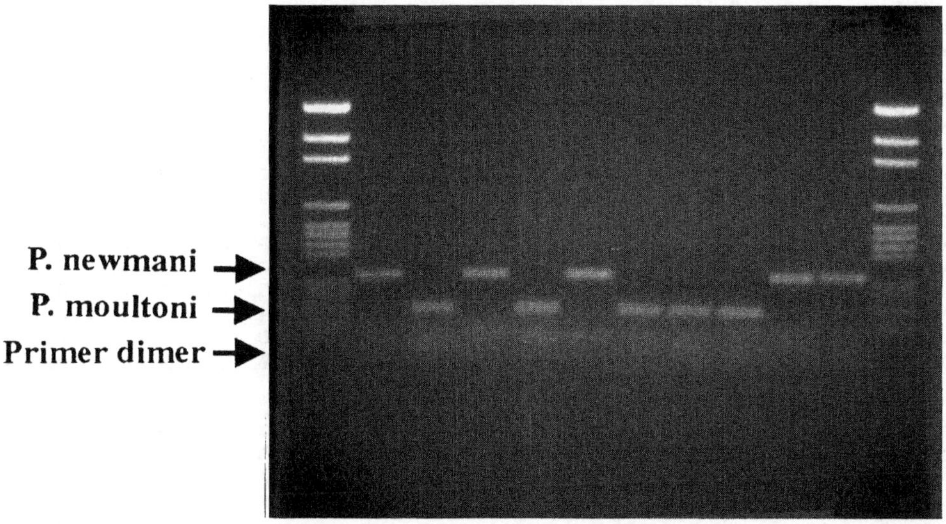

Figure 3. Gel photo showing species-specific PCR products for individual *Pseudocalanus* spp. First (#1) and last (#12; left to right) lanes are molecular size markers; lanes #2 – #11 include PCR products for individual copepods of both species. PCR products for *P. newmani* are larger (246 bp) and migrate less far; products for *P. moultoni* are smaller (125 bp) and migrate farther.

without respect to size (see Bucklin et al., 1997a; 1998a). Approximately 20 individuals in each sample were identified to species by competitive PCR. The relative abundance of each species was converted to estimated absolute abundance based on counts of female *Pseudocalanus* spp. in Broadscale Survey samples done by the URI Zooplankton Sorting Group (E.G. Durbin and P. Garrahan, personal communications).

Results

MtCOI sequence variation and phylogenetic reconstruction

A 300 bp region of mitochondrial COI was sequenced for eight species in three genera of calanoid copepods: *Calanus finmarchicus*, *C. glacialis* and *C. helgolandicus*; *Neocalanus cristatus*, *N. flemingeri* and *N. plumchrus*; and *Pseudocalanus moultoni* and *P. new-*

```
                                                                                        54
C.finmarchicus     TAT TTA ATG GCA GGG GCT TGA TCA GGA ATA GTT GGG ACA GGG CTC AGT ATA ATT
C.glacialis        ..C ... T.A ..T ..T ..G ... ... ... ..G A.C ..A ..C ... T.G ..G ... ...
C.helgolandicus    ... ... T.. ..C ..T ..G .AC ... ... ... A.C ..T ..G ..A ... ... ... ...
N.flemingeri       ..C ... T.A C.T ... ... ..G ... ..T ... A.. ..C ..T ... ..G ..G ... ..C
N.plumchrus        ..C ... T.A C.T ... ..C ..G ... ..T ... A.. ..T ..T ..A T.A .AA ... ..C
N.cristatus        ... ... TGA C.T ..T ..C ..G ... ..T ... A.. ..T ..G ..T T.A ..A ... ..C
P.newmani          ..C ... ..A ..T ..T ..A ... G.. ..T ... A.. ... ... ... T.A ..A ... ...
P.moultoni         ... ... ..A ..T ... ..A ..G G.. ..T ..G A.. ..T ... ..C T.A ..A ... ...
C.fin-pseudo       ..C A.. G.A TTT ..A ..A .TT ... ..T ..G ..A ..T ..T .CA T.A ... ..G T.A

                                                                                       108
C.finmarchicus     ATC CGA ATG GAG TTG GGC CAA GCA GGA AGA CTA ATT GGG GAT GAT CAG ATT TAC
C.glacialis        ..T ... T.A ..A ... ..T ..G ..C ..C TC. T.G ... ..A ... ... ... ... ...
C.helgolandicus    ..T ..T C.A ..A ..A ..T ... .CT GGG TCT TTA ATT GGA GAT G.. ..A G.A ..T
N.flemingeri       ... ... T.A ... ..A ..T ..G ... ..C TCT ..T ... ... ... ... ... ... ..T
N.plumchrus        ... ..C C.A ... ..A ..T ... ..T ..C TC. ..T ... ... ... ... ... ... ..T
N.cristatus        ..T ..G C.A ... C.C ..T ..G ..G ..C TC. T.. ..C ..T ... ... ... ... ..T
P.newmani          ..T ... ..A ..G TTA GGT CAG GCA GGA TC. T.G ... ..A ... ... ... ... ..T
P.moultoni         ..T ... ..A ..A ... ..T ..G ..T ... TC. CTA ATC GGG GAT GAC C.A ... ..T
C.fin-pseudo       ... A.. T.A ..A ..A .CA GG. C.. ..T GCT A.G T.. ..A ... ... ..T T.A ..T

                                                                                       162
C.finmarchicus     AAT GTC GTG GTT ACA GCT CAC GCA TTT ATT ATA ATT TTT TTC ATG GTT ATG CCT
C.glacialis        ..C ..A ..T ..A ..T ... ..T ..T ... ... ... ... ... ..T ... ... ... ...
C.helgolandicus    ..C ..T ..A ..A ..T ..A ... ... ... ... ... ... ... ..T ..A ... ... ...
N.flemingeri       ... ... ..A ... ..T ... ..T ... ... ... ... ... ... ..T ... ... ..A ..A
N.plumchrus        ... ..T ... ... ..T ... ... ... ... ..C ... ... ... ... ... ... ..A ..A
N.cristatus        ... ..A ..A ... ..T ... ... ... ... ..C ... ... ... ..T ... ..A ..A ...
P.newmani          ... ..A ..T ... ..T ..A ..T ... ... ... ... ... ... ... ... ..C ..A ..A
P.moultoni         ... ..A ..T ... ..T ..G ..T ..G ... ..C ... ... ... ..T ..A ... ..A ..A
C.fin-pseudo       ... ... A.A ..C ..T ..A ..T ..T ... G.C ..G ... ..C ..T T.A ..A ... ..A

                                                                                       216
C.finmarchicus     ATT CTT ATT GGG GGG TTT GGT AAC TGA TTA GTT CCT TTG ATA TTA GGG GCG GCG
C.glacialis        ... T.G ... ... ..A ... ..A ..T ... ... ..G ... ..A ... ... ..C ..A ..A
C.helgolandicus    ... T.A ... ..A ..A ... ..A ... ... ..G ..C ... ..A ... ..G ..T ..A ..A
N.flemingeri       ..C ... ... ... ... ... ... ... ..G ... ..G ... ..A ... C.G ..A ..T ..C
N.plumchrus        ..C ..A ... ..T ... ..C ... ..T ... C.T ..C ... ..A ... C.G ... ..A ..A
N.cristatus        ... T.A ... ..A ... ... ... ... ..G C.. C.G ..A C.A ... ..G ..T ... ..A
P.newmani          ..C T.A ... ... ..A ... ..C ... ... ... ..A ... C.A ... C.G ..T ... ..A
P.moultoni         ... T.A ... ..A ..T ... ... ..T ... C.. ..C ..C C.T ... C.. ... ..A ..A
C.fin-pseudo       G.C T.A ... ..A ..A ... ... ..T ... ... A.A CCA TTA TAC ATA GGA GCA C..

                                                                                       270
C.finmarchicus     GAT ATA GCT TTC CCC CGT ATA AAT AAT ATA AGG TTT TGG TTC TTA ATA CCA GCG
C.glacialis        ... ..G ..C ..T ..A ..A ... ... ... ... ..A ..C ... ... ... ..G ... ...
C.helgolandicus    ... ..G ..A ..T ..T ... ... ... ... ... ..A ..C ... ... ... ..G ... ..T
N.flemingeri       ... ..G ..A ..T ..A ... ... ... ... ... ... ..C ... ... ... ... ... ..T
N.plumchrus        ... ..G ..G ..T ..T ... ... ... ..C ... ... ... ... ..T ..G ... ... ..C
N.cristatus        ..C ..G ..A ... ... ... ... ... ... ... ... ..C ..A ... ..G ... ... ..A
P.newmani          ... ..G ... ..T ..A ... ... ... ... ... ... ..A ..T ... ... ... ... ..C
P.moultoni         ... ... ..C ..T ..T ... ... ... ..C ... ..A ... ..A ..T ... ..G ... ..C
C.fin-pseudo       ... ..G ..A ..T ..T A.A T.. ... ... T.. ..T ... ... ..A ... CC. ..T ..A

                                                       300
C.finmarchicus     TTG ATC ATG CTT CTA TCT AGG TCT CTA GTG
C.glacialis        ..A ... ..A ..G CTC TCA AGG TCA TTA GTC
C.helgolandicus    ..A ..T ..A ... T.G ..A ..A ... ..G ..T
N.flemingeri       ... ..T ... ... ..T ... ..A ... T.. ..T
N.plumchrus        ..A ..T ..A ... T.. ..G ..A ... T.. ..T
N.cristatus        ..A ..T ..A ..A T.. ... ..C ... T.. ...
P.newmani          ..A ..T ..A ... ..T ... ..A ... T.. ..A
P.moultoni         ... ... ... T.A ..T ..A ..A ..C T.. ..T
C.fin-pseudo       ..A T.A T.A T.A T.. GG. TCT ... T.. ..T
```

Figure 4. DNA sequence for a 300 bp region of the mitochondrial COI gene for eight species in three genera of calanoid copepods: *Calanus finmarchicus*, *C. glacialis* and *C. helgolandicus*; *Neocalanus cristatus*, *N. flemingeri* and *N. plumchrus*; and *Pseudocalanus moultoni* and *P. newmani*. Sequences shown are the consensus sequence of three or more individual sequences, except that the most frequent sequence was used for *Pseudocalanus* spp. Species-specific PCR primer positions for *Calanus* and *Pseudocalanus* spp. are indicated by bold, underlined type. Note that the primers were synthesized as the reverse complements of the sequences shown.

Table 2. Standard station numbers (STN), vertical depth strata [Depth (m)], coordinates of collection sites and numbers of female *Pseudocalanus moultoni* (P.m.) and *P. newmani* (P.n.) in samples collected during the April, 1997, Broadscale Survey cruise of the U.S. GLOBEC Georges Bank Study. *Pseudocalanus* spp. abundances were estimated from relative abundances of the two species among ~20 individuals in sub-samples of the original sample, and concentrations of both species in samples analyzed by the Georges Bank zooplankton group (personal communication, P. Garrahan & E. G. Durbin, University of Rhode Island). See Bucklin et al. (1998) for methods and original data

STN	Depth (m)	Latitude	Longitude	P.m.	P.n.
3	2–14	40° 32′	68° 26′	550	1376
3	15–36	"	"	2121	1306
4	1–14	41° 01′	68° 16′	669	0
7	1–13	40° 26′	67° 14′	3	59
9	2–14	40° 60′	67° 20′	0	2501
9	15–43	"	"	100	251
12	1–17	41° 24′	67° 33′	383	0
13	1–14	41° 17′	67° 04′	356	1008
17	1–12	41° 11′	66° 29′	153	8
17	13–38	"	"	0	415
18	1–10	41° 24′	66° 42′	112	48
18	11–38	"	"	0	1437
20	4–14	41° 44′	66° 32′	566	849
20	15–39	"	"	7017	9356
23	2–12	41° 48′	66° 12′	692	115
23	13–38	"	"	1253	368
25	1–14	42° 20′	65° 46′	0	859
25	15–41	"	"	0	3675
27	3–13	41° 55′	66° 42′	80	121
27	14–39	"	"	266	354
29	0–14	42° 18′	66° 54′	471	138
29	15–38	"	"	1293	216
30	15–40	41° 53′	67° 14′	102	0
34	2–16	41° 52′	68° 17′	5849	278
34	16–41	"	"	8833	0
36	1–12	41° 26′	68° 18′	3065	0
38	4–11	41° 29′	68° 58′	42	2
38	12–40	"	"	4666	0
39	1–14	42° 08′	66° 02'	1012	644
39	15–37	"	"	0	466

mani (Figure 4). The DNA sequences of congeneric species differed by 13 – 22% of the nucleotides; differences between the sequences of species of different genera ranged from 14 – 22% (Table 3). There were 1 – 10 non-synonymous substitutions and 37 – 56 synonymous substitutions (Table 4) between congeneric species. Species of different genera differed by 3 – 11 non-synonymous substitutions and 37 – 61 synonymous substitutions. The protein sequences for mitochondrial COI differed between congeneric species by 0 – 5 amino acids, with one to five amino acid substitutions between non-congeneric species (Table 5).

An aberrant sequence (putative 'pseudogene') for *C. finmarchicus* differed from all other COI sequences by 31 – 36% of the nucleotides (Table 3); by 43 – 59 non-synonymous substitutions and 45 – 50 synonymous substitutions (Table 4); and by 32 – 34 amino acids (Table 5). In most cases, the pseudogene was most different from the *C. finmarchicus* COI sequence.

The evolutionary relationships among the three genera were not well resolved by this region of the mitochondrial COI sequence (neither nucleotide

Table 3. Proportion of nucleotide differences between species of *Calanus, Neocalanus*, and *Pseudocalanus*, based on a 300 bp region of mitochondrial COI. Comparisons between congeneric species are indicated in bold-face type. Abbreviations are: *Calanus finmarchicus* (Cf), *C. glacialis* (Cg), *C. helgolandicus* (Ch), *Neocalanus flemingeri* (Nf), *N. plumchrus* (Np), *N. cristatus* (Nc), *Pseudocalanus newmani* (Pn), *P. moultoni* (Pm), *C. finmarchicus pseudogene* (Cf_p)

	Cf	Cg	Ch	Nf	Np	Nc	Pn	Pm
Cfp	0.36	0.32	0.31	0.31	0.31	0.34	0.31	0.33
Cf		**0.20**	**0.22**	0.17	0.21	0.22	0.18	0.22
Cg			**0.17**	0.18	0.22	0.21	0.17	0.19
Ch				0.20	0.20	0.20	0.19	0.22
Nf					**0.13**	**0.17**	0.14	0.20
Np						**0.16**	0.15	0.18
Nc							0.17	0.20
Pn								**0.16**

Table 4. Number of synonymous (above diagonal) and non-synonymous (below diagonal) differences between species of *Calanus, Neocalanus* and *Pseudocalanus* for a 100 codon sequence of mitochondrial COI. Comparisons between congeneric species are indicated in bold-face type. Abbreviations are given in legend to Table 3

Cfp	Cf	Cg	Ch	Nf	Np	Nc	Pn	Pm	
Cfp		59	50	49	45	45	55	43	48
Cf	49		**54**	**56**	44	53	56	48	58
Cg	46	**7**		**45**	50	61	58	47	54
Ch	45	**10**	**6**		53	50	52	50	56
Nf	48	6	3	8		**37**	**45**	37	54
Np	49	10	5	9	**3**		**43**	38	45
Nc	47	11	5	9	**6**	**5**		43	53
Pn	49	6	3	9	6	8	8		**47**
Pm	50	8	4	11	7	8	8	**1**	

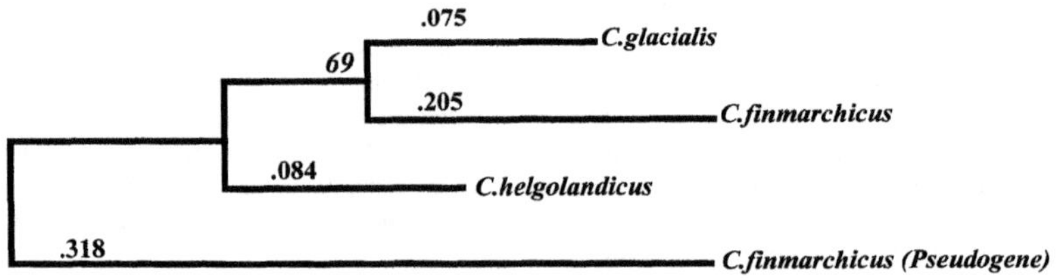

Figure 5. Phylogenetic reconstruction of evolutionary relationships among sibling species of *Calanus*, based on the COI nucleotide sequences (using all bases) as shown in Figure 4. The *C. finmarchicus* pseudogene sequence was used as an outgroup. Tree reconstruction was by Neighbor-Joining (Saitou & Nei, 1987) using Tamura/Nei distances (Tamura & Nei, 1993) and alpha = 2.0; the tree was bootstrapped 1000×. Distances are given on branches; bootstrap value is given at the branch point.

Table 5. Number of amino acid differences between species of *Calanus, Neocalanus* and *Pseudocalanus* based on 100 codons for mitochondrial COI. Comparisons between congeneric species are indicated in bold-face type. Abbreviations are given in legend to Table 3

	Cf	Cg	Ch	Nf	Np	Nc	Pn	Pm
Cfp	32	32	32	32	33	32	34	34
Cf		**3**	**5**	4	5	5	2	2
Cg			**2**	1	2	3	3	3
Ch				3	4	5	5	5
Nf					**1**	**2**	4	4
Np						**3**	5	5
Nc							5	5
Pn								**0**

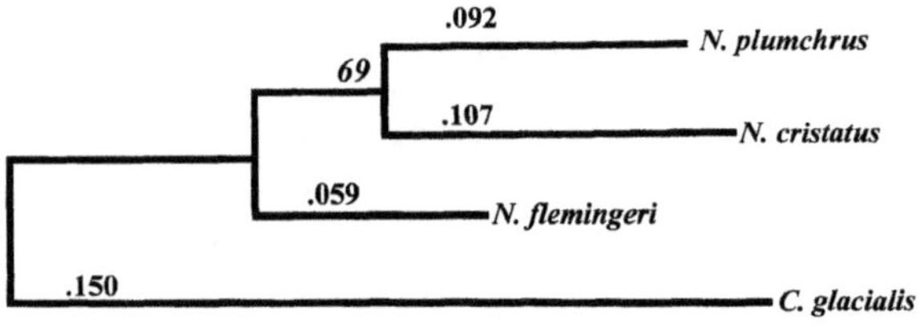

Figure 6. Gene tree showing the systematic relationships of three *Neocalanus* spp. based on the consensus COI nucleotide sequences. The COI sequence for *C. glacialis* is used as an outgroup. See the legend of Figure 5 for phylogeny reconstruction methods.

nor amino acid), probably because the genera were distantly related and the COI sequence variation was either highly saturated (third position bases) or relatively unvariable (first and second position bases). However, systematic relationships among both *Calanus* (Figure 5) and *Neocalanus* (Figure 6) sibling species were well resolved based on the nucleotide sequences (including all bases). Similar tree topologies were obtained for both *Calanus* spp. and *Neocalanus* spp. using the nucleotide sequences but omitting the third base position and using the amino acid sequences.

Discrimination of Calanus *and* Pseudocalanus *spp. by species-specific PCR*

Species-specific PCR was used to determine the relative abundances of three co-occurring *Calanus* spp. in Oslofjord, as a confirmation of morphologically-based identification of the distribution and abundance of *Calanus* spp. in Norwegian fjords. Molecular discrimination of several individuals in each sample confirmed the presence of three co-occurring species of *Calanus* in the Oslofjord and revealed that the most abundant species was *C. helgolandicus* (Table 1).

The relative abundances of the two sibling *Pseudocalanus* spp. in zooplankton samples collected during April, 1997 Broadscale Survey cruise on Georges Bank ranged from 100% *P. moultoni* to 100% *P. newmani* (Table 2). Estimated abundances of female *Pseudocalanus* spp. in these samples ranged from 44 m^{-2} to 16 373 m^{-2} (E.G. Durbin & P. Garrahan, University of Rhode Island, personal communications). The patterns of distribution of the two species on Georges Bank differed markedly: *P. moultoni* was predominant on the crest of the Bank and in the basins of the Gulf of Maine to the north of the Bank. In contrast, *P. newmani* predominated on Browns Bank, to the northeast of Georges Bank, and on the northeast peak and southern flank of Georges Bank (Figure 7). The pattern was clearest for samples collected at 15–40m; the species were more intermixed in surface waters.

Discussion

The mitochondrial COI gene showed sufficient variation in DNA sequence between species to easily

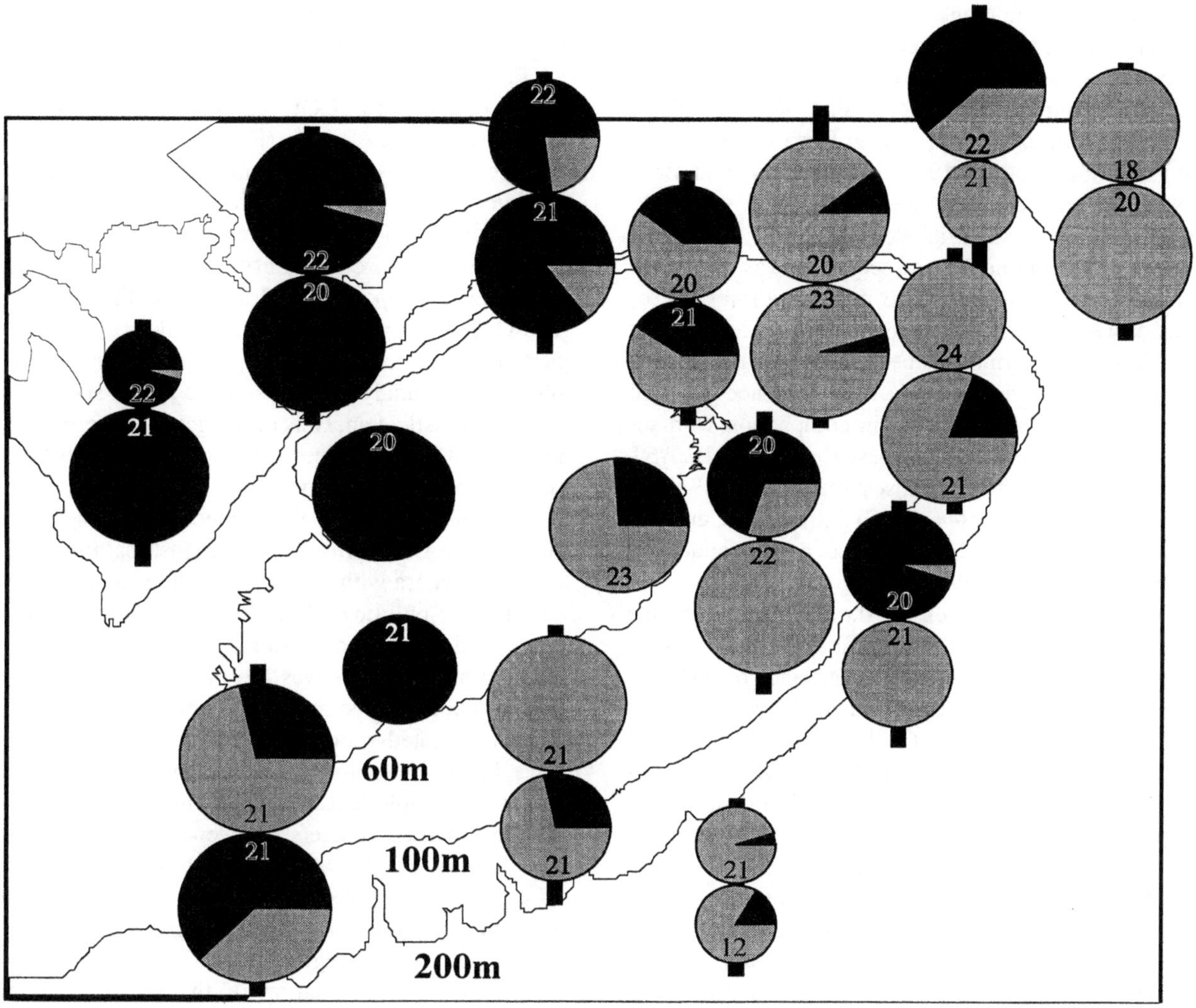

Figure 7. Estimated abundances of female *P. moultoni* (black shading) and *P. newmani* (gray shading) in samples collected at 0 – 15 m (top) and 15 – 40 m (bottom) from Georges Bank in the NW Atlantic in April 1997. Pie sizes indicate total abundances of female *Pseudocalanus* spp. based on counts. The three sizes reflect order-of-magnitude differences in abundance: 1 – 100 m^{-2} (small); 100 – 1000 m^{-2} (medium); and >1000 m^{-2} (large). Numbers on pies are sample sizes for relative species' abundances based on species-specific PCR. See Table 2 for station location and actual species' abundances.

resolve differences between the sibling species of calanoid copepods (Figure 4). This finding is consistent with previous studies using COI to examine variation within a harpactacoid copepod (Burton & Lee, 1994) and between species of insects (Brown et al., 1994; Juan et al., 1995; Pederson, 1996; Stauffer et al., 1997) and a diverse group of invertebrates (Folmer et al., 1994), although amounts of sequence differences associated with species comparisons differed among the taxa studied.

The percentage of sequence differences between congeneric species of *Calanus*, *Neocalanus* and *Pseudocalanus* for the 300 bp region of COI ranged from 13 to 22% (Table 3), and were thus similar to that for a 440 bp region of 16S rRNA between *Calanus* spp. (Bucklin et al., 1992, 1995), but about twice that observed between *Pseudocalanus* spp. (Bucklin et al., 1997a, 1998a). The great majority of nucleotide differences between species were synonymous substitutions, as opposed to non-synonymous substitutions (Table 4).

The aberrant COI sequence obtained for *C. finmarchicus* may be best explained by the presence of a pseudogene, presumably one encoded on the nuclear genome. Similar findings of mitochondrial elements in the nuclear genome have been reported in sea urchins (Jacobs and Grimes, 1986), molluscs (Harasewych et al., 1997), humans (Tsuzuki et al., 1983; Fukuda et al., 1985; Kamimura et al., 1989; Parfait et al., 1998) and rats (Zullo et al., 1991). In the published studies the nuclear sequences were 80%–92% similar to the original mitochondrial element. In the case of *C. finmarchicus*, the nucleotide sequence was only 64% similar to the *C. finmarchicus* mitochondrial COI sequence we obtained; the nucleotide sequences differed by 52 non-synonymous substitutions and the protein sequences by 32 amino acids. The sequence was still identified as a COI sequence in comparison with sequences of the GenBank database (based on a BLAST search). Interestingly, the closest match was 88% similarity to a *Murex troscheli* COI sequence (GenBank accession No. U86329) identified as a pseudogene (Harasewych et al., 1997).

Other possible causes for the aberrant, COI-like sequence from *C. finmarchicus* may include heteroplasmy, caused by bi-parental inheritance of mtDNA. This condition has been observed in bivalves (Quesada et al., 1996) and humans (Gocke et al., 1998), among other species. However, the large divergence, including numerous amino acid substitutions, of the sequence from the *C. finmarchicus* mtCOI suggests that the aberrant sequence is not transcribed. The lack of stop codons – a usual feature of pseudogenes – is not statistically significant (none occurring vs. <2 expected on the basis of random occurrence).

This level of divergence suggests that the origin of the aberrant COI sequence (whatever the mechanism) may predate the divergence of *Calanus* spp. (see Fukuda et al., 1985 and Zullo et al., 1991 for pseudogene dating methods; see Bucklin et al., 1992 for estimation of speciation rates of *Calanus*). However, the sequence provided an accurate and reliable marker of *C. finmarchicus* in comparison to the other *Calanus* spp. This sequence was the only product of amplification of *C. finmarchicus* by the primers LCO-1490 and HCO-2198. Our protocol testing (using individual copepods collected across the N. Atlantic Ocean) and systematic use of positive and negative controls in each PCR confirmed that of the three N. Atlantic *Calanus* spp., only *C. finmarchicus* was amplified by the species-specific primer designed from this sequence. Additional protocol testing will be required if the species-specific primer based on the pseudogene is to be used to determine the presence of *C. finmarchicus* in unsorted zooplankton samples, rather than to discriminate species of *Calanus*.

Phylogenetic reconstruction based on the nucleotide sequence for COI resolved the relationships among species of *Calanus* (Figure 5) and *Neocalanus* (Figure 6) with statistical confidence. The COI nucleotide sequence was an informative character for systematic relationships among sibling species and confirmed the close relationship of *N. plumchrus* and *N. cristatus* proposed by Bradford & Jillett (1974). The COI sequence is useful for studies of speciation patterns within calanoid copepod genera. In addition, phylogenetic analyses are useful to assess species diversity; statistical analysis of the gene trees provides a measure of distinctiveness of taxa and allows for easy recognition of cryptic species.

In contrast, neither the nucleotide nor the amino sequence for COI resolved relationships among the copepod genera with statistical confidence, regardless of the tree-building algorithm (including distance-based and maximum parsimony analyses). The most likely reason for this was the high degree of saturation in nucleotide substitutions among the relatively distantly-related genera. However, rates of evolution have been shown to vary along the COI sequence, and it is possible that other portions of the gene may provide useful characters for deeper phylogenies (see e.g. Lunt et al., 1996). Relationships among genera of calanoid copepods may also be investigated using sequences of more slowly-evolving genes, such as a nuclear rRNA gene region.

Despite some variation of the sequence within each species, the mitochondrial COI sequence is a useful molecular systematic character for identification and discrimination of species that lack definitive morphological characters. The sequences must differ sufficiently between species to allow identification of regions that are desirable for primer design (in terms of base composition and self-compatibility, see Rychlik, 1992) and are located at sufficiently distant positions along the sequence to allow multiplexing of the reaction. Only if the lengths of the amplified products are sufficiently different can the results of a single competitive species-specific PCR be discriminated by agarose gel electrophoresis. The sequenced COI gene portion included several short regions of sequence that differed between species by ~30% of the nucleotides, but exhibited little or no variation within species.

Carefully designed, species-specific oligonucleotide primers are valuable taxonomic tools for ecologists and oceanographers. Using species-specific PCR, hundreds to thousands of individuals can be identified to accurately characterize stage-specific patterns of distribution and abundance at sufficient spatial and temporal resolution. In this study, we provide preliminary evidence of the usefulness of this type of analysis by discriminating individuals of three co-occurring species of *Calanus* (Figures 1 and 2) and two co-occurring sibling species of *Pseudocalanus* (Figure 3) using species-specific PCR of individuals collected in zooplankton samples from two regions.

Discrimination of Calanus *spp. in the Oslofjord*

Studies of zooplankton communities in Norwegian fjords have occasionally assumed the presence of a single *Calanus* spp.; *C. finmarchicus* in the fjords of northern and western Norway, and *C. helgolandicus* in southern Norway. In fact, the communities may be more complex, with mixtures of two or even three different *Calanus* spp. In a study of population genetic structure of *C. finmarchicus* in the fjords of western Norway based on the mitochondrial 16S rRNA sequence variation, Bucklin & Kaartvedt (manuscript in preparation) found nearly equal mixtures of *C. finmarchicus, C. glacialis* and *C. helgolandicus* in at least one of the fjords.

Our goal in this study was to spot-check zooplankton samples collected as part of a time-series analysis of community dynamics in the Oslofjord for relative abundances of *Calanus* spp. This type of analysis will not determine the species' patterns of distribution and abundance, but it may alert oceanographers to unexpected species' diversity in their system. Alternatively, unsorted zooplankton samples may be homogenized, the DNA purified and the resultant 'assemblage DNA' analyzed for the presence of a number of targeted species by competitive PCR. The extent to which this type of analysis could be made quantitative – to determine the abundance as well as the presence of a particular species – is unclear.

Our limited assessment of the Oslofjord samples by species-specific PCR confirmed that the majority of *Calanus* in this community is *C. helgolandicus* and that *C. finmarchicus* and *C. glacialis* are present as well (Table 1). Larger sample sizes will need to be assayed in order to make quantitative conclusions about the species' relative abundances and patterns of distribution.

Molecular discrimination of Pseudocalanus *spp.*

Although the two sympatric sibling species are nearly indistinguishable based on morphological characters, the patterns of distribution and abundance of *P. moultoni* and *P. newmani* on Georges Bank suggest that the species are ecologically quite distinct. The estimated abundances of *Pseudocalanus* spp. varied widely in samples collected during the April, 1997 Broadscale Survey of Georges Bank: from 44 to 16 373 m^{-2} (P. Garrahan & E.G. Durbin, unpublished data; Table 2). The relative abundances of *P. moultoni* and *P. newmani*, discriminated by competitive species-specific PCR, also varied widely: from 100% *P. moultoni* on the crest of the Bank and in the basins of the Gulf of Maine, to 100% *P. newmani* on Browns Bank and on the southern flank of Georges Bank (Figure 7). The patterns of distribution and abundance of the two species suggest that *P. moultoni* prefers the crest of Georges Bank and the basins of the Gulf of Maine, while *P. newmani* predominates in offshore waters. Strong northeast winds, which just preceded the April, 1997 Broadscale Survey (R. Limeburner, Woods Hole Oceanographic Institution, personal communication), may be responsible for transporting *P. newmani* from Browns Bank, where it is persistent almost thoughout the year (McLaren et al., 1989), across the southern flank of Georges Bank.

Based on our observations, we hypothesize that populations of *P. moultoni* may be actively concentrated inside the 60 m isobath of Georges Bank, while *P. moultoni* outside of this region will be diluted by offshore populations of *P. newmani*. This hypothesis is consistent with the biogeographic distributions described for the two species by Frost (1989), who considered *P. moultoni* to be a coastal species and *P. newmani* to be a cosmopolitan, oceanic species. These results suggest that the life history of *P. moultoni* may be adapted to retention in the cyclonic gyre present for some portion of the year over Georges Bank, perhaps by virtue of life history, behavior, preferred habitat, or vertical distribution.

Conclusions

Molecular analysis of marine zooplankton is likely to continue to reveal taxonomically-significant genetic partitioning within species, including the discovery of cryptic and sibling species. This may be especially true for calanoid copepods, which are plagued by numerous sibling species assemblages that are discriminated by subtle morphological characters. Molecular assessment of sibling species of calanoid copepods has revealed significant genetic divergence between species. Sequence variation of mitochondrial DNA regions is particularly useful to: reconstruct the evolutionary and phylogenetic history of a speciose genus, infer the boundaries of species and genera and identify and discriminate species with few or subtle morphological characters. In particular, species-specific competitive PCR may be useful to identify the species of an individual copepod of any size or life stage. Competitive species-specific PCR may also be useful for the analysis of DNA purified from unsorted zooplankton samples to determine the presence or absence of particular species in the assemblage.

A caution for biological oceanographers is that our ability to understand the community dynamics of planktonic assemblages may depend on our ability to accurately measure species' diversity and discriminate individuals of morphologically- and systematically-similar species. The myth of 'functional groups' in marine zooplankton may mask a complex set of interactions between species that effectively partition the ocean along temporal and spatial boundaries that are unseen by typical analyses. Molecular systematic assessment using simple protocols may be useful and necessary to understand the population dynamics of target species and the community dynamics of planktonic assemblages.

Acknowledgements

This study would not have been possible without the collaboration of B. W. Frost (University of Washington, U.S.A.), who identified *Pseudocalanus moultoni* and *P. newmani* and sent samples of these and other species. F. Norrbin (University of Tromso, Norway) and A. Fosshagen (University of Bergen, Norway) also collected samples for this study. E. G. Durbin and P. Garrahan (University of Rhode Island) shared unpublished data on the distribution of *Pseudocalanus* spp. on Georges Bank. The samples of *Neocalanus* spp. were collected and identified by A. Tsuda (Hokkaido National Fisheries Research Institute, Japan) and C. B. Miller (Oregon State University). Technical assistance was provided by J. Conroy and S. B. Smolenack (both University of New Hampshire). Thanks are also due to the Captain and crew of the *R/V Oceanus* on the April, 1997 Broadscale Survey (OCE-302). Funding for this study was provided by the U.S. National Science Foundation as part of the U.S. GLOBEC Program (Grants No. OCE-9529100 and OCE-9632840). Contribution number 106 from the U.S. GLOBEC George Bank Study.

References

Avise, J. C., 1994. Molecular Markers, Natural History and Evolution, Chapman and Hall, New York, NY. 511 pp.

Banks, M., D. Hedgecock & C. Waters, 1993. Discrimination between closely related Pacific oyster species (*Crassostrea*) via mitochondrial DNA sequences coding for large sununit rRNA. Mol. Mar. Biol. Biotechnol. 2: 129–136.

Bradford, J. M. & J. B. Jillett, 1974. A revision of generic definitions in the Calanidae (Copepoda, Calanoida). Crustaceana 27: 5–16.

Bradford, J. M., 1988. Review of the taxonomy of the Calanidae (Copepoda) and the limits to the genus*Calanus*. Hydrobiologia 167/168: 73–81.

Brown, J. M., O. Pellmyr, J. N. Thompson & R. G. Harrison, 1994. Phylogeny of *Greya* (Lepidoptera: Prodoxidae), based on nucleotide sequence variation in mitochondrial cytochrome oxidase I and II: congruence with morphological data. Mol. Biol. Evol. 11: 128–141.

Bucklin, A., B. W. Frost & T. D. Kocher, 1992. DNA sequence variation of the mitochondrial 16S rRNA in *Calanus* (Copepoda; Calanoida): intra- and inter-specific patterns. Molec. Mar. Biol. Biotech. 1: 397–407.

Bucklin, A., B. W. Frost & T.D. Kocher, 1995. Molecular systematics of seven species of *Calanus* and three species of *Metridia* (Calanoida; Copepoda). Mar. Biol. 121: 655–664.

Bucklin, A., T. C. LaJeunesse, E. Curry, J. Wallinga & K. Garrison (1996a) Molecular genetic diversity of the copepod, *Nannocalanus minor:* genetic evidence of species and population structure in the N. Atlantic Ocean. J. mar. Res. 54: 285–310.

Bucklin, A., R. Sundt & G. Dahle, 1996b. Population genetics of *Calanus finmarchicus* (Copepoda; Calanoida) in the North Atlantic. Proceedings of an ICES Workshop on a *TransAtlantic Study of Calanus finmarchicus*. Ophelia 44: 29–45.

Bucklin, A., R. S. Hill, N. J. Mottola & A. M. Bentley, 1997a. Seasonal patterns of distribution and abundance of the copepods, *Pseudocalanus moultoni* and *P. newmani*, on Georges Bank: evidence for a dynamic balance between retention and loss. Internat. Cons. Expl. Seas Science Mtg., September, 1997. Background Paper T: 06.

Bucklin, A. , S. B. Smolenack, A. M. Bentley & P. H. Wiebe, 1997b. Gene flow patterns of the euphausiid, *Meganyctiphanes norvegica*, in the N. Atlantic based on DNA sequences for mitochondrial cytochrome oxidase I and cytochrome b. J. Plank. Res. 19: 1763–1781.

Bucklin, A., A. M. Bentley & S. P. Franzen, 1998a. Distribution and relative abundance of the copepods, *Pseudocalanus moultoni* and *P. newmani*, on Georges Bank based on molecular identification of sibling species. Mar. Biol. (in press).

Bucklin A., C. C. Caudill & M. Guarnieri, 1998b. Population genetics and phylogeny of marine planktonic copepods. Chapter 14. In: K. C. Cooksey (ed.). Molecular Approaches to the Study of the Ocean. London: Chapman & Hall: 303–318.

Burton, R. S. & B.-N. Lee, 1994. Nuclear and mitochondrial gene genealogies and allozyme polymorphism across a major phylogenetic break in the copepod *Tigriopus californicus*. Proc. natn. Acad. Sci. 91: 5197–5201.

Charlieu, J.-P., 1994. Distinction between almost-identical DNA sequences by polymerase chain reaction. Chapter 12. In H. G. Griffin & A. M. Griffin (eds), PCR Technology Current Innovations. CRC Press, Boca Raton, FL: 101–106.

Clary, D. O. & D. R. Wolstenholme, 1985. The mitochondrial DNA molecule of *Drosophila yakuba*: nucleotide sequence, gene organization and genetic code. J. molec. Evol. 22: 252–271.

Cunningham, C. W., N. W. Blackstone & L. W. Buss, 1992. Evolution of king crabs from hermit crab ancestors. Nature 355: 539–542.

Davis, C. S, 1987. Zooplankton Life Cycles. In: Backus R. H. (ed.). Georges Bank, MIT Press, Cambridge, MA: 256–267.

DeDecker, A. H. B., B. Z. Kaczmaruk & G. Marska, 1991. A new species of*Calanus* (Copepoda, Calanoida) from South African waters. Ann. S. Afr. Mus. 101: 27–44.

DeLong, E. F., G. S. Wickman & N. R. Pace, 1989. Phylogenetic strains: ribosomal RNA-based for the identification of single cells. Science 243: 1360–1363.

Dixon, D. R., D. A. S. B. Jollivet, L. R. J. Dixon, J. A. Nott & P. W. H. Holland, 1995. The molecular identification of early life-history stages of hydrothermal vent organisms. In L. M. Parson, C. L. Walker & D. R. Dixon (eds.), Hydrothermal Vents and Processes, Geol. Soc. Spec. Publ. 87: 343–350.

Engels, W., 1992. *Amplify.* Computer Freeware. Genetics Department, University of Wisconsin, Madison, WI 53706.

Fell, J. W., 1995. rDNA targeted oligonucleotide primers for the identification of pathogenic yeasts in a polymerase chain reaction. J. Ind. Microbiol. 14: 475–477.

Fleminger, A. & K. Hulsemann, 1977. Geographical range and taxomonic divergence in North Atlantic *Calanus* (*C. helgolandicus*, *C. finmarchicus* and *C. glacialis*). Mar. Biol. 40: 233–248.

Folmer, O., M. Black, W. Hoen, R. Lutz & R. Vrijenhoek, 1994. DNA primers for amplification of mitochondrial cytochrome c oxidase subunit I from diverse metozoan invertebrates. Molec. Mar. Biol. Biotech. 3: 294–299.

France, S. C. & T. D. Kocher, 1996. DNA sequencing of formalin-fixed crustaceans from archival research collections. Mol. Mar. Biol. Biotech. 5: 304–313.

Frost, B. W., 1971. Taxonomic status of *Calanus finmarchicus* and *C. glacialis* (Copepoda), with special reference to adult males. J. Fish. res. Bd. Can. 28: 23–30.

Frost, B. W., 1974. *Calanus marshallae*, a new species of calanoid copepod closely allied to the sibling species *C. finmarchicus* and *C. glacialis*. Mar. Biol. 26: 77–99.

Frost, B. W., 1989. A taxonomy of the marine calanoid copepod genus *Pseudocalanus*. Can. J. Zool. 67: 525–551.

Gibbs, R. A., P.-N. Nguyen & C. T. Caskey, 1989. Detection of single DNA base differences by competitive oligonucleotide priming. Nuc. Acids Res. 17: 2437–2448.

Gocke, C. D., F. A. Benko & P. K. Rogan, 1998. Transmission of mitochondrial DNA heteroplasmy in normal pedigrees. Hum. Genet. 102: 182–186.

Grainger, E. H., 1961. The copepods *Calanus glacialis* and *Calanus finmarchicus* (Gunnerus) in Canadian Arctic-Subarctic waters. J. Fish. Res. Bd. Can. 18: 663–678.

Harasewych, M. G., S. L. Adamkewicz, J. A. Blake, D. M. Saudek, T. Spriggs & C. J. Bult, 1997. Phylogeny and relationships of pleurotomariid gastropods (Mollusca: Gastropoda): an assessment based on partial 18S rRNA and cytochrome c oxidase I sequences. Mol. Mar. Biol. Biotechnol. 6: 1–20.

Hulsemann, K., 1991. *Calanus euxinus*, new name, a replacement name for *Calanus ponticus* Karavaev, 1894 (Copepoda: Calanoida). Proc. biol. Soc. Wash. 104: 620–621.

Jacobs, H. T. & B. Grimes, 1986. Complete nucleotide sequences of the nuclear pseudogenes for cytochrome oxidase subunit I and the large mitochondrial ribosomal RNA in the sea urchin *Strongylocentrotus purpuratus*. J. mol. Biol. 187: 509–527.

Jaschnov, W. A., 1955. Morphology, distribution and systematics of *Calanus finmarchicus* s.l. [Russ.] Zool. Zh. 34: 1210–1223.

Juan, C., P. Oromi & G. M. Hewitt, 1995. Mitochondrial DNA phylogney and sequential colonization of Canary Islands by darking beetles of the genus *Pimelia* (Tenebrionidae). Proc. R. Soc. Lond. B Biol. Sci. 261: 173–180.

Jukes, T. H. & C. R. Cantor, 1969. Evolution of protein molecules. In Munro, H. N. (ed.), Mammalian Protein Metabolism, Academic Press, New York: 21–31.

Knowlton, N., 1993. Sibling species in the sea. Ann. Rev. Ecol. Syst. 24: 189–216.

Kumar, S., K. Tamura & M. Nei, 1993. MEGA: Molecular Evolutionary Genetics Analysis, Version 1.0, Pennsylvania State University, University Park, PA 16802.

Lunt, D. H., D. X. Zhang, J. M. Szymura & G. M. Hewitt, 1996. The insect cytochrome oxidase I gene: evolutionary patterns and conserved primers for phylogenetic studies. Insect Mol. Biol. 5: 153–165.

McLaren, I. A., E. Laberge, C. J. Corkett & J.-M. Sevigny, 1989. Life cycles of four species of *Pseudocalanus* in Nova Scotia. Can. J. Zool. 67: 552–558.

Mackas, D. L., H. Sefton, C. B. Miller & A. Raich, 1993. Vertical habitat partitioning by large calanoid copepods in the oceanic subarctic Pacific during spring. Progr. Oceanogr. 32: 259–294.

Medeiros-Bergen, D. E., R. R. Olson, J. A. Conroy & T. D. Kocher, 1995. Distribution of holothurian larvae determined with species-specific genetic probes. Limnol. Oceanogr. 40: 1225–1235.

Miller, C. B., 1988. *Neocalanus flemingeri*, a new species of Calanidae (Copepoda; Calanoida) from the subarctic Pacific Ocean, with a comparative redescription of *Neocalanus plumchrus* (Marukawa). Progr. Oceanogr. 20: 263–273.

Olson, R. R., J. Runstadler & T. D. Kocher, 1991. Whose larvae? Nature 351: 357–358.

Palumbi, S. R. & J. Benzie, 1991. Large mitochondrial DNA differences between morphologically similar Penaeid shrimp. Molec. Mar. Biol. Biotech. 1: 27–34.

Parfait, B., P. Rustin, A. Munnich & A. Rotig, 1998. Co-amplification of nuclear pseudogenes and assessment of heteroplasmy of mitochondrial DNA mutations. Biochem. Biophys. Res. Commun. 247: 57–59.

Pedersen, B. V., 1996. A phylogenetic analysis of cuckoo bumblebees (Psithyrus, Lepeletier) and bumblebees (Bombus, Latreille) inferred from sequences of the mitochondrial gene cytochrome oxidase I. Mol. Phylogenet. Evol. 5: 289–297.

Quesada, H., D. A. Skibinski & D. O. Skibinski, 1996. Sex-biased heteroplasmy and mitochondrial DNA inheritance in the mussel *Mytilus galloprovincialis* Lmk. Curr. Genet. 29: 423–426.

Rychlik, W., 1992. OLIGO, Ver. 4.04. Computer software. National Biosciences, Inc, Plymouth, MN.

Saitou, N. & M. Nei, 1987. The neighbor-joining method: a new method for reconstructing phylogenetic trees. Mol. Biol. Evol. 4: 406–425.

Sameoto, D. D., L. O. Jaroszynski & W. B. Fraser, 1980. BIONESS, a new design in multiple net zooplankton samplers. Can. J. Fish. aquat. Sci. 37: 722–724.

Sevigny, J.-M., I. A. McLaren & B. W. Frost, 1989. Discrimination among and variation within species of *Pseudocalanus* based on the GPI locus. Mar. Biol. 102: 321–327.

Skjoldal, H. R. & F. Rey, 1989. Pelagic production and variability of the Barents Sea ecosystem. In K. Sherman & L. M. Alexander (eds.), Biomass Yields and Geography of Large Marine Ecosystems AAAS Publ, 241–286.

Stauffer, C., F. Lakatos & G. M. Hewitt, 1997. The phylogenetic relationships of seven European Ips (Scolytidae, Ipinae) species. Insect. Mol. Biol. 6: 233–240.

Swofford, D. L., 1993. Phylogenetic analysis using parsimony (PAUP), Ver. 3.1, University of Illinois, Champaign.

Tamura, K. & M. Nei, 1993. Estimation of the number of nucleotide substitutions in the control region of mitochondrial DNA in humans and chimpanzees. Mol. Biol. Evol. 10: 512–526.

Wiebe, P. H., A. W. Morton, A. M. Bradley, R. H. Backus, J. E. Craddock, V. Barber, T. J. Cowles & G. R. Flierl, 1985. New developments in the MOCNESS, an apparatus for sampling zooplankton and micronekton. Mar. Biol. 87: 313–323.

Wilson, A. C., R. L. Cann, S. M. Carr, M. George, U. B. Gyllensten, K. M. Helm-Bychowski, R. G. Higuchi, S. R. Palumbi, E. M. Proger, R. D. Sage & M. Stoneking, 1985. Mitochondrial DNA and two perspectives on evolutionary genetics. Biol. J. linn. Soc. 26: 375–400.

Hydrobiologia **401:** 255–264, 1999.
J.P. Zehr & M.A. Voytek (eds), Molecular Ecology of Aquatic Communities.

Ecological implications of molecular biomarkers: assaying sub-lethal stress in the midge *Chironomus tentans* using heat shock protein 70 (HSP-70) expression

N.K. Karouna-Renier[1,*] & J.P. Zehr[1,2]
[1]*Department of Biology, Rensselaer Polytechnic Institute, 110 8th Street, Troy, NY 12180, U.S.A.*
[2]*Current address: Institute of Marine Sciences, University of California Santa Cruz, Santa Cruz, CA 95064, USA*

Key words: HSP-70, *Chironomus tentans*, heat shock, biomarkers

Abstract

Aquatic community structure is a reflection of the changes in constituent populations and the complex interactions between these organisms and environmental stressors. Consequently, shifts in populations and community structure can be used to assess water quality. However, these indicators only reflect damage already sustained by an ecosystem and are not useful for prediction of potential ecological impacts. Molecular/biochemical indicators, such as heat shock proteins, can provide early indication of environmental stress on aquatic communities. The heat shock protein response involves the synthesis of an array of proteins that protect organisms from cellular damage resulting from exposure to a variety of stressors. Consequently, stress proteins have the potential of being an important screening tool indicating exposure to, and/or biological effects of environmental contaminants. The midge larva *Chironomus tentans* has been extensively used in bioassays of freshwater systems. However, investigations of stress proteins as environmental biomarkers in midges are lacking. To evaluate the potential use of HSP-70 as a biomarker of environmental stress, we completed a preliminary characterization of the stress protein response in *C. tentans* upon exposure to heat shock. Western immunoblotting indicated an increase in a 72 kD protein after larvae were exposed to 33 °C, 35 °C, and 37 °C. The observed induction was rapid, appearing within 5–10 min, and persisted for over 24 h after removal of the stressor. The results are discussed with regard to the use of the HSP-70 biomarker as an environmental screening tool. It is proposed that the HSP-70 biomarker is most applicable as a sublethal toxicity test endpoint indicative of the presence of biochemically significant levels of stress.

Introduction

The structure of aquatic communities is determined by the complex interactions among trophic levels and the responses of individual populations to environmental conditions. Environmental stressors impact aquatic communities by selectively targeting individual species (Simpson et al., 1985; Clements et al., 1988). The resulting changes in the abundance and diversity of organisms can have a cascading effect throughout a community by shifting trophic structure. These community level changes are therefore an expression of pollution effects at lower levels of organization (Attrill & Depledge, 1997).

* Author for correspondence

The distinct relationship between water quality and aquatic community structure has led to the extensive use of population or community-level assessments as measures of aquatic ecosystem health. *In situ* assessments of water quality using aquatic biota have historically focused on whole-animal and community responses, such as abundance and diversity (Johnson et al., 1993). Similarly, laboratory toxicity tests commonly utilize survival, growth and reproduction as monitoring endpoints. Although these methods are widely used and are fairly accurate at assessing the health of aquatic systems, they have several limitations:

1. they measure stress indirectly, that is, they rely on what are essentially secondary indicators of environmental perturbation, the results of stress on organisms, and

2. they lack the sensitivity necessary to detect subtle chemical changes in the environment (Dyer et al., 1993).

In recent years, attention has focused on the development and use of molecular biomarkers as ecotoxicological tools. Measurements of biotic health using biochemical/molecular parameters potentially provide indications of chronic toxicity earlier and more efficiently than monitoring changes at higher levels of biological organization (Dyer, 1991; Ryan & Hightower, 1994). The rationale for this is as follows:

1. changes in gene expression (i.e. protein synthesis) may occur at levels far below toxic thresholds determined by currently utilized bioassay endpoints; and
2. changes in protein expression may be more precisely quantified than standard bioassay results.

These responses at the molecular level can serve as indicators of both exposure and effect, and underlie the changes seen in populations or in community structure (Stegeman et al., 1992).

Upon exposure to environmental stresses, organisms have been shown to exhibit a generalized protective response known as the stress protein or heat shock protein (HSP) response (Sanders, 1993; Sanders et al., 1994). This cellular stress response, characterized by the rapid synthesis of a suite of proteins, was first observed in *Drosophila* in response to severe thermal shock (Ritossa, 1962). It has since become evident that this protein induction occurs as a result of exposure to a wide variety of stressors including elevated temperatures, trace metals, pesticides, oxidizing agents and other contaminants (Dyer, 1991; Sanders, 1993). This sensitivity to a multiplicity of stressors indicates the potential usefulness of these proteins as quantitative and qualitative biomarkers of environmental conditions.

Under normal cellular conditions, stress proteins are present at low levels. These constitutive proteins play a role in normal cellular functions such as protein synthesis, folding, and transport (Sanders et al., 1994). Under adverse environmental conditions, the synthesis of stress proteins increases, and the proteins take on the additional roles of repairing denatured proteins, which allows the organism to withstand the stress (Sanders et al., 1994). The most abundant and widely studied group of stress proteins is the 70 kD protein family, commonly known as HSP-70 (Ryan & Hightower, 1994). The structure of HSP-70 is highly conserved, with animals as diverse as humans and nematodes sharing similar antigenic binding domains (Sanders, 1993). The induction of proteins in this family has been observed in a wide range of organisms, including invertebrates, in response to various stressors, demonstrating their potential use as general marker of environmental stress.

The midge *Chironomus tentans* has been extensively used for assessing acute and sub-lethal toxicity associated with contaminated sediments and water (U.S. EPA, 1994; Sibley et al., 1997; Benoit et al., 1997). The advantages of using *C. tentans* as a test species include:

1. cultivability,
2. sensitivity to many contaminants,
3. common occurrence in many freshwater systems,
4. importance in many food webs, and
5. availability of standard protocols for assays (Burton & MacPherson, 1995; Benoit et al., 1997).

Traditionally, toxicity tests using *C. tentans* have focused on survival as an endpoint (acute toxicity) rather than the sublethal effects of contaminants. However, population or community-level impacts of contaminants more commonly result from chronic, sublethal changes in organisms, rather than mortality (Benoit et al., 1997). In recent years, organismal growth has become a commonly used endpoint in bioassessments of aquatic environments. However, the use of heat shock proteins as bioindicators of chemical environmental stress, has not been investigated in *C. tentans*.

Immunoassays based on antibodies raised against specific stress proteins have the best potential for environmental monitoring because the accumulation of the chosen biomarker can be quantified directly (Sanders, 1990; Stegeman et al., 1992). The separation of proteins using SDS-PAGE (Laemmli, 1970) followed by Western blotting has been the most commonly utilized technique for monitoring expression of HSP-70 in aquatic organisms (Dyer et al., 1993; Ryan & Hightower, 1994; Sanders et al., 1994). In the present study, we used immunological techniques for qualitative monitoring of HSP-70 induction in *C. tentans* as a sublethal biomarker of environmental stress. The main objective of this research was to undertake a basic characterization of the HSP-70 response in midges. Organisms were exposed to heat shock in a series of experiments to compare the HSP-70 response under different temperature regimes.

Organisms

Brood stock of *C. tentans* larvae were obtained from established cultures maintained by the New York State

Department of Environmental Conservation (Troy, NY) and Aquatic Research Organisms, Inc. (Wilmington, MA). The midges were reared in two 57-l aquaria at room temperature (23±2 °C) and a light:dark cycle of 16h:8h. Soft, reconstituted water (APHA, 1992) hardness 40–48 mg/l as $CaCO_3$, alkalinity 30–35 mg/l as $CaCO_3$ and pH 7.2–7.6, was used both for rearing and exposure experiments.

Larvae in the rearing tanks were fed Tetramin® (Tetra Werke, Melle, Germany) *ad libitum,* and supplemented by shredded plain white paper towels. The paper towels were prepared by soaking overnight in acetone to solubilize organic contaminants, rinsed and soaked overnight in deionized water (U.S. EPA, 1994).

To insure uniform age of larvae used in the assays, egg masses were isolated as follows. Adult *C. tentans* were collected from culture tanks and placed in a 1000 ml flask containing approximately 300 ml of distilled water. Females lay egg masses within 24 – 48 h of fertilization. Egg masses were removed daily and placed in separate, aerated rearing chambers containing culture water. Larvae were allowed to develop in the rearing chambers until early second instar approximately 8 d after hatching (U.S. EPA, 1994).

Methods

Heat exposures

Twenty-five second instar larvae were placed into aerated 50 ml Erlenmeyer beakers containing 40 ml of culture water. The bottom of each chamber was covered with a thin layer of acid washed glass beads (150–212 μm; Sigma Chemical Co.) to reduce stress and provide substrate for larval case building. Organisms were placed into each chamber and allowed to adjust to the conditions, overnight.

Experimental chambers for the heat shock experiments were immersed in a preheated, precision controlled water bath. All incubations were kept under constant aeration. Reference incubations were concurrently established and maintained at room temperature. The larvae were simultaneously subjected to different temperature-time combinations, and then returned to room temperature (23±2 °C) for a period of 15 min, unless noted otherwise. Mortality was assessed in each chamber at the end of all exposures. The criterion for death was the complete immobilization of a larva when subjected to gentle mechanical prodding. Surviving larvae were removed from their chambers, blotted dry on a paper towel and frozen in liquid nitrogen for analysis at a later time. Total transfer time for all 25 larvae per exposure chamber was approximately 7 min.

In the initial heat shock experiments, the HSP-70 response was assayed in organisms exposed to 33 °C, 35 °C and 37 °C for 1, 2, 4, 6, 12 and 24 h. Organisms used in reference incubations for these experiments remained at room temperature until 8 h after heat shock commenced. A separate experiment was conducted to further determine the rapidity of the heat shock response, wherein second instar larvae were exposed to brief heat shock (5, 10, 20, 30, 40, 50 and 60 min). Reference organisms for this experiment were frozen after 60 min at room temperature. A recovery experiment was also completed, which investigated the persistence of the HSP-70 response after exposure to 35 °C for 4 h. Organisms were returned to room temperature and sampled at 1, 4, 6, 10 and 24 h after exposure. Reference organisms remained at room temperature until the end of the heat exposure.

Sample preparation

Frozen larvae from each exposure were transferred as a bulk sample to a microcentrifuge tube and homogenized for 5 s using a teflon micro-pestle (Kontes, Inc.). SDS sample buffer (75 μl; 63 mm Tris-HCl pH 6.8, 10% Glycerol, 2% SDS, 0.0025% Bromophenol Blue) was added to each tube and samples were homogenized again for 30 s. A sub-sample was removed from each tube for determination of total protein using the Biorad DC Protein Assay (Biorad Laboratories, Hercules, CA). Dithiothreitol (DTT) (50 mm final concentration) was then added to each tube, the sample homogenized for <20 s, and heated at 95–100 °C for 10 min. The homogenates were centrifuged at 14 000 rpm for 20 min. Samples were stored at −80 °C until analysis.

Electrophoresis

The supernatant was analyzed for HSP-70 concentrations by loading 62.5 μg total protein per lane on a 10% SDS polyacrylamide gel with 4% stacking gel (Laemmli, 1970). The gels were run at 200V for approximately 1h (Biorad Mini-protein II) using a MOPS (3-(N-morpholino) propane sulfonic acid) SDS Running buffer (50 mM MOPS, 50 mM Tris Base, 3.5 mM SDS, 1 mM EDTA, pH 7.7; Novex Corporation, CA).

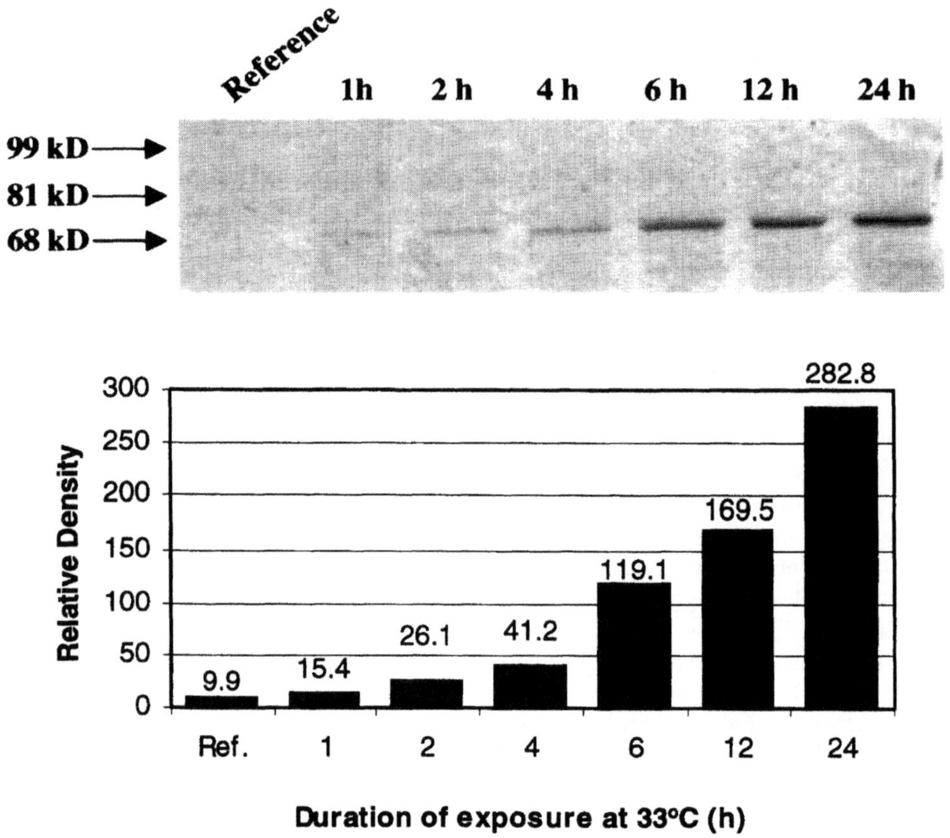

Figure 1. (TOP) Immunoblot analysis of HSP-70 in *C. tentans* after heat shock. Organisms were exposed to 33 °C for 1, 2, 4, 6, 12 and 24 h. Reference organisms were sampled after 8 h at room temperature. Equal amounts of protein extracts were resolved by SDS-PAGE, electroblotted onto nitrocellulose, and probed with an anti-HSP-70 monoclonal antibody (mAb 3A3). The reactions were visualized colorimetrically using horseradish peroxidase conjugated secondary antibodies. The positions of standard molecular weight markers are shown in kD. (BOTTOM) Relative intensity of the 72 kD protein bands detected by the anti-HSP-70 antibody.

Immunoblotting

After electrophoresis, gels were equilibrated for 15 min in transfer buffer (1% SDS, 0.192 M glycine, 0.025 M Tris and 20% methanol) and electroblotted onto nitrocellulose at 125 V for 75 min. Membranes were blocked using 5% skim milk in Tris-buffered saline (TBS) with 0.5% Tween-20 and 0.05% Triton X-100 (TBSTT) for 45 min followed by brief rinsing with TBSTT. The membranes were incubated overnight with monoclonal anti-HSP-70 raised against recombinant human stress protein 70 expressed in *E. coli* (mAb 3A3, Affinity Bioreagents, CO) at 1:2000 in TBSTT. This antibody reacts with both the constitutive and inducible forms of HSP-70.

Unbound antibody was removed by a 45 min wash in TBSTT, and the membrane was incubated in a peroxidase-conjugated, secondary antibody (horse anti-mouse, Vector Laboratories, CA; 1:100 dilution), overnight. Excess secondary antibody was removed by a 30 min wash in TBSTT, followed by 15 min in TBS. HSP-70 was visualized using the horse radish peroxidase (4-chloro-1-napthol; Biorad, CA) reaction.

Analysis

Immunoblots were scanned using a Hewlett Packard Scanjet 4C. To aid in the observation of trends and show changes in HSP-70 expression, relative densities of bands were determined using the Gel Plot add-in of NIH Image (National Institutes of Health, U.S.A.). The actual intensities are arbitrary units of relative optical density.

Results

Western blot analysis showed that only one protein band of an approximate molecular weight of 72 kD is recognized by HSP-70 clone 3A3 in *C. tentans*.

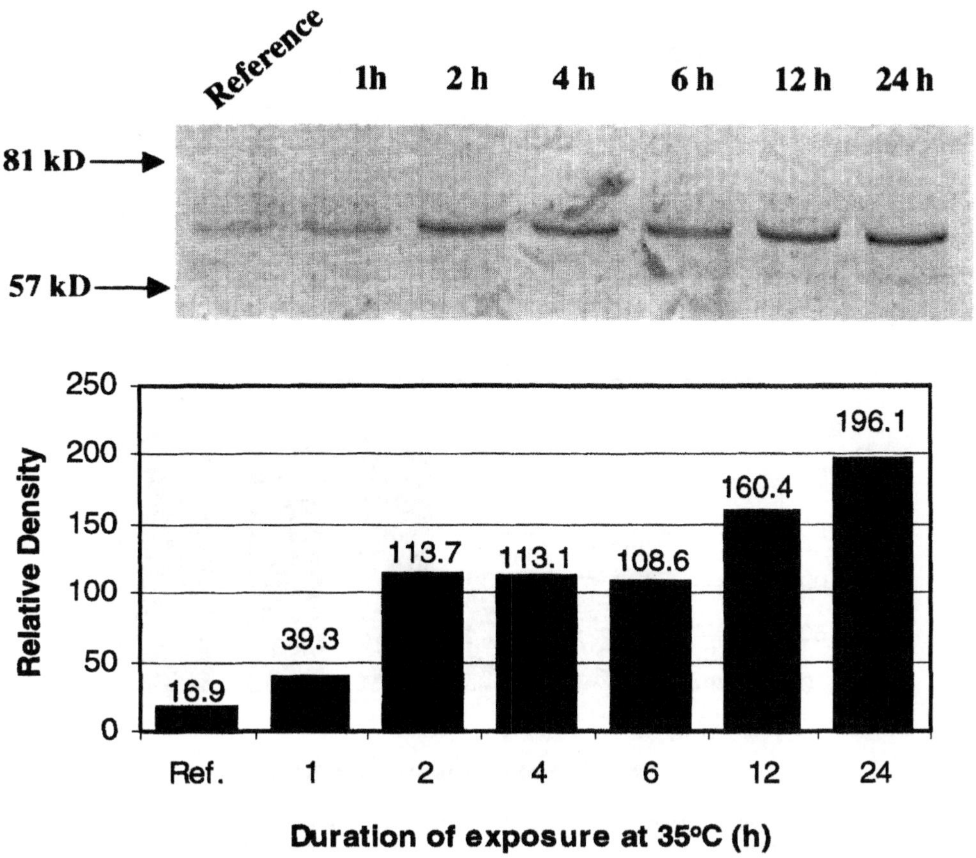

Figure 2. (TOP) Western blot analysis of HSP-70 levels in *C. tentans* after a 35 °C heat shock over a 24 h period. Protein samples represent exposures of 1, 2, 4, 6, 12 and 24 h to heat. Reference organisms were sampled after 8 h at room temperature. Stress protein levels were visualized using a monoclonal anti-HSP-70 antibody (mAb 3A3). The positions of standard molecular weight markers are shown in kD. (BOTTOM) Relative intensity of the 72 kD protein bands visualized on the Western blot.

The band is present at extremely low levels even in unstressed animals but its concentration increases dramatically in the heat shocked organisms.

Survival after the 24 h exposures to 33 °C and 35 °C was 90–100%. At 33 °C, HSP-70 levels in second instar larvae continued to increase over the 24 h exposure period, relative to the reference (Figure 1). However, at 35 °C, HSP-70 levels appear to peak more rapidly and after approximately 4h of exposure, only minimal increases in concentration are seen (Figure 2).

After 4 h of exposure to a 37 °C heat shock, the larvae began to exhibit physical signs of stress (lethargy, mortality; 79% survival). At 12 h of exposure, 50% of the organisms had died, and 100% mortality was observed at 24 h. However, surviving organisms continued to exhibit elevated levels of HSP-70 at all sampling time points (Figure 3). A similar heat shock experiment was conducted at 40 °C, although 100% of the second instar larvae died within 1 h of exposure to this elevated temperature.

Results of the short-term exposure experiments indicated that the HSP-70 response was extremely rapid (Figure 4). Levels of the protein in *C. tentans* larvae increased within 5–10 min of exposure to 35 °C, relative to reference samples. A time-dependent response was observed in the recovery of HSP-70 to near pre-stressed levels after a 4h exposure to 35 °C. High levels of HSP-70 were maintained through 10 h of recovery at room temperature (23±2 °C). By 24 h, protein levels had decreased substantially, although they were still elevated relative to reference larvae (Figure 5).

Discussion

Previous studies of *C. tentans* and *C. thummi* using metabolic labeling, reported that several newly synthesized proteins from the HSP-70 family, with molecular weights of 76 and 73 kD, and 76, 74, and

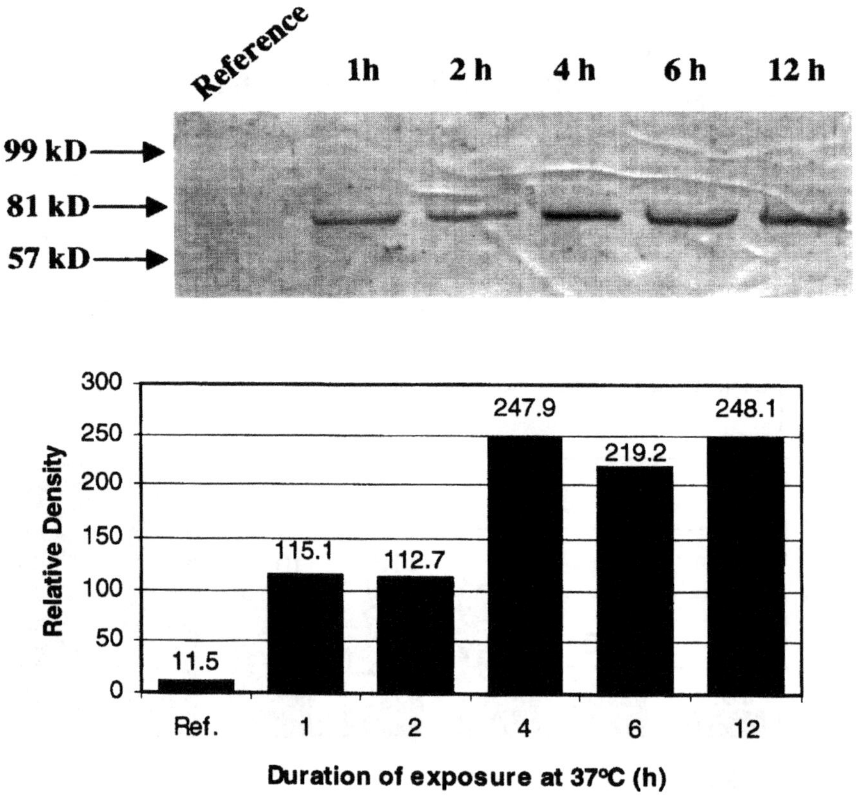

Figure 3. (TOP) Western blot analysis of *C. tentans* exposed to a 37 °C heat shock for increasing duration of time (1, 2, 4, 6 and 12 h). Reference organisms were sampled after 8 h at room temperature. After 4 h of exposure to a heat shock of 37 °C, *C. tentans* exhibited increased mortality. However, surviving organisms maintained high levels of HSP-70. Stress protein levels were visualized using a monoclonal anti-HSP-70 antibody (mAb 3A3). The positions of standard molecular weight markers are shown in kD. (BOTTOM) Relative intensity of the 72 kD bands detected by Western immunoblotting.

70 kD, respectively, can be seen in response to heat shock (Tanguay & Vincent, 1980; Barettino et al., 1988; Carretero et al., 1991). In the present study, the monoclonal antibody clone 3A3 cross-reacted with a 72 kD protein band. Studies of other invertebrate species using one-dimensional PAGE have indicated that this antibody recognizes a single band of either 70 or 72 kD (Sanders et al., 1994). Western blot analysis using two-dimensional SDS-Page would be required to identify the number of isoforms of HSP-70 present in each band visualized in the one-dimensional blots (Smerdon, 1995).

The present results indicate that the heat shock response in *C. tentans* was rapid at all temperatures examined. An increase in HSP-70 concentration could be detected within 5–10 min of exposure to heat stress of 35 °C. This demonstrates the sensitivity of this biomarker to thermal stress and indicates that even short exposure to this stressor is sufficient to stimulate this defense system. Lezzi et al. (1981) reported an increase in salivary gland chromosomal puffs within 5 min of transferring *C. tentans* from 18 °C to 37 °C. Chromosomal puffing in heat shocked organisms has been related to an increase in the synthesis of distinct RNA species, which results in the production of heat shock proteins (Lezzi et al., 1981). The rapidity of the HSP-70 response in *C. tentans* is an important characteristic in the evaluation of this protein as a biomarker of environmental stressors.

In our experiments, after 24 h of recovery at room temperature from a 4 h, 35 °C heat shock, HSP-70 concentrations in *C. tentans* were substantially reduced from their peak levels, although they did not completely return to pre-stressed levels. Morcillo & Diez (1996) found that heat shock puffs on the chromosomes of *C. thummi* persist until about 7 h of recovery at 18 °C, indicating that synthesis may be active after removal of the stressor. After *de novo* synthesis of the heat shock proteins has stopped, a

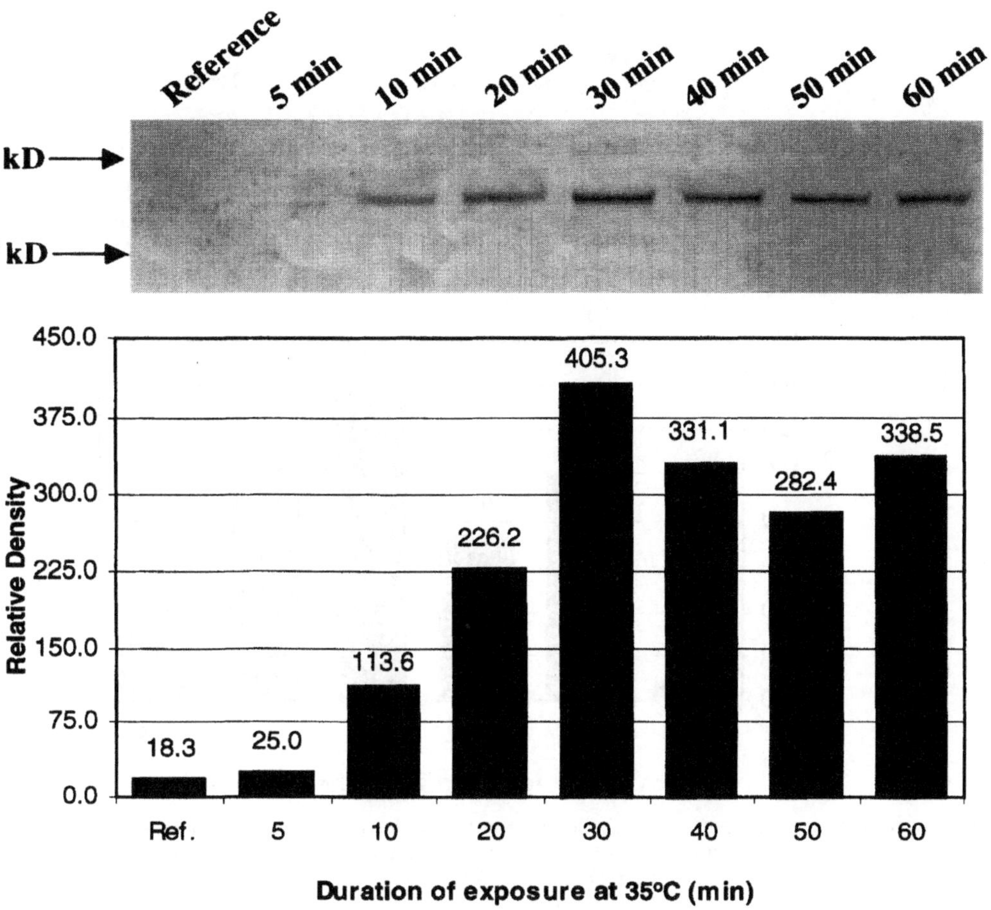

Figure 4. (TOP) Immunblot detection of HSP-70 in *C. tentans* exposed to a 35 °C heat shock for 5, 10, 20, 30, 40, 50 and 60 min. Reference organisms for this experiment were frozen after 60 min at room temperature. The membrane was probed with an anti-HSP-70 monoclonal antibody (mAb 3A3). The positions of standard molecular weight markers are shown in kD. (BOTTOM) Relative intensity of the 72 kD bands detected on the Western Blot.

gradual decrease in HSP levels occurs as the proteins are broken down (Sanders, 1990).

However, it appears that the stress protein response may be overwhelmed at acutely lethal stressor levels and exposure times. Mortality significantly increased in the present study after 4 h of exposure to 37 °C. Similarly, no organisms survived after a 1 h heat shock at 40 °C. Tanguay & Vincent (1980) found that after 20–40 min heat shock at 39 °C, and 10 min heat shocks at 41 °C and 43 °C, heat shock protein synthesis is reduced or entirely shut down. At temperatures above 40 °C, many of the chromosomal puffs observed by Lezzi et al. (1981) fail to form. This may indicate that as a progressive inactivation of the stress protein response occurs at extreme thermal stress, the organisms' defense against the stressor is eliminated, leading to death.

Linking levels of biological organization

Organismal growth and reproduction integrate numerous biochemical, physiological and behavioral processes that are affected by environmental conditions (Sibley et al., 1997). A strong theoretical link exists between changes in these processes in individual organisms, and population structure. Organisms stressed by environmental contaminants have less energy available for growth and reproduction. Sibley et al. (1997) reported that a reduction in the growth of *C. tentans* larvae is associated with a proportional decline in reproductive output of adult females. The resulting reduction in the number of larvae in succeeding generations could have severe impacts on stressed populations (Sibley et al., 1997). In turn, this decline can affect the structure of the community, which is controlled by variations in its constituent populations

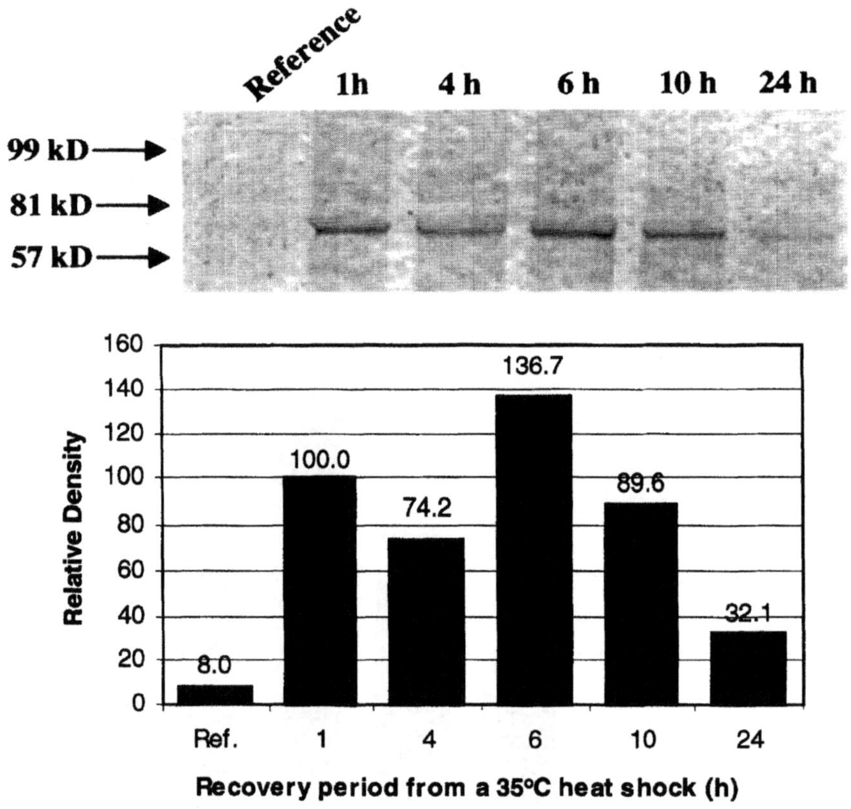

Figure 5. (TOP). Western blot analysis of the recovery time course of HSP-70 levels in *C. tentans* after exposure to a 4 h, 35 °C heat shock. Samples were taken 1, 4, 6, 10 and 24 h after the return of organisms to room temperature. Reference organisms remained at room temperature until the end of the heat exposure. The membrane was probed with an anti-HSP-70 monoclonal antibody (mAb 3A3). The positions of standard molecular weight markers are shown in kD. (BOTTOM) Relative intensity of the HSP-70 bands detected on the Western Blot.

and their response to environmental stress (Attrill & Depledge, 1997).

Although studies of the correlations of population and community level changes with stress protein levels are few, several relationships with physiological indicators of pollutant-induced stress have been reported. Sanders et al. (1991) found that in the mussel *Mytilus edulis*, exposed to environmentally realistic Cu levels, an increase in HSP-60 concentrations corresponded with inhibition of growth, as expressed by scope-for-growth measurements. The HSP-60 response was observed at an order of magnitude lower Cu concentration than that which caused inhibition of organismal growth. Similarly, Köhler et al. (1996) found that elevated HSP-70 levels in the slug *Derocerus reticulatum* were induced by metal concentrations that are known to cause adverse effects on survival, fecundity and offspring rate. Clearly, further studies are needed to elucidate the link between the stress protein response in *C. tentans* and detrimental effects at higher levels of biological organization.

Environmental monitoring using the HSP-70 biomarker

The use of the HSP-70 biomarker is most applicable as a diagnostic screening tool that could indicate the presence of biochemically significant levels of contaminants, which correlate with decreased physiological function and survival of organisms. Similarly to the manner that traditional toxicity test endpoints such as growth, survival and fecundity are used today, the stress protein response may be used as a rapid indicator of the presence of sublethal stress and to suggest that further study of the aquatic community in the affected location is warranted. More research is needed to characterize the HSP-70 response in *C. tentans* to different stressors before this biomarker can be applied generally. Additionally, it is not currently possible to conclude from a stress protein assay which stressor is responsible for the observed effects. Therefore, for a comprehensive assessment of environmental quality, analytical chemistry methods and community biomonitoring techniques must still accompany these studies.

Several critical issues regarding the use of HSP-70 as a biomarker of environmental stress should not be overlooked. Care must be taken when applying the stress protein response as a biomarker of environmental contamination, such that normal or seasonal background levels of the proteins are not misinterpreted as an indicator of contaminant stress (Fader et al., 1994). Yu et al. (1994) and Fader et al. (1994) demonstrated that fish and other aquatic organisms maintain detectable concentrations of constitutive HSP-70. In the present study, we also found that low levels of HSP-70 are maintained in unstressed *C. tentans* larvae. Moreover, Fader et al. (1994) reported that natural fish populations from uncontaminated streams exhibit seasonal variations in background levels of HSP-70. These results illustrate one of the major difficulties in using heat shock protein levels from organisms collected *in situ*. Further confounding this issue is the difficulty in finding a true control or reference site. Unstressed organisms must be available for comparison and a better understanding of normal variability is necessary before assays using wild populations can be successful. Although similar problems or protein patterns may be present in laboratory testing, better controls can be established by using cultured organisms of the same age and from the same background conditions.

Several studies have indicated that basic characteristics of the HSP-70 response may differ based on the organism studied and type of stressor that is applied. Sanders et al. (1995) reported that in the fathead minnow *Pimephales promelas*, heat shock elicits the stress protein response much more rapidly than do elevated Cu concentrations. Similarly, the subcellular distribution of HSP-70 also differs based on the stressor type. These variations were attributed to fundamental differences in the targets of the stressors (Sanders et al., 1995). Stressor specific effects have also been observed under conditions of continuous, moderate stress. Sanders (1993) reported that the HSP-70 response is sustained for over six weeks in *Mytilus edulis* exposed to 10 °C increase in temperature. However, Veldhuizen-Tsoerkan et al. (1990, 1991) demonstrated the transient nature of the HSP-70 response under long-term exposure of *M. edulis* to Cd. Under these conditions, the stress protein response may act in concert or sequentially with other detoxification mechanisms. Consequently, it may be necessary to re-evaluate the proper application of the HSP-70 biomarker in environmental monitoring, specifically focusing on exposure duration.

Studies of the effects of environmental perturbation on aquatic communities can benefit from application of biomarkers that facilitate detection of stressors before changes in community structure. A number of proteins are involved in response of eukaryotes to stress, including heat shock proteins, metallothioneins, and P450 cytochromes. These proteins are expressed in response to various types of environmental stressors, and thus, provide biomarkers for stresses on individual populations. Preferred characteristics of such biomarkers are a well-characterized response (either a general response to environmental stress, or specific response to stressor of interest), and well-characterized time-response, which allows for detection of the response within reasonable periods.

In the present study, the HSP-70 response was investigated in *C. tentans*, in order to determine the feasibility of using HSP70 as an assay for sublethal stress levels. The midge *C. tentans* has been extensively used in bioassays of freshwater systems and is therefore a good candidate for the development of this biomarker. Further studies are needed to demonstrate a cause and effect relationship between this biomarker and higher level indicators of environmental contamination.

Acknowledgements

The authors would like to thank Y. Chen and M. Mellon for technical assistance. Support for this work was provided in part by the United States Geological Survey, Water Resources Division, Troy, NY. Dr J. J. H. Ciborowski provided many valuable comments on an earlier draft of this manuscript.

References

APHA, 1992. Standard Methods for the Examination of Water and Wastewater. American Public Health Association, Washington, D.C.

Attrill, M. J. & M. H. Depledge, 1997. Community and population indicators of ecosystem health: targeting links between levels of biological organization. Aquat. Toxicol. 38: 183–197.

Barettino, D., G. Morcillo & J. L. Diez, 1988. Induction of the heat shock response by carbon dioxide in *Chironomus thummi*. Cell Differentiation 23: 27–36.

Benoit, D. A., P. K. Sibley, J. L. Juenemann & G. T. Ankley, 1997. *Chironomus tentans* life-cycle test: design and evaluation for use in assessing toxicity of contaminated sediments. Environ. Toxicol. Chem. 16: 1165–1176.

Burton, G. A. & C. MacPherson, 1995. Sediment toxicity testing issues and methods. In Hoffmann D. A., B. A. Rattner, G.

A. Burton & J. Cairns (eds), Environmental Toxicology. Lewis Publishers, CRC Press, Inc., Boca Raton (FL): 70–103.

Carretero, M. T., M. J. Carmona, & J. L. Diez, 1991. Thermotolerance and heat shock proteins in *Chironomus*. J. Insect Physiol. 37: 239–246.

Clements, W. H., D. S. Cherry & J. Cairns, Jr., 1988. Impact of heavy metals on insect communities in streams: a comparison of observational and experimental results. Can. J. Fish. aquat. Sci. 45: 2017–2025.

Dyer, S. D. 1991. Evaluation of the efficacy of the stress protein response as a biochemical water quality monitoring method. University of North Texas, Denton, TX.

Dyer, S. D., G. L. Brooks, K. L. Dickson, B. M. Sanders & E. G. Zimmerman, 1993. Synthesis and accumulation of stress proteins in tissues of arsenite-exposed fathead minnows (*Pimephales promelas*). J. Toxicol. Chem. 12: 913–924.

Fader, S. C., Z. Yu & J. R. Spotila, 1994. Seasonal variation in heat shock proteins (HSP70) in stream fish under natural conditions. J. Therm. Biol. 19: 335–341.

Johnson, R. K., T. Wiederholm & D. M. Rosenberg, 1993. Freshwater biomonitoring using individual organisms, populations, and species assemblages of benthic macroinvertebrates. In Rosenberg D. M. & V. H. Resh (eds), Freshwater Biomonitoring and Benthic Macroinvertebrates. Chapman & Hall, Inc., New York, NY: 40–125.

Köhler, H.-R., B. Rahman, S. Graff, M. Berkus & R. Triebskorn, 1996. Expression of the stress-70 protein family due to heavy metal contamination in the slug, *Derocerus reticulatum*: an approach to monitor sublethal stress conditions. Chemosphere 33: 1327–1340.

Laemmli, U. K., 1970. Cleavage of structural proteins during the assembly of the head of bacteriophage T4. Nature 227: 680–685.

Lezzi, M., B. Meyer & R. Mahr, 1981. Heat shock phenomena in *Chironomus tentans* I. In vivo effects of heat, overheat and quenching on salivary chromosome puffing. Chromosoma (Berl.) 83: 327–339.

Morcillo, G., & J. L. Diez, 1996. Telomeric puffing induced by heat shock in *Chironomus thummi*. J. Biosci. 21: 247–257.

Ritossa, F. M., 1962. A new puffing pattern induced by heat shock and DNP in *Drosophila*. Experimentia 18: 571–573.

Ryan, J. A. & L. E. Hightower, 1994. Evaluation of heavy-metal ion toxicity in fish cells using a combined stress protein and cytotoxicity assay. Environ. Toxicol. Chem. 13: 1231–1240.

Sanders, B., 1990. Stress proteins: potential as multitiered biomarkers. In Shugart L., J. McCarthy (eds), Biomarkers of Environmental Contamination. Lewis Publishers, Boca Raton (FL): 165–191.

Sanders, B. M., 1993. Stress proteins in aquatic organisms: an environmental perspective. CRC Crit. Rev. Toxicol. 23: 49–75.

Sanders, B. M., L. S. Martin, P. A. Nakagawa, D. A. Hunter & S. Miller, 1994. Specific cross-reactivity of antibodies raised against two major stress proteins, stress 70 and chaperonin 60, in diverse species. Environ. Toxicol. Chem. 13: 1241–1249.

Sanders, B. M., L. S. Martin, W. G. Nelson, D. K. Phelps & W. Welch, 1991. Relationships between accumulation of a 60-kDa stress protein and scope-for-growth in Mytilus edulis exposed to a range of copper concentrations. Mar. envir. Res. 31: 81–97.

Sanders, B. M., J. Nguyen, L. S. Martin, S. R. Howe & S. Coventry, 1995. Induction and subcellular localization of two major stress proteins in response to copper in the fathead minnow *Pimephales promelas*. Comp. Biochem. Physiol. 112C: 335–343.

Sibley, P. K., D. A. Benoit & G. T. Ankley, 1997. The significance of growth in *Chironomus tentans* sediment toxicity tests: relationship to reproduction and demographic endpoints. Environ. Toxicol. Chem. 16: 336–345.

Simpson, K. W., R. W. Bode & J. R. Colquhoun, 1985. The macroinvertebrate fauna of an acid-stressed headwater stream system in the Adirondack Mountains, New York. Freshwat. Biol. 15: 671–681.

Smerdon, G. R., J. P. Chapple & A. J. S. Hawkins, 1995. The simultaneous immunological detection of four Stress-70 protein isoforms in *Mytilus edulis*. Mar. envir. Res. 40: 399–407.

Stegeman, J. J., M. Brouwer, R. T. Di Giullo, L. Forlin, B. A. Fowler, B. M. Sanders & P. A. Van Veld, 1992. Molecular responses to environmental contamination: enzyme and protein systems as indicators of chemical exposure and effect. In Huggett R. J., R. A. Kimerle, P. M. Mehrle Jr. & H. L. Bergman (eds), Biomarkers: Biochemical, Physiological, and Histological Markers of Anthropogenic Stress. Lewis Publishers, Inc., Ann Arbor (MI): 235–335.

Tanguay, R. M. & M. Vincent, 1980. Biosynthesis and characterization of heat shock proteins in *Chironomus tentans* salivary glands. Can. J. Biochem. 59: 67–73.

U.S. EPA (United States Environmental Protection Agency), 1994. Methods for measuring the toxicity and bioaccumulation of sediment associated contaminants using freshwater invertebrates. EPA 600-R94-024. Technical Report. Washington, DC.

Veldhuizen-Tsoerkan, M. B., D. A. Holwerda, C. A. van der Mast & D. I. Zandee, 1990. Effects of cadmium exposure and heat shock on protein synthesis in gill tissue of the sea mussel *Mytilus edulis* L. Comp. Biochem. Physiol. C. 96: 419–426.

Veldhuizen-Tsoerkan, M. B., D. A. Holwerda, C. A. van der Mast & D. I. Zandee, 1991. Synthesis of stress proteins under normal and heat shock conditions in gill tissue of sea mussels (*Mytilus edulis*) after chronic exposure to cadmium. Comp. Biochem. Physiol. 100: 699–706.

Yu, Z., W. E. Magee & J. R. Spotila, 1994. Monoclonal antibody ELISA test indicates that large amounts of constitutive HSP-70 are present in salamanders, turtle and fish. J. therm. Biol. 19: 41–53.

Hydrobiologia **401:** 265–277, 1999.
J.P. Zehr & M.A. Voytek (eds), Molecular Ecology of Aquatic Communities.

RNA–DNA ratio and other nucleic acid-based indicators for growth and condition of marine fishes

L. Buckley[1], E. Caldarone[2] & T.-L. Ong[3]
[1]*URI/NOAA CMER Program, Graduate School of Oceanography, University of Rhode Island, Narragansett, Rhode Island 02882-1197, U.S.A.*
[2]*NOAA, National Marine Fisheries Service, Northeast Fisheries Science Center, Narragansett, Rhode Island 02882-1199, U.S.A.*
[3]*Graduate School of Oceanography, University of Rhode Island, Narragansett, Rhode Island 02882-1197, U.S.A.*

Key words: nucleic acids, fish larvae, RNA/DNA, condition, growth

Abstract

Nucleic acid analysis has provided useful tools to study recent growth and mortality of young fishes and their responses to environmental variability. The ratio of RNA–DNA (R/D) has been shown to respond to changes in feeding conditions and growth after periods as short as 1–3 days in a variety of fish species. The earliest studies used primarily UV-based methods, but most investigators now use more sensitive, fluorometric dye-binding assays to estimate RNA and DNA in individual larvae. These newer methods are very sensitive to procedural details and choice of standards. Analytical methods, normalization and calibration procedures to optimize information obtained from nucleic acid analysis are discussed. We present examples illustrating the technique's utility, and problems encountered when applying nucleic acid-based indices to fish larvae and early juveniles. The wide use of R/D analysis in studies of fish early life stages, together with a proliferation of analytical methods, demands a major intercalibration exercise.

Introduction

Organisms respond to their environment at all levels of organization, ranging from molecular to population and community levels. These responses are frequently difficult to measure or predict with an acceptable level of certainty. Given limited resources, sampling constraints, analytical problems, predictive capability and a wide variety of other considerations, the choice of appropriate indicators of environmental response is critical. Frequently, it is information at the population level that is the most needed but the hardest to acquire. Birth, mortality and growth rates are essential to understanding the demographics of populations and the effects of environmental variability. Marine fish are a good example in point.

Temperate marine fish typically spawn large numbers of small eggs. Each year, an individual female may produce thousands to millions of eggs with a dry weight of 10–1000 μg, depending upon the species and other factors. Egg production is relatively easy to estimate from assessments of the adult population, and egg and larval surveys. Estimation of growth and mortality rates is more problematic, particularly over short time intervals (days to weeks). Mortality is high during the first several months of life and it is the rare individual that survives to reproduce. Relatively small differences in growth and mortality rates (on the order of a few % d^{-1}) in the first year of life can result in large differences (1–2 orders of magnitude) in the number of individuals entering the reproductive stage annually (Houde, 1987). Knowledge of the effects of the environment on mortality and growth rates during the first year of life is essential to understanding the role of the environment in regulating the abundance of marine fishes. Furthermore, methods that provide data on short temporal and small spatial scales are particularly informative in linking environmental conditions at the time of sampling to variability in growth and survival.

In this paper, we describe how nucleic acid levels have been used to estimate recent growth and mortality of young fishes and their responses to environmental variability. We trace the development of nucleic acid-based indices from their first application to fish larvae in the late 1970s through contemporary recruitment fisheries oceanography studies. We do not present an exhaustive review of the topic, but rather have selected several studies illustrative of the utility and problems encountered in working with fish larvae and early juveniles. Many of the examples presented come from our own work with a variety of species over two decades. The emphasis has been placed on the effects of natural environmental variability on growth and its estimation using nucleic acid levels. The foundation of much of this work is the concept that the amount of DNA per cell is relatively constant, while the amount of RNA varies with physiological status, the requirement for protein synthesis, and growth.

Initial UV method-based studies

Early work (circa 1950–1975) on nucleic acid levels in fish and their relation to feeding and growth focused almost exclusively on older juvenile and adult fishes (see Bulow, 1970 and 1987 for reviews). Total RNA and total DNA were estimated in either whole fish or isolated tissues using a variety of colorimetric- and UV-based techniques. In most instances, RNA levels were reported to increase with increased feeding and growth. Sex and stage of gonad maturation, among other factors, were reported to affect nucleic acid levels in the different tissues examined.

The earliest work on fish larvae used similar UV-based methods on pools of whole larvae (Buckley, 1979, 1980). These studies with cold water marine species showed a rapid decline in the RNA content (μg larva^{-1}) and concentration (μg mg dry weight^{-1}), and in the ratio of RNA–DNA (R/D) after periods of starvation as short as two days. In winter flounder (*Pseudopleuronectes americanus* Walbaum) reared at 8 °C, starved and fed larvae were clearly separated on the basis of R/D at age 9 d, only 2 d after feeding initiation in fed groups (Buckley, 1980). At age 28 d, fed and starved winter flounder larvae could be differentiated clearly 2 d after removal of food. Gut contents of whole larvae had little effect on R/D estimates since values for actively feeding larvae were unchanged 18 h after removal from food. Clemmesen (1996) reexamined the effect of gut contents in herring larvae (*Clupea harengus*) and found no measurable effect on larval R/D. In Atlantic cod (*Gadus morhua* Linnaeus), differences in R/D of fed and starved larvae were observed 7 d after hatch (Buckley, 1979), corresponding to the time of complete yolk absorption. The magnitude of the decrease in RNA content of starved larvae, exceeding 50% in 4 d, indicated that ribosomal RNA (rRNA) and possibly other species of RNA were rapidly degraded during starvation. Working with striped bass (*Morone saxatilis*) larvae reared at 20 °C, Wright & Martin (1985) reported significant differences in R/D within a day of a change in feeding status.

In cod larvae reared at 7 °C, R/D of pooled samples was directly related to growth of the cohort over periods of 2–8 d before or after the nucleic acid measurement (Buckley, 1979). Growth rate was estimated from the change in protein content of pooled samples taken at intervals. The relationship was not significant for intervals longer than 8 d and could not be tested for intervals <2 d, since a minimum of 2 d was required to obtain reliable estimates of protein growth from sequential samples. In a study of winter flounder larvae reared at three temperatures (5, 7, 10 °C), a direct linear relation was observed between R/D and recent growth (Buckley, 1982). Addition of temperature to the regression, as a second independent variable, significantly reduced the residual mean square. These data suggested that the increased growth rate observed at higher temperatures was not accomplished or accompanied by an increase in R/D at the higher temperatures, but rather by increased growth at a given R/D. A similar model relating R/D, temperature and growth was reported for sand lance larvae (*Ammodytes americanus* Dekay) reared at 3 temperatures (5, 7 and 9 °C) (Buckley et al., 1984).

A general model relating R/D, temperature and larval growth was developed from data on eight species of temperate marine fishes reared under a variety of feeding conditions at temperatures between 2 and 20 °C (Buckley, 1984). All larvae were reared within the range of tolerance of the individual species. Growth increased with both temperature and R/D according to the equation:

$$G_{pi} = 0.93T + 4.75(\mathrm{R/D}) - 18.18$$
$$(n = 23,\ r^2 = 0.92)$$

where G_{pi} is the mean protein growth rate of a tank population in % d^{-1}, T is the water temperature and R/D is the mean for the population measured at intervals over the course of the experiments. RNA and DNA were estimated using a modification of the

Schmidt-Thannhauser UV method on pooled samples (Buckley, 1979). Analysis of residuals, the variety of species included in the study and the similarity of the general model to species-specific models for winter flounder and sand lance (Buckley, 1982, 1984) suggested that the general model was widely applicable to the larvae of temperate marine fish.

These data on a variety of species indicated that bulk nucleic acid analysis could provide information on recent growth and feeding condition of fish larvae that was difficult to obtain using other methods.

Contemporary fluorometric-based studies

In the late 1980s, a number of investigators began using more sensitive fluorometric procedures to analyze RNA and DNA levels in individual larvae rather than pooled samples (Clemmesen, 1988; Robinson & Ware, 1988; Westerman & Holt, 1988). With few exceptions, a large number of studies on a wide variety of species using several different analytical techniques have confirmed the initial findings of distinct differences in R/D values of fed and starved larvae after periods of starvation as short as 1–2 d (Clemmesen, 1988; Robinson & Ware, 1988; Westerman & Holt, 1988; Rooker & Holt, 1996). These data revealed a wealth of variability at the individual level. However, as the number of research groups applying nucleic acid analysis to fish larvae increased, the number of different analytical techniques used proliferated and it rapidly became obvious that results from the fluorometric assays were very sensitive to procedural details and choice of standards (Caldarone & Buckley, 1991; McGurk & Kusser, 1992; Clemmesen, 1993). Clearly, if the UV-based models described earlier are used to calculate growth from fluorometric estimates of R/D, then the two analytical methods (UV and specific fluorometric assay) must be calibrated and a conversion factor applied to the fluorometric values, as has recently been done by Gronkjaer et al. (1997) for Baltic cod (*Gadus morhua*). Also, intercalibration among methods and laboratories is an essential prerequisite to meaningful comparisons of findings among research groups.

While early work demonstrated the potential for using bulk nucleic acid levels in whole fish larvae to estimate nutritional condition and recent growth, application to field samples required that a range of practical and theoretical questions be considered including:

1. What are the effects of the analytical methods used on the estimates of nucleic acid levels, condition, growth and mortality?
2. How do nucleic acid levels change with larval growth and development, and what is the best method to normalize RNA content to larval size?
3. How do environmental factors affect nucleic acid levels, and the relations among feeding, growth and nucleic acid levels?
4. What is the significance of variability in R/D levels observed at the individual level?

Each of these questions has been considered in a fair degree of detail and a variety of answers has been offered by different investigators. Although there is some degree of consensus among researchers, each question remains open. The answers can have profound effects on interpretation of field results.

Analytical methods

A variety of analytical methods has been used to estimate RNA and DNA content of fish larvae (Buckley & Bulow, 1987; Clemmesen, 1988; Caldarone & Buckley, 1991), making direct comparisons among laboratories tenuous. The most commonly used UV-based nucleic acid analysis method is fairly straightforward (Buckley & Bulow, 1987). Through a series of steps, RNA and DNA are hydrolyzed to nucleotides and their concentration is calculated from their absorbance at 260 nm. Because an established extinction coefficient is used, a researcher can be reasonably assured of obtaining nucleic acid values comparable to those of other laboratories. This method thus has the advantage of allowing the researcher to directly use the published UV-based growth-temperature-R/D models (Buckley, 1982, 1984) when analyzing larval fish. The disadvantages are that analysis time is lengthy, small larvae must be pooled to obtain sufficient sample size, and errors due to incomplete digestion of either RNA or DNA, or contamination of the fractions by polypeptides, can occur.

Fluorometric methods are more sensitive, more rapid, and equally as precise as the UV method; however, they require more attention to procedural details. Currently there are two prevalent fluorometric methods for determining nucleic acids. The 'two dye' method uses ethidium bromide (EB) to measure total nucleic acids (i.e. both DNA and RNA) in the extracted sample. Hoechst H33258 is used to measure the amount of DNA, and RNA is then calculated by subtraction (Clemmesen, 1988; Caldarone & Buckley, 1991; Canino & Caldarone, 1995). The 'one dye' method uses EB to measure total nucleic acid fluorescence. RNA is then enzymatically digested by RNase and the remaining fluorescence is attributed to DNA (Raae et al. 1988; Clemmesen, 1993). Alternatively, DNase is also added to distinguish between DNA and residual fluorescence (Westerman & Holt, 1988).

All of the fluorometric methods rely on a series of dilutions of RNA and DNA to construct standard curves. The choice of commercial nucleic acid preparation, which has historically differed among laboratories, is critical because final estimates of RNA and DNA content can vary by a factor of two or more simply by changing the source of the standards (Caldarone & Buckley, 1991). Electrophoretic gels of commercial DNA preparations from herring sperm and testes, and RNA preparations from *E. coli*, calf liver, and brewer's yeast indicated extensive degradation and a preponderance of low molecular weight fragments (Caldarone, unpub.). Recently, products manufactured specifically for molecular biological research purposes have proven to be very pure and of consistently good quality, e.g. highly polymerized calf thymus DNA and 18S, 28S rRNA (eukaryotic rRNA). The main criteria for a DNA standard should be the absence of RNA contamination and a high molecular weight. Since rRNA makes up the majority of total RNA (85–94%) in a cell (Young, 1970), 18S, 28S rRNA is a good choice for a RNA standard. Batch to batch consistency is also an important consideration in selecting a standard. Use of common standards among laboratories would significantly advance comparisons of data between laboratories and reduce the need for time-consuming intercalibrations.

When preparing tissues for fluorometric analysis, it is necessary to include a step that will dissociate proteins from nucleic acids. Histones will block binding of EB and Hoechst by as much as 40% (Brodie et al., 1975; Morgan et al., 1979). As with standards, there is a wide variety of dissociation methods used by different laboratories. Any procedure that dissociates proteins completely and reproducibly and provides acceptable blanks is satisfactory. We have found extraction with *n*-lauroylsarcosine to be a very simple, effective, and highly reproducible technique (Caldarone & Buckley, 1991). Recoveries of nucleic acid standards from *n*-lauroylsarcosine-extracted homogenates are similar to those reported for more labor-intensive solvent extractions (Clemmesen, 1993). Again, standardization of extraction and dissociation procedures would facilitate comparisons among laboratories.

At a minimum, for each different species, stage, and tissue type, the following factors should be evaluated before a specific fluorometric method is routinely used:

1. Recovery of spikes: A set of samples should be spiked with RNA and DNA before the protein dissociation step to determine whether the recovery rates of the nucleic acids are consistent and within an acceptable range.
2. Precision: Replicate samples of the same homogenate should be extracted and analyzed to determine whether the reproducibility of the method is within acceptable limits.
3. Residual fluorescence: Both RNase and DNase should be added to a portion of the sample to determine the remaining fluorescence. If the remaining fluorescence is significant, DNase (with the appropriate cofactors of Mg^{2+} and Ca^{2+}) should be routinely added to the samples to determine the true DNA content.
4. Subsampling: If the fish is susbsampled (a specific portion dissected out), it must be done consistently. The R/D of different portions of a larva can vary by a factor of two (Caldarone, unpub.).

Other procedural details that should be observed are:

1. The fluorescence yield of all newly prepared standards should be checked against previously prepared standards; the individual slopes and their ratio should be comparable to those of previous standards.
2. As a quality control measure, a large homogenate can be aliquoted into smaller portions, frozen, and then run along with each set of samples. The RNA and DNA concentration and R/D ratio of these replicates should fall within an acceptable range each time the sample is run.

Our standard protocol is a one-dye (EB) RNase-only procedure which uses *n*-lauroylsarcosine to dissociate the proteins, and highly polymerized calf thymus DNA (Sigma D4764) and 18S+28S calf liver rRNA (Sigma R0889) as standards. This method is very rapid because no homogenization of larval tissue is required and the fluorescence of 80 samples is recorded in less than two minutes in a 96-well plate reader (Wagner et al., 1998).

Normalization to larval size/developmental stage

The larval stage of marine fishes is characterized by rapid exponential growth. Tissues, organs and physiological functions are typically rudimentary at hatching and develop rapidly during the larval period. Hatching is followed by a yolk-sac stage lasting for a period of days to weeks, depending on the species and temperature. Yolk-sac larvae are dependent on endogenous reserves for energy and substrates. During this period, larvae typically increase in length but lose mass, primarily due to the use of protein as an energy source. Exogenous feeding is initiated during or shortly after the yolk-sac stage with the transition to feeding on phytoplankton, microflagellates and small zooplankton. Once exogenous feeding has been established, a period of rapid increase in weight follows. During this period, standard length, dry weight, protein, RNA and DNA content are all highly correlated (Richard et al., 1991; Folkvord et al., 1996).

In actively feeding larvae, nucleic acid content (μg larva^{-1}) increases with larval age and size. Estimation of recent growth or condition from larval nucleic acid content requires normalization or adjustment for larval age, development or size. Accurate estimation of age requires microscopic examination of daily rings in the otoliths, a procedure considerably more labor intensive than estimation of nucleic acid content. Since age is difficult to establish in field-caught larvae, normalization to age is usually not practical (although see Gronkjaer et al., 1997). Normalization to larval size has generated some controversy and several different procedures have been proposed. By far the most widely used and extensively tested procedure has been to normalize the RNA content to the DNA content and express the data as the ratio of RNA to DNA. Use of R/D has several advantages, but presents several concerns. RNA and DNA are relatively easy to measure concurrently, although good analytical technique and rigid quality control are essential. Net degradation of RNA occurs during periods of restricted feeding and growth, while DNA is conserved, resulting in a decrease in RNA concentration and an increase in DNA concentration (due to the decrease in tissue dry weight). This effectively amplifies the R/D signal and helps to explain why R/D is frequently the most sensitive index of starvation examined (Buckley, 1980; Martin & Wright, 1987; Ferron & Leggett, 1994). Normalization of RNA to dry weight, carbon or protein is less satisfactory since, like RNA, they decrease during periods of starvation, effectively dampening the signal. Determination of larval dry weight presents some practical problems since very small larvae tend to be delicate and may break during handling, and results are affected by ambient humidity (Bergeron et al., 1997). Typically when R/D, RNA/protein, protein/DNA and RNA concentration are correlated with recent growth, R/D explains the largest part of the observed variability in growth (Malloy & Targett, 1994; Caldarone & Buckley, in prep., Buckley, unpub.).

Some concerns with the use of R/D are related to the particular analytical methods employed or, more generally, to the use of a ratio and the compounding of errors (Hovenkamp & Witte, 1991; Suthers et al., 1996). Estimation of RNA by difference and the differential yield of RNA and DNA with fluorochrome dyes may further compound the error. These considerations and the sensitivity of the fluorometric assays to procedural details and standards place a very high premium on careful analytical technique with rigorous quality control.

Normalization of RNA to DNA does not completely remove the effects of larval size or development stage, since persistent patterns in R/D are observed during growth and development. For example, until the larva's endogenous yolk reserves are exhausted, R/D and most other measures of larval condition are relatively poor indicators of available food levels. Bergeron (1997) has questioned the utility of R/D analysis to provide information on nutritional condition of early stage larvae based in part on the time required to observe differences in fed and starved groups after hatching. However, during much of this time larvae rely on yolk reserves and may not feed, even in the presence of suitable prey. Several studies have demonstrated significant differences in R/D of fed and starved larvae within 1–3 days after yolk absorption (Buckley, 1980; Clemmesen, 1987, 1989).

In culture, R/D levels are typically stable through the yolk-sac stage and decrease for a period of days after yolk absorption, although in red drum (*Sciaenops ocellatus* Linnaeus), R/D levels decreased during the yolk-sac stage (Westerman & Holt, 1994). DNA is conserved through the period of falling R/D and changes in the ratio are driven by a loss of RNA. When exogenous feeding and growth are established, R/D increases for a period of days to weeks and then is relatively stable through the remainder of the larval period (Buckley, 1980; Clemmesen, 1987; Rooker & Holt, 1996; Gronkjaer et al., 1997). Mortality can be very high after yolk absorption, and the die-off of weaker larvae may contribute to the observed increase in R/D upon initiation of feeding and growth.

While trends have been observed in R/D during early larval growth and development, Buckley (1982) and Rooker & Holt (1996) have demonstrated that the relation between R/D and growth was constant during periods when R/D increased rapidly with length or age. For Atlantic cod (*G. morhua*), a single regression equation adequately described the relationship among growth, temperature and R/D for the period between complete yolk-sac absorption and the transition to the pelagic juvenile stage (roughly 10–38 d at temperatures between 3 and 9 °C) (Caldarone & Buckley, in prep.). An examination of residuals indicated no trend with larval size or development stage after yolk sac absorption. However, inclusion of yolk-sac larvae resulted in a decrease in the r^2 and a pattern in the residuals. The model consistently overestimated growth of yolk-sac larvae. These data suggest that the models should not be used with yolk-sac larvae and that comparisons of R/D between yolk-sac stage and older larvae should be avoided.

Field studies have not consistently indicated a correlation between standard length and R/D in fish larvae (Lough et al., 1996; Chicharo et al., 1998). Even when a correlation between R/D and standard length is observed (Canino et al., 1991; Folkvord et al., 1996; Rooker & Holt, 1996), the benefit of using standard length as an independent variable to adjust treatment or site mean R/D values in an analysis of covariance should be carefully considered. When a covariate or independent variable (length) was influenced by the treatment, then adjustment results in removal of part of the treatment effect (Steel & Torrie, 1960, p. 307). Clearly, prey availability affects length, through its effect on growth, as well as R/D. Since after yolk absorption the relationship between R/D and growth appears constant through the larval period, adjustment of R/D for larval size should only be considered when one wants to adjust for differences in growth rate. A secondary consideration is that length estimates of field-caught larvae are subject to considerable error, due to differential shrinkage during capture, death and preservation.

A variety of studies using different indices of condition have established that the incidence of starvation is higher in first-feeding larvae than in older, larger larvae that have established feeding and growth (Hewitt et al., 1985; Theilacker et al., 1996; Clemmesen et al., 1997). Adjustment of R/D for age or size during the larval period could mask real differences in condition and growth of large versus small larvae. A better approach may be to acknowledge the difference in condition or growth rate, and examine the possible causes (developmental stage, age, prey availability and size, temperature etc.).

Clemmesen (1994, 1996) suggested that the critical R/D (the lowest point to which R/D can fall and larvae still recover) determined in culture as a function of larval size and development stage could be used to estimate the percentage of starved field-caught larvae for each size/developmental stage. In herring larvae (*Clupea harengus* Linnaeus), the critical R/D indicative of starving larvae increased as a linear function of larval size (Clemmesen, 1994). This approach is relatively insensitive to errors in estimation of standard length, since the critical R/D appears to be a linear function of length with a slope not much greater than 0 (e.g. Atlantic herring slope 0.03, Clemmesen, 1994). The problem of direct comparisons of R/D values in larvae of different size/developmental stages is avoided. Clemmesen et al. (1997) used this approach to estimate the percentage of anchovy (*Engraulis anchoita*) larvae in starving condition on the Brazilian and Argentinean shelves. They stressed that care should be taken in comparing larvae from different temperatures, since temperature influences larval metabolism, development, and the relationship between length and developmental stage. Also, the effect of temperature on the relationship between larval size/developmental stage and the critical R/D indicative of starved larvae is largely unknown. There is, however, some evidence suggesting that the critical R/D may decrease with increasing temperature (Buckley, 1984; Robinson & Ware, 1988; Caldarone & Buckley, in prep.). Robinson & Ware (1988) defined the critical R/D as the value corresponding to zero growth in the Buckley (1994) multi-species growth-temperature-R/D model. This implies that the critical R/D decreases with increasing temperature.

Laboratory calibration studies with juvenile fish (Foster et al., 1992; Mathers et al., 1992; Malloy & Targett, 1994; Rooker & Holt, 1996) indicate that the results from work with larvae should not be extrapolated to juveniles without critical evaluation. During the juvenile period, concentrations of both RNA and DNA decrease while concentrations of protein, lipid and ash increase with development and ossification (Love, 1970). During this same period, weight-specific growth begins to decrease. As the mass of the organism increases, a larger portion of the total RNA pool is involved with turnover of enzymes and structural proteins, and net synthesis of new protein decreases as a percent of body weight (Houlihan et al., 1988). Foster et al. (1992), working with juvenile cod (*G. morhua*) acclimated to two temperatures, observed a positive linear relationship between tissue fractional protein synthesis rates and tissue RNA concentration and a negative trend when compared with tissue R/D. Rooker & Holt (1996) determined that the relative decline in R/D in starved individuals decreased with increasing age in early juvenile red drum (*Sciaenops ocellatus*). When comparing individuals of different size in the late juvenile period, RNA concentration (μg mg^{-1} dry weight) or RNA/protein may be a better indicator of growth. Eventually, as the fish matures, seasonal cycles of temperature and gonad development have large effects on nucleic acid levels in different tissues. As the fish become larger and it becomes practical to dissect out individual tissues for analysis, care should be taken in the selection of tissues for analysis. The appropriate tissue will depend upon the questions addressed. These considerations suggest that the growth-temperature-R/D relation should be determined for each stage under consideration and trends (deviations) with age and development examined carefully. Until this has been accomplished, direct comparisons of R/D levels among larvae and juveniles of widely different size should not be considered.

Other approaches to normalization of nucleic acid levels

Relative (residual) RNA content has been proposed recently as an alternative to using R/D for estimation of larval growth and feeding condition (Suthers et al., 1996). RNA content (μg larva^{-1}) is regressed against standard length and deviations from the regression line are used as indicators of larval condition and growth. This method suffers from several severe limitations. The observed increase in relative RNA content of larvae in the terminal stages of starvation (Suthers et al., 1996) is a particular concern, since levels approach those of well-fed larvae. This increase in relative RNA content of starving larvae may be due to osmoregulation failure. An increase in larval density (Sclafani et al., 1997) and shrinkage in standard length would result from loss of water to the hypertonic environment. Relative RNA content is very sensitive to errors in measurement of standard length, since RNA content increases with roughly the cube of length. Although live standard length can be accurately measured, calculating an accurate live length from the length of a dead field-caught larva is extremely difficult, due to differential shrinkage. Field-caught and preserved larvae can shrink to as little as 60% of their initial live length. The few rigorous studies published on larval shrinkage have shown that larvae shrink upon capture in a net and shrink further upon preservation (Theilacker, 1980; Radtke, 1989; Jennings, 1991). Furthermore, larvae shrink to varying degrees depending on a wide variety of factors including species, size, condition, temperature, towing speed, towing time, washdown technique, processing time, processing methods, preservation medium and preservation time. Even when standard length is estimated from video images taken prior to preservation on shipboard, the degree of shrinkage can be very difficult to control or estimate.

Other problems in using relative RNA content arise from the choice of a regression line. Use of a laboratory-derived regression for fieldwork is problematic due to the difficulty in estimating length as discussed above. Fitting a regression line to field data is not satisfactory since no a priori information on the condition of larvae is available. A line fit to a group of fish in poor condition will yield both positive and negative deviations. The same is true when fitting a line to a group of fish in good condition. The method would be of some utility, however, when it was desirable to obtain information on relative condition of larvae.

The use of DNA concentration expressed as μg mg dry weight^{-1} (DNA/DW) and more recently μg mg carbon^{-1} (DNA/C) has been proposed as a index of condition in larvae (Bergeron et al., 1991, 1997; Bergeron, 1997). While promising, this method requires additional investigation in a variety of species and a wider range of feeding levels and temperatures. This method may be most useful in newly-hatched through first-feeding larvae.

Environmental factors

Temperature and food availability are the two environmental variables most often correlated with larval growth rate. Temperature is a controlling factor in that it governs the rate of chemical reactions, pacing metabolic requirements, digestion and the rate of growth (Fry, 1971). Temperature must be considered in any study of larval growth and mortality. Only within very narrow temperature ranges ($<$2 °C) can RNA concentration and R/D be compared directly, without correction for temperature. The growth-temperature-R/D models (Buckley, 1984) imply that for a given R/D, growth increases about 1% d^{-1} for every 1 °C increase in water temperature. In several studies, RNA concentration and R/D have been shown to decrease with increasing temperature, even while growth rate increased. Goolish et al. (1984) have suggested that the elevated R/D levels observed at low temperatures are part of a compensatory mechanism for reduced RNA activity at low temperatures.

The portion of the observed variability in growth explained by temperature will depend upon the temperature range considered and the state of limiting factors such as food availability (Buckley, 1984). RNA concentration and R/D are good indicators of limiting factors such as food availability in post-yolk sac larvae. Larval R/D and food availability are highly correlated in a wide variety of species in both laboratory and field studies (Buckley, 1980; Buckley et al., 1984; Buckley & Lough, 1987; Theilacker et al., 1996). In winter flounder (*Pseudopleuronectes americanus* Walbaum) larvae reared at 5, 7 and 10 °C and fed a high density of wild plankton (>2000 plankters l^{-1}), temperature alone and R/D alone explained 40% and 53% of the observed variability in growth, respectively. Temperature and R/D together explained 72% of the variability in growth (Buckley, 1982). In sand lance (*Ammodytes americanus*) larvae reared at different temperatures (2, 4, 6, 7 °C) and prey levels (200, 500 and 1000 rotifers l^{-1}), temperature alone explained little of the observed variability in growth, while R/D alone explained 55%. Temperature and R/D together explained 61% of the observed variability in growth. Addition of food availability, as a third independent variable to a regression model that already included R/D and temperature, explained very little of the remaining variability in growth (Buckley et al., 1984). An analysis of data on eight species of temperate marine fish larvae reared over a wide range of temperatures (2–20 °C) and prey levels (excess to starvation) showed that temperature and R/D explained 46 and 24%, respectively, of the observed variability in growth. Temperature and R/D together explained 92% of the observed variability (Buckley, 1984). The high percentage of the variability in growth explained in the eight-species model may be due in part to the use of average R/D and growth values for individual cultures.

Diel cycles

In most of the early laboratory calibration work on R/D in fish larvae, sampling frequency was at most once per day. Recently, variable results have been obtained when diel patterns in bulk nucleic acid levels of fish larvae have been examined. Clemmesen (1996) found no diel pattern in R/D of herring larvae at 30, 37 or 42 d after hatch. In 20-day-old larval red drum, Rooker & Holt (1996) reported a pronounced diel pattern in R/D. Values were highest during daytime periods (0800–2000) and reduced at night (0000–0400). In 14-day-old Atlantic cod sampled every 4 h, R/D values of fed larvae were relatively stable through a 72-h period (Figure 1). A dramatic decline in R/D was observed in starved larvae 40 h after removal of food (Ong et al., in prep.). R/D of Atlantic cod and haddock larvae sampled on the southern flank of Georges Bank was unrelated to time of capture (Lough et al., 1996). However, sampling time did have a significant effect on R/D of cod larvae sampled in the Baltic (Gronkjaer et al., 1997), and on R/D and residual RNA content of sardine larvae collected off northern Spain (Chicharo et al., 1998). Differences in metabolic rates, food requirements and digestion times may explain the presence of a strong diel pattern in some warm water species and the apparent absence of a strong diel pattern in some cold water species. When a significant diel pattern is observed, care should be taken in sampling design and data analysis to account for this additional source of variability.

Individual variability and calibration of nucleic acid-based indices

The primary motivation for changing from UV to fluorometric analytical methods was to examine variability among individual larvae (Clemmesen, 1988). Analysis of individuals has provided a wealth of information on variability at the individual level in a wide variety of species both in the field and culture. However, the high variability in R/D sometimes observed in culture among individuals within a tank or between tanks in replicate treatments has caused some to question the utility of the approach (Bergeron, 1997).

The observed variability raises two fundamental questions:

1. What are the sources of the observed variability? and

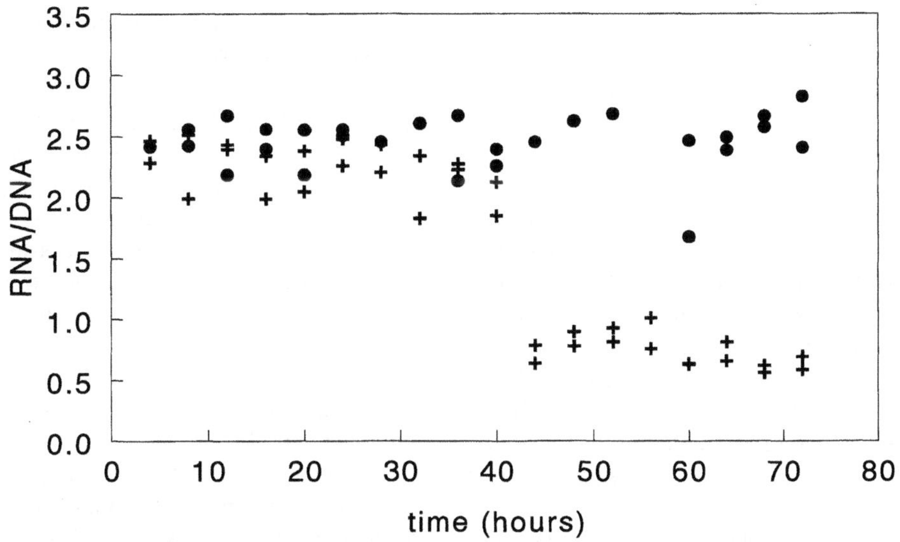

Figure 1. RNA/DNA values of starved (+) and fed (●) 14 d Atlantic cod larvae reared at 8 °C. The experiment began 14 d after hatching with fed larvae. Data points are RNA/DNA values for duplicate samples of 5 larvae each taken from starved and fed tanks. Larvae were sampled every 4 h for 72 h.

2. What can the variability tell us about the biochemistry, physiology and ecology of fish larvae?

Identifying the sources of variability is essential to answering the second question. Variability can arise from several sources including analytical methods, collection or culture techniques, developmental and genetic differences, and chance or stochastic processes. Careful analytical technique is the only assurance that the observed or reported variability is real. Unfortunately, there are some examples in the literature where analytical methods must be questioned since factors such as precision, absence of residual fluorescence, recoveries or effects of using only a portion of the larva were not evaluated. Although bulk nucleic acid levels in fish have proven fairly stable after collection, care should be taken to minimize towing and processing times, and to chill samples during sorting and processing.

Due to their small size and requirement for live prey, culture of temperate marine fishes presents many problems and pitfalls. Culture problems can readily complicate laboratory experiments designed to test or calibrate nucleic acid-based estimators of growth and feeding condition. Some of these problems were pointed out by Folkvord & Moksness (1995) commenting on a paper on nucleic acid levels in fed and starved herring (*Clupea harengus*) (Mathers et al., 1994). Culture problems can originate from two sources:

1. poor initial egg or larval quality, and
2. culture conditions.

Culture conditions include initial stocking density, water quality, temperature, gas saturation, lighting, and the quantity/quality of prey offered. Addition of prey to a tank does not insure that all larvae will initiate feeding and grow rapidly, or even that a significant portion of larvae will initiate feeding and survive to metamorphosis. Often mortality is high in both fed and starved groups shortly after yolk absorption. Microscopic examination of individual larvae in fed groups often indicates that some larvae never initiate feeding while others may feed (indicated by the presence of food in the gut) but not grow. A large portion of the variability in R/D observed among individuals within a tank or among replicate tanks can be explained by the variability in performance (growth and mortality) of individual larvae or replicates. High mortality or slow growth of fed groups indicates problems with larval culture that will compromise calibration experiments. Conclusions drawn from observations on variability among individuals within a group (tank) and differences or lack of differences between groups should take into consideration the performance of the individual larvae in terms of both growth and survival.

The change to fluorometric methods and analysis of individuals has raised some new considerations when establishing the relation among R/D, temperature, and growth. Calibration of the relationships among nucleic acid levels, temperature and growth requires independent estimates of larval growth. Specific growth is usually estimated using the equations:

$$G_i = (\ln Wt_2 - \ln Wt_1)/(t_2 - t_1)$$

$$G(\%/d) = 100\,(e^{G_i} - 1)$$

where G_i is the instantaneous or specific growth rate, W is the dry weight or protein content, t is time in days, and G is the growth in % per day. Since R/D appears to reflect feeding and recent growth over a period of a few days, care should be exercised in selecting the sampling interval. The interval must be long enough to obtain an accurate estimate of growth in length, weight or protein content, but short enough so that the R/D estimate will be reflective of feeding and growth over the entire interval.

Now that most investigators analyze individual larvae for nucleic acid content, choice of a method to estimate growth presents new problems. Typically, growth is estimated from the weight, protein content, or length of the larvae at the time of sampling (t_2) and the mean value of the population at an earlier sampling time (t_1). Use of an individual value and a population mean can introduce error since an individual larva may have varied considerably from the population mean at the earlier sampling point. Alternatively recent growth can be estimated from the width of the outer otolith increments (Clemmesen & Doan, 1996), or for larvae reared in individual containers where growth in live length can be estimated from recorded images of the same larva over time (Bertram et al., 1996). Although promising, both of these approaches have their own limitations and can introduce error to the growth estimate.

Mesocosm and field studies have also revealed considerable variability in R/D and growth of individual larvae, even when variability in size and/or age is removed. Other approaches such as otolith microstructure analysis have indicated that larval growth and size at age are highly variable even within a cohort. The challenge in both laboratory and field studies is to identify the sources of variability observed at the individual level, and to determine the relationship between variability at the individual level and longer-term growth and survival of the cohort. If careful attention is paid to analytical, sampling and/or culture methods, variability among individuals and replicates should not be dismissed as an artifact.

Use of nucleic acid analysis in recruitment fisheries oceanography

Nucleic acid-based indicators of larval growth and starvation have been employed in two large, multi-year, recruitment fisheries oceanography studies off the northwest and northeast coasts of North America: Fisheries Oceanography Coordinated Investigation (FOCI) and U.S. Global Ocean Ecosystems Dynamics (GLOBEC). These programs illustrate the utility of nucleic acid-based indicators of larval growth and starvation when coupled with a broad spectrum of physical and biological measurements made over extended periods of time.

Estimation of mortality during the larval and juvenile periods represents a major impediment to understanding the factors responsible for the observed variability in recruitment of marine fishes. Predation and starvation appear to be the two major sources of mortality. The two are interrelated since starvation and reduced growth due to food limitation can make individuals more vulnerable to predators by diminishing escape responses and by extending the duration of vulnerability to the predator. R/D has been used extensively in the field to estimate the percentage of larvae or juveniles in starving condition (Buckley & Lough, 1987; Clemmesen, 1988; Robinson & Ware, 1988; Clemmesen et al., 1997). However, the growth rate of healthy larvae and the time that fish remain in starving condition before death are needed to estimate the starvation mortality rate in % d^{-1} (Theilacker et al., 1996). This approach was used in FOCI to estimate condition and starvation mortality rates in walleye pollock (*Theragra chalcogramma*) in the western Gulf of Alaska. Estimates of mortality rates were based on histological criteria however, while condition estimates based on R/D and histological criteria were in good agreement (Theilacker et al., 1996). A strong positive association was observed between microzooplankton prey concentrations, larval gut contents and condition estimated from R/D or histological criteria. Larval walleye pollock appeared most vulnerable to starvation for a 2-week period after first feeding (Theilacker et al., 1996).

An alternate approach for estimating starvation mortality is to first estimate growth rate from R/D and temperature, and then estimate starvation mortality from the relation between growth and starvation mortality. Buckley et al. (1993) demonstrated an inverse relation between growth rate of survivors and mortality of cohorts for six of nine species reared in

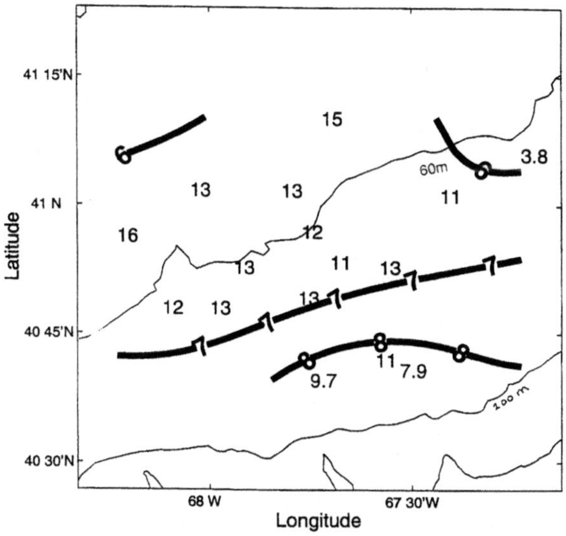

Figure 2. Mean water temperature and growth of Atlantic cod larvae sampled on the southern flank of Georges Bank in May 1993. Protein growth rates plotted (range 3.8–15% d^{-1}) are the mean values for larvae collected at each station with a bongo net towed obliquely from the surface to within 10 m of the bottom. Stations yielding less than three larvae for biochemical analysis are not shown. Recent protein growth rates were estimated from water temperature and RNA/DNA using a growth-temperature-RNA/DNA model derived from laboratory culture of Atlantic cod at 3 temperatures and 3 prey levels. Bold contour lines and numbers are the mean water temperature (°C) in the top 60 m of the water column. The 60 m and 100 m bottom depth contours are also shown.

the laboratory. Groups of larvae of each species were reared at several discrete prey levels at a single temperature near the species optimum. A general model relating growth rate, temperature and mortality in the nine species indicated that mortality increased with temperature and decreased with growth, suggesting that a full model relating temperature, growth, and mortality should be developed for each species of interest.

As part of the U.S. GLOBEC – Northwest Atlantic Georges Bank Program, R/D analysis has been used to examine recent larval growth on spatial scales ranging from tens of meters in depth to hundreds of kilometers horizontally (Lough et al., 1996). A major goal of this program has been to understand the effects of environmental variability, specifically global climate change, on the dynamics of cod and haddock populations. Efforts have been concentrated on the larval and early pelagic-juvenile periods, since the environment acting upon these early-life stages may determine year-class strength. Growth, mortality and retention of larvae are key factors under investigation. Recent larval growth rates were estimated from R/D of individual larvae and water temperature using a model developed from data on cod larvae reared at three temperatures and three prey levels (Caldarone & Buckley, in prep.). Figure 2 maps larval Atlantic cod growth rates and water temperature on the southern flank of Georges Bank in May 1993. The most rapid larval growth was observed in water of about 7 °C in the northwest portion of the larval distribution. Larval growth was slower in water that was either warmer or colder than 7 °C.

Future directions

There is sufficient interest in nucleic acid analysis of early life history stages of fish to warrant a major international effort to intercalibrate current analytical methods. Considerable additional work is necessary to calibrate nucleic acid-based methods for estimation of larval and juvenile condition, growth and mortality. The relationship between growth and R/D has been reported for relatively few species (Buckley, 1980; Rooker & Holt, 1996). Full growth-temperature-R/D models have been reported for only winter flounder and sand lance (Buckley, 1982; Buckley et al., 1984). Standardization of analytical methods and intercalibration between laboratories is essential for efficient use of resources and comparisons among species and systems. Comparisons among sites, species and systems are most effectively made using derived estimates of growth or starvation rather than by comparing R/D directly, due to the effect of temperature on RNA activity, changes in R/D with development, and differences in analytical techniques.

Some of the problems that arise with methods based on bulk RNA result from the heterogeneous nature of RNA. Ribosomal RNA (rRNA), messenger RNA (mRNA), and transfer RNA (tRNA) have different structures, functions, and half-lives. They respond at different times in different ways during development, feeding and growth. Since rRNA makes up the majority of total RNA, changes in total RNA or R/D primarily reflect changes in rRNA. The concentration of individual RNAs can be estimated using a variety of techniques, most based on hybridization of labeled probes for unique nucleotide sequences. It is routine now to produce specific probes for almost any target RNA. The challenge is to identify specific RNAs that provide the desired information on metabolic pathways or vital rate processes of interest on the desired time scales. While changes in bulk RNA and mRNA levels can be observed within 1–2 d of a change in food availability, measurable changes in certain mRNA may be observed within hours of a change in the environment. Heat-shock protein mRNA was induced within 2 h of exposure of *Fundulus heteroclitus* to elevated temperatures (Koban et al., 1991). In Atlantic cod larvae, metallothionein mRNA increases four-fold within six days of removal from food (McNamara & Buckley, 1994). Also, a dramatic decrease in 18S rRNA titer to 3–5% of time-zero values was observed in 14-day-old Atlantic cod larvae 40 h after removal of food (Ong et al., in prep.). These studies suggest that specific nucleic acid-based biomarkers, indicative of condition and growth rate of fish larvae, will provide new insight into the factors affecting growth, survival and mortality of fish.

Acknowledgements

We thank Jeanne Burns and anonymous reviewers for their thoughtful review of the manuscript. This effort was partially funded by NOAA award NA66RG0303 and National Science Foundation award OCE-9730712.

References

Bergeron, J. P., 1997. Nucleic acids in ichthyoplankton ecology: a review, with emphasis on recent advances for new perspectives. J. Fish Biol. 51 (Suppl. A): 284–302.

Bergeron, J. P., M. Boulhic & R. Galois, 1991. Effet de la privation de nourriture sur la teneur en ADN de la larve de sole (*Solea solea* L.) ICES J. mar. Sci. 48: 127–134.

Bergeron, J. P., J. Person-Le Ruyet & C. Koutsikopoulos, 1997. Use of carbon rather than dry weight to assess the DNA content and nutritional condition index in sole larvae. ICES J. mar. Sci. 54: 148–151.

Bertram, D. F., T. J. Miller & W. C. Leggett, 1996. Individual variation in growth and development during the early life stages of winter flounder, *Pleuronectes americanus*. Fish. Bull. U.S.A. 95: 1–10.

Brodie, S., J. Giron & S. A. Latt, 1975. Estimation of accessibility of DNA in chromatin from fluorescence measurements of electronic excitation energy transfer. Nature 253: 470–471.

Buckley, L. J., 1979. Relationships between RNA-DNA ratio, prey density and growth rate in cod (*Gadus morhua* L.) larvae. J. Fish Res. Bd Can. 36: 1497–1502.

Buckley, L. J., 1980. Changes in RNA, DNA, and protein content during ontogenesis of the winter flounder (*Pseudopleuronectes americanus*) and the effect of starvation. Fish. Bull. U.S.A. 77: 703–708.

Buckley, L. J., 1982. Effects of temperature on growth and biochemical composition of larval winter flounder, *Pseudopleuronectes americanus*. Mar. Ecol. Progr. Ser. 8: 181–186.

Buckley, L. J., 1984. RNA-DNA ratio: an index of larval fish growth in the sea. Mar. Biol. 80: 291–298.

Buckley, L. J. & F. J. Bulow, 1987. Techniques for estimation of RNA, DNA, and protein in fish. In Summerfelt R. C. and G. E. Hall, (eds), Age and Growth of Fish. Iowa State University Press, Ames, Iowa. 345–354.

Buckley, L. J. & R. G. Lough, 1987. Recent growth, biochemical composition, and prey field of larval haddock (*Melanogrammus aeglefinus*) and Atlantic cod (*Gadus morhua*) on Georges Bank. Can. J. Fish. aquat. Sci. 44: 14–25.

Buckley, L. J., S. I. Turner, T. A. Halavik, A. S. Smigielski, S. M. Drew, & G. C. Laurence, 1984. Effects of temperature and food availability on growth, survival and RNA–DNA ratio of larval sand lance (*Ammodytes americanus*). Mar. Ecol. Progr. Ser. 15: 91–97.

Buckley, L. J., A. S. Smigielski, T. A. Halavik, B. R. Burns & G. C. Laurence, 1993. Growth and survival of the larvae of three species of temperate marine fishes at discrete prey densities. II. Cod (*Gadus morhua*), winter flounder (*Pseudopleuronectes americanus*), and silver hake (*Merluccius bilinearus*). In Walther, B. T. & H. J. Fyhn, (eds), Physiological and Biochemical Aspects of Fish Development. Univ. of Bergen, Norway.

Bulow, F. J., 1970. RNA-DNA ratios as indicators of recent growth rates of a fish. J. Fish Res. Bd Can. 27: 2343–2349.

Bulow, F. J., 1987. RNA-DNA ratios as indicators of growth in fish. In R. C. Summerfelt and G.E. Hall (eds), The Age and Growth of Fish. Iowa State University Press, Ames, Iowa. 45–64.

Caldarone, E. M. & L. J. Buckley, 1991. Quantitation of DNA and RNA in crude tissue extracts by flow injection analysis. Anal. Biochem. 199: 137–141.

Canino, M. F. & E. M. Caldarone, 1995. Modification and comparison of two fluorometric techniques for determining nucleic acid contents of fish larvae. Fish. Bull. U.S.A. 93: 158–165.

Canino, M. F., K. M. Bailey & L. S. Incze, 1991. Temporal and geographic differences in feeding and nutritional condition of walleye pollock larvae *Theragra chalcogramma* in Shelikof Strait, Gulf of Alaska. Mar. Ecol. Prog. Ser. 79: 27–35.

Chicharo, M. A., L. Chicharo, L. Valdes, E. Lopez-Jamar & P. Re, 1998. Estimation of starvation and diel variation of the RNA/DNA ratios in field-caught *Sardina pilchardus* larvae off the north of Spain. Mar. Ecol. Prog. Ser. 164: 273–283.

Clemmesen, C., 1987. Laboratory studies on RNA/DNA ratios of starved and fed herring (*Clupea harengus*) and turbot (*Scophthalmus maximus*) larvae. J. Cons. perm. int. Explor. Mer. 43: 122–128.

Clemmesen, C., 1988. An RNA and DNA fluorescence technique to evaluate the nutritional condition of individual marine fish larvae. Meeresforch. 32: 134–143

Clemmesen, C., 1989. RNA/DNA ratios of laboratory-reared and wild herring larvae determined with a highly sensitive fluorescence method. J. Fish Biol. 35 (Suppl. A): 331–333.

Clemmesen, C., 1993. Improvements in the fluorimetric determination of the RNA and DNA content in individual marine fish larvae. Mar. Ecol. Prog. Ser. 100: 177–183.

Clemmesen, C., 1994. Importance and limits of RNA/DNA ratios as a measure of nutritional condition in fish larvae. Proc. International Workshop: Survival Strategies in Early Life Stages of Marine Resources. 67–82. Yokohama, Japan. 11–14 October 1994.

Clemmesen, C., 1996. Importance and limits of RNA/DNA ratios as a measure of nutritional condition in fish larvae. In Y. Watanabe, Y. Yamashita & Y. Ooseki (eds), Survival Strategies in Early Life Stages of Marine Resources. A.A. Balkema. Rotterdam: 67–82.

Clemmesen, C. & T. Doan, 1996. Does otolith structure reflect the nutritional condition of a fish larva? Comparison of otolith structure and biochemical index (RNA/DNA ratio) determined on cod larvae. Mar. Ecol. Prog. Ser. 138: 33–39.

Clemmesen, C., R. Sanchez & C. Wongtschowski, 1997. A regional comparison of the nutritional condition of S.W. Atlantic anchovy larvae, *Engraulis anchoita*, based on RNA/DNA ratios. Arch. Fish. Mar. Res. 45: 17–43.

Ferron, A. & W. C. Leggett, 1994. An Appraisal of Condition Measures for Marine Fish Larvae. Adv. Mar. Biol. 30: 217–303.

Folkvord, A. & E. Moksness, 1995. RNA/DNA ratios and growth of herring larvae. Mar. Ecol. Prog. Ser. 121: 311–312.

Folkvord, A., L. Ystanes, L. Johannessen & E. Moksness, 1996. RNA:DNA ratios and growth of herring (*Clupea harengus*) larvae reared in mesocosms. Mar. Biol. 126: 591–602.

Foster, A. R., D. F. Houlihan, S. J. Hall, & L. J. Burren. 1992. The effects of temperature acclimation on protein synthesis rates and nucleic acid content of juvenile cod (*Gadus morhua* L.). Can. J. Zool. 70: 2095–2102.

Fry, J. E. J., 1971. The effects of environmental factors on the physiology of fish. In Hoar, W. S. & D. J. Randall (eds), Fish Physiology, Academic Press, New York, 6: 1–98.

Goolish, E. M., M. G. Barron & I. R. Adelman, 1984. Thermoclimatory response of nucleic acid and protein content of carp muscle tissue: influence of growth rate and relationship to glycine uptake by scales. Can. J. Zool. 62: 2164–2170.

Gronkjaer, P., C. Clemmesenn & M. St. John, 1997. Nutritional condition and vertical distribution of Baltic cod larvae. J. Fish Biol. 51 (Suppl. A): 352–369.

Hewitt, R. P., G. H. Theilacker & N. C. H. Lo, 1985. Causes of mortality in young jack mackerel. Mar. Ecol. Prog. Ser. 26: 1–10.

Houde, E. D., 1987. Fish Early Life Dynamics and Recruitment Variability. Amer. Fish. Society. Symp. 2: 17–29.

Houlihan, D. F., S. J. Hall, C. Gray & B. S. Noble, 1988. Growth rates and protein turnover in Atlantic cod, *Gadus morhua*. Can. J. Fish. aquat. Sci. 45: 951–964.

Hovenkamp, F. & J. Witte, 1991. Growth, otolith growth and RNA/DNA ratios of larval plaice *Pleuronectes platessa* in the North Sea 1987–1989. Mar. Ecol. Prog. Ser. 70: 105–116.

Jennings, S., 1991. The effects of capture, net retention and preservation upon lengths of larval and juvenile bass, *Dicentrarchus labrax* (L.). J. Fish Biol. 38: 349–357.

Koban, M., A. A. Yup, L. B. Agellon & D. A. Powers. Molecular adaptation to environmental temperature: heat-shock response of the eurythermal teleost *Fundulus heteroclitus*. Molec. mar. Biol. Biotechnol. 1: 1–17.

Lough, R. G., E. M. Caldarone, T. K. Rotunno, E. A. Broughton, B. R. Burns & L. J. Buckley, 1996. Vertical distribution of cod and haddock eggs and larvae feeding and condition in stratified and mixed waters on southern Georges Bank May 1992. Deep Sea Research 43: 1875–1904.

Love, R. M., 1970. The Chemical Biology of Fishes. Academic Press, New York.

Malloy, K. D. & T. E. Targett, 1994. The use of RNA:DNA ratios to predict growth limitation of juvenile summer flounder (*Paralichthys dentatus*) from Delaware and North Carolina estuaries. Mar. Biol. 118: 367–375.

Martin, F. D. & D. A. Wright, 1987. Nutritional state analysis and its use in predicting striped bass recruitment: laboratory calibration. Am. Fish. Soc. Symp. 2: 109–114.

Mathers, E. M., D. F.Houlihan & M. J. Cunningham, 1992. Nucleic acid concentrations and enzyme activities as correlates of growth rate of the saithe *Pollachius virens*: growth-rate estimates of open-sea fish. Mar. Biol. 112: 363–369.

Mathers, E. M., D. F. Houlihan & L. J. Burren, 1994. RNA, DNA, and protein concentrations in fed and starved herring *Clupea harengus* larvae. Mar. Ecol. Prog. Ser. 107: 223–231.

McGurk M. & M. W. Kusser, 1992. Comparison of three methods of measuring rRNA and DNA concentrations of individual Pacific herring, *Clupea pallasi*, larvae. Can. J. Fish. aquat. Sci. 49: 967–974.

McNamara, P. T. & L. J. Buckley, 1994. Identification and characterization of metallothionein cDNA from mRNA transcripts induced by starvation in Atlantic cod (*Gadus morhua*). Molec. Mar. Biol. Biotechnol. 3: 252–260.

Morgan, A. R., D. H. Evans, J. S. Lee & D. E. Pulleyblallnk, 1979. Review: Ethidium fluorescence assay. Part II: Enzymatic studies and DNA-protein interactions. Nucleic Acids Res. 7: 571–594.

Ong, T. L., R. Armstrong & L. Buckley. Diel study of RNA productions associated with starvation in Atlantic cod *Gadus morhua* larvae. (In prep.).

Raae, A. J., I. Opstad, P. Kvenseth & B. T. Walther, 1988. RNA, DNA and protein during early development in feeding and starved cod (*Gadus morhua* L.) larvae. Aquaculture 73: 247–259.

Radtke, R. L., 1989. Larval fish age, growth, and body shrinkage: information available from otoliths. Can. J. Fish. aquat. Sci. 46: 1884–1894.

Richard, P., J.-P. Bergeron, M. Boulhic, R. Galois & J. P.-L. Ruyet, 1991. Effect of starvation on RNA, DNA and protein content of laboratory-reared larvae and juveniles of *Solea solea*. Mar. Ecol. Prog. Ser. 72: 69–77.

Robinson, S. M. C. & D. M. Ware, 1988. Ontogenetic development of growth rates in larval Pacific herring, *Culpea harengus pallasi*, measured with RNA-DNA ratios in the Strait of Georgia, British Columbia. Can. J. Fish. aquat. Sci. 45: 1422–1429.

Rooker, J. R. & G. J. Holt, 1996. Application of RNA:DNA ratios to evaluate the condition and growth of larval and juvenile red drum (*Sciaenops ocellatus*). Mar. Freshwat. Res. 47: 283–290.

Sclafani, M., G. Sterling & W. C. Leggett. 1997. Osmoregulation, nutritional effects and buoyancy of marine larval fish: a bioassay for assessing density changes during the earliest life-history stages. Mar. Biol. 129: 1–9.

Steel, R. G. D. & J. H. Torrie, 1960. Principles and Procedures of Statistics, McGraw-Hall, 481 pp.

Suthers, I. M., J. J. Cleary, S. C. Battaglene & R. Evans, 1996. Relative RNA content as a measure of condition in larval and juvenile fish. Mar. Freshwater Res. 47: 301–307.

Theilacker, G. H., 1980. Changes in body measurements of larval northern anchovy, *Engraulis mordax*, and other fishes due to handling and preservation. Fish. Bull. U.S.A. 78: 685–692.

Theilacker, G. H., K. M. Bailey, M. F. Canino & S. M. Porter, 1996. Variations in larval walleye pollock feeding and condition: a synthesis. Fisheries Oceanography 5 (Suppl. 1): 112–123.

Wagner, M., L. J. Buckley, A. Durbin and T. Durbin, 1998. RNA:DNA ratios as indicators of nutritional condition in the copepod *Calanus finmarchicus*. Marine Ecol. Prog. Ser. 162: 173–181.

Westerman, M. E. & G. J. Holt, 1988. The RNA-DNA ratio: measurement of nucleic acids in larval *Sciaenops ocellatus*. Contrib. Mar. Sci. 30: 117–124.

Westerman, M. E. & G. J. Holt, 1994. RNA:DNA ratio during the critical period and early larval growth of the red drum *Sciaenops ocellatus*. Mar. Biol. (Berlin) 121: 1–9.

Wilder, I. B. & J. G. Stanley, 1983. RNA-DNA ratio as an index to growth in salmonid fishes in the laboratory and in streams contaminated by carbaryl. J. Fish Biol. 22: 165–172.

Wright, D. A. & F. D. Martin, 1985. The effect of starvation on RNA:DNA and growth of larval striped bass, *Morone saxatilis*. J. Fish Biol. 27: 479–485.

Young, V. R. 1970. The role of skeletal and cardiac muscle in the regulation of protein metabolism. In Munro H.H. (ed), Mammalian protein metabolism, Academic Press, London. 585–674.

Index

GPSR Compliance

The European Union's (EU) General Product Safety Regulation (GPSR) is a set of rules that requires consumer products to be safe and our obligations to ensure this.

If you have any concerns about our products, you can contact us on ProductSafety@springernature.com

In case Publisher is established outside the EU, the EU authorized representative is:

Springer Nature Customer Service Center GmbH
Europaplatz 3
69115 Heidelberg, Germany

Batch number: 10800215

Printed by Printforce, the Netherlands